Fractal Analysis of the Binding and Dissociation Kinetics for Different Analytes on Biosensor Surfaces

Fractal Analysis of the Binding and Dissociation Kinetics for Different Analytes on Biosensor Surfaces

Ajit Sadana

Department of Chemical Engineering, and National Center for Computational Hydroscience and Engineering, University of Mississippi, University, MS, USA

AND

Neeti Sadana, MD

Department of Anesthesiology, University of Miami, Miami, Florida, USA

ELSEVIER

Amsterdam • Boston • Heidelberg • London • New York • Oxford
Paris • San Diego • San Francisco • Singapore • Sydney • Tokyo

Elsevier
Radarweg 29, PO Box 211, 1000 AE Amsterdam, The Netherlands
Linacre House, Jordan Hill, Oxford OX2 8DP, UK

First edition 2008

Library of Congress Cataloging-in-Publication Data
A catalog record for this book is available from the Library of Congress

British Library Cataloguing in Publication Data
A catalogue record for this book is available from the British Library

ISBN: 978-0-444-53010-3

For information on all Elsevier publications
visit our website at books.elsevier.com

Transferred to Digital Printing 2010

Contents

Preface

Applications for biosensors in diverse areas of application continue to increase. The initial application for biosensors was for the effective management of diabetes mellitus (DM) by the detection of glucose levels in blood. However, since then medical and other applications for biosensors are gradually increasing. The similarities between the detection of harmful pathogens brought into focus by implied and realistic terrorist threats and biomarkers for the early detection of disease is being exploited in the development of biosensors for different applications. The present book in the series of books on analyte–receptor interaction kinetics on biosensor surfaces has a co-author (Neeti Sadana, MD) to emphasize the medical applications of biosensors, and to provide a better perspective of their applications. The initial chapter now provides not only a balance for the whole book, but also eases one into what to expect in the following chapters.

Fractal mathematics provides a convenient means to provide novel physical insights into the kinetics of analyte–receptor reactions occurring on biosensor surfaces. It is hoped that eventually this will help improve the biosensor performance parameters such as sensitivity, selectivity, response time, stability, etc. with the eventual goal of, for example, (a) the early detection of biomarkers for harmful and insidious diseases and (b) providing emergency personnel that extra few minutes to help move large sections of human population from terrorist threat affected areas.

Chapter 1 focuses on the vast medical implications of fractals in modern medical practice. The cost of current health care on society and diseased individuals is examined. Common illnesses such as cardiovascular disease, cancer, and diabetes mellitus are explored to find practical uses for fractal theory.

Chapter 2 outlines the basic fractal theory to model the binding and the dissociation (if applicable) kinetics used. A simple single-fractal model to analyze the kinetics is initially presented. This is then followed by dual- and triple-fractal models to analyze complex binding and dissociation kinetics when a single-fractal analysis did not provide an adequate fit. In Chapter 3 different examples are presented wherein the biosensor performance parameters have been enhanced by experimental modification and validated by the fractal theory kinetics presented. This includes enhancement of biosensor sensitivity, immobilization, reproducibility, resolution performance, etc.

The detection of harmful bacteria, toxins, and pathogens on biosensor surfaces is outlined in Chapter 4. TV (CNN) news in the United Sates in February 2007 indicated the concerns that consumers expressed over food items available in grocery stores, and the number of illnesses and also deaths resulting from pathogenic bacteria-contaminated food (particularly different types of meat). Suggestions were made to increase the testing for these pathogenic bacteria (such as *Salmonella*, *Listeria monocytogenes*, and *Escherichia coli)* by the governmental agencies responsible. One of the major themes for the book is the application of biosensors for the early detection of insidious diseases, specifically their biomarkers. This has been brought into proper perspective in the first chapter by an expert

MD. In Chapter 5 we analyze the binding kinetics of different biomarkers for the early detection of diseases. Some of the biomarkers analyzed include C-reactive protein (MI, myocardial infarction) and glutamic acid decarboxylase (marker for IDDM, insulin-dependent diabetes mellitus).

Chapter 6 analyzes the binding and dissociation kinetics of analytes that help control diseases on biosensor surfaces. One example analyzed includes drugs that help control blood pressure. Biosensors are being used extensively nowadays in drug discovery. Chapter 7 analyzes the binding and dissociation kinetics of small molecules involved in drug discovery on biosensor surfaces.

Alzheimer's disease is an insidious disease and generally affects the older generation. Prions are generally recognized to be involved in the onset and in the progression of this disease. Chapter 8 analyzes the binding and dissociation kinetics of prion-related interactions occurring on biosensor surfaces. Chapter 9 analyzes the binding and dissociation of DNA–analyte interactions occurring on biosensor surfaces. Some of the interactions analyzed include T7 DNA polymerase/DNA, DNA/histones, and hybridization of a molecular beacon (probe)/complement with and without micro-bubble activation.

Chapter 10 analyzes the binding and dissociation kinetics of protein–analyte interactions on biosensor surfaces. Chapter 11 analyzes the binding and dissociation kinetics of different compounds on biosensor surfaces. The examples presented in this chapter were taken at random from the literature, and placed together in this chapter with apparently no common theme amongst them.

Chapter 12 analyzes the binding and dissociation kinetics of environmental contaminants on biosensor surfaces. The kinetics of binding and dissociation (if applicable) of atrazine, acetylcholine, and catechol on biosensor surfaces is analyzed. The last chapter, Chapter 13, is the capstone chapter which analyzes the economics involved in setting up a biosensor industry, and the markets for biosensors in the different areas of application. This type of information is difficult to get free of charge in the open literature, besides both of the authors have experience only in the academic areas. One may expect to pay a few thousand dollars for reports that provide this type of information available from private sources. However, these reports too are rapidly outdated due to the quick changing landscape of biosensor economics. Nevertheless, the different economic aspects are presented, albeit from an academic viewpoint, along with projected estimates of the biosensor markets in the different areas of application.

The co-author, Ajit Sadana expresses his appreciation to Dr. Kai-Fong Lee, Dean, School of Engineering at the University of Mississippi for his continued support and encouragement for research that facilitates the writing of treatises like these.

Neeti Sadana, MD, Miami, Florida
Ajit Sadana, Oxford, Mississippi

– 1 –

INTRODUCTION

COST AND IMPLICATIONS OF MEDICAL CARE OF SOME COMMON DISEASES

The burgeoning cost of health care has been a topic of debate in local, regional, national, and most importantly world politics. Heart disease, diabetes mellitus (DM), and cancer together make up the top three causes of death worldwide. The economic and personal burden of heart disease, which includes primarily myocardial infarctions, cerebrovascular disease, hypertension, and the broad term coronary artery disease cost the European Union 169 billion euros in 2003 a new study recently found. That figure comes out to 230 euros for every man, woman, and child in the European Union regardless of location and previous medical history. The United Kingdom spent the largest portion of this budget. Not surprisingly, the UK has a diet and lifestyle most similar to the United States. This study also found that cardiovascular disease (CVD) accounted for 268.5 million working days lost and that the two million deaths from CVD cost 24.4 billion euros and involved 2.18 million lost working years (BBC News article, 2006).

What about the United States? Heart disease is the number one killer of Americans with no prejudices against men or women of all races, cultures, and backgrounds. The American Heart Association (2007) estimates 1.1 million people in the U.S. will suffer a heart attack this year. 600,000 will suffer from some form of cerebrovascular accident or stroke. Of these 600,000, 30% will die within a year from the incident and almost as many will have some form of permanent disability as a sequelae of the stroke. Studies estimate that by age 65, almost more than half of Americans will have some form of CVD. The American Heart Association (2007) also reveals that heart disease and stroke cost the U.S. more than $350 billion in 2003 and that expense keeps rising. Of the estimated $350 billion spent on the annual health care budget, $209 billion was in direct medical costs, $32 billion for lost productivity, and even more startling $110 billion for loss of future productivity due to premature death primarily from heart disease and its counterpart cerebrovascular disease (American Heart Association, 2007).

When talk of medical cost occurs, the need to discuss Diabetes is not far behind. Diabetes itself places people at a higher risk for heart disease. Chronic diabetics set themselves up for such complications as blindness, kidney disease and failure requiring dialysis, and extremity amputation. DM comes in two flavors. The juvenile onset type 1 DM and the more prevalent adult onset type 2 DM. The American Diabetes Association claims direct medical and indirect expenditures attributable to diabetes in 2002 were thought to be in the way of $132 billion. Direct medical costs made up $91.8 billion and included diabetes care, chronic complications as discussed above, and the superimposed medical conditions associated with DM. Essential for understanding the economics of DM patients is in-patient hospital days, nursing home care and office visits for dental care, optometry care, and the use of licensed dieticians. The indirect cost of DM includes lost workdays, days of decreased activity secondary to disease, mortality, and permanent disability. The cost was estimated to be $39.8 billion. The cost per person of the $865 billion U.S. heath care expenditure included $13,243 per capita for people with diabetes and $2560 for people without diabetes. Not surprisingly, diabetics had expenditures 3–4 times that of nondiabetics. More frightening is the prospect of future generations with an increased incidence of DM. If prevalence rates remain constant, the Census Bureau estimates the number of people diagnosed with diabetes could increase to 14.5 million by 2010 and 17.4 million by the year 2020. This would increase the cost of diabetes to $156 billion by 2010 and $192 billion by 2020 (Diabetes Care, 2003).

Cancer seems to be that disease that we as humans have much sympathy towards. Cancer is the human body's way of attacking its own machinery and only rarely is a direct result of inactivity, overeating, and lethargy as the above mentioned CVD and DM. Nevertheless, cancer is a formidable opponent in the health care crisis that we face today. Breast cancer is the second major cause of cancer death in American women with death rates over 44,000 per year. While ovarian cancer, the evil cousin of breast cancer, accounts for much fewer deaths then breast cancer, it represents approximately 4% of female cancers. Even more unfortunate is the genetic link between breast and ovarian cancer. An enormous breakthrough in cancer research occurred when two breast cancer susceptibility genes BRCA 1 and BRCA 2 were identified in the early 1990s (Science, 1994). The human chromosome 17q21 contains the BRCA 1 gene and physical mapping and study of this gene led to the discovery that women with a mutation of this gene or its sister BRCA 2 on chromosome 13q12-q13 were at an increased risk for breast and/or ovarian cancer at some time in life.

BRCA mutations conferred a high risk for disease at an early age. Disease at an earlier age conferred a poorer prognosis. Even more frightening was the elucidation that BRCA mutations tend to run in families, many times in an autosomal dominant pattern, affecting generations of women. The function of these genes only recently was discovered as participation in DNA repair via radiation-induced breaks. Mutations in BRCA 1 and BRCA 2 lead to errors in DNA replication and eventually abnormal cancerous growth. Testing for BRCA 1 and BRCA 2 remain for tertiary care university centers due to cost. It would be enormously beneficial to make the testing for genetic susceptibility available for the masses.

Why did we take the time to outline the cost and implications of heart disease, DM and cancer? Simple. To emphasize the fact that while we cannot eradicate these diseases, it is possible to control the level of associated morbidity and mortality by early detection. Biomarkers have been elucidated to detect various aspects of heart disease, DM, and certain cancers. Specifically, biosensors are microprocessors that allow the recognition and conversion of certain biologic elements to be converted to chemical, electrical, magnetic, optical, and thermal signals. These signals are then processed and amplified for display and interpretation. The first biosensors were used to detect glucose levels in diabetics. Early and more accurate glucose levels can lead to tighter control and titration of oral hypoglycemic drugs. By early detection and better glucose control, the use of insulin may be delayed.

Delayed use of insulin adds years to life as well as decreases the complications of long-term insulin therapy and certainly improves quality of life. Biosensors are becoming more sophisticated as will be elucidated further in this book and allow us greater reliability, greater accuracy, and ease of use. But more money and research are still required to take this technology to the level of bedside use and becoming the standard of care in medicine.

Newer uses of fractals and disease-related analytes include an endothelial biosensing system. This system recognizes early or prehypertension leading to the earlier diagnoses and implementation of diet and exercise regimes followed by earlier use of antihypertensives. Again, hypertension is a clinical predictor for both heart disease and coronary artery disease, which are major sources of morbidity and mortality worldwide as previously mentioned.

Detection of C-reactive protein, a nonspecific marker for inflammation in human plasma, has already been used to detect, stratify, and treat accordingly myocardial infarctions. As current cardiac biomarkers troponin I and creatine kinase MB are performed routinely every 8 h as a blood assay times 3 before a myocardial infarction is ruled out or an acute coronary syndrome is ruled in. The management of an acute MI is vastly different from acute coronary syndrome such that the accuracy of these markers is truly essential. Lactate dehydrogenase levels and the change in ratio of LDH 1 > LDH 2 is also now used for late presentations of MI while troponins mentioned above are elevated for days. Fractal models provide for earlier and more accurate levels of these biomarkers and are part of the standard of care workup for patients coming to emergency rooms throughout the world. Further research of these biomarkers with fractal theory could only further medical practice and accuracy. Would not it be nice to know sooner rather than later that the pain felt in one's chest is angina verses an acute MI? The implication of this technology is vast.

A related biomarker is the use of a fluorescent coagulation assay using a fiber optic evanescent wave sensor for the detection of thrombin. Thrombin leads to the production of fibrin via both the intrinsic and extrinsic coagulation pathways. These pathways make a soft clot that thrombin further makes stronger by cross-linking molecules to make the clot resistant to dissolution. Clot formation is essential in normal hemostasis as well as bleeding tendencies that result in life-threatening coagulation disorders.

The above mentioned biosensor examples are only a preview of the vast implications of fractal models in disease control. Many more will be further elucidated in this book along with the mechanism of fractal action. It is important to realize that taking relatively new technology and designing it to be used at the bedside for rapid and accurate patient care is well within reach. New applications for biosensors are being discovered at an enormous pace, but do require adequate research and funding. In addition, cooperation and understanding by the medical community at large is essential for the appropriate use of this fascinating and practical knowledge.

REFERENCES

American Heart Association, Investment in Research Saves Lives and Money. Volume No. 2, 2007.

BBC News article, 'Heart Disease Costs EU Billions'. Wednesday February 22, 2006.

Diabetes Care, *Economic Costs of Diabetes in the U.S. in 2002*, 2003, 26(3), 917–932.

Science, *Localization of a Breast Cancer Susceptibility Gene, BRCA 2, to Chromosome 13q12-13*, 1994, September 30, 265(5181), 2088–2090.

– 2 –

Modeling and Theory

2.1 INTRODUCTION

In a biosensor based assay the molecule to be detected (analyte) is present in solution and the appropriate receptor is immobilized on a solid surface. The interaction between the analyte and the receptor on the solid biosensor surface is detected either by a change in the refractive index (in SPR—surface plasmon resonance—instruments) or by changes in the fluorometric intensity, ultraviolet light intensity, etc. The SPR biosensor protocol analyzes the binding (and dissociation where applicable) kinetic curves using classical saturation models involving analyte–receptor binding using 1:1, 1:2, etc. ratios, generally under diffusion-free conditions and assuming that the receptors are homogeneously distributed over the sensor surface. Computer programs and software that come with the equipment provide values of the binding (and the dissociation) rate coefficients. Though a careful analysis and experimental protocol may eliminate or minimize the influence of diffusional limitations; realistically speaking, it is more appropriate to include a heterogeneous distribution on the sensing surface. Heterogeneity on the sensing surface and in the biosensor systems itself may be due to other reasons, such as, non-specific binding, inherent irregularities on the sensing surface, mixture of receptors on the surface, and mixture of analytes in solution which includes the analyte of interest.

Two factors need to be addressed whilst analyzing the analyte–receptor binding and dissociation kinetics. The system by its design is heterogeneous. For example, and as indicated above, the receptors immobilized on the biosensor surface may exhibit some heterogeneity, that is, surface roughness. No matter how careful one is in immobilizing the receptors on the biosensor surface, there will be some degree of heterogeneity on the surface.

Henke *et al.* (2002) have used the atomic force microscopy (AFM) technique to determine the effects of the cleaning of fused silica and glass on surface roughness. This is for biosensor use. Note that prior to the immobilization of receptors on the surface, the surface needs to be cleaned to remove contaminants, and to create surface attachment sites, for example, for hydroxyl groups.

For the analyte–receptor binding (and dissociation) to take place the analyte by the diffusion process must come within the 'proximity' of the active site on the receptor. Mass transport limitations may be minimized or eliminated if the system is either properly designed or properly operated or both. In most cases, however, both diffusional effects and heterogeneity aspects will be present in biosensor systems, and their influence on binding and dissociation kinetics needs to be determined. Ideally, one would like to determine the influence of each of these separately on the binding and dissociation kinetics. In the theoretical analysis presented below (the Havlin (1989) analysis) the effects of diffusion and heterogeneity are presented coupled together. One possible way of accounting for the presence of diffusional limitations and the heterogeneity that exists on the surface is by using fractals. Ideally, and as indicated above, one would prefer to decouple the influence of diffusion and heterogeneity. Presumably, an approach other than fractal analysis is required to decouple these two effects.

A characteristic feature of fractals is self-similarity at different levels of scale. Fractals exhibit dilatational symmetry. Fractals are disordered systems, and the disorder is described by non-integral dimensions (Pfeifer and Obert, 1987). Fractals have non-integral dimensions, and are smaller than the dimension they are embedded in. In other words, the highest value that a fractal can have is three. In our case, an increase in the degree of heterogeneity on the biosensor surface would lead to an increase in the value of the fractal dimension. Another way of looking at the fractal dimension is its 'space filling' capacity. The more the space a surface fills, the higher is its fractal dimension. The fractal dimension cannot have a negative value, and very low values of the fractal dimension on the surface indicate that the surface exists as a Cantor like dust.

Kopelman (1988) indicates that surface diffusion-controlled reactions that occur on clusters or islands are expected to exhibit anomalous and fractal-like kinetics. These kinetics exhibit anomalous reactions orders and time-dependent (e.g., binding) rate coefficients. As long as surface irregularities show scale invariance they can be characterized by a single number, the fractal dimension. Later on in the book we will characterize the surfaces of the biosensors used in the different examples by a fractal dimension. More specifically, we will characterize the heterogeneity present on these biosensor surfaces by a fractal dimension.

The fractal dimension is a global property, and it is insensitive to structural or morphological details (Pajkossy and Nyikos, 1989). Markel *et al.* (1991) indicate that fractals are scale self-similar mathematical objects that possess non-trivial geometrical properties. Furthermore, these authors indicate that rough surfaces, disordered layers on surfaces, and porous objects all possess fractal structure. A consequence of the fractal nature is a power-law dependence of a correlation function (in our case the analyte–receptor on the biosensor surface) on a coordinate (e.g., time).

Pfeifer (1987) indicates that fractals may be used to track topographical features of a surface at different levels of scale. Lee and Lee (1995) indicate that the fractal approach permits a predictive approach for transport (diffusion related) and reaction processes occurring on catalytic surfaces. This approach may presumably be extended to diffusion-limited analyte–receptor reactions occurring on biosensor surfaces.

The binding of an analyte in solution to a receptor attached to a solid (albeit flow cell or biosensor surface) is a good example of a low dimension reaction system in which the distribution tends to be 'less random' (Kopelman, 1988), and a fractal analysis would provide novel physical insights into the diffusion-controlled reactions occurring at the surface. Also, when too many parameters are involved in a reaction, which is the case for these analyte–receptor reactions on a solid (e.g., biosensor surface), a fractal analysis provides a useful lumped parameter. It is appropriate to pay particular care to the design of such systems and to explore new avenues by which further insight or knowledge may be obtained in these biosensor systems. The fractal approach is not new and has been used previously in analyzing different phenomena on lipid membranes.

Fatin-Rouge *et al.* (2004) have recently presented a summary of cases where the analysis of diffusion properties in random media have led to significant theoretical and experimental interest. These cases include soils (Sahimi, 1993), gels (Starchev *et al.*, 1997; Pluen *et al.*, 1999), bacteria cytoplasm (Berland *et al.*, 1995; Schwille *et al.*, 1999), membranes (Saffman and Delbruck, 1975; Peters and Cherry, 1982; Ghosh and Webb, 1988), and channels (Wei *et al.*, 2000). Coppens and Froment (1995) have analyzed the geometrical aspects of diffusion and reaction occurring in a fractal catalyst pore. In this chapter, and in this book as a whole, we are extending the analysis to analyte–receptor binding (and dissociation) on biosensor surfaces.

Fatin-Rouge *et al.* (2004) indicate that in most real systems disorder may exist over a finite range of distances. Harder *et al.* (1987) and Havlin (1989) indicate that in this range the diffusion process cannot be characterized by the classical Fick's law. In this range, anomalous diffusion applies. Fatin-Rouge *et al.* (2004) emphasize that at larger distances than in the above window range, the effects of disorder on diffusion may be very small due to statistical effects, and may cancel each other.

Prior to presenting the Havlin (1989) analysis modified for the analyte–receptor binding occurring on biosensor surfaces, it is appropriate to present briefly the analysis presented by Fatin-Rouge *et al.* (2004) on size effects on diffusion processes within agarose gels, and apply it to analyte–receptor binding and dissociation for biosensor kinetics. This analysis provides some insights into general fractal-related processes. Fatin-Rogue *et al.* (2004) have considered diffusion within a fractal network of pores. They indicate that fractal networks such as percolating clusters may be characterized by a power law distribution (Havlin, 1989):

$$M \propto (L)^{D_f} \tag{2.1}$$

Here M is the average number of empty holes in the (gel) space characterized by a linear size, L. The exponent, D_f is the mass fraction dimension. Fatin-Rogue *et al.* (2004) emphasize that in the general case of fractals, D_f is smaller than the dimension of space of interest. Furthermore, the independence of D_f on scale is also referred to as self-similarity, and is an important property of rigorous fractals.

Havlin and Ben-Avraham (1987) indicate that the diffusion behavior of a particle within a medium can be characterized by its mean-square displacement, $r^2(t)$ versus time, t, which is written as:

$$r^2(t) = \gamma\, t^{(2/D_w)} \tag{2.2a}$$

Here γ is the transport coefficient, and D_w is the fractal dimension for diffusion. Normal or regular diffusion occurs when D_w is equal to 2. In this case, $r^2(t)$ is equal to γt. In other words,

$$r^2(t) = 2dDt \tag{2.2b}$$

Here d is the dimensionality of space, and D is the diffusion coefficient.

Harder *et al.* (1987) and Havlin (1989) describe *anomalous diffusion* wherein the particles sense obstructions to their movement. This is within the fractal matrix, or in our case due to heterogeneities on the biosensor surface, perhaps due to irregularities on the biosensor surface. Fatin-Rogue *et al.* (2004) are careful to point out that anomalous diffusion may also occur due to nonelastic interactions between the network and the diffusing particles in a gel matrix (Saxton, 2001). Furthermore, Fatin-Rouge *et al.* (2004) indicate that *anomalous diffusion* is different from *trapped diffusion* wherein the particles are permanently trapped in holes, and are unable to come out of these holes. When the particles (analyte in our case) are in these trapped holes, then as time, $t \rightarrow \infty$, the mean-square displacement, $r^2(t)$ tends to a constant value. Fatin-Rouge *et al.* (2004) emphasize that in real heterogeneous porous media anomalous diffusion of particles occurs over a limited length- or time-scales since the structure is only fractal over a limited size scale. In other words, there is a lower bound and an upper bound over which the fractal structure applies. Similarly, in our case, the anomalous diffusion of the analyte on the biosensor surface occurs over a limited range of length- or time-scales.

For anomalous diffusion, one may combine the right-hand sides of eqs. (2.2a) and (2.2b). Then, the diffusion coefficient, D is given by (Fatin-Rouge *et al.*, 2004):

$$D(t) = \left(\frac{1}{4}\right)\gamma\, t^{[(2/D_{\mathrm{w}})-1]} \tag{2.3}$$

Due to the temporal nature of $D(t)$, it is better to characterize the diffusion of the analyte in our case by D_{w}. If we were still talking about the medium and gels, then D_{w} would refer to the diffusing medium.

We will now develop the theory for the analyte–receptor binding and dissociation on biosensor surfaces. We will use the (Havlin, 1989) approach.

2.2 THEORY

We present now a method of estimating fractal dimension values for analyte–receptor binding and dissociation kinetics observed in biosensor applications. The following chapters will present the different examples of data that have been modeled using the fractal analysis. The selection of the binding and dissociation data to be analyzed in the later chapters is constrained by whatever is available in the literature.

2.2.1 Variable rate coefficient

Kopelman (1988) has indicated that classical reaction kinetics are sometimes unsatisfactory when the reactants are spatially constrained on the microscopic level by either walls, phase boundaries, or force fields. Such heterogeneous reactions, for example bioenzymatic reactions, that occur at interfaces of different phases, exhibit fractal orders for elementary reactions and rate coefficients with temporal memories. In such reactions, the rate coefficient exhibits a form given by:

$$k_1 = k' t^{-b} \qquad 0 \le b \le 1\, (t \ge 1) \tag{2.4}$$

In general, k_1 depends on time whereas $k' = k_1(t = 1)$ does not. Kopelman (1988) indicates that in three dimensions (homogeneous space) $b = 0$. This is in agreement with the results obtained in classical kinetics. Also, with vigorous stirring, the system is made homogeneous and b again equals zero. However, for diffusion-limited reactions occurring in fractal spaces, $b \geq 0$; this yields a time-dependent rate coefficient.

Antibodies may form fractal clusters on biosensor surfaces. These antibodies or receptors on the biosensor surface may consist of islands of highly organized or disorganized antibodies. This has similarity to the growth of crystalline structures. It is quite possible that a cooperative effect may arise due to this tightly organized fractal structures. This is one possibility that could lead to an increase in the binding rate coefficient with an increase in the fractal dimension or the degree of heterogeneity on the biosensor surface.

The diffusion-limited binding kinetics of antigen (or antibody or analyte or substrate) in solution to antibody (or antigen, or receptor, or enzyme) immobilized on a biosensor surface has been analyzed within a fractal framework (Sadana and Beelaram, 1994; Sadana *et al.*, 1995). One of the findings, for example, is that an increase in the surface roughness or fractal dimension leads to an increase in the binding rate coefficient. Furthermore, experimental data presented for the binding of HIV virus (antigen) to the antibody immobilized on a surface displays characteristic ordered 'disorder' (Anderson, 1993). This indicates the possibility of a fractal-like surface.

A biosensor system (wherein either the antigen, antibody, analyte, or substrate is attached to the surface), along with its different complexities, which include heterogeneities on the surface and in solution, diffusion-coupled reaction, time-varying adsorption or binding rate coefficients, etc., can be characterized as a fractal system.

The diffusion of reactants towards fractal surfaces has been analyzed (De Gennes, 1982; Pfeifer *et al.*, 1984a,b; Nyikos and Pajkossy, 1986). Havlin (1989) has briefly reviewed and discussed these results. The diffusion is in the Euclidean space surrounding the fractal surface (Giona, 1992). Havlin (1989) presents an equation that may be utilized to describe the build-up of the analyte–receptor on a biosensor surface during the binding reaction. The receptor is immobilized on the biosensor surface. This equation is given below. In all fairness, at the outset, it is appropriate to indicate that the biosensor surface is assumed to be fractal, or possibly so.

Ideally, it is advisable to provide independent proof or physical evidence for the existence of fractals in the analysis of analyte–receptor reactions occurring on biosensor surfaces. Also, and as indicated earlier, if the diffusion effects can be separated from the heterogeneity effects, then one may better understand the effects of each of these on analyte–receptor reactions occurring on biosensor surfaces. In general, diffusion effects may be minimized either by increasing flow rates or by immobilizing less amounts of receptors on the biosensor surface.

In general, to demonstrate fractal-like behavior log–log plots of distribution of molecules $M(r)$ as a function of the radial distance (r) from a given molecule are required. This plot should be close to a straight line. The slope of the log $M(r)$ *versus* log(r) plot determines the fractal dimension. In our case, one could try to obtain a log–log plot of two variables, k and time, t and perform a least squares fit in this parameter space to find the slope of the curve. A regression coefficient at this stage could be beneficial in understanding the efficacy of this metric. However, an easier method, without the use of the required log-log plots, is presented below.

This is the equation developed by Havlin (1989) for diffusion of analytes towards fractal surfaces.

2.2.2 Single-fractal analysis

In the literature some authors refer to binding as comprising of two phases, an association phase and a dissociation phase. In this chapter and in the book, we will refer to binding as just binding. The dissociation phase is separate.

Binding rate coefficient

Havlin (1989) indicates that the diffusion of a particle (analyte) from a homogeneous solution to a solid surface (e.g., receptor-coated surface) on which it reacts to form a product (analyte–receptor complex) is given by:

$$(\text{Analyte} - \text{Receptor}) \sim \begin{cases} t^{(3-D_{\mathrm{f,bind}})/2} = t^{p} & (t < t_{\mathrm{c}}) \\ t^{1/2} & (t > t_{\mathrm{c}}) \end{cases} \tag{2.5a}$$

where the analyte–receptor represents the association (or binding) complex formed on the surface. Here $p = -b$, and D_{f} is the fractal dimension of the surface. Havlin (1989) states that the crossover value may be determined by $r_{\mathrm{c}}^2 \sim t_{\mathrm{c}}$. Above the characteristic length, r_{c}, the self-similarity of the surface is lost and the surface may be considered homogeneous. Eq. (2.5a) indicates that the concentration of the product [analyte–receptor] on a solid fractal surface scales at short and intermediate times as analyte–receptor $\sim t^{p}$ with the coefficient $p = (3 - D_{\mathrm{f}})/2$ at short time scales and $p = 1/2$ at intermediate time scales. Note that D_{f}, $D_{\mathrm{f,assoc}}$, and $D_{\mathrm{f,bind}}$ are used interchangeably. This equation is associated with the short-term diffusional properties of a random walk on a fractal surface. Note that, in perfectly stirred kinetics on a regular (nonfractal) structure (or surface), the binding rate coefficient, k_1 is a constant, that is, it is independent of time. In other words, the limit of regular structures (or surfaces) and the absence of diffusion-limited kinetics leads to k_1 being independent of time. In all other situations, one would expect a scaling behavior given by $k_1 \sim k' t^{-b}$ with $-b = p < 0$. Also, the appearance of the coefficient, p different from $p = 0$ is the consequence of two different phenomena, that is, the heterogeneity (fractality) of the surface and the imperfect mixing (diffusion-limited) condition. Finally, for a homogeneous surface where D_{f} is equal to 2, and when only diffusional limitations are present, $p = 1/2$ as it should be. Another way of looking at the $p = 1/2$ case (where $D_{\mathrm{f,bind}}$ is equal to two) is that the analyte in solution views the fractal object, in our case the receptor-coated biosensor surface, from a 'large distance.' In essence, in the association process, the diffusion of the analyte from the solution to the receptor surface creates a depletion layer of width $(+t)^{1/2}$ where + is the diffusion constant. This gives rise to the fractal power law,

$$(\text{Analyte–Receptor}) \sim t^{(3 - D_{\mathrm{f,bind}})/2}$$

The values of the parameters k (binding rate coefficient), p, and D_{f} in eq. (2.5a) may be obtained for analyte–receptor association kinetics data. This may be done by a regression analysis using, for example, Corel Quattro Pro (1997) along with eq. (2.5a) where (analyte–receptor) $= kt^p$ (Sadana and Beelaram, 1994; Sadana *et al.*, 1995). The fractal dimension may be obtained from the parameter p. Since $p = (3 - D_{f,bind})/2$, $D_{f,bind}$ is equal to $(3 - 2p)$. In general, low values of p would lead to higher values of the fractal dimension, $D_{f,bind}$. Higher values of the fractal dimension would indicate higher degrees of 'disorder' or heterogeneity or inhomogeneity on the surface.

Another way of looking at the diffusive process is that it inherently involves fluctuations at the molecular level that may be described by a random walk (Weiss, 1994). This author indicates that the kinetics of transport on disordered (or heterogeneous) media itself needs to be described by a random-walk model. When both of these are present, that is, the diffusion phenomena as well as a fractal surface, then one needs to analyze the interplay of both of these fluctuations. In essence, the disorder on the surface (or a higher fractal dimension, D_f) tends to slow down the motion of a particle (analyte in our case) moving in such a medium. Basically, according to Weiss (1994) the particle (random walker analyte) is trapped in regions in space, as it oscillates for a long time before resuming its motion.

Havlin (1989) indicates that the crossover value may be determined by $r_c^2 \sim t_c$. Above the characteristic length, r_c, the self-similarity of the surface of the surface is lost. Above t_c, the surface may be considered homogeneous, and 'regular' diffusion is now present. One may consider the analysis to be presented as an intermediate 'heuristic' approach in that in the future one may also be able to develop an autonomous (and not time-dependent) model of diffusion-limited kinetics in disordered media.

It is worthwhile commenting on the units of the association and the dissociation rate coefficient(s) obtained for the fractal analysis. In general, for SPR biosensor analysis, the unit for the analyte–receptor complex on the biosensor surface is RU (resonance unit). One thousand resonance units is generally 1 ng/(mm)^2 (of surface), or one resonance unit is 1 pg/(mm)^2. Here, ng and pg are nanogram and picogram, respectively. Then, to help determine the units for the binding coefficient, k, from eq. (2.5a):

$$\text{(Analyte} - \text{Receptor), pg/(mm)}^2 = kt^p = kt^{(3-D_{f,bind})/2}$$

This yields a unit for the binding rate coefficient, k as $(\text{pg})(\text{mm})^{-2}(\text{sec})^{(D_{f,bind}-3)/2}$. Note that the unit of dependence in time exhibited by the association (or binding) rate coefficient, k changes slightly depending on the corresponding fractal dimension obtained in the binding phase, $D_{f,bind}$. The fractal dimension value is less than or equal to three. Three is the highest value of the fractal dimension, since the system is embedded in a three-dimensional system. k and k_{bind}, and D_f, and $D_{f,bind}$ are used interchangeably in this chapter and in the book.

It should be indicated that different laboratories use different technologies or different experimental designs to analyze the binding affinity of ligands to target proteins (or analytes) of interest (or to determine the rate coefficients for association and dissociation kinetics for binding). The comparison of data between different technologies and experimental designs and conclusions thereof should be made with great caution. The fractal analysis is of value in that it provides the pros and cons of different *in vitro* technologies (or more precisely, in this case, analysis procedures). It makes the user of the technology aware of the quality of data generated and what can be done to improve the analysis.

One might very reasonably question the utility of the approach considering the different dimensions, and subsequently the units one may obtain even for the same interactions. It would be difficult to compare this technique with other approaches for different interactions.

Nevertheless, the inclusion of the surface effects is essential, albeit difficult. This is especially true, if the rate coefficients for association and dissociation for binding are very significantly dependent on the nature of the surface. Unless, a simpler and alternate approach is suggested that includes the surface effects, it is reasonable for the present, to follow the present approach. Hopefully, modifications, to this approach may be suggested that permit the comparison for different interactions as well as with other approaches.

It would be useful to specify what the carrier of fractal properties is. It could either be the analyte surface, the receptor surface, or the immobilizing (in our case, the biosensor) surface. There is a considerable body of work on fractal surface properties of proteins (Li *et al.*, 1990; Federov *et al.*, 1999; Dewey and Bann, 1992; Le Brecque, 1992). Le Brecque (1992) indicates that the active sites (in our case the receptors on the biosensor surface) may themselves form a fractal surface. Furthermore, the inclusion of non-specific association sites on the surface would increase the degree of heterogeneity on the surface, thereby leading to an increase in the fractal dimension of the surface. At present, we are unable to specify what the carrier of the fractal properties is. This is exacerbated by our re-analysis of kinetic data available in the literature. Presumably, it is due to a composite of some or all of the factors mentioned above. No evidence of fractality is presented.

Dissociation rate coefficient

The diffusion of the dissociated particle (receptor or analyte) from the solid surface (e.g., analyte–receptor complex coated surface) into solution may be given as a first approximation by:

$$\begin{aligned}(\text{Analyte} - \text{Receptor}) &\sim -t^{(3-D_{\text{f,diss}})/2} \qquad (t > t_{\text{diss}}) \\ &= -k_{\text{diss}} t^{(3-D_{\text{f,diss}})/2}\end{aligned} \tag{2.5b}$$

Here $D_{\text{f,diss}}$ is the fractal dimension of the surface for the dissociation step. t_{diss} represents the start of the dissociation step. This corresponds to the highest concentration of the analyte–receptor complex on the surface. Henceforth, its concentration only decreases. $D_{\text{f,bind}}$ may or may not be equal to $D_{\text{f,diss}}$ k_{d} and k_{diss}, and $D_{\text{f,d}}$ and $D_{\text{f,diss}}$ are used interchangeably in this chapter and in this book.

One may obtain a unit for the dissociation rate coefficient, k_{d} in a similar manner as done for the binding rate coefficient. In this case, the units for the binding and the dissociation rate coefficient are the same. The unit for the dissociation rate coefficient, k_{d} is $(\text{pg})(\text{mm})^{-2}(\text{sec})^{(D_{\text{f,diss}}-3)/2}$. Once again, note that the unit dependence on time exhibited by k_{d} changes slightly due to the dependence on $D_{\text{f,diss}}$.

2.2.3 Dual-fractal analysis

Binding rate coefficient

The single-fractal analysis we have just presented is extended to include two fractal dimensions. At present, the time ($t = t_1$) at which the first fractal dimension 'changes' to the second fractal dimension is arbitrary and empirical. For most part it is dictated by the data analyzed and the experience gained by handling a single-fractal analysis. The r^2 (regression coefficient) value obtained is also used to determine if a single-fractal analysis is sufficient, or one needs to use a dual-fractal analysis to provide an adequate fit. Only if the r^2 value is less than 0.97 for a single-fractal analysis, do we use a dual-fractal model.

In this case, the analyte–receptor complex is given by:

$$(\text{Analyte}-\text{Receptor}) \sim \begin{cases} t^{(3-D_{\text{f1,bind}})/2} = t^{p_1} & (t < t_1) \\ t^{(3-D_{\text{f2,bind}})/2} = t^{p_2} & (t_1 < t < t_2 = t_c) \\ t^{1/2} & (t > t_c) \end{cases} \tag{2.5c}$$

In analyte–receptor binding the analyte–receptor binds with the active site on the surface and the product is released. In this sense the catalytic surface exhibits an unchanging fractal surface to the reactant in the absence of fouling and other complications. In the case of analyte–receptor association the biosensor surface exhibits a changing fractal surface to the analyte in solution. This occurs as each association reaction takes place, smaller and smaller amounts of 'association' sites or receptors are available on the biosensor surface to which the analyte may bind. Furthermore, as the reaction proceeds, there is an increasing degree of heterogeneity on the biosensor surface for some reaction systems. This is manifested by two degrees of heterogeneity, or two fractal dimensions on the biosensor surface. In the theoretical limit one might envisage a temporal fractal dimension wherein there is a continuous change in the degree of heterogeneity on the surface; though of course, such situations would be very rare, if at all.

Surfaces exhibit roughness, or a degree of heterogeneity at some scale. This degree of heterogeneity on the surface may be due to fracture or erosion. In our case of biosensors, this may arise due to: (a) the inherent roughness of the biosensor surface; or (b) the immobilization or deposition of the receptors on the biosensor surface. The method of deposition of the receptors on the surface would also lead to different degrees of heterogeneity on the surface. The binding reaction takes place between the analyte in solution and the receptors on the surface through chemical bond formation and subsequent molecular association. The geometric nature (or parameter) of the surface will significantly influence these reactions. The influence of surface morphology and structure has been analyzed (Lee and Lee, 1994; Chaudhari *et al.*, 2002, 2003). It would be of interest to determine the scale of these roughness heterogeneities. Are these at the Angstrom level or lower? With the current emphasis on nanotechnology and nanobiotechnology these types of questions are becoming more and more relevant and of significance. The nature of surfaces in general, and of biosensors in particular (our case) should exhibit a fractal nature at the molecular level. Furthermore, one of the reasons for the emphasis on nanotechnology is that as one goes down in scale, the properties of some substances change, sometimes for the better. It is these beneficial changes that one wishes to exploit in nanotechnology and nanobiotechnology. Hopefully, similar parallels can be drawn on analyzing the fractal nature of biosensor surfaces. Do they exhibit self-similarity; and if they do what are their limits? In other words, what are their lower and upper bounds?

Furthermore, each binding event need not result in the formation of an analyte–receptor complex on the biosensor surface. All of the receptors on the biosensor surface are presumably not, and do not exhibit the same activity. In other words, their active sites should comprise of presumably a probability distribution in 'activity.' *In lieu* of any prior information, it is reasonable to assume a bell-shaped Gaussian (or normal) distribution of active sites on the surface. A probabilistic approach is more realistic here. Such sort of analyses have presumably not been performed (at least this author is unaware of this) for analyte–receptor reactions occurring on biosensor surfaces. Thus, the fractal analysis is a convenient method of providing a lumped parameter analysis of analyte–receptor reactions occurring on biosensor surfaces.

Note that, at present, the dual-fractal analysis does not have a basis at the molecular level. This represents two different levels of heterogeneity on the biosensor surface. But, in some of the examples presented, a single-fractal analysis is clearly inadequate to model the data. Only in these cases does one resort to a dual-fractal analysis. The binding rate coefficients, k_1 and k_2 in the dual-fractal analysis have the same units $(\mathrm{pg})(\mathrm{mm})^{-2}(\mathrm{sec})^{(D_{\mathrm{f1,bind}}-3)/2}$ and $(\mathrm{pg})(\mathrm{mm})^{-2}(\mathrm{sec})^{(D_{\mathrm{f2,bind}}-3)/2}$, respectively, as the association rate coefficient, k, in the single-fractal analysis.

Dissociation rate coefficient

In this case the dissociation rate coefficient is given by:

$$(\mathrm{Ab}\cdot\mathrm{Ag}) \approx \begin{cases} -t^{(3-D_{\mathrm{f1,diss}})/2} & (t_{\mathrm{diss}} < t < t_{\mathrm{d1}}) \\ -t^{(3-D_{\mathrm{f2,diss}})/2} & (t_{\mathrm{d1}} < t < t_{\mathrm{d2}}) \end{cases} \tag{2.5d}$$

Here $D_{\mathrm{f,diss}}$ is the fractal dimension of the surface for the dissociation step. t_{diss} represents the start of the dissociation step. This corresponds to the highest concentration of the analyte–receptor on the surface. Henceforth, its concentration only decreases. $D_{\mathrm{f,bind}}$ or $D_{\mathrm{f,assoc}}$ may or may not be equal to $D_{\mathrm{f,diss}}$. The dissociation rate coefficients, k_{d1} and k_{d2} in the dual-fractal analysis have the same units $(\mathrm{pg})(\mathrm{mm})^{-2}(\mathrm{sec})^{(D_{\mathrm{fd1}}-3)/2}$ and $(\mathrm{pg})(\mathrm{mm})^{-2}(\mathrm{sec})^{(D_{\mathrm{fd2}}-3)/2}$, respectively, as the dissociation rate coefficient, k_{d}, in the single-fractal analysis.

2.2.4 Triple-fractal analysis

One resorts to a triple-fractal analysis if the dual-fractal analysis does not provide a reasonable fit. As will be shown later on in the book, one resorts to a triple-fractal analysis when the dual-fractal analysis does not provide an adequate fit. The equation for the fractal analysis equation is generic in nature, and one may easily extend the single- and the dual-fractal analysis equations (eqs. (2.5a) and (2.5c)) to describe the binding (and/or the dissociation) kinetics for a triple fractal analysis. In fact, in the extreme case, n fractal dimensions may be present. In this case, the degree of heterogeneity, D_{f} or the fractal dimension is continuously changing on the biosensor surface, and the surface needs to be represented by $D_{\mathrm{f}i}$ where i goes from 1 to n. Similarly, we have n binding rate coefficients on the biosensor surface. A similar representation may also be or made for the dissociation phase.

It is perhaps appropriate to at least mention one more approach that has been used to model the binding kinetics on surfaces.

2.2.5 Pfeifer's fractal binding rate theory

Pfeifer (1989) has suggested an alternate form of the binding rate theory. In the equation given in this reference N is the number of complexes, N_0 is the number of receptors on the solid surface, D is the diffusion coefficient of the analyte, L is the receptor diameter, and λ is the mean distance between two neighboring receptors. This equation may also be used to analyze the analyte–receptor binding kinetics. The problem, however, is that it may not be possible in all instances to estimate *a priori* all of the parameters described in the equation (not given here). In that case, one may have to approximate or assume certain values, and this will affect the accuracy and reliability of the analysis. The suggested equation does have an advantage compared to the fractal analysis described above in that it does include a pre-factor necessary to convert the time interval over which fractal scaling is observed into a length interval. It also provides an expression for t_c ($=L^2/D$), which separates the short-term regime from the long-term regime. The short-term regime is the one in which the anomalous diffusion applies. At the end of the short-term interval ($t = t_c$), the self-similarity of the system is lost, the surface is homogeneous, and regular diffusion applies.

Pfeifer (1989) states that the application of the above equation is contingent on:

(a) The analyte is uniformly distributed in the solution at time, t equal to zero.
(b) Binding is irreversible and first-order (N equals the number of analyte particles that have reached the receptors).
(c) Binding occurs whenever an incoming analyte particle hits a receptor surface for the first time. In other words, the 'sticking' probability is one.

It is very difficult to imagine perhaps any one or all of these conditions being satisfied for analyte–receptor binding interactions occurring in continuous-flow reactors.

Given the extremely small volume of the flow channels there is a high probability that the mixing of the analyte is not proper. This in turn may lead to analyte depletion in the flow channel. Also, the binding cannot be assumed to be irreversible in all instances. There may be cases of extremely fast binding and dissociation, especially for analytes with low affinity which can dissociate in the continuously flowing buffer without any regeneration reagent.

Condition (c) may be satisfied, however, it does not include the 'sticking' probability, in that each collision leads to a binding event. Also, the presence of non-specific binding, avidity effects, and binding with reactions or binding of dissociated analytes may interfere with condition (c) being satisfied. Furthermore, the equation makes assumptions about the number of active sites, and the immobilized receptors. For example it states that the analyte binds to one specific active site. The receptor cannot bind to more than one analyte molecule at one time (1:1 binding).

The equilibrium dissociation rate coefficient, $K_D = k_{diss}/k_{assoc}$ can be calculated using the above models. The K_D value is frequently used in analyte–receptor reactions occurring on biosensor surfaces. The ratio besides providing physical insights into the analyte–receptor system is of practical importance since it may be used to help determine (and possibly enhance) the regenerability, reusability, stability, and other biosensor performance parameters. K_D has the unit $(\text{sec})^{[D_{f,diss}-D_{f,assoc}]/2}$. This applies to both single- as well as dual-fractal analyses. For example, for a single-fractal analysis, K_D has the units $(\text{sec})^{[D_{fd}-D_f]/2}$. Similarly, for a dual-fractal analysis, the affinity, K_{D1} has the units $(\text{sec})^{[D_{fd1}-D_{fassoc1}]/2}$ and K_{D2} has the units $(\text{sec})^{[D_{fd2}-D_{fassoc2}]/2}$. Note the difference in the units of the equilibrium dissociation rate coefficient obtained for the classical as well as the fractal-type kinetics. Though the definition of the equilibrium dissociation rate coefficient is the same in both types of kinetics (ratio of the dissociation rate coefficient to the association rate coefficient), the difference(s) in the units of the different rate coefficients eventually leads to a different unit for the equilibrium dissociation rate coefficient in the two types of kinetics. This is not entirely unexpected since the classical kinetic analysis does not include the characteristics of the surface in the definition of the equilibrium dissociation rate coefficient whereas the present fractal analysis does. Thus, one may not be able to actually compare the equilibrium dissociation rate coefficient affinities in these two types of systems. This is a significant difference in the kinetic analysis of binding and dissociation reactions on biosensor surfaces from what is available in the literature.

REFERENCES

Anderson, J, *NIH Panel Review Meeting*, Case Western Reserve University, Cleveland, Ohio, July, 1993.

Berland, KM, PTC So, and E Gratton, Two-photon fluorescence correlation spectroscopy: method and application to the intracellular environment. *Biophysical Journal*, 1995, **68**, 694–701.

Chaudhari, A, CC Yan, and SL Lee, Multifractal analysis of diffusion-limited reactions over surfaces of diffusion-limited aggregates. *Chemical Physics Letters*, 2002, **207**, 220–226.

Chaudhari, A, CC Yan, and SL Lee, *Journal of Physics A*, 2003, **36**, 3757.

Coppens, MO and GF Froment, Diffusion and reaction in a fractal catalyst pore. 1. Geometrical aspects. *Chemical Engineering Science*, 1995, **50**(6), 1013–1026.

Corel Quattro Pro, Corel Corporation Limited, Ottawa, Canada, 1997.

De Gennes, PG, Diffusion-controlled reactions in polymer melts. *Radiation Physics & Chemistry*, 1982, **22**, 193.

Dewey, TG and JG Bann, Diffusion-controlled reaction in polymer melts. *Biophysical Journal*, 1992, **63**, 594.

Fatin-Rouge, N, K Starchev, and J Buffle, Size effects on diffusion process with agarose gels. *Biophysical Journal*, 2004, **86**, 2710–2719.

Federov, BA, BB Federov, and PW Schmidt, An analysis of the fractal properties of globular proteins. *Journal of Chemical Physics*, 1999, **99**, 4076–4083.

Ghosh, RN and WW Webb, Results of automated tracking of low density lipoprotein receptors on cell surfaces. *Biophysical Journal*, 1988, **53**, A352.

Giona, M, First-order reaction-diffusion in complex fractal media. *Chemical Engineering Science*, 1992, **47**, 1503–1515.

Harder, FH, S Havlin, and A Bunde, Diffusion in fractals with singular waiting-time distribution. *Physics Reviews B*, 1987, **36**, 3874–3879.

Havlin, S, Molecular diffusion and reaction, in *The Fractal Approach To Heterogeneous Chemistry: Surfaces, Colloids, Polymers* (ed. D. Avnir), Wiley, New York, 1989, pp. 251–269.

Havlin, S and D Ben-Avraham, Diffusion in disordered media. *Advances in Physics*, 1987, **36**, 695–798.

Henke, L, N Nagy, and UJ Krull, An AFM determination of the effects of surface roughness caused by cleaning of fused silica and glass substrates in the process of optical biosensor preparation. *Biosensors & Bioelectronics*, 2002, **17**, 547–555.

Kopelman, R, Fractal reaction kinetics. *Science*, 1988, **241**, 1620–1624.

Le Brecque, M, *Mosaic*, 1992, **23**, 12–15.

Lee, CK and SL Lee, Multifractal scaling analysis of autocatalytic and autopoisoning reactions over DLA surfaces. *Chemical Physics Letters*, 1994, **228**, 539–546.

Lee, CK and SL Lee, Multifractal scaling analysis of reactions over fractal surfaces. *Surface Science*, 1995, **325**, 294–310.

Li, HL, S Chen, and H Zhao, Fractal mechanisms for the allosteric effects of proteins and enzymes. *Biophysical Journal*, 1990, **58**, 1313–1320.

Markel, VA, LS Muratov, MI Stockman, and TF George, Theory and numerical simulation of optical properties of fractal clusters. *Physics Reviews B*, 1991, **43**(10), 8183.

Nyikos, L and T Pajkossy, Diffusion to fractal surfaces. *Electrochimica Acta*, 1986, **31**, 1347.

Pajkossy, T and L Nyikos, Diffusion to fractal surfaces. II. Verification of theory. *Electrochimica Acta*, 1989, **34**, 171.

Peters, R and RJ Cherry, Lateral and rotational diffusion of bacteriorhodopsin in lipid bilayers: experimental test of the Saffman-Delbruck equations. *Proceedings of the National Academy of Sciences of the United States of America*, 1982, **79**, 4317–4321.

Pfeifer, P. (1987), Characterization of surface irregularity, Chapter 2. In: Preparative Chemistry Using Supported Reagents, Academic Press, San Diego.

Pfeifer, P, D Avnir, and DJ Farin, Molecular fractal surfaces. *Nature (London)*, 1984a, **308**, 261.

Pfeifer, P, D Avnir, and DJ Farin, Surface geometric irregularity of particulate materials. The fractal approach, *Journal of Colloid & Interface Science*, 1984b, **103**(1), 112.

Pfeifer, P and M Obert, Fractals: Basic concepts and terminology in *The Fractal Approach to Heterogeneous Chemistry: Surfaces, Colloids, Polymers* (ed. D. Avnir), Wiley, New York, 1989, pp. 251–269.

Pluen, A, PA Netti, KJ Rakesh, and DA Berk, Diffusion of macromolecules in agarose gels: comparison of linear and globular configurations. *Biophysical Journal*, 1999, **77**, 542–552.

Sadana, A, JP Alarie, and T Vo-Dinh, A ß-cyclodextrin based fiber-optic chemical sensor: a fractal analysis. *Talanta*, 1995, **42**, 1567.

Sadana, A and A Beelaram, Fractal analysis of antigen-antibody binding kinetics: biosensor applications. *Biotechnology Progress*, 1994, **9**, 45.

Saffman, PG and M Delbruck, Brownian motion in biological membranes. *Proceedings of the National Academy of Sciences of the United States of America*, 1975, **72**, 3111–3113.

Sahimi, M, Flow phenomena in rocks: From continuum models to fractals, percolation, cellular automata and simulated annealing. *Reviews in Modern Physics*, 1993, **65**, 1393–1534.

Saxton, MJ, Anomalous diffusion due to binding: a Monte Carlo study. *Biophysical Journal*, 2001, **77**, 2251–2265.

Schwille, P, U Haupts, S Maiti, and WW Webb, Molecular dynamics in living cells observed by fluorescence correlation spectroscopy with one- and two-photon excitation. *Biophysical Journal*, 1999, **77**, 2251–2265.

Starchev, K, J Sturm, G Weill, and CH Brogen, Brownian motion and electrophoretic transport in agarose gels studied by epifluorescence microscopy and simple particle tracking analysis. *Journal of Physical Chemistry*, 1997, **101**, 5659–5663.

Wei, QH, C Bechinger, and P Leiderer, Single-file diffusion of colloids in one-dimensional channels. *Science*, 2000, **287**, 625–627.

Weiss, GH, *Fractals in Science*, Springer-Verlag, p. 119, Berlin, 1994.

– 3 –

Biosensor Performance Parameters and their Enhancement

3.1 INTRODUCTION

Biosensor research and application is gradually becoming an established area of widespread usage. As indicated elsewhere in the literature and in later chapters in this book and in other books the areas of biosensor application are increasing continuously.

This is because of the simplicity of their usage for different applications. Besides, often the analytical results are obtained near real time or at real time itself. One aspect of biosensor application that needs careful study is the enhancement of biosensor performance parameters. Quite often, researchers especially in the academic arena are content with developing a biosensor for a novel application. Little interest or resources are expended on the ways and means of improving biosensor performance parameters such as sensitivity, reproducibility, validation, enhanced response, stability, improvement in resolution, detection time, limit of detection (LOD), etc.

The industrial sector, which is under constant pressure due to intense competition to improve biosensor performance parameters is particularly interested in this area of biosensor development. However, as expected, they will guard this type of information very carefully, especially from their competitors. Now that we recognize that biosensors are here to stay, it behooves one to explore different avenues by which one may be able to enhance the different biosensor performance parameters. One recognizes that if modifications are made for the biosensor to enhance a particular performance parameter, then it is quite possible that another biosensor parameter may exhibit a decrease in its performance characteristic. For example, if one enhances the sensitivity of a biosensor by some type of modification, then perhaps it is quite possible that after this modification, the biosensor may exhibit a decrease in its stability characteristic.

It would be useful to relate the biosensor performance parameter(s) to a common variable, such as the degree of heterogeneity or fractal dimension of the biosensor surface. Then, by changing the degree of heterogeneity to help improve a particular biosensor performance parameter, one does have *a priori* an estimate of the influence of this change in heterogeneity on other biosensor performance parameters. This seems like an optimization problem. Surely, experience in the biosensor R&D field is irreplaceable here. Simple as well as complex mathematical expressions can and will continue to provide insights in the biosensor R&D arena. Sometimes, theory will lead the way for further experimentation, and vice versa. However, practical experience in the biosensor R&D field is irreplaceable as far as improving biosensor performance parameters are concerned.

In this chapter we analyze the influence of the degree of heterogeneity or the fractal dimension, D_f on the biosensor surface on:

(a) the enhancement of fluorescence for the hybridization of a molecular beacon to a target DNA in the presence of site-specific DNA nickase (Zheleznaya *et al.*, 2006). Tyagi and Kramer (1996) indicate that molecular beacons are synthetic DNA and RNA probes that fluoresce on hybridization.

(b) A 10-fold improvement in resolution performance using a novel surface plasmon resonance (SPR) biosensor with Au (silver) nanoclusters when compared with conventional SPR biosensor during hybridization studies (Hu *et al.*, 2004). These authors indicate that there is a requirement to develop ultrahigh resolution SPR biosensors, which are capable of detecting very low biomolecular interactions at low concentrations. These low biomolecular interactions may be at levels less than 200 Daltons (Da). Apparently, the use of nanoparticles leads to an increase in the roughness of the Au film used, which in turn leads to a significant enhancement in biosensor performance (via light scattering and energy adsorption phenomena). This is consistent with the emphasis in the chapter and in the other chapters in the book on the fractal dimension, D_f (roughness) or the degree of heterogeneity on the biosensor chip surface.

(c) Sensitivity and specificity enhancements for the SPR biosensor detection of prostate-specific antigen (PSA)-α_1-antichymotrypsin (Cao *et al.*, 2006). Different self-assembled monolayers (SAMs) were used to improve the sensitivity of the biosensor. A sandwich strategy using an intact PSA polyclonal antibody was used as an amplifying agent to enhance the signal. Chapman *et al.* (2000) and Ciu *et al.* (2003) indicate that the antibody–antigen–antibody sequence in a sandwich strategy exhibits a higher sensitivity than that exhibited by an antibody–antigen sequence.

(d) Reproducibility studies for a fluorescent sensor for imidazole derivatives (Zhang *et al.*, 2004). This sensor is based on the monomer–dimer equilibrium of a zinc porphyrin complex in a polymer film. 2.5×10^{-6}M and 2.5×10^{-5}M solutions were switched and passed over the sensor chip.

(e) SPR biosensor sensitivity enhancement of small molecules using progesterone as a model compound (Mitchell *et al.*, 2005). Sequential binding formats led to signal enhancement.

(f) Nonspecific binding (NSB) studies of bovine serum albumin (BSA) and immunoglobulin (IgG) on a 1:9 (molar ratio of EG_6–COOH/EG_3–OH) sensor chip surface (Choi *et al.*, 2005). Here EG_6-COOH is

$$HS(CH_2)_{11}(OCH_2CH_2)_6OCHCOOH \quad \text{and} \quad EG_3-OH$$
$$\text{is } HS(CH_2)_{11}(OCH_2CH_2)_3OH$$

Binding and dissociation (wherever applicable) kinetics are analyzed for the above-mentioned analyte–receptor interactions occurring on biosensor surfaces. Hopefully, valuable insights will be obtained that may be used to help increase or change biosensor performance parameters either singly, or simultaneously in desired directions. Once again, this analysis will not replace practical experience in the development of biosensors, but hopefully it should complement it.

3.2 THEORY

Havlin (1987) has reviewed and analyzed the diffusion of reactants towards fractal surfaces. The details of the theory and the equations involved for the binding and the dissociation phases for analyte–receptor binding are available (Sadana, 2001). The details are not repeated here; except that just the equations are given to permit an easier reading. These equations have been applied to other biosensor systems (Sadana, 2001, 2005; Ramakrishnan and Sadana, 2001). For most applications, a single- or a dual-fractal analysis is often adequate to describe the binding and the dissociation kinetics. Peculiarities in the values of the binding and the dissociation rate coefficients, as well as in the values of the fractal dimensions with regard to the dilute analyte systems being analyzed will be carefully noted, if applicable.

In this chapter we analyze the binding and dissociation kinetics for the enhancement of fluorescence during the hybridization of a molecular beacon to a target DNA in the presence of site-specific DNA nickase (Zheleznaya *et al.*, 2006), a 10-fold improvement in resolution performance using a novel SPR biosensor with Au (gold) nanoclusters when compared with conventional SPR biosensor during hybridization studies (Hu *et al.*, 2004), sensitivity and specificity enhancements for the SPR biosensor detection of PSA-α_1-antichymotrypsin (Cao *et al.*, 2006), reproducibility studies for a fluorescent sensor for imidazole derivatives (Zhang *et al.*, 2004), SPR biosensor sensitivity enhancement of small molecules using progesterone as a model compound (Mitchell *et al.*, 2005), binding kinetics of antiglutamic acid decarboxylase (anti-GAD) in solution to GAD immobilized on four different SAMs on a BIAcore sensor chip surface (Choi *et al.*, 2006), and NSB studies of BSA and IgG on a 1:9 (molar ratio of EG_6–COOH/EG_3–OH) sensor chip surface (Choi *et al.*, 2006).

3.2.1 Single-fractal analysis

Binding rate coefficient

Havlin (1989) indicates that the diffusion of a particle (analyte [Ag]) from a homogeneous solution to a solid surface (e.g., receptor [Ab]-coated surface) on which it reacts to form a product (analyte–receptor complex; (Ab.Ag)) is given by

$$(\text{Ab.Ag}) \approx \begin{cases} t^{(3-D_{\text{f,bind}})/2} = t^{p} & t < t_{\text{c}} \\ t^{1/2} & t > t_{\text{c}} \end{cases} \tag{3.1}$$

Here $D_{\text{f,bind}}$ or D_{f} (used later on in the chapter) is the fractal dimension of the surface during the binding step; t_{c} is the cross-over value. Havlin (1989) indicates that the cross-over value may be determined by $r_{\text{c}}^2 \sim t_{\text{c}}$. Above the characteristic length, r_{c}, the self-similarity of the surface is lost and the surface may be considered homogeneous. Above time, t_{c} the surface may be considered homogeneous, since the self-similarity property disappears, and "regular" diffusion is now present. For a homogeneous surface where D_f is equal to 2, and when only diffusional limitations are present, $p = 1/2$ as it should be. Another way of looking at the $p = 1/2$ case (where $D_{f,bind}$ is equal to two) is that the analyte in solution views the fractal object, in our case, the receptor-coated biosensor surface, from a "large distance." In essence, in the association process, the diffusion of the analyte from the solution to the receptor surface creates a depletion layer of width $(Ðt)^{1/2}$ where $Ð$ is the diffusion constant. This gives rise to the fractal power law, (Analyte·Receptor) ~ $t^{(3-D_{f,bind})/2}$. For the present analysis, t_c is arbitrarily chosen and we assume that the value of the t_c is not reached. One may consider the approach as an intermediate "heuristic" approach that may be used in the future to develop an autonomous (and not time-dependent) model for diffusion-controlled kinetics.

Dissociation rate coefficient

The diffusion of the dissociated particle (receptor [Ab] or analyte [Ag]) from the solid surface (e.g., analyte [Ag]-receptor [Ab]) complex coated surface) into solution may be given, as a first approximation by

$$(\text{Ab.Ag}) \approx -t^{(3-D_{f,diss})/2} = -t^{p}\,(t > t_{diss}) \tag{3.2}$$

Here $D_{f,diss}$ is the fractal dimension of the surface for the dissociation step. This corresponds to the highest concentration of the analyte–receptor complex on the surface.

Henceforth, its concentration only decreases. The dissociation kinetics may be analyzed in a manner "similar" to the binding kinetics.

3.2.2 Dual-fractal analysis

Binding rate coefficient

Sometimes, the binding curve exhibits complexities and two parameters (k, D_f) are not sufficient to adequately describe the binding kinetics. This is further corroborated by low values of r^2 factor (goodness of fit). In that case, one resorts to a dual-fractal analysis (four parameters; k_1, k_2, D_{f1}, and D_{f2}) to adequately describe the binding kinetics. The single-fractal analysis presented above is thus extended to include two fractal dimensions. At present, the time ($t = t_1$) at which the "first" fractal dimension "changes" to the "second" fractal dimension is arbitrary and empirical. For the most part, it is dictated by the data analyzed and experience gained by handling a single-fractal analysis. A smoother curve is obtained in the "transition" region, if care is taken to select the correct number of points for the two regions. In this case, the product (antibody–antigen; or analyte–receptor complex, Ab·Ag or analyte·receptor) is given by

$$(\text{Ab·Ag}) \approx \begin{cases} t^{(3-D_{f1,bind})/2} = t^{p1} & (t < t_1) \\ t^{(3-D_{f2,bind})/2} = t^{p2} & (t_1 < t < t_2 = t_c) \\ t^{1/2} & (t > t_c) \end{cases} \tag{3.3}$$

In some cases, as mentioned above, a triple-fractal analysis with six parameters (k_1, k_2, k_3, D_{f1}, D_{f2}, and D_{f3}) may be required to adequately model the binding kinetics. This is when the binding curve exhibits convolutions and complexities in its shape due to perhaps to the very dilute nature of the analyte (in some of the cases to be presented) or for some other reasons. Also, in some cases, a dual-fractal analysis may be required to describe the dissociation kinetics.

3.3 RESULTS

The fractal analysis will be applied to different analyte–receptor reactions occurring on biosensor chip surfaces with the specific purpose of trying to help improve the different biosensor performance parameters such as selectivity, sensitivity, reproducibility, specificity, LOD, etc. Attempts will be made to relate particularly changes in the fractal dimension on the biosensor chip surface with the changes in the different biosensor performance parameters.

At the outset it should be indicated that alternate expressions for fitting the binding and dissociation data are available that include saturation, first-order reaction, and no diffusional limitations, but these expressions are deficient in describing the heterogeneity that inherently exists on the surface. It is this heterogeneity on the biosensor surface that one is attempting here to relate to the different biosensor performance parameters. More specifically the question we wish to answer is that how may one change the heterogeneity or the fractal dimension, D_f on the biosensor chip surface in order that one may be able to enhance the different biosensor performance parameters.

Other modeling attempts also need to be mentioned. One might justifiably argue that appropriate modeling may be achieved by using a Langmuirian or other approach. The Langmuirian approach may be used to model the data presented if one assumes the presence of discrete classes of sites, for example double exponential analysis as compared with the single-fractal analysis. Lee and Lee (1995) indicate that the fractal approach has been applied to surface science, for example, adsorption and reaction processes. These authors emphasize that the fractal approach provides a convenient means to represent the different structures and morphology at the reaction surface. These authors also emphasize using the fractal approach to develop optimal structures and as a predictive approach. Another advantage of the fractal technique is that the analyte–receptor association is a complex reaction, and the fractal analysis via the fractal dimension and the rate coefficient provide a useful lumped parameter analysis of the diffusion-limited reaction occurring on a heterogeneous surface.

In a classical situation, to demonstrate fractality, one should make a log-log plot, and one should definitely have a large amount of data. It may be useful to compare the fit to some other forms, such as exponential, or involving saturation, etc. At present, we do not present any independent proof or physical evidence of fractals in the examples presented. Nevertheless, we still use fractals and the degree of heterogeneity on the biosensor surface to gain insights into enhancing the different biosensor performance parameters. The fractal approach is a convenient means (since it is a lumped parameter) to make the degree of heterogeneity that exists on the surface more quantitative. Thus, there is some arbitrariness in the fractal approach to be presented. The fractal approach provides additional information about interactions that may not be obtained by a conventional analysis of biosensor data. In this chapter as mentioned above, we are attempting to relate the fractal dimension, D_f or the degree of heterogeneity on the biosensor surface with the different biosensor performance parameters. More specifically, we are interested in finding out how changes in the fractal dimension or the degree of heterogeneity on the biosensor chip surface affect the different biosensor parameters of interest.

Unless specifically mentioned there is no nonselective adsorption of the analyte. In other words, NSB is ignored. Nonselective adsorption would skew the results obtained very significantly. In these types of systems, it is imperative to minimize this nonselective adsorption. We also do recognize that, in some cases, this nonselective adsorption may not be a significant component of the adsorbed material and that the rate of association, which is of a temporal nature would depend on surface availability. Later on in the chapter an example of nonspecific adsorption is analyzed (Choi *et al.*, 2005).

Zheleznaya *et al.* (2006) recently indicate that molecular beacons are being increasingly used in different applications that include quantitative polymerase chain reaction (PCR), single polymorphism (SNP) studies, and for the detection of DNA binding proteins (Broude, 2002). Zheleznaya *et al.* (2006) indicate that site-specific DNA nickases recognize short-specific sequences on double-stranded DNA. These authors emphasize that these nickases unlike restriction endonucleases cleave just a single strand. They noted a significant enhancement of fluorescence on hybridization of a molecular beacon to a target DNA. This was observed in the presence of a site-specific DNA nickase. These authors emphasize that their nickase molecular beacon (NMB) permits a two order of magnitude increase in the fluorescent signal.

Figure 3.1a shows the binding of the target 0.5 nM ss DNA of M13mp19 phage in solution to the DNA molecular beacon St7 with the nickase in the absence of calf thymus DNA. St7 is an oligonucleotide, and has a stem length of 7 bp. Its stem arms are underlined. The oligonucleotide sequence of St7 is 5′-FAM-GGCATCTTCTAGAGTCGACC-CTGCAGGCATGAGATGCC-BHQ1-3′. A single-fractal analysis is adequate to describe the binding kinetics. The values of the binding rate coefficient, k and the fractal dimension, D_f are given in Tables 3.1a and 3.1b, respectively. The values of the binding of the binding rate coefficient presented in Table 3.1a were obtained from a regressions analysis using Corel Quattro Pro 8.0 (Corel Quattro Pro 8.0, 1997) to model the data using eqs. (3.1a) and (3.1b), wherein (Analyte · Receptor) = $kt^{(3-D_f)/2}$ for a single-fractal analysis for the binding phase, and (Analyte Receptor) = $kt^{(3-D_{f1})/2}$ and $kt^{(3-D_{f2})/2}$ for a dual-fractal analysis. The binding rate coefficient values presented in Table 3.1a are within 95% confidence values. For example, for the binding (hybridization) of 0.5 nM target in solution in the absence of calf thymus DNA to a site-specific DNA nickase (molecular beacon with a DNA nickase recognition site) the binding rate coefficient, k is equal to 0.0029 ± 0. Since the error is ± 0, the (95% confidence limit indicates that the k values lie between $0.0028 - 0$ and $0.0029 + 0$, and the values are precise and significant.

Figures 3.1b and c show the binding rate curve of the target at 5.0 and 50 nM ss DNA of M13mp19 phage in solution, respectively to the molecular beacon St7 with the nickase in the absence of calf thymus DNA. A dual-fractal analysis is required to adequately describe the binding in both of these cases. The values of (a) the binding rate coefficient,

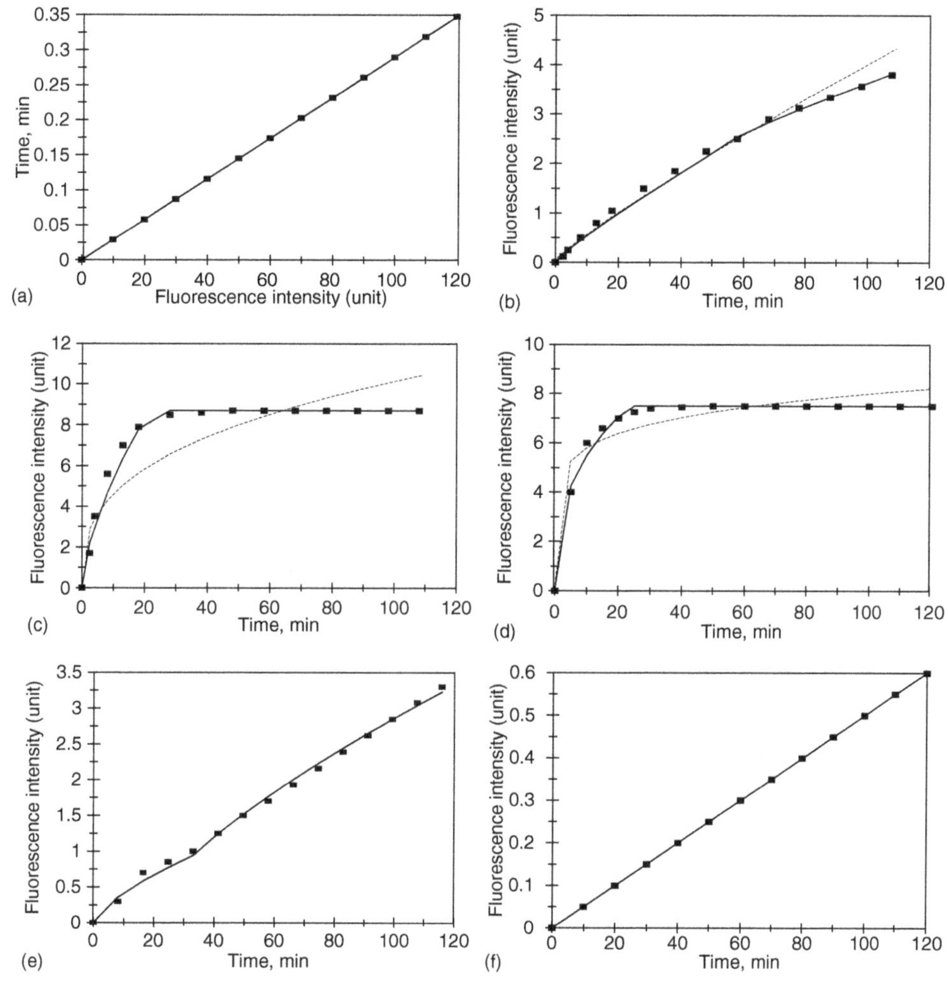

Figure 3.1 Binding of different concentrations in nM of the target, ss DNA of M13mp19 phage in solution to the molecular beacon St7 (0.5 μM) with the nickase. Hybridization studies (Zheleznaya *et al.*, 2006): (a) 0.5, (b) 5, (c) 50. Effect of calf thymus on the nickase-molecular beacon (NMB) assay. Binding of different concentrations in nM of the target ss DNA of M13mp19 phage in solution to the molecular beacon St7 (0.5 μM) with the nickase. Hybridization studies (Zheleznaya *et al.*, 2006): (d) 50 nM of target + 200 μg/mL of calf thymus DNA; (e) 5 nM of target + 200 μg/mL of calf thymus DNA; (f) 0.5 nM of target + 200 μg/mL of calf thymus DNA.

k, and the fractal dimension, D_f, for a single-fractal analysis, and (b) the binding rate coefficients, k_1 and k_2, and the fractal dimensions, D_{f1} and D_{f2} for a dual-fractal analysis are given in Tables 3.1a and 3.1b. It is of interest to note that as the target concentration in solution increases from 0.5 to 5.0 and 50 nM, the binding is adequately described by a single- and a dual-fractal analysis, respectively. In other words, there is a change in the binding mechanism as one goes from the lower analyte (target) concentration (0.5 nM) to the higher (5 and 50 nM) target concentrations in solution.

Table 3.1a

Binding (hybridization) rate coefficients for different concentrations of ss DNA of M13mp19 phage (target DNA) in solution to a site-specific DNA nickase (molecular beacon with a DNA nickage recognition site) in the absence and in the presence of 200 μg/mL calf thymus DNA

Target DNA concentration in solution (nM)	k	k_1	k_2
(a) In the absence of calf thymus DNA			
0.5	0.0029 ± 0.0	na	na
5.0	0.07353 ± 0.0103	0.06118 ± 0.0081	0.1849 ± 0.0033
50	2.1984 ± 0.6146	1.2260 ± 0.2877	8.7 ± 0.0
(b) In the presence of 200 μg/mL calf thymus DNA			
0.5	0.005 ± 0.0	na	na
5.0	0.1209 ± 0.0108	na	na
50	1.4223 ± 0.441	2.3610 ± 0.1818	7.5 ± 0.0

Note: The oligonucleotide St7 (stem length of 7 bp) used (5′-FAM-GGCATCTTCTAGAGTCGACCT-GCAGGCATGAGATGCC-BHQ1-3′) (Zheleznaya *et al.*, 2006).

Table 3.1b

Fractal dimensions for the binding (hybridization) of different concentrations of ss DNA of M13mp19 phage (target DNA) in solution to a site-specific DNA nickase (molecular beacon with a DNA nickage recognition site) in the absence and in the presence of 200 μg/mL calf thymus DNA

Target DNA concentration in solution (nM)	D_f	D_{f1}	D_{f2}
(a) In the absence of calf thymus DNA			
0.5	1.0 ± 1.88E-16	na	na
5.0	1.2606 ± 0.05968	1.0834 ± 0.08944	1.7068 ± 0.06676
50	2.3232 ± 0.1118	1.7188 ± 0.2054	3.0 − 6E-15
(b) In the presence of 200 μg/mL calf thymus DNA			
0.5	1.0 ± 8.5E-16	na	na
5.0	1.6276 ± 0.04808	na	na
50	2.7178 ± 0.0566	2.2682 ±0.1168	3.0 − 7.2E-15

Note: The oligonucleotide St7 (stem length of 7 bp) used (5′-FAM-GGCATCTTCTAGAGTCGACCT-GCAGGCATGAGATGCC-BHQ1-3′) (Zheleznaya *et al.*, 2006).

Note also that for the binding at 5.0 and 50 nM target concentration in solution where a dual-fractal analysis is used to describe the binding kinetics an increase in the fractal dimension or the degree of heterogeneity on the biosensor chip surface leads to an increase in the binding rate coefficient. For example, for the binding of 5.0 nM target in solution, an increase in the fractal dimension by 57.5% from a value of D_{f1} equal to 1.0834 to D_{f2} equal to 1.7068 leads to an increase in the binding rate coefficient by a factor of 2.793 from a value of k_1 equal to 0.06118 to k_2 equal to 0.1849. An increase in the degree of heterogeneity on the biosensor chip surface leads to an increase in the binding rate coefficient at least for this case. A similar result is also obtained at the higher, 50 nM analyte (target) concentration in solution.

Figure 3.1d shows the binding of 50 nM of the target ss DNA of the M13mp19 phage in the presence of 200 μg/ML calf thymus DNA in solution to the DNA molecular beacon St7. A dual-fractal analysis is required to adequately describe the binding kinetics. The values of (a) the binding rate coefficient, k and the fractal dimension, D_f for a single-fractal analysis, and (b) the binding rate coefficients, k_1 and k_2, and the fractal dimensions, D_{f1} and D_{f2} are given in Tables 3.1a and 3.1b.

It is of interest to compare the binding rate coefficients and the fractal dimensions for the binding of 50 nM target in solution in the absence (Case A) and in the presence of 200 μg/mL calf thymus DNA (Case B). As one goes from case A to Case B the fractal dimension D_{f1} increases by 31.2% from a value of 1.7188 to 2.2682, and the binding rate coefficient, k_1 increases by a factor of 1.926 from a value of k_1 equal to 1.2260 to 2.3610. Changes in the binding rate coefficient and in the fractal dimension on the biosensor surface are in the same direction.

Figure 3.2a shows the increase in the binding rate coefficient, k of the target ss DNA of M13mp19 phage in solution in the 0.5–50 nM concentration range in the absence and in

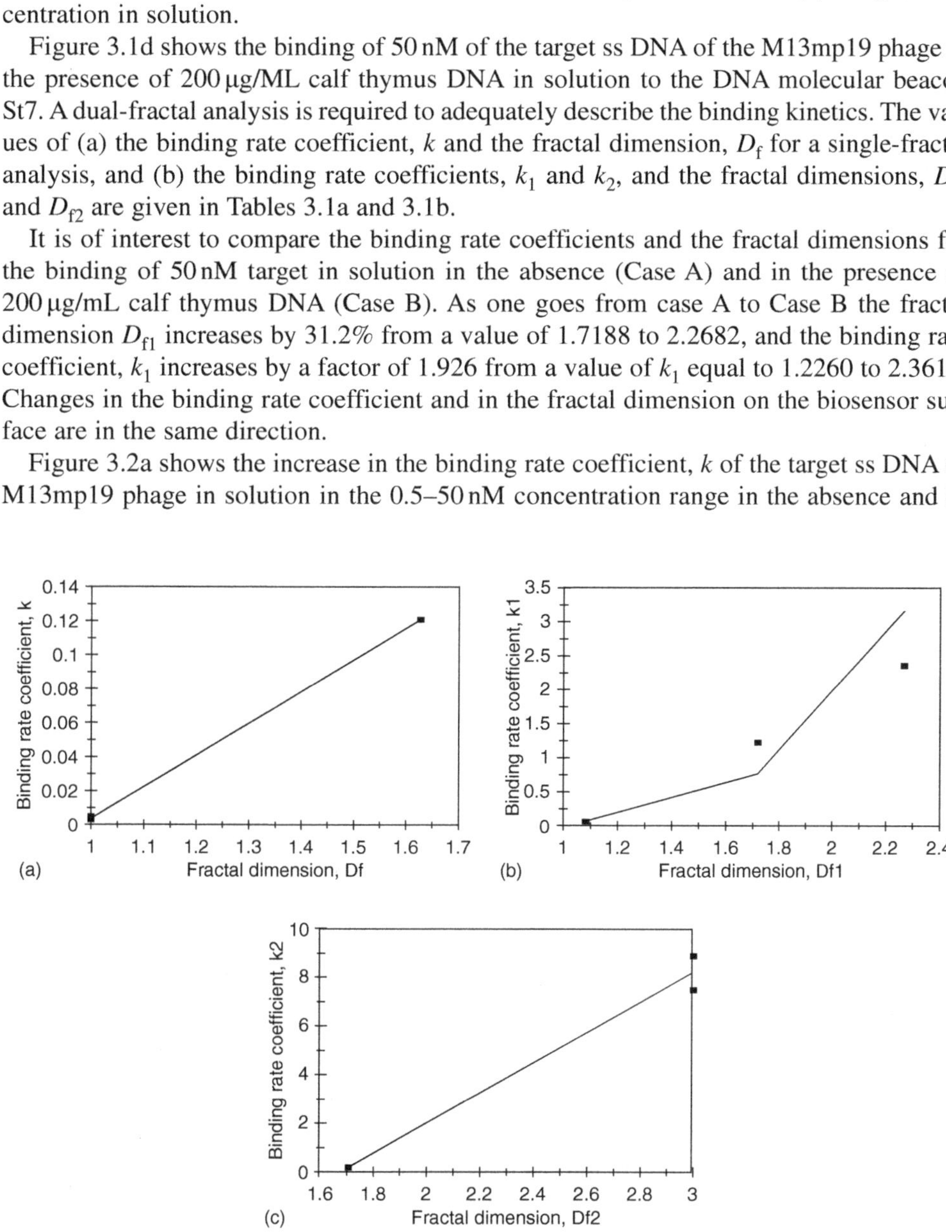

Figure 3.2 (a) Increase in the binding rate coefficient, k with an increase in the fractal dimension, D_f. (b) Increase in the binding rate coefficient, k_1 with an increase in the fractal dimension, D_{f1}. (c) Increase in the binding rate coefficient, k_2 with an increase in the fractal dimension, D_{f2}.

the presence of 200 μg/mL calf thymus DNA in solution and when a single-fractal analysis applies. The binding rate coefficient, k is given by

$$k = (0.003808 \pm 0.00179) D_f^{7.0988 \pm 0.9685} \tag{3.2a}$$

The fit is good. However, only three data points are available. The availability of more data points would lead to a more reliable fit. The binding rate coefficient, k is very sensitive to the fractal dimension, D_f or the degree of heterogeneity that exists on the biosensor chip surface as noted by the higher than seventh (equal to 7.0988) order of dependence exhibited.

Figure 3.2b shows the increase in the binding rate coefficient, k_1 of the target ss DNA of M13mp19 phage in solution in the 0.5–50 nM concentration range in the absence and in the presence of 200 μg/mL calf thymus DNA in solution and when a dual-fractal analysis applies. The binding rate coefficient, k_1 is given by

$$k_1 = (0.04847 \pm 0.03793) D_{f1}^{5.1012 \pm 1.0959} \tag{3.2b}$$

The fit is good. However, only three data points are available. The availability of more data points would lead to a more reliable fit. The binding rate coefficient, k_1 is very sensitive to the fractal dimension, D_{f1} or the degree of heterogeneity that exists on the biosensor chip surface as noted by the higher than fifth (equal to 5.1012) order of dependence exhibited.

Figure 3.2c shows the increase in the binding rate coefficient, k_2 of the target ss DNA of M13mp19 phage in solution in the 0.5–50 nM concentration range in the absence and in the presence of 200 μg/mL calf thymus DNA in solution and when a dual-fractal analysis applies. The binding rate coefficient, k_2 is given by

$$k_2 = (0.005097 \pm 0.000656) D_{f2}^{6.7171 \pm 0.2628} \tag{3.2c}$$

The fit is good. However, only three data points are available. The availability of more data points would lead to a more reliable fit. The binding rate coefficient, k_2 is very sensitive to the fractal dimension, D_{f2} or the degree of heterogeneity that exists on the biosensor chip surface as noted by the higher than sixth (equal to 6.7171) order of dependence exhibited.

Hu *et al.* (2004) have presented the binding (hybridization) kinetics of a target DNA(15 bp mer) in solution to a complementary probe (15 bp mer) (5′-GTTACCACACGGATG-3′) immobilized on a conventional SPR biosensor surface and on a novel nanocluster-enhanced SPR biosensor surface. These authors indicate that a simple co-sputtering method using a multitarget sputtering system is used to fabricate the Au nanocluster-embedded dielectric film. Also, the roughness of the film is increased by the inclusion of the colloidal Au nanoparticles in the gold film. This they indicate leads to an increase in light scattering and in the energy adsorption phenomena.

Figure 3.3a shows the binding of the target DNA (15 bp mer) in solution to the complementary probe (15 bp mer) (5′-GTTACCACACGGATG-3′) immobilized on a conventional SPR biosensor. A dual-fractal analysis is required to adequately describe the binding

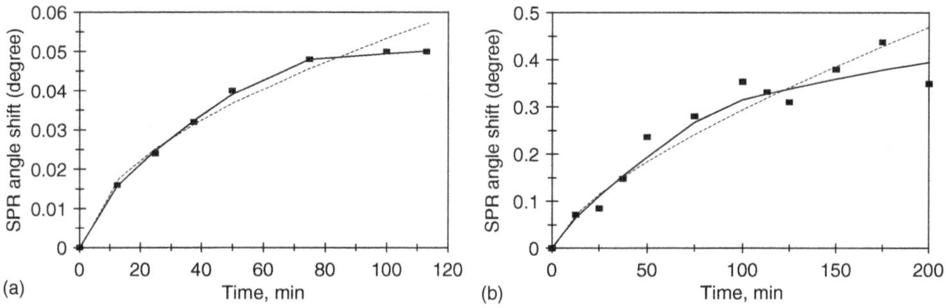

Figure 3.3 (a) Binding (hybridization) of the target, cDNA in solution to the probe DNA on a conventional surface plasmon resonance (SPR) biosensor (Hu *et al.*, 2004). (b) Binding (hybridization) of the target, cDNA in solution to the probe DNA on a novel nanocluster-enhanced SPR biosensor (Hu *et al.*, 2004).

Table 3.2a

Binding rate coefficients for DNA hybridization on (a) a conventional SPR biosensor, and (b) on a nanocluster-enhanced SPR biosensor (Hu *et al.*, 2004)

Analyte in solution/receptor on surface	k	k_1	k_2
Target cDNA (15 bp mer)/ probe DNA (15 bp mer) (5′-GTTACCACACGGATG-3′) on a conventional SPR biosensor surface	0.004331 ± 0.000386	0.002993 ± 0.000096	0.030329 ± 0.000295
Target cDNA (15 bp mer)/ probe DNA (15 bp mer) (5′-GTTACCACACGGATG-3′) on a nanocluster-enhanced SPR biosensor surface	0.013066 ± 0.002654	0.00854 ± 0.00158	0.06932 ± 0.00928

kinetics. The values of (a) the binding rate coefficient, k, and the fractal dimension, D_f, for a single-fractal analysis, and (b) the binding rate coefficients, k_1 and k_2, and the fractal dimensions, $D_{f1}$and D_{f2} for a dual-fractal analysis are given in Tables 3.2a and 3.2b.

Figure 3.3b shows the binding of the target DNA (15 bp mer) in solution to the complementary probe (15 bp mer) (5′-GTTACCACACGGATG-3′) immobilized on a nanocluster-enhanced SPR biosensor surface. Here too, a dual-fractal analysis is required to adequately describe the binding kinetics. The values of (a) the binding rate coefficient, k, and the fractal dimension, D_f, for a single-fractal analysis, and (b) the binding rate coefficients, k_1 and k_2, and the fractal dimensions, D_{f1} and D_{f2} for a dual-fractal are given in Tables 3.2a and 3.2b. It is of interest to compare the binding (hybridization) kinetics on the conventional SPR biosensor surface and on the nanocluster-enhanced biosensor surface. Note that for the dual-fractal analysis used to model the binding kinetics for both cases, (a) the fractal dimension, D_{f1} decreases by 14% from a value of 1.6868 for the conventional

TABLE 3.2b

Fractal dimensions for DNA hybridization on (a) a conventional SPR biosensor, and (b) on a nanocluster-enhanced SPR biosensor (Hu *et al.*, 2004)

Analyte in solution/receptor on surface	D_f	D_{f1}	D_{f2}
Target cDNA (15 bp mer)/ probe DNA (15 bp mer) (5′-GTTACCACACGGATG-3′) on a conventional SPR biosensor surface	1.9202 ± 0.08802	1.6868 ± 0.0608	2.7856 ± 0.06504
Target cDNA (15 bp mer)/ probe DNA (15 bp mer) (5′-GTTACCACACGGATG-3′) on a nanocluster-enhanced SPR biosensor surface	1.6520 ± 0.1331	1.4050 ± 0.1751	2.3426 ± 0.5356

SPR biosensor to a value of 1.4050 for the nanocluster-enhanced SPR biosensor surface, and (b) the fractal dimension, D_{f2} decreases by 15.9% from a value of 2.7856 for the conventional SPR biosensor to a value of 2.3426 for the nanocluster-enhanced SPR biosensor surface. Note, however, that both of the binding rate coefficients, k_1 and k_2 exhibit increases as one goes from the conventional SPR biosensor to the nanocluster-enhanced SPR biosensor. The binding rate coefficient, k_1 is higher by a factor of 2.853 for the nanocluster-enhanced SPR biosensor surface when compared to the conventional SPR biosensor surface. Similarly, the binding rate coefficient, k_2 value is higher by a factor of 2.89 for the nanocluster-enhanced SPR biosensor surface when compared with the conventional SPR biosensor surface. Note that though the nanocluster-enhanced SPR biosensor surface exhibits lower fractal dimension values (lower degrees of heterogeneity or roughness on the biosensor surface) than the conventional SPR biosensor surface, the corresponding binding rate coefficients k_1 and k_2 are higher for the nanocluster-enhanced SPR biosensor surface. This is presumably due to the "nano" nature of the surface. In this case, the changes in the binding rate coefficient and in the degree of heterogeneity on the biosensor surface are in opposite directions.

Choi *et al.* (2005) recently developed an enhanced performance SPR immunosensor for diagnosing type I diabetes by using modifications of mixed SAMs. The analyte these authors wanted to detect was monoclonal anti-GAD. They indicate that GAD plays a key role in the initial immunological events that lead to the destruction of pancreatic β cells which produce insulin (Baekkeskov *et al.*, 1990). Choi *et al.* (2005) were able to obtain an enhancement in the detection of GAD by reducing the steric hindrance on using SAMs of heterogeneous lengths. Their technique also permitted them to reduce the NSB. These authors used the following SAMs:

SAM1 (10:1 ratio of 3-MPA to 3-MUA)
3-MPA is 3-mercaptopropionic acid
11-MUA is 11-mercaptoundecanoic acid

SAM2 (10:1 ratio of 3-MPOH to 11-MUA)
3-MPOH is 3-mercaptopropanol
SAM3 (10:1 ratio of 11-MUOH to 11-MUA)
11-MUOH is 11-mercaptoundecanol
SAM4 (10:1 ratio of 3-MPOH to 3-MUA)

Figure 3.4a shows the binding of 450 nM monoclonal anti-GAD 65 (purified mouse IgG) to SAM4 immobilized on a BIAcore 2000 sensor chip surface using strepravidin. A dual-fractal analysis is required to adequately describe the binding kinetics. The values of (a) the binding rate coefficient, k, and the fractal dimension, D_f, for a single-fractal analysis, and (b) the binding rate coefficients, k_1 and k_2, and the fractal dimensions, D_{f1} and D_{f2} for a dual-fractal are given in Tables 3.3a and 3.3b.

Figure 3.4b shows the binding of 450 nM monoclonal anti-GAD 65 (purified mouse IgG) to SAM3 immobilized on a BIAcore 2000 sensor chip surface using strepravidin. A dual-fractal analysis is required to adequately describe the binding kinetics. The values of (a) the binding rate coefficient, k, and the fractal dimension, D_f, for a single-fractal analysis, and (b) the binding rate coefficients, k_1 and k_2, and the fractal dimensions, D_{f1} and D_{f2} for a dual-fractal are given in Tables 3.3a and 3.3b.

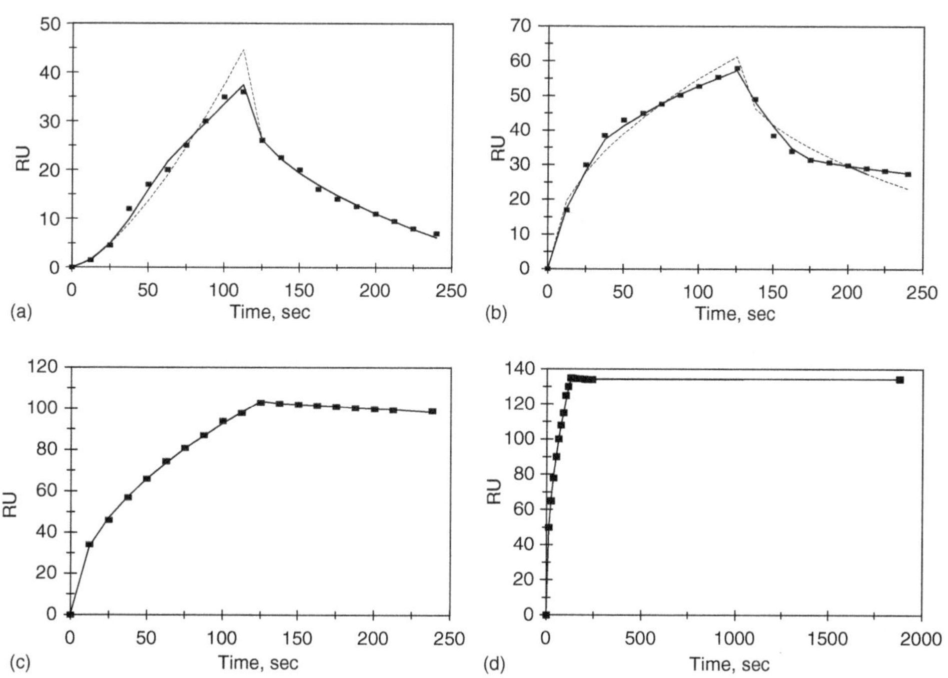

Figure 3.4 Binding kinetics of anti-GAD (glutamic acid decarboxylase) antibody (450 nM) in solution to GAD immobilized on four different SAMs (self-assembled monolayers) on a BIAcore 2000 sensor chip surface (Choi *et al.*, 2005): (a) SAM 4, (b) SAM 3, (c) SAM 1, (d) SAM 2.

Figure 3.4c shows the binding of 450 nM monoclonal anti-GAD 65 (purified mouse IgG) to SAM3 immobilized on a BIAcore 2000 sensor chip surface using strepravidin. In this case, a single-fractal analysis is required to adequately describe the binding kinetics. The values of the binding rate coefficient, k, and the fractal dimension, D_f, for a single-fractal analysis, are given in Tables 3.3a and 3.3b.

Figure 3.4d shows the binding of 450 nM monoclonal anti-GAD 65 (purified mouse IgG) to SAM2 immobilized on a BIAcore 2000 sensor chip surface using streptavidin. In this case, a single-fractal analysis is required to adequately describe the binding kinetics. The values of the binding rate coefficient, k, and the fractal dimension, D_f, for a single-fractal analysis, are given in Tables 3.3a and 3.3b.

Zhang *et al.* (2004) have developed a fluorescent sensor for imidazole derivatives. This sensor is based on monomer–dimer equilibrium of a zinc porphyrin complex in a polymeric film. The bridging interaction of the imidazole ring of the analyte with the zinc (II) center of the prophyrin was the molecular recognition step. Pyrene excimer fluorescence was the transduction signal for the recognition process. The sensor was applied for the fluorescence assay of histidine in aqueous solution by immobilizing the sensing material in a plasticized polyvinyl chloride (PVC) membrane.

Imidazoles are frequently found in a large number of natural products and pharmacologically active molecules. Imidazole and its derivatives may also be used as corrosion inhibitors and adhesion promoters (Larsen *et al.*, 2000; Micanovic *et al.*, 1994). Zhang *et al.* (2004) state that ion-selective electrodes for imidazole derivatives based on liquid- and solid-state membranes have been developed (Tastuma and Watnable, 1995; Pihel *et al.*, 1995; Amini *et al.*, 1999).

Table 3.3a

Binding and dissociation rate coefficients for antiglutamic acid decarboxylase (GAD) antibody in solution to different self-assembled monolayers (SAMs) immobilized on a SPR biosensor surface (Choi *et al.*, 2005)

Analyte in solution/ SAM on SPR biosensor surface	k	k_1	k_2	k_d	k_{d1}	k_{d2}
450 nM monoclonal anti-GAD 65 (purified mouse IgG)/SAM4	0.04426 ± 0.0088	0.02133 ± 0.00370	0.4489 ± 0.0209	3.0036 ± 0.0859	na	na
450 nM monoclonal anti-GAD 65 (purified mouse IgG)/SAM3	5.7513 ± 0.05313	3.1421 ± 0.2142	10.4457 ± 0.0553	3.3715 ± 0.5862	1.3168 ± 0.1940	13.5371 ± 0.0650
450 nM monoclonal anti-GAD 65 (purified mouse IgG)/SAM1	9.6993 ± 0.1149	na	na	0.04469 ± 0.0021	na	na
450 nM monoclonal anti-GAD 65 (purified mouse IgG)/SAM2	16.3429 ± 0.3267	na	na	0.03598 ± 0.0025	na	na

Table 3.3b

Fractal dimensions for the binding and the dissociation phase for antiglutamic acid decarboxylase (GAD) antibody in solution to different self-assembled monolayers (SAMs) immobilized on a SPR biosensor surface (Choi *et al.*, 2005)

Analyte in solution/SAM on SPR biosensor surface	D_f	D_{f1}	D_{f2}	D_{fd}	D_{fd1}	D_{fd2}
450 nM monoclonal anti-GAD 65 (purified mouse IgG)/SAM4	0.0780 + 0.1794	0.0	1.1274 ± 0.3010	0.0530 ± 0.0255	na	na
450 nM monoclonal anti-GAD 65 (purified mouse IgG)/SAM3	2.0200 ± 0.0804	1.6346 ± 0.1267	2.293 ± 0.0288	2.0156 ± 0.1569	1.4154 ± 0.2638	2.6594 ± 0.0139
450 nM monoclonal anti-GAD 65 (purified mouse IgG)/SAM1	2.0188 ± 0.0107	na	na	1.0664 ± 0.04852	na	na
450 nM monoclonal anti-GAD 65 (purified mouse IgG)/SAM2	2.1278 ± 0.0168	na	na	1.5664 ± 0.06934	na	na

Zhang *et al.* (2004) indicate that in the presence of the bridging ligand, histidine, the ligation of the imidazole residue to the Zn (II) center of the porphyrin causes the monomer species to be converted to the dimmer species. This in turn yields the strong excimer emission of pyrene at 454 nm with a little increase in the monomer fluorescence. These authors presented reproducibility studies of the measurements obtained with the optode membrane M1 by switching 2.5×10^{-6}M and 2.5×10^{-5}M histidine. They emphasize that their membrane exhibits good reproducibility and reversibility.

Figure 3.5a shows for Run #1 the binding and dissociation of the 2.5×10^{-5}M and 2.5×10^{-6}M histidine solutions that were switched alternatively. A dual-fractal analysis is required to adequately describe the binding kinetics. The dissociation kinetics may be adequately described by a single-fractal analysis. The values (a) the binding rate coefficient, k and the fractal dimension, D_f for a single-fractal analysis, and (b) the binding rate coefficients, k_1 and k_2 and the fractal dimensions, D_{f1} and D_{f2} for a dual-fractal analysis, and (c) the dissociation rate coefficient, k_d and the fractal dimension, D_{fd} for a single-fractal analysis are given in Tables 3.4a and 3.4b.

Figure 3.5b shows for Run #2 the binding and dissociation of the 2.5×10^{-5}M and 2.5×10^{-6}M histidine solutions that were switched alternatively. A dual-fractal analysis is required to adequately describe the binding and the dissociation kinetics. The values (a) the binding rate coefficient, k and the fractal dimension, D_f for a single-fractal analysis, and (b) the binding rate coefficients, k_1 and k_2 and the fractal dimensions, D_{f1} and D_{f2} for a dual-fractal analysis, (c) the dissociation rate coefficient, k_d and the fractal dimension, D_{fd} for a single-fractal analysis, and (d) the dissociation rate coefficients, k_{d1} and k_{d2} and the fractal dimensions, D_{fd1} and D_{fd2} for a dual-fractal analysis are given in Tables 3.4a and 3.4b.

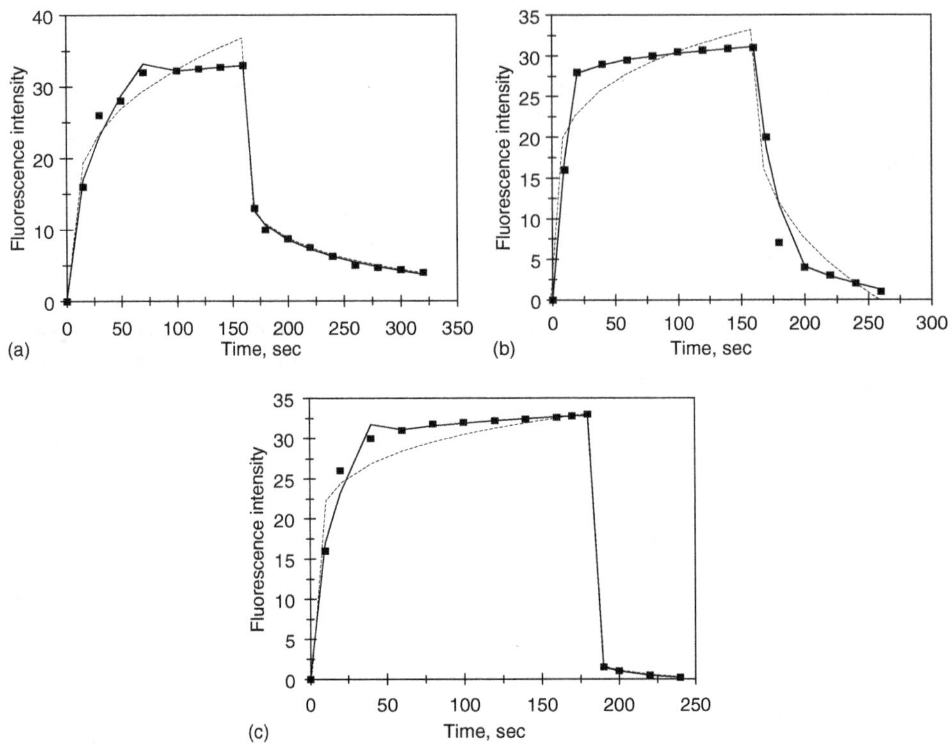

Figure 3.5 Binding of histidine in solution to optode M1 plasticized PVC membrane. 2.5×10^{-6}M and 2.5×10^{-5}M solutions were switched. Reproducibility studies (Zhang *et al.*, 2004): (a) Run #1, (b) Run #2, (c) Run #3 in chronological order.

Table 3.4a

Binding and dissociation rate coefficients for imidazole derivatives in solution to zinc (II) porphyrin conjugate with an appended pyrene subunit immobilized in a plasticized PVC membrane.Reproducibility studies

Run number	k	k_1	k_2	k_d	k_{d1}	k_{d2}
One	8.9876 ± 1.0429	5.2698 ± 0.5673	25.7495 ± 0.0207	15.0834 ± 0.2473	na	na
Two	13.1730 ± 1.8381	6.5027 ± 1.5528	23.9236 ± 0.07721	7.0058 ± 1.4258	2.7645 ± 0.8583	17.9823 ± 0.135
Three	15.9342 ± 2.4934	5.9629 ± 0.8946	25.3826 ± 0.0701	29.9075 ± 0	na	na

Note: 2.5×10^{-6}M and 2.5×10^{-5}M histidine solutions were switched on an optode M1 membrane (Zhang *et al.*, 2004); na: not applicable.

Table 3.4b

Fractal dimensions for the binding and dissociation phase for imidazole derivatives in solution to zinc (II) porphyrin conjugate with an appended pyrene subunit immobilized in a plasticized PVC membrane.

Run number	D_f	D_{f1}	D_{f2}	D_{fd}	D_{fd1}	D_{fd2}
One	2.4536 ± 0.1004	2.1338 ± 0.1761	2.9025 ± 0.00458	2.7386 ± 0.0122	na	na
Two	2.6332 ± 0.09816	2.1422 ± 0.4368	2.8650 ± 0.007886	2.3564 ± 0.1511	1.7046 ± 0.5518	2.7818 ± 0.0223
Three	2.7270 ± 0.07706	2.0932 ± 0.2852	2.9003 ± 0.01095	2.9549 ± 1.04E-04	na	na

Note: Reproducibility studies. 2.5×10^{-6}M and 2.5×10^{-5}M histidine solutions were switched on an optode M1 membrane (Zhang *et al.*, 2004); na: not applicable.

Figure 3.5c shows for Run #3 the binding and dissociation of the 2.5×10^{-5}M and 2.5×10^{-6}M histidine solutions that were switched alternatively. A dual-fractal analysis is required to adequately describe the binding kinetics. The dissociation kinetics may be adequately described by a single-fractal analysis. The values (a) the binding rate coefficient, k and the fractal dimension, D_f for a single-fractal analysis, and (b) the binding rate coefficients, k_1 and k_2 and the fractal dimensions, D_{f1} and D_{f2} for a dual-fractal analysis, and (c) the dissociation rate coefficient, k_d and the fractal dimension, D_{fd} for a single-fractal analysis are given in Tables 3.4a and 3.4b.

It is of interest to compare the binding rate coefficients, k_1 and k_2 and the fractal dimensions, D_{f1} and D_{f2} when a dual-fractal analysis applies for the chronological Run #s 1, 2, and 3. The average values of k_1, k_2, D_{f1}, and D_{f2} for these three runs are 5.912, 25.02, 2.123, and 2.889, respectively. The binding rate coefficient, k_1 is within 10.8% of its average value. The binding rate coefficient, k_2 is within 2.9% of its average value.

The fractal dimension, D_{f1} is within 1.4% of its average value. The fractal dimension, D_{f2} is within 0.831% of its average value. This indicates that these runs exhibit reproducible results in accordance with a similar statement made by the original authors (Zhang *et al.*, 2004).

Figure 3.6a shows the increase in the binding rate coefficient, k_2 with an increase in the fractal dimension, D_{f2} when a dual-fractal analysis applies for the reproducibility runs (#1, 2, and 3) presented in Figures 3.5a–c. The binding rate coefficient, k_2 is given by

$$k_2 = (0.0909 \pm 0.00067)D_{f2}^{5.294 \pm 0.7091} \tag{3.3a}$$

The fit is good. Only three data points are available. The availability of more data points would lead to a more reliable fit. The binding rate coefficient, k_2 is very sensitive to the degree of heterogeneity present on the biosensor chip surface as noted by the higher than fifth (equal to 5.294) order of dependence exhibited.

Figure 3.6b shows the increase in the affinity, K_i $(= k_i/k_{di})$ with an increase in the ratio of the fractal dimensions in the binding and in the dissociation phases, D_{fi}/D_{fdi} when a

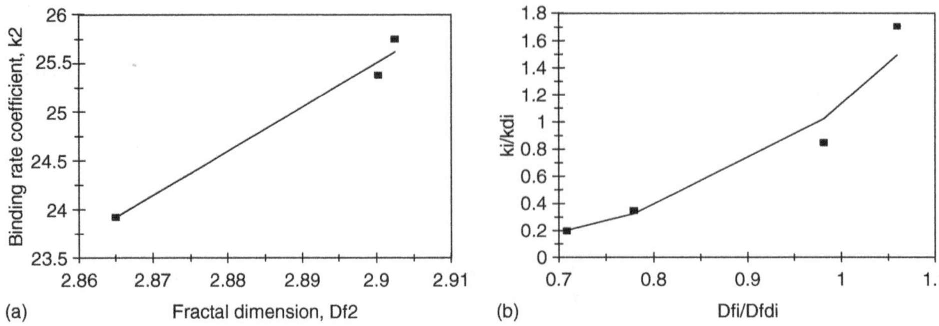

Figure 3.6 (a) Increase in the binding rate coefficient, k_2 with an increase in the fractal dimension, D_{f2}. (b) Increase in the affinity, $K_i(=k_i/k_{di})$ with an increase in the fractal dimension ratio, D_{fi}/D_{fdi}.

dual-fractal analysis applies for the reproducibility runs (#1, 2, and 3) presented in Figures 3.5a–c. The affinity, k_i is given by

$$K_i\left(=\frac{k_i}{k_{di}}\right)=(1.1204\pm0.2068)\left(\frac{D_{fi}}{D_{fdi}}\right)^{4.9546\pm0.5159} \quad (3.3b)$$

The fit is very good. Only four data points are available. The availability of more data points would lead to a more reliable fit. The affinity, K_i is very sensitive to the degree of heterogeneity present on the biosensor chip surface as noted by the close to fifth (equal to 4.9546) order of dependence exhibited.

Cao *et al.* (2006) indicate that prostate cancer is a major cause of death for the male population. Savage and Waxman (1996) indicate that this disease is increasing rapidly, and it is estimated to be the leading cause of cancer in men by the year 2010. Cao *et al.* (2006) emphasize the importance of the early and accurate detection of prostate cancer. PSA can be used to detect prostate cancer in its early stages. Armbuster (1993) and Savage and Waxman (1996) indicate that PSA is a premier tumor marker for prostate cancer. Cao *et al.* (2006) have developed a strategy that enhances sensitivity and specificity using the SPR biosensor for the detection of PSA- α_1-antichymotrypsin detection.

PSA-α_1-antichymotrypsin (PSA-ACT complex) was detected by Cao *et al.* (2006) in HBS buffer and in human serum. Various oligo (ethylene glycol) mixtures were used to minimize steric hindrance and NSB. These authors were able to enhance the sensitivity and specificity of PSA-ACT complex detection by employing a simple sandwich strategy. Figure 3.7a shows the binding of 1000 ng/mL of PSA-ACT complex in solution to the biotinylated PSA-ACT mAb (monoclonal antibody) immobilized on a 1:9 sensor chip surface (primary response). The clear gold chips were immersed into 0.5 mM ethanol solutions containing 1:09 molar ratio of EG_6–COOH/EG_3–OH to form a mixed SAM (self-assembled monolayer). EG_6-COOH is $HS(CH_2)_{11}(OCH_2CH_2)_6\ OCH_2COOH$ and EG_3-OH is $HS(CH_2)_{11}(OCH_2CH_2)_3$–OH. A dual-fractal analysis is required to adequately describe the binding kinetics. The values of (a) the binding rate coefficient, k and the fractal dimension, D_f for a single-fractal analysis and (b) the binding rate coefficients, k_1 and k_2 and the

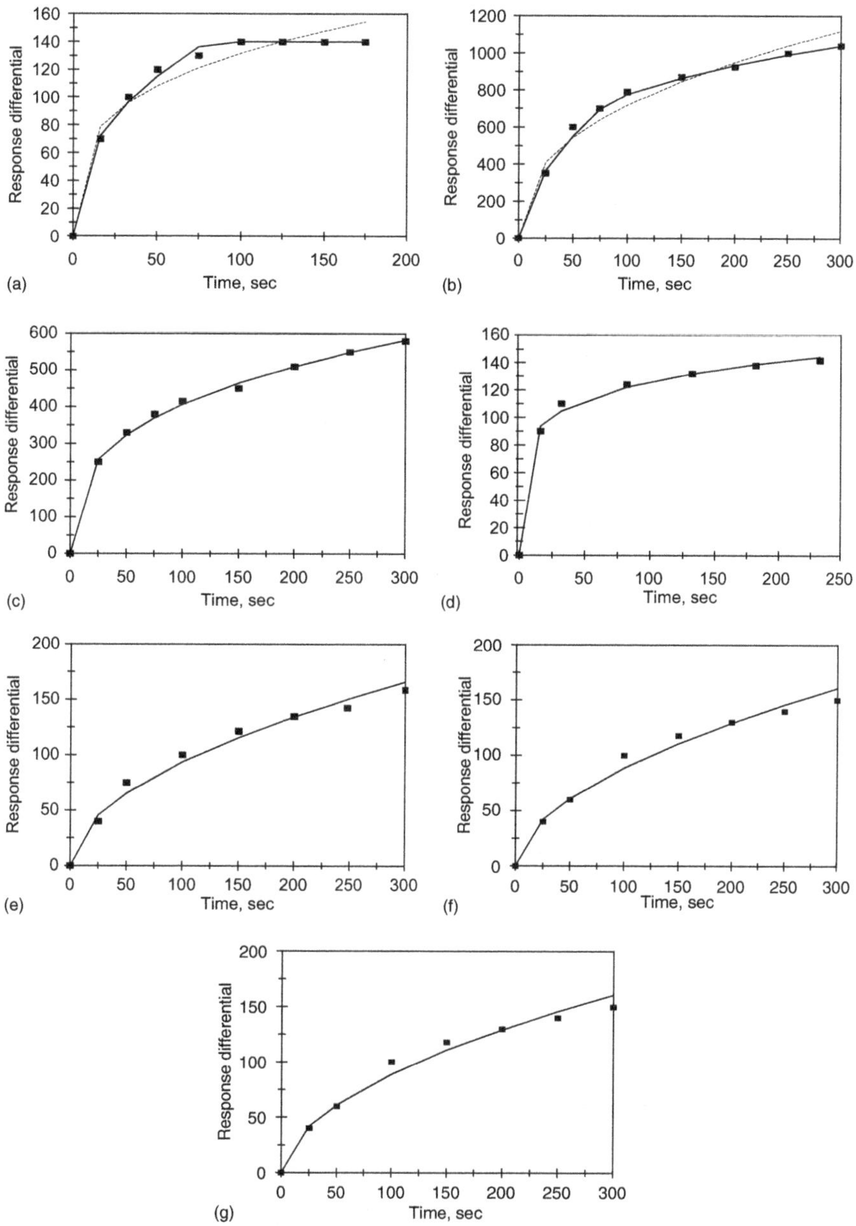

Figure 3.7 Binding of different concentrations of the PSA-ACT (prostate-specific antigen-α_1-chymotrypsin) complex in solution to the biotinylated PSA-ACT mAb layer immobilized on a surface plasmon resonance (SPR) 1:9 sensor chip surface (Cao *et al.*, 2006): (a) 1000 ng/mL PSA-ACT, primary response; (b) 1000 ng/mL PSA-ACT, enhanced response; (c) 500 ng/mL PSA-ACT, enhanced response; (d) 500 ng/mL PSA-ACT, enhanced response; (e) 100 ng/mL PSA-ACT, enhanced response; (f) 10 ng/mL PSA-ACT, enhanced response.

Table 3.5

Binding rate coefficients and fractal dimensions for the primary response and the enhanced response for prostate-specific antigen-α_1-antichymotrypsin detection by surface plasmon resonance (Cao *et al.*, 2006)

Analyte in solution/ biotinylated PSA-ACT mAb layer	Type of response: primary or enhanced	k	k_1	k_2	D_f	D_{f1}	D_{f2}
1000 ng/mL	Primary	35.486 ± 3.311	22.231 ± 1.277	140 ± 0.0	2.4304 ± 0.08272	2.1600 ± 0.09980	3.0 ± 0.0
1000 ng/mL	Enhanced	112.07 ± 11.885	55.044 ± 4.359	228.998 ± 2.297	2.1924 ± 0.0892	1.8234 ± 0.1464	2.4690 ± 0.03856
500 ng/mL	Enhanced	89.270 ± 2.182	na	na	2.3418 ± 0.02136	na	na
500 ng/mL	Primary	59.917 ± 2.050	na	na	2.6770 ± 0.02870	na	na
100 ng/mL	Enhanced	8.595 ± 0.89	na	na	1.9622 ± 0.0886	na	na
10 ng/mL	Enhanced	7.424 ± 0.571	na	na	1.9212 ± 0.0664	na	na

fractal dimensions, D_{f1} and D_{f2} for a dual-fractal analysis are given in Table 3.5. For a dual-fractal analysis it is of interest to note that as the fractal dimension increase by 38.8% from a value of D_{f1} equal to 2.16 to D_{f2} equal to 3.0 (maximum value), the binding rate coefficient increases by a factor of 6.30 from a value of k_1 equal to 22.231 to k_2 equal to 140.

Figure 3.7b shows the binding of 1000 ng/mL of PSA-ACT in solution to the biotinylated PSA-ACT mAb (monoclonal antibody) immobilized on the 1:9 sensor chip surface ((secondary response). In this case as indicated previously a sandwich-type assay was used. A dual-fractal analysis is once again required to adequately describe the binding kinetics. The values of (a) the binding rate coefficient, k and the fractal dimension, D_f for a single-fractal analysis and (b) the binding rate coefficients, k_1 and k_2 and the fractal dimensions, D_{f1} and D_{f2} for a dual-fractal analysis are given in Table 3.5. For a dual-fractal analysis it is of interest to note that as the fractal dimension increases by 35.4% from a value of D_{f1} equal to 1.8234 to D_{f2} equal to 2.4690, the binding rate coefficient increases by a factor of 4.16 from a value of k_1 equal to 55.044 to k_2 equal to 228.98.

It is of interest to note that though the fractal dimension, D_{f1} and D_{f2} values for the enhanced response are lower than the corresponding ones for the primary response, the binding rate coefficients, k_1 and k_2 for the enhanced response (as expected) are higher than those of the primary response. Presumably, it is the sandwich-type assay in the enhanced response that leads to the higher binding rate coefficients in spite of the corresponding lower fractal dimension values when compared to the primary response.

Figure 3.7c shows the binding of 500 ng/mL of PSA-ACT complex in solution to the biotinylated PSA-ACT mAb (monoclonal antibody) immobilized on a 1:9 sensor chip surface (primary response). A single-fractal analysis is required to adequately describe the binding kinetics. The values of the binding rate coefficient, k and the fractal dimension, D_f for a single-fractal analysis are given in Table 3.5.

Figure 3.7d shows the binding of 500 ng/mL of PSA-ACT complex in solution to the biotinylated PSA-ACT mAb (monoclonal antibody) immobilized on a 1:9 sensor chip surface (enhanced response). A single-fractal analysis is required to adequately describe the binding kinetics. The values of the binding rate coefficient, k and the fractal dimension, D_f for a single-fractal analysis are given in Table 3.5.

It is of interest to note once again that though the fractal dimension, D_f value for the enhanced response is lower than the corresponding ones for the primary response, the binding rate coefficients, k for the enhanced response (as expected) is higher than that of the primary response. Once again, as indicated above, presumably, it is the sandwich-type assay in the enhanced response that leads to the higher binding rate coefficients in spite of the corresponding lower fractal dimension values when compared to the primary response.

Figure 3.8a shows the increase in the binding rate coefficient, k_i (k, k_1, or k_2) with an increase in the fractal dimension, D_{fi}(D_f, D_{f1}, or D_{f2}) for the primary response. For the data given in Table 3.5 and in Figure 3.8a, the binding rate coefficient, k_i is given by

$$k_i = (0.3129 \pm 0.0577) D_{fi}^{5.4747 \pm 0.7177} \tag{3.4a}$$

Only three data points are available. The fit is good. The availability of more data points would lead to a more reliable fit. The binding rate coefficient, k_i is extremely sensitive to the fractal dimension, D_{fi} or the degree of heterogeneity that exists on the surface as noted by the close to five and one-half (equal to 5.4747) order of dependence exhibited.

Figure 3.8b shows the increase in the binding rate coefficient, k_i (k, k_1, or k_2) with an increase in the fractal dimension, D_{fi} (D_f, D_{f1}, or D_{f2}) for the enhanced response. For the data given in Table 3.5 and in Figure 3.8b, the binding rate coefficient, k_i is given by

$$k_i = (5.128 \pm 3.583) D_{fi}^{3.833 \pm 2.314} \tag{3.4b}$$

Only three data points are available. There is scatter in the data. Note that the binding rate coefficient, k for a single-fractal analysis, and the binding rate coefficients, k_1 and k_2 for a dual-fractal analysis are plotted on the same graph. Part of the scatter could be due to this. The availability of more data points would lead to a more reliable fit. The binding rate coefficient, k_i is sensitive to the fractal dimension, D_{fi} or the degree of heterogeneity that exists on the surface as noted by the close to fourth (equal to 3.833) order of dependence exhibited. Note that the enhanced response exhibits a lower order of dependence on the degree of heterogeneity that exists on the biosensor chip surface than the primary response (order equal to 5.4747). This indicates that the enhanced response is less sensitive than the primary response on the degree of heterogeneity that exists on the biosensor chip surface.

Figure 3.8c shows the increase in the binding rate coefficient, k with an increase in the PSA concentration in the 10–500 ng/mL range in solution when a single-fractal analysis is used for the enhanced response. For the data given in Table 3.5 and in Figure 3.8c, the binding rate coefficient, k is given by

$$k = (1.315 + 2.519)[\text{PSA concentration in ng/mL}]^{0.5694 \pm 0.3838} \tag{3.4c}$$

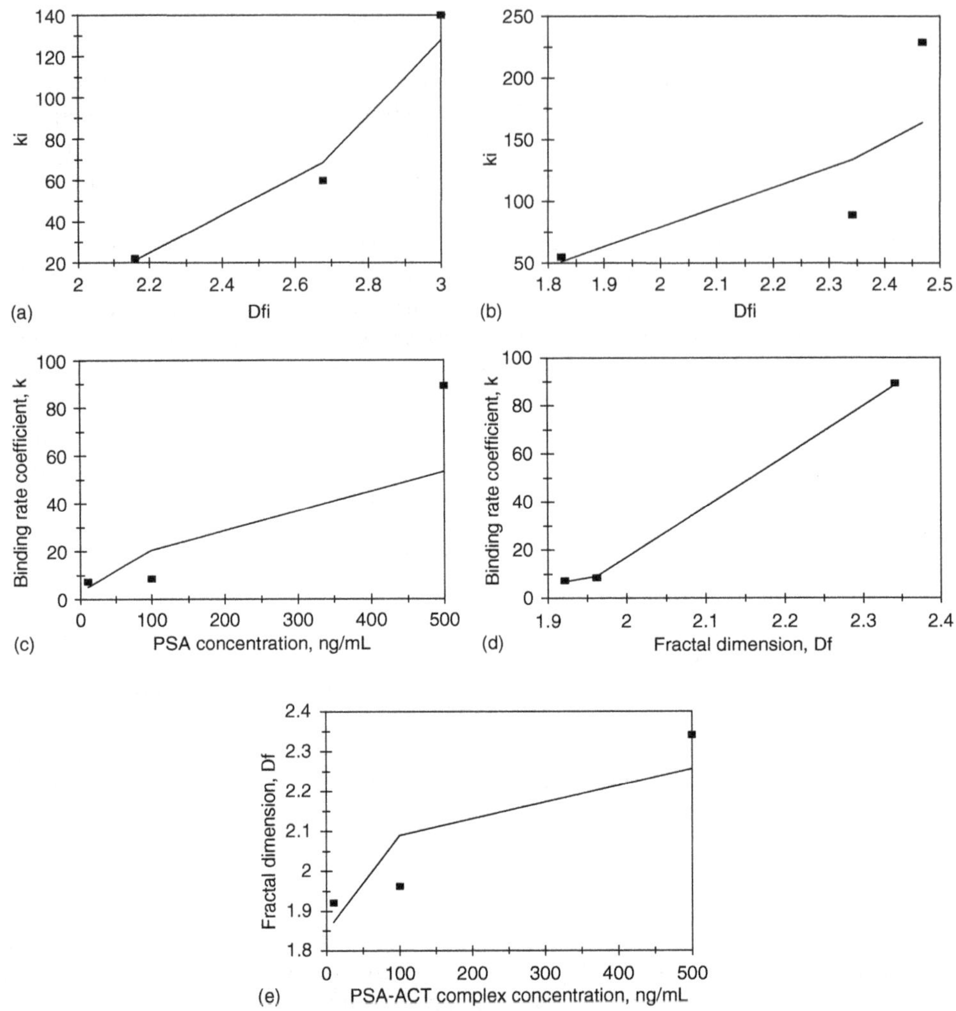

Figure 3.8 (a) Increase in the binding rate coefficient, k_i with an increase in the fractal dimension, D_{fi} (primary response). (b) Increase in the binding rate coefficient, k_i with an increase in the fractal dimension, D_{fi} (enhanced response). (c) Increase in the binding rate coefficient, k with an increase in the PSA concentration (in ng/mL) in solution. (d) Increase in the binding rate coefficient, k with an increase in the fractal dimension, D_f. (e) Increase in the fractal dimension, D_f with an increase in the PSA-ACT complex concentration (in ng/mL) in solution.

Only three data points are available. There is considerable scatter in the data as noted by the large error in the binding rate coefficient. Only the positive error is presented, since the binding rate coefficient cannot be negative. The availability of more data points would lead to a more reliable fit. The binding rate coefficient, k for the enhanced response exhibits an order of dependence between one-half and first (equal to 0.5963) on the PSA concentration in solution. The non-integer order of dependence exhibited by the binding rate coefficient, k on the PSA concentration in solution lends support to the fractal nature of the system.

Figure 3.8d shows the increase in the binding rate coefficient, k with an increase in the fractal dimension, D_f for the enhanced response. For the data given in Table 3.5 and in Figure 3.8d, the binding rate coefficient, k_k is given by

$$k = (0.001618 \pm 0.000149) D_f^{12.823 \pm 0.5744} \tag{3.4d}$$

Only three data points are available. The fit is very good. The availability of more data points would lead to a more reliable fit. The binding rate coefficient, k is very sensitive to the fractal dimension, D_f or the degree of heterogeneity that exists on the surface as noted by the close to thirteenth (equal to 12.823) order of dependence exhibited.

Figure 3.8e shows the increase in the fractal dimension, D_f with an increase in the PSA concentration in the 10–500 ng/mL range in solution when a single-fractal analysis is used for the enhanced response. For the data given in Table 3.5 and in Figure 3.8c, the fractal dimension, D_f is given by

$$D_f = (1.6771 \pm 0.1351)[\text{PSA-ACT concentration in ng/mL}]^{0.04775 \pm 0.02787} \tag{3.4e}$$

The fit is reasonable. Only three data points are available. The availability of more data points would lead to a more reliable fit. The fractal dimension, D_f is only very slightly dependent on the PSA-ACT complex concentration in solution as noted by the very low (equal to 0.04775) order of dependence exhibited.

Mitchell *et al.* (2005) have recently developed an SPR biosensor format that exhibits sensitivity enhancement using progesterone as a model compound. Gold nano particles were used, and progesterone was immobilized to a dextran surface of a Biacore biosensor. Both label prebinding and sequential binding formats were used. Figure 3.9a shows the binding of the primary monoclonal antibody (mAb, 25 μg/mL) to immobilized progesterone on the biosensor chip surface. A dual-fractal analysis is required to adequately describe the binding kinetics. The values of (a) the binding rate coefficient, k and the fractal dimension, D_f for a single-fractal analysis, and (b) the binding rate coefficients, k_1 and k_2 and the fractal dimensions, D_{f1} and D_{f2} for a dual-fractal analysis are given in Table 3.6.

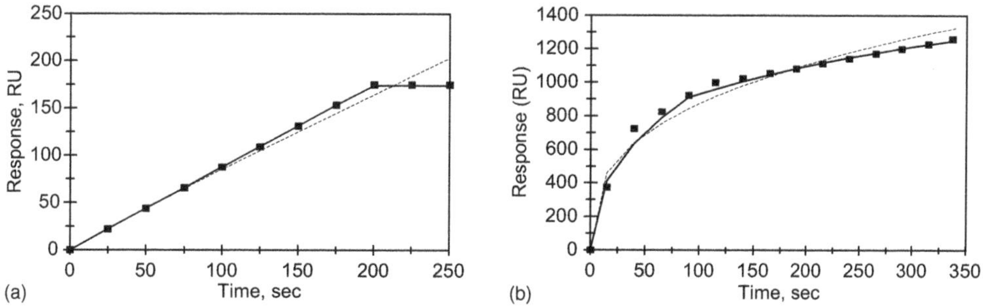

Figure 3.9 Binding of (a) primary monoclonal antibody (mAb; 25 μg/mL) and (b) secondary antibody enhancement (800 μg/mL) to progesterone (used as a model compound) immobilized on a dextran surface of a Biacore SPR biosensor chip surface (Mitchell *et al.*, 2005).

Table 3.6

Binding rate coefficients and fractal dimensions for (a) primary monoclonal antibody (mAb) binding and (b) secondary binding enhancement (Mitchell *et al.*, 2005)

Analyte in solution/ progesterone (model compound) immobilized on sensor chip surface	k	k_1	k_2	D_f	D_{f1}	D_{f2}
Primary monoclonal antibody, 25 μg/mL/ progesterone	1.0913 ± 0.0786	0.876 ± 0.0	175 ± 0.0	1.9076 ± 0.0632	1.0 ± 5.6 E-15	3.0–2.6E-13
Secondary antibody enhancement, 800 μg/ mL/progesterone	182.94 ± 16.80	122.57 ± 11.92	301.03 ± 1.564	2.3204 ± 0.055	2.1076 ± 0.101	2.512 ± 0.146

It is of interest to note that for the binding of the primary monoclonal antibody (mAb) for a dual-fractal analysis, as the fractal dimension increases by a factor of three from D_{f1} equal to 1.0 to D_{f2} equal to 3.0, the binding rate coefficient increases by a factor of 199.77 from a value of k_1 equal to 0.876 to k_2 equal to 175.

Figure 3.9b shows the binding of the secondary antibody enhancement (800 μg/mL) to immobilized progesterone on the biosensor chip surface. A dual-fractal analysis is once again required to adequately describe the binding kinetics. The values of (a) the binding rate coefficient, k and the fractal dimension, D_f for a single-fractal analysis, and (b) the binding rate coefficients, k_1 and k_2 and the fractal dimensions, D_{f1} and D_{f2} for a dual-fractal analysis are given in Table 3.6.

Once again, it is of interest to note that for the binding of the secondary antibody enhancement for a dual-fractal analysis, as the fractal dimension increases by a factor of 1.192 from D_{f1} equal to 2.1076 to D_{f2} equal to 2.511, the binding rate coefficient increases by a factor of 2.46 from a value of k_1 equal to 122.572 to k_2 equal to 301.033.

For the initial binding phase, note that (a) the fractal dimension, D_{f1} values are 1.0 and 2.1076 for the primary antibody binding and the secondary antibody enhancement, and (b) the (corresponding) binding rate coefficients are 0.876 and 122.572, respectively. As expected, when one compares the binding rate coefficients for the second phase of binding also, the k_2 value is higher for the secondary antibody enhancement case (equal to 301.033) than that of the primary monoclonal antibody binding case (equal to 175). Note the very significant (greater than three orders of magnitude) increase in the binding rate coefficient, k_1, as one goes from the primary monoclonal antibody binding to that of the secondary antibody enhancement case.

Cao *et al.* (2006) indicate that for clinical applications it is essential to minimize NSB. In other words, one needs to minimize the interactions between human fluid proteins and the biosensor surface. These authors analyzed the NSB between the 1:9 biosensor surface and the serum proteins such as albumin (BSA), IgG, and fibrinogen. These authors indicate that the oligo(ethylene glycol) (OEG) used SAMs were very effective in reducing NSB. These authors emphasize that the SAM surface of the OEG used provided a template for water

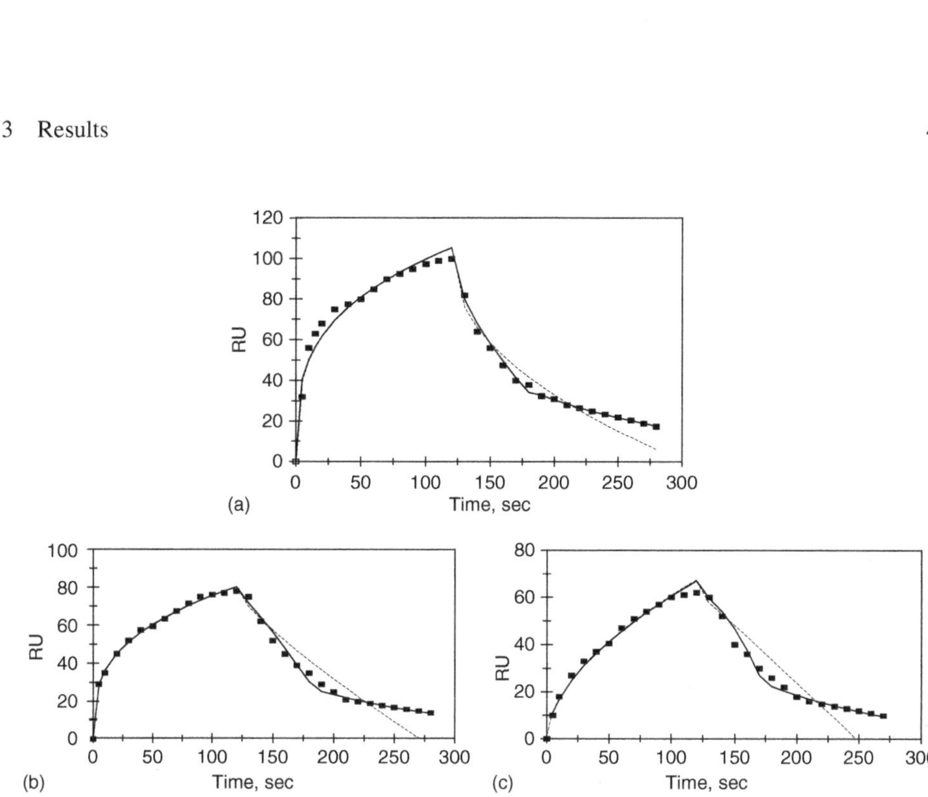

Figure 3.10 Nonspecific binding (NSB) of different concentrations of BSA (bovine serum albumin) in solution to the 1:9 SPR sensor chip surface (Choi *et al.*, 2006): (a) 10 μg/mL, 18.7 RU; (b) 20 μg/mL, 13.1 RU; (c) 40 μg/mL, 11.6 RU.

nucleation. This interfacial water layer according to these authors prevented direct contact between the surface and proteins (Silin *et al.*, 1997; Wang *et al.*, 1997).

Figure 3.10a shows the NSB of 10 μg/mL BSA in solution (18.7 RU) to the 1:9 (EG_6–COOH/EG_3–OH; molar ratio) mixed SAM gold-coated biosensor surface. A single-fractal analysis is adequate to describe the binding kinetics. A dual-fractal analysis is required to adequately describe the dissociation kinetics. The values of (a) the binding rate coefficient, k and the fractal dimension, D_f for a single-fractal analysis, (b) the dissociation rate coefficient, k_d for and the fractal dimension for the dissociation phase, D_{fd} for a single-fractal analysis, and (c) the dissociation rate coefficients, k_{d1} and k_{d2}, and the fractal dimensions, D_{fd1} and D_{fd2} for a dual-fractal analysis are given in Tables 3.7a and 3.7b.

Figure 3.10b shows the NSB of 20 μg/mL BSA in solution (13.1 RU) to the 1:9 (EG_6–COOH/EG_3–OH; molar ratio) mixed SAM gold-coated biosensor surface. A single-fractal analysis is adequate to describe the binding kinetics. A dual-fractal analysis is required to adequately describe the dissociation kinetics. The values of (a) the binding rate coefficient, k and the fractal dimension, D_f for a single-fractal analysis, (b) the dissociation rate coefficient, k_d for and the fractal dimension for the dissociation phase, D_{fd} for a single-fractal analysis, and (c) the dissociation rate coefficients, k_{d1} and k_{d2}, and the fractal dimensions, D_{fd1} and D_{fd2} for a dual-fractal analysis are given in Tables 3.7a and 3.7b.

It is of interest to note that as one goes from the 10 μg/mL BSA in solution to the 20 μg/mL BSA in solution, (a) the fractal dimension, D_f decreases by 1.9% from a value of 2.400 to 2.3542, and (b) the binding rate coefficient, k decreases by 31.9% from a value of

Table 3.7a

Binding and dissociation rate coefficients for nonspecific binding (NSB) of different concentrations of bovine serum albumin (BSA) and IgG in solution to the 1:9 (EG_6–COOH/EG_3–OH; molar ratio) OEG [(oligo)nucleotide] mixture on bare gold surface (Cao *et al.*, 2006)

Analyte in solution, RU	k	k_1	k_2	k_d	k_{d1}	k_{d2}
(a) BSA 10 μg/mL (NSB: 18.7 RU)	25.104 ± 2.436	na	na	7.798 ± 0.926	4.431 ± 0.429	23.645 ± 0.099
20 μg/mL (NSB: 13.1 RU)	17.092 ± 0.3174	na	na	1.2954 ± 0.3053	0.4446 ± 0.0829	18.975 ± 1.365
40 μg/mL (NSB: 11.6 RU)	4.6879 ± 0.3090	na	na	0.4090 ± 0.1639	0.06499 ± 0.0217	11.2771 ± 0.2243
(b) IgG 10 μg/mL; NSB 15.8 RU	20.754 ± 2.173	12.474 ± 0.461	98.668 ± 1.432	9.3114 ± 2.8424	2.1130 ± 0.4888	87.998 ± 1.548
20 μg/mL; NSB 13.6 RU	17.534 ± 2.1524	9.6840 ± 0.4039	152.342 ± 0.608	13.972 ± 1.7665	na	na
40 μg/mL; NSB 10.8 RU	31.674 ± 17.301	4.3626 ± 0.2405	49.784 ± 0.573	6.4565 ± 1.529	1.9534 ± 0.308	29.932 ± 1.254

Table 3.7b

Fractal dimensions for the binding and dissociation phase for nonspecific binding (NSB) of different concentrations of bovine serum albumin (BSA) and IgG in solution to the 1:9 (EG_6–COOH/EG_3–OH; molar ratio) OEG [(oligo)nucleotide] mixture on bare gold surface (Cao *et al.*, 2006)

Analyte in solution, RU	D_f	D_{f1}	D_{f2}	D_{fd}	D_{fd1}	D_{fd2}
(a) BSA 10 μg/mL (NSB: 18.7 RU)	2.400 ± 05238	na	na	2.0286 ± 0.07344	1.6826 ± 0.1102	2.5086 ± 0.01253
20 μg/mL (NSB: 13.1 RU)	2.3542 ± 0.0109	na	na	1.3678 ± 0.1425	0.7172 ± 0.2038	2.5196 ± 0.0392
40 μg/mL (NSB: 11.6 RU)	1.8884 ± 0.037780	na	na	0.9336 ± 0.2302	0.0 + 0.3896	2.3862 ± 0.0537
(b) IgG 10 μg/mL; NSB 15.8 RU	1.9332 ± 0.09054	1.6212 ± 0.0574	2.6286 ± 0.04984	1.6630 ± 0.2358	0.7282 ± 0.3996	2.64248 ± 0.06856
20 μg/mL; NSB 13.6 RU	1.8972 ± 0.1053	1.5342 ± 0.0643	2.8561 ± 0.02994	1.8996 ± 0.1067	na	na
40 μg/mL; NSB 10.8 RU	2.2776 ± 0.3298	1.2714 ± 0.08444	2.4468 ± 0.05668	1.6060 ± 0.1881	0.8538 ± 0.2812	2.2862 ± 0.09464

25.104 to 17.092. Note that changes in the fractal dimension, D_f or the degree of heterogeneity on the surface and in the binding rate coefficient, k are in the same direction.

Figure 3.10c shows the NSB of 40 μg/mL BSA in solution (11.6 RU) to the 1:9 (EG_6–COOH/EG_3–OH; molar ratio) mixed SAM gold-coated biosensor surface. A single-fractal analysis is adequate to describe the binding kinetics. A dual-fractal analysis is

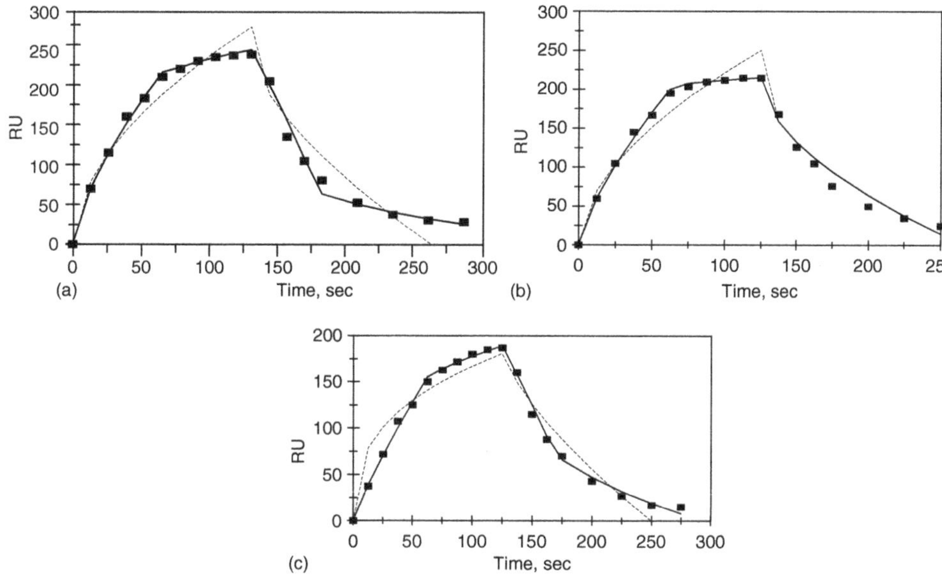

Figure 3.11 Nonspecific binding (NSB) of different concentrations of IgG in solution to the 1:9 SPR sensor chip surface (Choi *et al.*, 2006): (d) 10 μg/mL, 15.8 RU; (e) 20 μg/mL, 13.6 RU; (f) 40 μg/mL, 10.8 RU.

required to adequately describe the dissociation kinetics. The values of (a) the binding rate coefficient, k and the fractal dimension, D_f for a single-fractal analysis, (b) the dissociation rate coefficient, k_d for and the fractal dimension for the dissociation phase, D_{fd} for a single-fractal analysis, and (c) the dissociation rate coefficients, k_{d1} and k_{d2}, and the fractal dimensions, D_{fd1} and D_{fd2} for a dual-fractal analysis are given in Table 3.7a and 3.7b.

It is of interest to note that as one goes from the 10 μg/mL BSA in solution to the 40 μg/mL BSA in solution, (a) the fractal dimension, D_f decreases by 21.3% from a value of 2.400 to 1.8884, and (b) the binding rate coefficient, k decreases by 81.3% from a value of 25.104 to 4.6879. Note once again that changes in the fractal dimension, D_f or the degree of heterogeneity on the surface and in the binding rate coefficient, k are in the same direction.

Figure 3.11a shows the NSB of 10 μg/mL IgG in solution (15.8 RU) to the 1:9 (EG_6–COOH/EG_3–OH; molar ratio) mixed SAM gold-coated biosensor surface. A dual-fractal analysis is required to adequately describe the binding and the dissociation kinetics. The values of (a) the binding rate coefficient, k and the fractal dimension, D_f for a single-fractal analysis, (b) the binding rate coefficients, k_1 and k_2 and the fractal dimensions, D_{f1} and D_{f2} for a dual-fractal analysis, (c) the dissociation rate coefficient, k_d and the fractal dimension for the dissociation phase, D_{fd}, for a single-fractal analysis, and (d) the dissociation rate coefficients, k_{d1} and k_{d2}, and the fractal dimensions, D_{fd1} and D_{fd2} for a dual-fractal analysis are given in Tables 3.7a and 3.7b.

It is of interest to note that that for a dual-fractal analysis (a) as the fractal dimension for NSB increases by 62.1% from a value of D_{f1} equal to 1.6212 to D_{f2} equal to 2.6286, the binding rate coefficient increases by a factor of 7.91 from a value of k_1 equal to 12.474 to k_2 equal to 98.668, and (b) as fractal dimension for NSB in the dissociation phase increases by a factor of 3.63 from a value of D_{fd1} equal to 0.7282 to D_{fd2} equal to 2.6424, the dissociation rate coefficient for NSB increases by a factor of 41.65 from a value of k_{d1} equal to 2.1130 to k_{d2} equal to 87.998.

Figure 3.11b shows the NSB of 20 μg/mL IgG in solution (13.6 RU) to the 1:9 (EG_6–COOH/EG_3–OH; molar ratio) mixed SAM gold-coated biosensor surface. A dual-fractal analysis is required to adequately describe the binding kinetics, and a single-fractal analysis is adequate describe the dissociation kinetics. The values of (a) the binding rate coefficient, k and the fractal dimension, D_f for a single-fractal analysis, (b) the binding rate coefficients, k_1 and k_2 and the fractal dimensions, D_{f1} and D_{f2} for a dual-fractal analysis, and (c) the dissociation rate coefficient, k_d and the fractal dimension for the dissociation phase, D_{fd}, for a single-fractal analysis are given in Tables 3.7a and 3.7b.

Once again, it is of interest to note that that for a dual-fractal analysis (a) as the fractal dimension for NSB increases by 86.2% from a value of D_{f1} equal to 1.5342 to D_{f2} equal to 2.8561, the binding rate coefficient increases by a factor of 15.73 from a value of k_1 equal to 9.6840 to k_2 equal to 152.342.

Figure 3.11c shows the NSB of 40 μg/mL IgG in solution (10.8 RU) to the 1:9 (EG_6–COOH/EG_3–OH; molar ratio) mixed SAM gold-coated biosensor surface. A dual-fractal analysis is required to adequately describe the binding and the dissociation kinetics. The values of (a) the binding rate coefficient, k and the fractal dimension, D_f for a single-fractal analysis, (b) the binding rate coefficients, k_1 and k_2 and the fractal dimensions, D_{f1} and D_{f2} for a dual-fractal analysis, (c) the dissociation rate coefficient, k_d and the fractal dimension for the dissociation phase, D_{fd}, for a single-fractal analysis, and (d) the dissociation rate coefficients, k_{d1} and k_{d2}, and the fractal dimensions, D_{fd1} and D_{fd2} for a dual-fractal analysis are given in Table 3.7a.

A indicated above, once again, note that that for a dual-fractal analysis (a) as the fractal dimension for NSB increases by 92.4% from a value of D_{f1} equal to 1.2714 to D_{f2} equal to 2.4468, the binding rate coefficient increases by a factor of 11.41 from a value of k_1 equal to 4.3626 to k_2 equal to 49.784, and (b) as fractal dimension for NSB in the dissociation phase increases by a factor of 2.68 from a value of D_{fd1} equal to 1.2714 to D_{fd2} equal to 2.4468, the dissociation rate coefficient for NSB increases by a factor of 15.32 from a value of k_{d1} equal to 1.9534 to k_{d2} equal to 29.932.

Figure 3.12a shows the increase in the binding rate coefficient, k with an increase in the fractal dimension, D_f for the NSB of BSA in solution to the 1:9 biosensor chip surface. For the data given in Table 3.7a and 3.7b, and plotted in Figure 3.12a, the binding rate coefficient, k is given by:

$$k = (0.0731 \pm 0.0147) D_f^{6.527 \pm 0.9735} \tag{3.5a}$$

The fit is good. Only three data points are available. The availability of more data points would lead to a more reliable fit. The binding rate coefficient, k for the NSB of BSA in

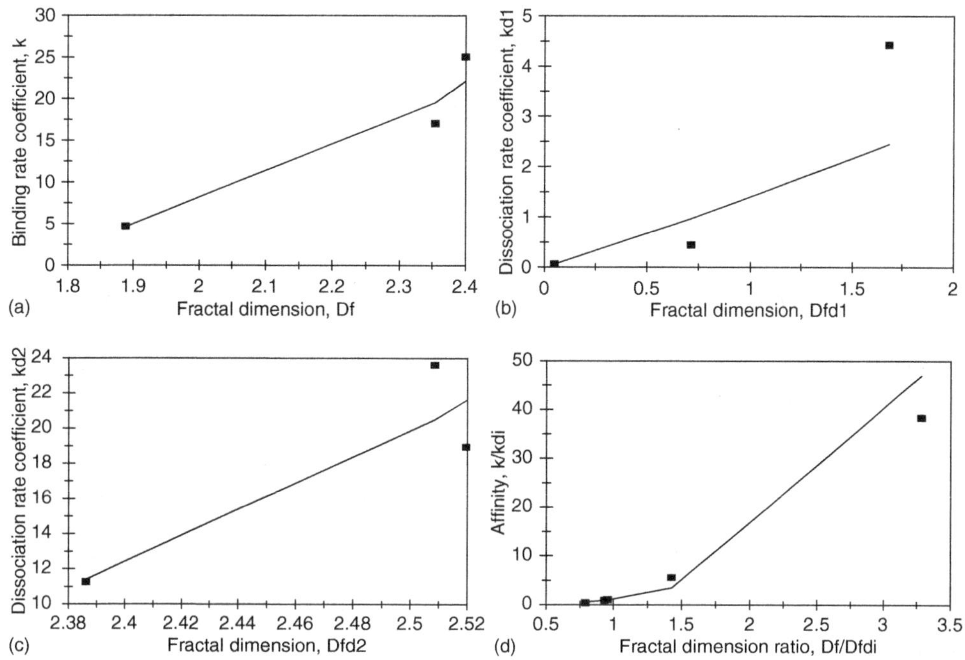

Figure 3.12 (a) Increase in the binding rate coefficient, k with an increase in the fractal dimension, D_f. (b) Increase in the dissociation rate coefficient, k_{d1} with an increase in the fractal dimension, D_{fd1}. (c) Increase in the dissociation rate coefficient, k_{d2} with an increase in the fractal dimension, D_{fd2}. (d) Increase in the affinity, $K(=k/k_{di})$ with an increase in the ratio of the fractal dimensions in the binding and in the dissociation phase (D_f/D_{fdi}).

solution to the 1:9 biosensor chip surface is very sensitive to the degree of heterogeneity or the fractal dimension, D_f that exists on the surface as noted by the very close to six and one-half order of dependence (equal to 6.527) exhibited.

Figure 3.12b shows the increase in the dissociation rate coefficient, k_{d1} with an increase in the fractal dimension, D_{fd1} for the NSB of BSA in solution to the 1:9 biosensor chip surface. This is when a dual-fractal analysis is required to adequately describe the dissociation kinetics. For the data given in Tables 3.7a and 3.7b, and plotted in Figure 3.12b, the dissociation rate coefficient, k_{d1} is given by:

$$k_{d1} = (1.393 + 2.386)D_{fd1}^{1.086 \pm 0.385} \tag{3.5b}$$

The fit is not good. There is considerable scatter in the data, as is also noted by the error in the dissociation rate coefficient presented. Only the positive error is presented, since the dissociation rate coefficient cannot have a negative value. The dissociation rate coefficient, k_{d1} for the NSB (and dissociation) of BSA in solution to the 1:9 biosensor chip surface exhibits close to a first (equal to 1.086) order of dependence on the degree of heterogeneity or fractal dimension, D_{fd1} that exists on the biosensor chip surface in the dissociation phase.

Figure 3.12c shows the increase in the dissociation rate coefficient, k_{d2} with an increase in the fractal dimension, D_{fd2} for the NSB of BSA in solution to the 1:9 biosensor chip surface. For the data given in Table 3.7a and 3.7b, and plotted in Figure 3.12c, the dissociation rate coefficient, k_{d2} is given by

$$k_{d2} = (0.000417 \pm 0.00099) D_{fd2}^{11.746 \pm 4.502} \tag{3.5c}$$

The fit is reasonable. The dissociation rate coefficient, k_{d2} for the NSB (and dissociation) of BSA in solution to the 1:9 biosensor chip surface is very sensitive to the degree of heterogeneity or the fractal dimension, D_{fd2} that exists on the surface as noted by the close to twelfth (equal to 11.746) order of dependence exhibited.

Affinity values are of interest to practicing biosensorists. Define affinity, K as the ratio of the binding rate coefficient to the dissociation rate coefficient. Thus, for a single-fractal analysis, the affinity, $K = k/k_d$ where k is the binding rate coefficient, and k_d is the dissociation rate coefficient. Similarly, when a dual-fractal analysis applies for the binding and the dissociation phase, then the affinity, $K_1 = k_1/k_{d1}$, and $K_2 = k_2/k_{d2}$, etc. Figure 3.12d shows the increase in the affinity with an increase in the ratio of fractal dimensions present in the binding and in the dissociation phase. Both K_1 and K_2 are plotted on the same graph to provide more reliability for the relation, as otherwise the number of data points available for each of them is very few. For the data presented in Figure 3.12d and given in Tables 3.7a and 3.7b, the affinity, K_i is given by

$$K_i\left(=\frac{k}{k_{di}}\right) = (1.5557 \pm 0.4784)\left(\frac{D_f}{D_{fdi}}\right)^{3.117 \pm 0.3022} \tag{3.5d}$$

The fit is good. Only five data points are available. The availability of more data points would lead to a more reliable fit. The affinity, K_i for the NSB (and dissociation) of BSA in solution to the 1:9 biosensor chip surface is sensitive to the degree of heterogeneity or the fractal dimension, D_{fi} that exists on the biosensor chip surface as noted by the close to third (equal to 3.117) order of dependence exhibited on the ratio of fractal dimensions present in the binding and in the dissociation phase.

This represents one possible way of manipulating the affinity value on the biosensor chip by changing the fractal dimension values present in the binding and in the dissociation phases. This can prove to be quite challenging since by making a modification on the sensor chip surface one may, in most cases, inadvertently or otherwise change the fractal dimension values for both the binding and the dissociation phases. Note that one is interested in the ratio of the fractal dimensions.

Figure 3.13a shows the increase in the binding rate coefficient, k_1 with an increase in the fractal dimension, D_{f1} for the NSB of IgG in solution to the 1:9 biosensor chip surface. For the data given in Tables 3.7a and 3.7b, and plotted in Figure 3.13a, the binding rate coefficient, k_1 is given by

$$k_1 = (1.5495 \pm 0.0179) D_{f1}^{4.302 \pm 0.06375} \tag{3.6a}$$

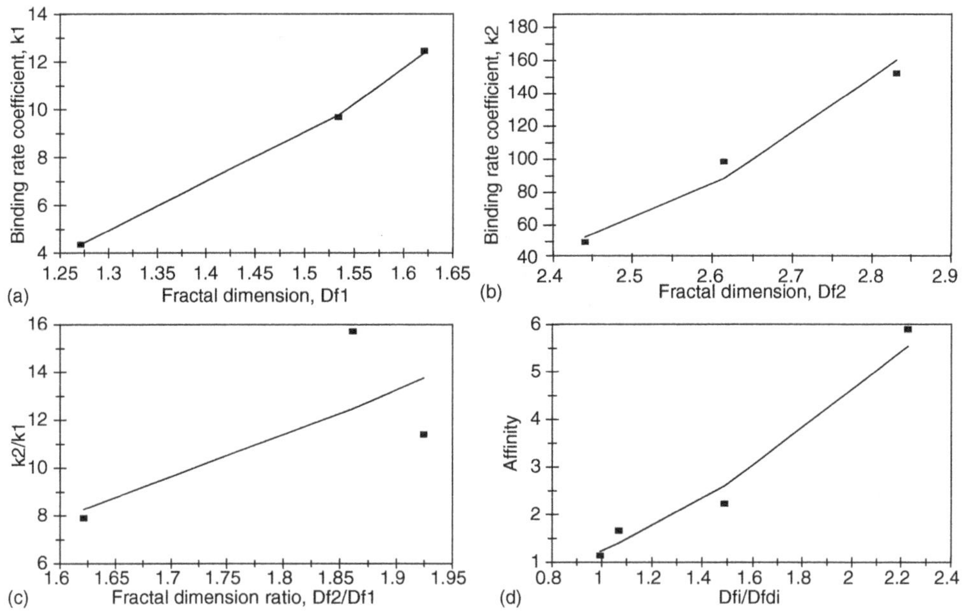

Figure 3.13 (a) Increase in the binding rate coefficient, k_1 with an increase in the fractal dimension, D_{f1}. (b) Increase in the binding rate coefficient, k_2 with an increase in the fractal dimension, D_{f2}. (c) Increase in the ratio of binding rate coefficients, k_2/k_1 with an increase in the ratio of fractal dimensions, D_{f2}/D_{f1}. (d) Increase in the affinity, $K_i(=k_i/k_{di})$ with an increase in the ratio of the fractal dimensions, D_{fi}/D_{fdi}.

The fit is good. Only three data points are available. The availability of more data points would lead to a more reliable fit. The binding rate coefficient, k for the NSB of IgG in solution to the 1:9 biosensor chip surface is very sensitive to the degree of heterogeneity or the fractal dimension, D_f that exists on the surface as noted by the close to fourth (equal to 4.302) order of dependence exhibited.

Figure 3.13b shows the increase in the binding rate coefficient, k_2 with an increase in the fractal dimension, D_{f2} for the NSB of IgG in solution to the 1:9 biosensor chip surface. For the data given in Tables 3.7a and 3.7b, and plotted in Figure 3.13b, the binding rate coefficient, k_2 is given by:

$$k_2 = (0.08575 \pm 0.01242) D_{f2}^{7.1785 \pm 1.2358} \tag{3.6b}$$

The fit is good. Only three data points are available. The availability of more data points would lead to a more reliable fit. The binding rate coefficient, k_2 for the NSB of IgG in solution to the 1:9 biosensor chip surface is very sensitive to the degree of heterogeneity or the fractal dimension, D_{f2} that exists on the surface as noted by the greater than seventh (equal to 7.1785) order of dependence exhibited.

Figure 3.13c shows for a dual-fractal analysis the increase in the ratio of the binding rate coefficients, k_2/k_1 with an increase in the ratio of fractal dimensions, D_{f2}/D_{f1}. For the data given in Table 3.7a and 3.7b, and plotted in Figure 3.13c, the ratio of the binding rate coefficients, k_2/k_1 is given by

$$\frac{k_2}{k_1} = (1.9712 \pm 0.6945)\left(\frac{D_{f2}}{D_{f1}}\right)^{2.968 \pm 2.348} \tag{3.6c}$$

The fit is not good. There is scatter in the data. This is also reflected in the error exhibited in the order of dependence on the ratio of the fractal dimensions, D_{f2}/D_{f1}. Only three data points are presented. The availability of more data points would lead to a more reliable fit. Nevertheless, the ratio of the binding rate coefficients, k_2/k_1 exhibits close to a third (equal to 2.968) order of dependence on the ratio of fractal dimensions, D_{f2}/D_{f1}.

Figure 3.13d shows the increase in the affinity with an increase in the ratio of fractal dimensions present in the binding and in the dissociation phase. Both K_1 and K_2 are plotted on the same graph to provide more reliability for the relation, as otherwise the number of data points for each of them is very few. For the data presented in Figure 3.13d and given in Tables 3.7a and 3.7b, the affinity, K_i is given by

$$K_i\left(=\frac{k_i}{k_{di}}\right) = (1.2403 \pm 0.2379)\left(\frac{D_{fi}}{D_{fdi}}\right)^{1.8703 \pm 0.2753} \tag{3.6d}$$

The fit is good. Only four data points are available. The availability of more data points would lead to a more reliable fit. The affinity, K_i for the NSB (and dissociation) of IgG in solution to the 1:9 biosensor chip surface is sensitive to the ratio of the fractal dimensions in the binding and in the dissociation phases that exists on the biosensor chip surface as noted by the close to second (equal to 1.8703) order of dependence exhibited.

Finally, it would be of interest to obtain an affinity plot versus the ratio of fractal dimensions present in the binding phase and in the dissociation phase for the NSB (and dissociation) of both BSA and IgG for the 1:9 biosensor chip surface. This would help determine if the 1:9 biosensor chip surface is a major factor in help determining the affinity exhibited. Figure 3.14 shows the increase in the affinity, K_i $(=k_i/k_{di})$ as the ratio of the fractal

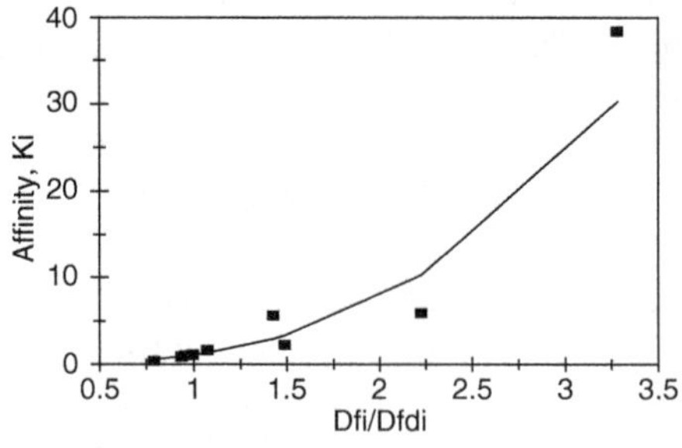

Figure 3.14 Increase in the affinity, K_i with an increase in the ratio of the fractal dimensions, D_{fi}/D_{fdi}. Data for BSA and IgG shown on the same plot.

dimensions in the binding and in the dissociation phase (D_{fi}/D_{fdi}) increases. For the data presented in Figure 3.14 and given in Tables 3.6a and 3.6b and in Tables 3.7a and 3.7b, the affinity, K_i is given by

$$K_i = (1.0940 \pm 0.533)\left(\frac{D_{fi}}{D_{fdi}}\right)^{2.7937 \pm 0.3016} \tag{3.6e}$$

The fit is good. This indicates that the 1:9 biosensor chip is a major determining factor in the affinity values exhibited during the NSB of either (10–40 μg/mL; 18.7–11.6 RU) BSA or (10–40 μg/mL; 15.8–10.8 RU) IgG in solution on the 1:9 biosensor chip surface. The affinity, K_i (= k_i/k_{di}) is sensitive to the ratio of fractal dimensions, (D_{fi}/D_{fdi}) as noted by the close to third (equal to 2.7937) order exhibited.

3.4 CONCLUSIONS

A fractal analysis is presented for the binding of different analytes in solution to different receptors immobilized on biosensor surfaces. The binding as well as the dissociation kinetics is described by either a single- or a dual-fractal analysis. The dual-fractal analysis is used only when the single-fractal analysis did not provide an adequate fit (sum of least squares less than 0.97). This was by regression analysis provided by Corel Quattro Pro 8.0 (1989).

The intent of this chapter is to provide (a) binding rate coefficient and fractal dimension values for the different analyte–receptor reactions occurring on biosensor surfaces, and (b) possibly relate the fractal dimension values obtained to the biosensor efficiency parameters. More specifically, how do changes in the fractal dimension on the biosensor surface affect changes in the above-mentioned performance parameters.

For the binding (hybridization) of 50 nM ss DNA of M13mp19 phage (target DNA) in solution in the absence (Case A) and in the presence of 200 μg/mL calf thymus DNA (Case B) and when a dual-fractal analysis applies, an increase in the fractal dimension, D_{f1} by 31.2% from a value of 1.7188 to 2.2682 as one goes from Case A to Case B leads to an increase in the binding rate coefficient, k_1 by a factor of 1.926 from a value of k_1 equal to 1.2260 to 2.3610. Predictive relations are also presented for the binding rate coefficients, k_1 and k_2 as a function of the fractal dimensions, D_{f1} and D_{f2}, respectively.

For the binding (hybridization) of the target DNA (15 bp mer) in solution to the complementary probe (15 bp mer) (5′-GTTACCACAGGATG-3′) immobilized on a conventional SPR biosensor surface (Case A) and on a nanocluster-enhanced SPR biosensor surface (Case B) (Hu *et al.*, 2004), one notes that (a) the fractal dimension, D_{f1} decreases by 14% from a value of 1.6868 for the conventional SPR biosensor to a value of 1.40405 for the nanocluster-enhanced SPR biosensor surface, and (b) the fractal dimension, D_{f2} decreases by 15.9% from a value of 2.7856 for the conventional SPR biosensor to a value of 2.3426 for the nanocluster-enhanced SPR biosensor surface. In this case, however, due to the “nano” nature of the nanocluster-enhanced SPR biosensor surface, both binding rate coefficients, k_1 and k_2 increase as one goes from the conventional SPR biosensor to the nanocluster-enhanced biosensor. The binding rate coefficient, k_1 is higher by a factor of 2.853 for the nanocluster-enhanced SPR biosensor when compared to the conventional SPR biosensor. Similarly, the binding rate coefficient, k_2 value is higher by a factor of 2.89 for the nanocluster-enhanced SPR biosensor surface when compared with the conventional SPR biosensor surface.

For the binding of 450 nM monoclonal anti-GAD 65 (purified mouse IgG) in solution to SAM1 and SAM2 immobilized on a SPR biosensor surface and where a single-fractal analysis applies, an increase in the fractal dimension by 5.4% from a value of D_f equal to 2.0188 to D_f equal to 2.1278 as one goes from the SAM1 case to the SMM2 case, leads to an increase in the binding rate coefficient, k by a factor of 1.685 from a value of k equal to 9.6993 to k equal to 16.3429.

Other examples of changes in the fractal dimension on the biosensor surface and how they affect the different biosensor performance parameters are also presented. At the outset of this book this chapter outlines and demonstrates via different examples, not only the importance of the degree of heterogeneity on the biosensor surface on the binding rate coefficients, but also how the changes in the degree of heterogeneity on the biosensor surface due to different experimental conditions would affect the different biosensor performance parameters. It would behoove practicing biosensorists to pay more and more attention to the nature of the biosensor surface, and how it may affect the different biosensor performance parameters.

REFERENCES

Amini, MK, S Shahrokhian, and S Tangenstaminejad, *Analytical Chemistry*, 1999, **71**, 2502—2505.

Armbuster, DA, Prostate specific antigen: biochemistry, analytical methods, and clinical application. *Clinical Chemistry*, 1993, **39**, 181–195.

Baekkeskov S, HJ Aanstoot, S Christgau, A Reetz, M Solimena, M Cascalho, F Folli, H Richter-Olesen, P De Camilli, and PD Camilli, Identification of the 64K autoantigen insulin-dependent diabetes as the GABA-synthesizing enzyme glutamic acid decarboxylase. *Nature*, 1990, **347**, 151–156.

Broude NE, Stem-Loop oligonucleotides: a robust tool for molecular biology and biotechnology. *Trends in Biotechnology*, 2002, **20**, 249–256.

Cao, C, JP Kim, BW Kim, H Chae, HC Yoon, SS Yang, and SJ Kim, A strategy for sensitivity and specificity enhancements in prostate specific antigen-α_1- antichymotrypsin detection based on surface plasmon resonance. *Biosensors & Bioelectronics* 2006, **21**, 2106–2113.

Chapman, RG, E Ostuni, L Yan, and G Whitesides, Preparation of mixed self-assembled monolayers (SAMs) that resist adsorption of proteins using the reaction of amines with a SAM that presents interchain carboxylic anhydride groups. *Langmuir* 2000, **16**, 6927–6936.

Choi, SH, JW Lee and SJ Sun, Enhanced performance of a surface plasmon resonance immunosensor for detecting Ab-GAD antibody based on the modified self-assembled monolayers. *Biosensors & Bioelectronics* 2005, **21**, 378–383.

Ciu, X, F Yang, Y Sha, and X Yang, Real time immunoassay of ferritin using surface plasmon resonance biosensor. *Talanta*, 2003, **60**, 53–61.

Corel Quattro Pro 8.0, Corel Corporation, Ottawa, Canada, 1989.

Corel Quattro Pro 8.0, Corel Corporation Limited, Ottawa, Canada, 1997.

Havlin, S, Molecular diffusion and reactions in *The Fractal Approach to Heterogeneous Chemistry: Surfaces, Colloids, Polymers* (ed. D. Avnir), Wiley, New York, 1989, pp. 251–269.

Hu, WP, KT Huang, JH Hsu, WY Chen, GL Chang and KA Lai, A novel ultrahigh-resolution surface plasmon resonance biosensor with an Au nanocluster-embedded dielectric film. *Biosensors & Bioelectronics*, 2004, **19**, 1465–1471.

Larsen, AS, JD Holbrey, FS Tham, and CA Reed, *Journal of the American Chemical Society*, 2000, **122**, 7264–7272.

Lee, CK, and SL Lee, Multi-fractal scaling analysis of reactions over fractal surfaces. *Surface Science*, 1995, **325**, 294–310.

Micanovic, R, R Procyk, WH Lin, and GR Matsueda, *Journal of Biological Chemistry*, 1994, **269**, 9190–9194.

Mitchell, JS, Y Wu, CJ Cook, and L Main, Sensitivity enhancement of surface plasmon resonance biosensing of small molecules. *Analytical Biochemistry*, 2005, **343**, 125–135.

Pihel, K, S Hsieh, JW Jorgenson, and MR Wightman, *Analytical Chemistry*, 1995, **67**, 4514–4521.

Ramakrishnan, A and A Sadana, A single fractal analysis of cellular analyte-receptor binding kinetics using biosensors. *BioSystems*, 2001, **59**, 35–51.

Sadana, A, A fractal analysis for the evaluation of hybridization kinetics in biosensors. *Journal of Colloid and Interface Science*, 2001, **151**(1), 166–177.

Sadana, A, *Fractal Binding and Dissociation Kinetics for Different Biosensor Applications*, Elsevier, Amsterdam, 2005.

Savage, P, and J Waxman, PSA and prostate cancer diagnosis. *European Journal of Cancer*, 1996, **32A**, 1097–1099.

Silin, V., H Weekall, and DJ Vanderah, SPR studies of the non-specific absorption kinetics of human IgG and BSA on gold surfaces modified by self-assembled monolayers (SAMs). *Journal of Colloid and Interface Science*, 1997, **185**, 94–103.

Tastuma, T, and T Watnable, *Analytical Chemistry*, 1995, **67**, 4514–4521.

Tyagi, S, and FR Kramer, Molecular beacons: probes that fluoresce upon hybridization. *Nature Biotechnology*, 1996, **14**, 303–308.

Wang, RLC, HJ Kreuzer, and M Gruze, Molecular information and solvation of oligo (ethylene glycol)-terminated self-assembled monolayers and their resistance to protein adsorption. *Journal of Physical Chemistry B*, 1997, **101**, 9767–9773.

Zhang, Y, R Yang, F Liu, and KA Li, Fluorescent sensor for imidazole derivatives based on monomer-dimer equilibrium of a zinc porphyrin complex in a polymer film. *Analytical Chemistry*, 2004, **76**, 7336–7345.

Zheleznaya, LA, DS Kopein, EA Rogulin, SI Gubanov and NI Matvienko, Significant enhancement of fluorescence on hybridization of a molecular beacon to a target DNA in the presence of site-specific DNA nickase. *Analytical Biochemistry*, 2006, **348**, 123–126.

– 4 –

Fractal Analysis of Harmful Bacteria, Toxins, and Pathogen Detection on Biosensor Surfaces

4.1 INTRODUCTION

The detection of harmful bacteria, toxins, and pathogens is an important area of investigation. The unintentional ingestion or becoming afflicted by the above three may lead to diseases which may require medical treatment or even hospitalization. The deliberate distribution amongst a human population of 'weaponized' forms of harmful bacteria, toxins, and pathogens is what is recognized nowadays as a biological or chemical threat. The current geopolitical climate emphasizes that countries need to be prepared to help minimize, control, and contain these types of threats if they are to occur in the future. Unintentional outbreaks of a disease due to harmful bacteria (such as *Escherichia coli*; mildly harmful), toxins, and pathogens can presumably be managed effectively by hospitals and other emergency personnel.

Subramanian *et al.* (2006) have recently used polyethylene glycol terminated alkanethiol mixed self-assembled monolayers on a SPR (surface plasmon resonance) biosensor to detect *E. coli* O157:H7. These authors state that in the year 2003, ERS (2003) estimated the cost associated with five major pathogens such as *Salmonella* and *E. coli* to be around US$6.9 billion annually. These pathogens also lead to illnesses, hospitalizations, and sometimes death. Regulations for safe levels of pathogens in for example, food materials have been passed in different countries. These give rise to the effective detection of different food contaminants by techniques that include biosensors.

E. coli O157:H7 has also been detected by antibody-immobilized biconal tapered fiber sensors (Rijal *et al.*, 2005), by self-excited PZT (piezoelectric)-glass microcantilevers (Campbell and Mutharasan, 2005), by a proteomic biosensor (Horner *et al.*, 2006), and by an integrating waveguide biosensor (Zhu *et al.*, 2005). Ko and Grant (2006) have used a novel fluorescence resonance energy transfer (FRET)-based optical fiber biosensor for the rapid detection of *Salmonella typhimurium*. According to these authors this is a portable biosensor that permits the on-site analysis of samples and should help reduce the economic impact caused by food recalls. Ngundi *et al.* (2006) have used monosaccharide arrays for the detection of bacterial toxins. They used arrays of *N*-acetylneuraminicacid (Neu5Ac) derivatives immobilized on planar waveguides. These derivatives served as the receptors for the protein toxins, such as cholera toxin and tetanus toxin. Balasubramanian *et al.* (2006) have recently developed a lytic phage as a specific and selective probe for the detection of *Staphylococcus aureus* using a surface plasmon resonance spectroscopic procedure. These authors were able to label-free detect *S. aureus* using lytic phage and a surface plasmon resonance-based SPREETA^{TM} biosensor. Their detection limit was 10^4cfu/ml. Finally, Moats and Sullivan (2004) have proposed a combinatorial approach to increase the performance of a multi-pathogen biosensor.

In this chapter we use fractal analysis to analyze the binding kinetics of: (a) serotype-specific detection of Dengue virus RNA using a microfluidic biosensor (Zaytseva *et al.*, 2005); (b) detection of *E. coli* O157:H7 by a mixed self-assembled monolayer based surface plasmon immunosensor (Subramanian *et al.*, 2006); (c) detection of bacteria using multiarray sensors (Karasinski *et al.*, 2005); (d) detection of cholera toxin using an array biosensor (Ngundi *et al.*, 2004); (e) detection of whole *Listeria monocytogenes* cells in contaminated samples using a surface plasmon resonance biosensor (Leonard *et al.*, 2004); (f) detection of *S. aureus* using surface plasmon resonance spectroscopy (Balasubramanian *et al.*, 2006); and (g) detection of bacteria using a disposable optical leaky waveguide biosensor (Zourob *et al.*, 2005). Fractal dimension and binding and dissociation (if applicable) rate coefficient values are presented. The fractal analysis to be presented may be viewed as an alternate analysis.

4.2 THEORY

Havlin (1987) has reviewed and analyzed the diffusion of reactants toward fractal surfaces. The details of the theory and the equations involved for the binding and the dissociation phases for analyte–receptor binding are available (Sadana, 2001). The details are not repeated here; except that just the equations are given to permit an easier reading. These equations have been applied to other biosensor systems (Sadana, 2001, 2005; Ramakrishnan and Sadana, 2001). For most applications, a single- or a dual-fractal analysis is often adequate to describe the binding and the dissociation kinetics. Peculiarities in the values of the binding and the dissociation rate coefficients, as well as in the values of the fractal dimensions with regard to the dilute analyte systems being analyzed will be carefully noted, if applicable.

4.2.1 Single-fractal analysis

Binding rate coefficient

Havlin (1989) indicates that the diffusion of a particle (analyte [Ag]) from a homogeneous solution to a solid surface (e.g., receptor [Ab]-coated surface) on which it reacts to form a product (analyte–receptor complex; Ab·Ag) is given by:

$$(\mathrm{Ab}\cdot\mathrm{Ag}) \approx \begin{cases} t^{(3-D_{f,bind})/2} = t^{p} & t < t_c \\ t^{1/2} & t > t_c \end{cases} \tag{4.1}$$

Here $D_{f,bind}$ or D_f (used later on in the book) is the fractal dimension of the surface during the binding step. t_c is the cross-over value. Havlin (1989) indicates that the cross-over value may be determined by $r_c^2 \sim t_c$. Above the characteristic length, r_c, the self-similarity of the surface is lost and the surface may be considered homogeneous. Above time, t_c the surface may be considered homogeneous, since the self-similarity property disappears, and 'regular' diffusion is now present. For a homogeneous surface where $D_f = 2$, and when only diffusional limitations are present, $p = 1/2$ as it should be. Another way of looking at the $p = 1/2$ case (where $D_{f,bind} = 2$) is that the analyte in solution views the fractal object, in our case, the receptor-coated biosensor surface, from a 'large distance.' In essence, in the association process, the diffusion of the analyte from the solution to the receptor surface creates a depletion layer of width $(Đt)^{1/2}$ where $Đ$ is the diffusion constant. This gives rise to the fractal power law, (Analyte–Receptor) $\sim t^{(3-D_{f,bind})/2}$. For the present analysis, t_c is arbitrarily chosen and we assume that the value of the t_c is not reached. One may consider the approach as an intermediate 'heuristic' approach that may be used in the future to develop an autonomous (and not time-dependent) model for diffusion-controlled kinetics.

Dissociation rate coefficient

The diffusion of the dissociated particle (receptor [Ab] or analyte [Ag]) from the solid surface (e.g., analyte [Ag]–receptor [Ab]) complex coated surface) into solution may be given, as a first approximation by:

$$(\text{Ab}\cdot\text{Ag}) \approx -t^{(3-D_{f,diss})/2} = -t^{p}(t > t_{diss}) \quad (4.2)$$

Here $D_{f,diss}$ is the fractal dimension of the surface for the dissociation step. This corresponds to the highest concentration of the analyte-receptor complex on the surface. Henceforth, its concentration only decreases. The dissociation kinetics may be analyzed in a manner 'similar' to the binding kinetics.

4.2.2 Dual-fractal analysis

Binding rate coefficient

Sometimes, the binding curve exhibits complexities and two parameters (k, D_f) are not sufficient to adequately describe the binding kinetics. This is further corroborated by low values of r^2 factor (goodness-of-fit). In that case, one resorts to a dual-fractal analysis (four parameters; k_1, k_2, D_{f1}, and D_{f2}) to adequately describe the binding kinetics. The single-fractal analysis presented above is thus extended to include two fractal dimensions. At present, the time ($t = t_1$) at which the 'first' fractal dimension 'changes' to the 'second' fractal dimension is arbitrary and empirical. For the most part, it is dictated by the data analyzed and experience gained by handling a single-fractal analysis. A smoother curve is obtained in the 'transition' region, if care is taken to select the correct number of points for the two regions. In this case, the product (antibody–antigen; or analyte–receptor complex, Ab·Ag or analyte–receptor) is given by:

$$(\mathrm{Ab{\cdot}Ag}) \approx \begin{cases} t^{(3-D_{f1,\mathrm{bind}})/2} = t^{p1} & (t < t_1) \\ t^{(3-D_{f2,\mathrm{bind}})/2} = t^{p2} & (t_1 < t < t_2 = t_c) \\ t^{1/2} & (t > t_c) \end{cases} \tag{4.3}$$

In some cases, as mentioned above, a triple-fractal analysis with six parameters (k_1, k_2, k_3, D_{f1}, D_{f2}, and D_{f3}) may be required to adequately model the binding kinetics. This is when the binding curve exhibits convolutions and complexities in its shape due to perhaps to the very dilute nature of the analyte (in some of the cases to be presented) or for some other reasons. Also, in some cases, a dual-fractal analysis may be required to describe the dissociation kinetics.

4.3 RESULTS

The fractal analysis will be applied for the detection of bacteria, viruses, toxins, and pathogens.

Zaytseva *et al.* (2005) have developed a microfluidic biosensor for the serotype-specific detection of Dengue virus RNA. These authors indicate that approximately 2.5 billion people worldwide are at risk from Dengue virus infections (WHO, 2000; Gubler, 1997). Zaytseva *et al.* (2005) indicate that the Dengue fever leads to an illness with mild febrile symptoms. Infection is transmitted by the *Aedes* mosquito. Since no Dengue vaccine is presently available these authors emphasize the need for rapid and reliable diagnostic procedures.

Zaytseva *et al.* (2005) have used a magnetic bead-based sandwich hybridization system in conjugation with liposome amplification. Their microfluidic biosensor system based on fluorescence detection was employed for the specific detection of nucleic acids. Figure 4.1a shows the binding of Dengue virus serotype 1 RNA in solution to the probe immobilized on the sensor chip surface in the presence of dextran sulfate in the hybridization buffer. The authors emphasize that dextran sulfate is a molecular crowding agent and in its presence the rate of probe–target hybridization is accelerated. A single-fractal analysis is adequate to describe the binding kinetics. The values of the binding rate coefficient, k and the fractal dimension, D_f are given in Table 4.1.

The values of the binding rate coefficient, k and the fractal dimension, D_f presented in Table 4.1 were obtained from a regression analysis using Corel Quattro Pro 8.0 (1997) to model the data using eq. (4.1) wherein $(\mathrm{Ab{\cdot}Ag}) = kt^{(3-D_f)/2}$. The binding rate coefficients presented in Table 4.1 are within 95% confidence limits. For example, for the binding of Dengue virus serotype 1 RNA in the presence of dextran sulfate in the hybridization buffer to the serotype specific DNA probe complementary to separate regions in the Dengue viral RNA immobilized on the magnetic bead surface the binding rate coefficient, k for a single-fractal analysis is given by $k = 34.015 \pm 4.173$. The 95% confidence limit indicates that the k value lies between 29.842 and 38.188. This indicates that the value is precise and significant.

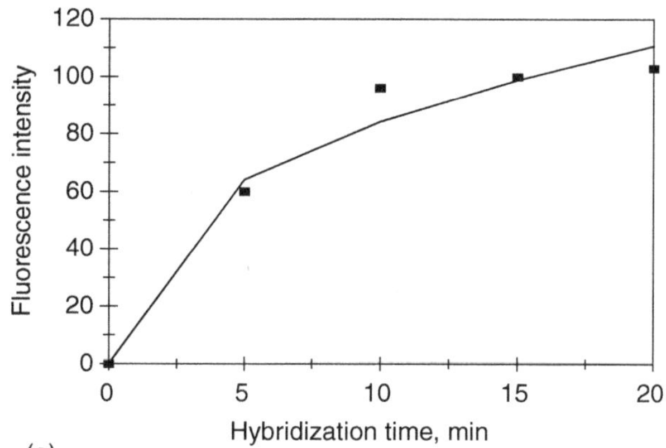

(a)

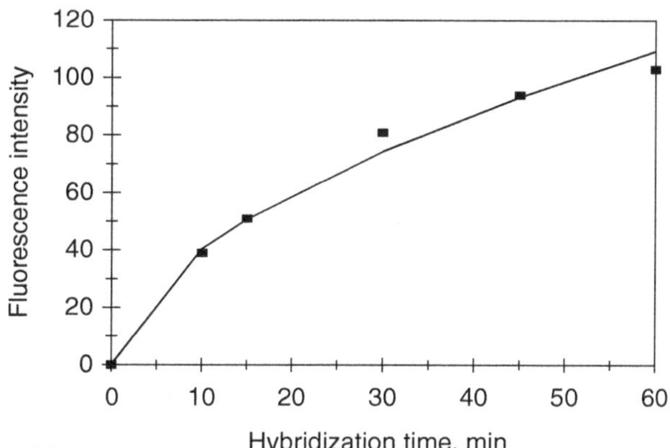

(b)

Figure 4.1 Binding of Dengue virus serotype 1 RNA in solution to the probe immobilized on the sensor chip surface (a) in the presence and (b) in the absence of the hybridization buffer (Zaytseva *et al.*, 2005).

Figure 4.1b shows the binding of Dengue virus serotype 1 RNA in solution to the probe immobilized on the sensor chip surface in the absence of dextran sulfate in the hybridization buffer. Once again, a single-fractal analysis is adequate to describe the binding kinetics. The values of the binding rate coefficient, k and the fractal dimension, D_f are given in Table 4.1. The molecular crowding agent, dextran sulfate is not present in this case, and the binding rate coefficient, k is lower by a factor of three (equal to 11.31) than when dextran sulfate is present (34.015). Also, note that the fractal dimension, D_f is higher (equal to 2.2116) when dextran sulfate is present in solution compared to when it is absent (equal to 1.8918). Note that changes in the fractal dimension and in the binding rate coefficient are in the same direction. A 16.9% increase in the fractal dimension from D_f equal to 1.8918 to D_f equal to 2.2116 on going from the case when dextran sulfate is absent to when it is

Table 4.1

Binding rate coefficients and fractal dimensions for (a) the binding (hybridization) of Dengue virus serotype 1 RNA in the absence and in the presence of dextran sulfate (Zaytseva *et al.*, 2005) and (b) effect of polyclonal antibody (PAb; anti-*E. coli* O157:H7) in a sandwich assay for *E. coli* detection (Subramanian *et al.*, 2006)

Analyte in solution/Receptor no biosensor surface	k	k_1	k_2	D_f	D_{f1}	D_{f2}	Reference
Dengue virus serotype 1 RNA in the presence of dextran sulfate in the hybridization buffer/ serotype-specific DNA probes complementary to separate regions in the Dengue viral RNA	34.015 ± 4.173	na	na	2.2116 ± 0.222	na	na	Zaytseva et al. (2005)
Dengue virus serotype 1 RNA in the absence of dextran sulfate in the hybridization buffer/ serotype-specific DNA probes complementary to separate regions in the Dengue viral RNA	11.311 ± 0.745	na	na	1.8918 ± 0.0852	na	na	Zaytseva *et al.* (2005)
E. coli O157:H7/10 μg/ml anti-*E. coli* polyclonal antibody immobilized on a sensor chip surface	1247.47 ± 132.22	na	na	2.5778 ± 0.0469	na	na	Subramanian *et al.* (2006)
E. coli O157:H7/20 μg/ml anti-*E. coli* polyclonal antibody immobilized on a sensor chip surface	2105.29 ± 256.43	767.03 ± 220.32	3092.14 ± 29.39	2.6354 ± 0.0534	2.0152 ± 0.5152	2.7754 ± 0.0696	Subramanian *et al.* (2006)
E. coli O157:H7/30 μg/ml anti-*E. coli* polyclonal antibody immobilized on a sensor chip surface	2808.86 ± 368.63	906.67 ± 248.10	4397.89 ± 64.73	2.6174 ± 0.5744	1.9232 ± 0.4934	2.7802 ± 0.0086	Subramanian *et al.* (2006)

present in solution leads to an increase in the binding rate coefficient, k by a factor of three from a value of k equal to 11.31 to k equal to 34.015.

Subramanian *et al.* (2006) have recently used a mixed self-assembled monolayer (SAM)-based surface plasmon resonance immunosensor to detect *E. coli* O157:H7 on an activated sensor chip. They used a direct and sandwich-type immunoassay to detect *E. coli* O157:H7. These authors emphasize that one of their intentions is to demonstrate the applicability of the mixed alkanethiol SAM based SPR biosensor platform as an alternative to the expensive Biacore system to analyze the kinetics of antibody–antigen reactions.

Figure 4.2a shows the binding of *E. coli* O157:H7 in solution to 10 μg/ml PAb (anti-*E. coli* O157:H7) immobilized on an activated SAM surface. A single-fractal analysis is adequate to describe the binding kinetics. The values of the binding rate coefficient, k and the fractal dimension, D_f for a single-fractal analysis are given in Table 4.1.

Figure 4.2b shows the binding of *E. coli* O157:H7 in solution to 20 μg/ml PAb immobilized on an activated SAM surface. A dual-fractal analysis is adequate to describe the binding kinetics. The values of (a) the binding rate coefficient, k and the fractal dimension, D_f for a single-fractal analysis, and (b) the binding rate coefficients, k_1 and k_2, and the fractal dimensions, D_{f1} and D_{f2} for a dual-fractal analysis are given in Table 4.1. Note that for a dual-fractal analysis, an increase in the fractal dimension by 37.7% from a value of D_{f1} equal to 2.0152 to D_{f2} equal to 2.7754 leads to an increase in the binding rate coefficient by a factor of 4.03 from a value of k_1 equal to 767.03 to k_2 equal to 3092.14. An increase in the degree of heterogeneity on the surface (increase in the fractal dimension) leads to an increase in the binding rate coefficient value. Note the change in the binding mechanism as one goes from the 10 μg/ml PAb to the 20 μg/ml PAb. At the lower concentration a single-fractal analysis is adequate to describe the binding kinetics, whereas at the higher concentration a dual-fractal analysis is required to adequately describe the binding kinetics.

Figure 4.2c shows the binding of *E. coli* O157:H7 in solution to 30 μg/ml PAb immobilized on an activated SAM surface. A dual-fractal analysis is adequate to describe the binding kinetics. The values of (a) the binding rate coefficient, k and the fractal dimension, D_f for a single-fractal analysis, and (b) the binding rate coefficients, k_1 and k_2, and the fractal dimensions, D_{f1} and D_{f2} for a dual-fractal analysis are given in Table 4.1 Note that for a dual-fractal analysis, an increase in the fractal dimension by 44.5% from a value of D_{f1} equal to 1.9232 to D_{f2} equal to 2.7802 leads to an increase in the binding rate coefficient by a factor of 4.85 from a value of k_1 equal to 906.67 to k_2 equal to 4397.89. Once again, an increase in the degree of heterogeneity on the surface (increase in the fractal dimension) leads to an increase in the binding rate coefficient value.

Karasinski *et al.* (2005) have developed multiarray sensors for the detection, classification, and differentiation of bacteria at subspecies and strain levels. A 96-well-type electrode array (DOX-dissolved oxygen sensor) is used along with principal component assay (PCA) for the rapid and routine classification of bacteria. These authors indicate that their system is based on the hypothesis that under identical experimental conditions different bacteria consume oxygen at different rates and are affected by selected antibiotics in different ways.

Figure 4.3a shows the increased absorbance due to cell replication of *Bacillus globiggi* 9372 on a microarray (DOX-dissolved oxygen) sensor (Karasinski *et al.*, 2005). The bacteria may be differentiated by the DOX-PCA system via the direct monitoring of oxygen. As the bacteria grow, oxygen in the medium is consumed with time due to their respiration.

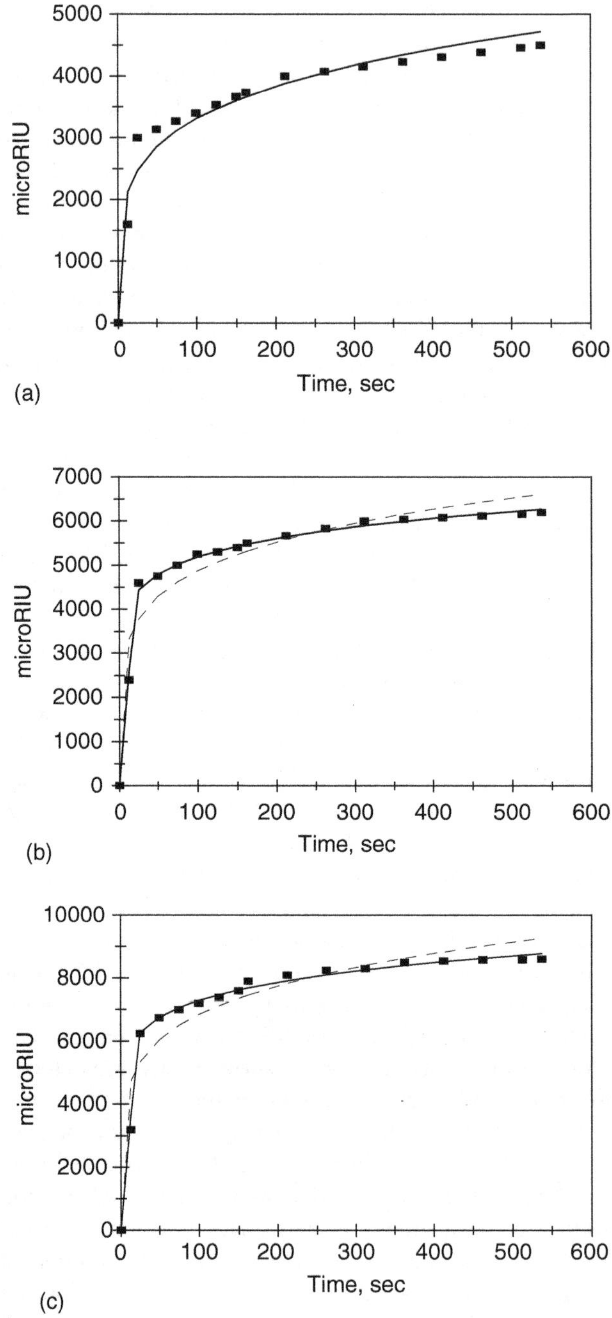

Figure 4.2 Binding of *E. coli* O157:H7 in solution to different concentrations (in μg/ml) of PAb (anti-*E. coli* O157:H7) immobilized on an activated SAM surface (Subramanian *et al.*, 2006). When both a dashed (- - -) and a solid (——) line are used, then the solid line represents a dual-fractal analysis, and the dashed line represents a single-fractal analysis. In this case, the solid line is the best fit line.

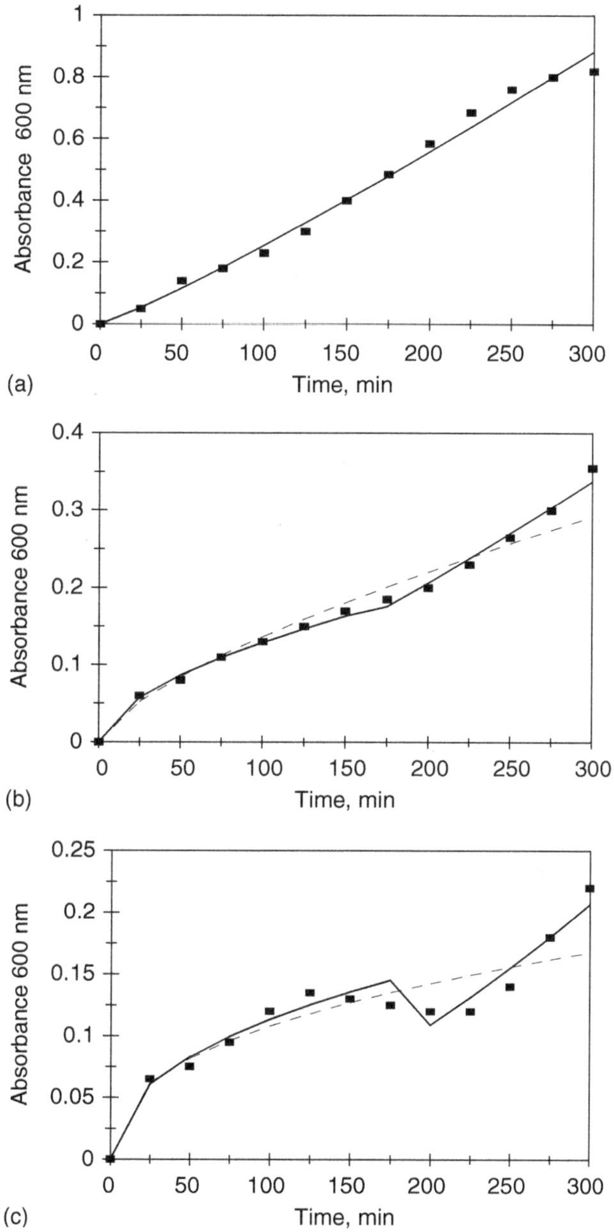

Figure 4.3 (a) Increase in the absorbance due to cell replication of *B. globiggi* 9372 on a microarray (DOX-dissolved oxygen) sensor (Karasinski *et al.*, 2005). (b) Increase in the absorbance due to cell replication of *B. globiggi* 9372 on a microarray (DOX-dissolved oxygen) sensor in the presence of the antibiotic, chloramphenicol (Karasinski *et al.*, 2005). (c) Increase in the absorbance due to cell replication of *B. globiggi* 9372 on a microarray (DOX-dissolved oxygen) sensor in the presence of the antibiotic, tetracycline (Karasinski *et al.*, 2005). When both a dashed (---) and a solid (——) line are used, then the solid line represents a dual-fractal analysis, and the dashed line represents a single-fractal analysis. In this case, the solid line is the best fit line.

Table 4.2

Binding rate coefficients and fractal dimensions for pathogen (*B. globigii* 9372) detection and for commonly used antibiotics (chloramphenicol, tetracycline) on a microarray (DOX-dissolved oxygen) sensor (Karasinski *et al.*, 2005)

Analyte in solution	k	k_1	k_2	D_f	D_{f1}	D_{f2}
B. globigii 9372	0.001393 ± 0.000124	na	na	0.7372 ± 0.06816	na	na
Chloramphenicol	0.005626 ± 0.000581	0.00891 ± 0.000465	0.000332 ± 0.000015	1.6156 ± 0.07842	1.8382 ± 0.0802	0.5718 ± 0.1996
Tetracycline	0.01708 ± 0.00274	0.01458 ± 0.00127	2.7E–05 ± 0.3E–05	2.1988 ± 0.1183	2.1096 ± 0.1126	0 + 0.6376

A single-fractal analysis is adequate to describe the binding kinetics. The values of the binding rate coefficient, k and the fractal dimension, D_f for a single-fractal analysis are given in Table 4.2.

Figure 4.3b shows the increased absorbance due to cell replication of *B. globigii* 9372 on a microarray (DOX-dissolved oxygen) sensor in the presence of the antibiotic, chloramphenicol (Karasinski *et al.*, 2005). A dual-fractal analysis is required to adequately describe the binding kinetics. The values of (a) the binding rate coefficient, k and the fractal dimension, D_f for a single-fractal analysis, and (b) the binding rate coefficients, k_1 and k_2 and the fractal dimensions, D_{f1} and D_{f2} for a dual-fractal analysis are given in Table 4.2. Note that as the fractal dimension or the degree of heterogeneity decreases by a factor of 3.2 from a value of D_{f1} equal to 1.8382 to D_{f2} equal to 0.5718, the binding rate coefficient decreases by a factor of 26.84 from a value of k_1 equal to 0.00891 to k_2 equal to 0.00032. Changes in the degree of heterogeneity in the DOX-PCA system and in the binding rate coefficient are once again in the same direction.

Figure 4.3c shows the increased absorbance due to cell replication of *B. globigii* 9372 on a microarray (DOX-dissolved oxygen) sensor in the presence of the antibiotic, tetracycline (Karasinski *et al.*, 2005). A dual-fractal analysis is once again required to adequately describe the binding kinetics. The values of (a) the binding rate coefficient, k and the fractal dimension, D_f for a single-fractal analysis, and (b) the binding rate coefficients, k_1 and k_2 and the fractal dimensions, D_{f1} and D_{f2} for a dual-fractal analysis are given in Table 4.2. Note that as the fractal dimension or the degree of heterogeneity decreases from a value of D_{f1} equal to 2.1096 to D_{f2} equal to ~0, the binding rate coefficient decreases by a factor of 540 from a value of k_1 equal to 0.0148 to k_2 equal to 2.7E–05. Once again, changes in the degree of heterogeneity in the DOX-PCA system and in the binding rate coefficient are in the same direction.

Ngundi *et al.* (2006) have recently used an array biosensor to determine the kinetic parameters for the binding of cholera toxin (CT) to immobilized Neu5Ac (*N*-Acetylneuraminic acid, a member of the sialic acid family) sialic acid and to anti-CT antibody (as a reference). These authors determined association rate coefficient values. However, their analysis like quite a few other analyses available in the literature ignored the effect of heterogeneities on the biosensor surface (for example, the fractal dimension which is the main focus of this book) on the kinetic parameters. The data presented by Ngundi *et al.* (2006) is re-analyzed to include surface effects to help determine the kinetic parameters such as the association (binding) rate coefficients. These authors do indicate that mechanistically viruses and toxins use protein–carbohydrate reactions to recognize host cells. Furthermore, Ngundi *et al.* (2006) indicate that quite a few of these protein toxins use one of their peptide chains to recognize and to bind carbohydrate receptors on cell surfaces before the toxins are internalized (Zang *et al.*, 1995; Emsley *et al.*, 2000; Fotinou *et al.*, 2001).

Figure 4.4a shows the binding of 0.1 μg/ml Cy5–CT (cholera toxin) in solution to anti-CT mAb (monoclonal antibody) covalently immobilized on a maleimide-activated planar waveguide using a thiol-terminated linker (Ngundi *et al.*, 2006). Cy5 is a bisfunctional dye. A dual-fractal analysis is required to adequately describe the binding kinetics. The values of (a) the binding rate coefficient, k and the fractal dimension, D_f for a single-fractal analysis, and (b) the binding rate coefficients, k_1 and k_2, and the fractal dimensions, D_{f1} and D_{f2} for a dual-fractal analysis are given in Table 4.3. It is of interest to note that for a dual-fractal analysis an increase in the fractal dimension by a factor of 1.479 from a value of D_{f1} equal to 1.0 to D_{f2} equal to 1.4794 leads to an increase in the binding rate coefficient by a factor of 5.70 from a value of k_1 equal to 0.3428 to k_2 equal to 1.9531. Note that changes in the binding rate coefficient and in the degree of heterogeneity on the surface or the fractal dimension are in the same direction.

Figure 4.4b shows the binding of 0.5 μg/ml Cy5–CT in solution to anti-CT mAb covalently immobilized on a maleimide-activated planar waveguide using a thiol-terminated linker (Ngundi *et al.*, 2006). A dual-fractal analysis is once again required to adequately describe the binding kinetics. The values of (a) the binding rate coefficient, k and the fractal dimension, D_f for a single-fractal analysis, and (b) the binding rate coefficients, k_1 and k_2, and the fractal dimensions, D_{f1} and D_{f2} for a dual-fractal analysis are given in Table 4.3. It is of interest to note that for a dual-fractal analysis an increase in the fractal dimension by a factor of 1.531 from a value of D_{f1} equal to 1.0004 to D_{f2} equal to 1.5322 leads to an increase in the binding rate coefficient by a factor of 7.02 from a value of k_1 equal to 0.6291 to k_2 equal to 4.4154. Note that changes in the binding rate coefficient and in the degree of heterogeneity on the surface or the fractal dimension are once again in the same direction.

Figure 4.4c shows the binding of 5.0 μg/ml Cy5–CT in solution to anti-CT mAb covalently immobilized on a maleimide-activated planar waveguide using a thiol-terminated linker (Ngundi *et al.*, 2006). A dual-fractal analysis is once again required to adequately describe the binding kinetics. The values of (a) the binding rate coefficient, k and the fractal dimension, D_f for a single-fractal analysis, and (b) the binding rate coefficients, k_1 and k_2, and the fractal dimensions, D_{f1} and D_{f2} for a dual-fractal analysis are given in Table 4.3. It is of interest to note that for a dual-fractal analysis an increase in the fractal dimension by a factor of 1.17 from a value of D_{f1} equal to 2.2820 to D_{f2} equal to 2.6742 leads to an increase in the binding rate coefficient by a factor of 3.94 from a value of k_1 equal to 220.77 to k_2 equal to 869.48.

Figure 4.5a and Table 4.3 show for a dual-fractal analysis the increase in the binding rate coefficient, k_1 with an increase in the Cy5–CT in the 0.1–5.0 μg/ml range in solution. For the data presented in Figure 4.5a the binding rate coefficient, k_1 is given by:

$$k_1 = (543 \pm 147.87)\ [\text{Cy5}-\text{CT}\ \mu\text{g/mL}]^{1.021 \pm 0.0865} \tag{4.4a}$$

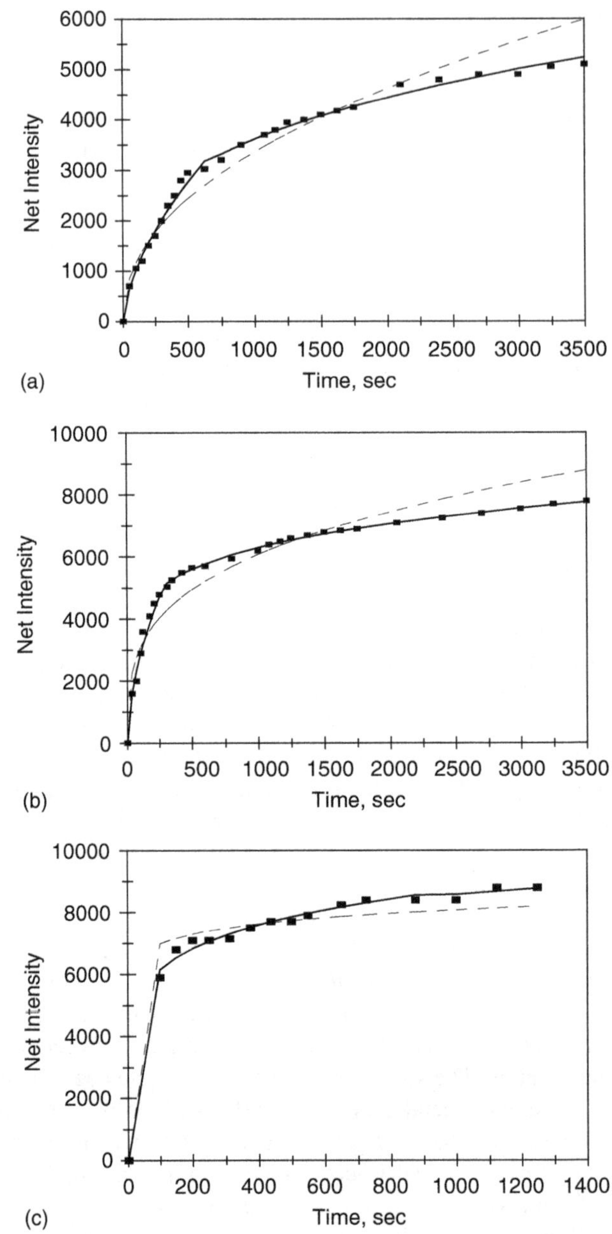

Figure 4.4 Binding of different concentrations (in μg/ml) of Cy5–CT in solution to anti-CT covalently immobilized on a malemide-activated planar waveguide using a thiol-terminated linker (Ngundi *et al.*, 2006): (a) 0.1; (b) 0.5; (c) 5.0. When both a dashed (---) and a solid (——) line are used, then the solid line represents a dual-fractal analysis, and the dashed line represents a single-fractal analysis. In this case, the solid line is the best fit line.

Table 4.3

Binding rate coefficients and fractal dimensions for 0.1–0.5 μg/ml Cy5-labeled CT (cholera toxin) in solution to (a) anti-CT mAb (monoclonal antibody) and (b) Neu5Ac (*N*-acetylneuraminic acid) immobilized onto an amine-functionalized waveguide via a crosslinker (6.8 Å, spacer arm length) (Ngundi *et al.*, 2006)

Analyte in solution/ receptor on surface	k	k_1	k_2	D_f	D_{f1}	D_{f2}
0.1 μg/ml Cy5-labeled CT/anti-CT mAb	0.4341 ± 0.0271	0.3428 ± 0.00003	1.9531 ± 0.0476	1.0828 ± 0.0212	1.0 ± 1.36E–14	1.4794 ± 0.0432
0.5 μg/ml Cy5-labeled CT/anti-CT mAb	0.7991 ± 0.0487	0.6291 ± 0.0011	4.4154 ± 0.0981	1.0852 ± 0.02094	1.0004 ± 0.00116	1.5322 ± 0.0423
5.0 μg/ml Cy5-labeled CT/anti-CT mAb	387.74 ± 25.71	220.77 ± 2.58	869.48 ± 44.68	2.4614 ± 0.03434	2.2820 ± 0.01934	2.6742 ± 0.0571
0.1 μg/ml Cy5-labeled CT/Neu5Ac	138.76 ± 15.92	58.046 ± 3.553	467.74 ± 10.05	2.0772 ± 0.0381	1.7544 ± 0.04946	2.4076 ± 0.0237
0.5 μg/ml Cy5 labeled CT/ Neu5Ac	789.86 ± 116.79	220.25 ± 20.87	1965.33 ± 17.93	2.4092 ± 0.04258	1.8910 ± 0.0851	2.6630 ± 0.00696
5.0 μg/ml Cy5 labeled CT/Neu5Ac	5226.09 ± 472.45	3049.07 ± 79.04	4338.93 ± 72.46	2.8738 ± 0.0407	2.6952 ± 0.02564	2.8027 ± 0.0776

The fit is good. Only three data points are available. The availability of more data points would lead to a more reliable fit. The binding rate coefficient, k_1 exhibits close to a first (equal to 1.021) order of dependence on the Cy5–CT concentration in solution in the 0.1–5.0 μg/ml range. The non-integer order of dependence exhibited on the Cy5–CT concentration in solution lends support to the fractal nature of the system.

Figure 4.5b and Table 4.3 show for a dual-fractal analysis the increase in the binding rate coefficient, k_2 with an increase in the Cy5–CT in the 0.1–5.0 μg/ml range in solution. For the data presented in Figure 4.5b the binding rate coefficient, k_2 is given by:

$$k_2 = (2048.60 \pm 1074.36)[\text{Cy5} - \text{CT}\ \mu\text{g/mL}]^{0554 \pm 0.152} \tag{4.4b}$$

The fit is good. Only three data points are available. The availability of more data points would lead to a more reliable fit. The binding rate coefficient, k_2 exhibits close to a one-half (equal to 0.554) order of dependence on the Cy5–CT concentration in solution in the 0.1–5.0 μg/ml range. Note that the order of dependence on the Cy5–CT concentration in solution decreases by almost a factor of two as one goes from k_1 (equal to 1.02) to k_2 (equal to 0.554). In other words, the binding rate coefficient, k_1 is almost twice as sensitive as k_2 to the Cy5–CT concentration in solution in the 0.1 to 5.0 μg/ml in solution.

Figure 4.6a and Table 4.3 show for a dual-fractal analysis the increase in the binding rate coefficient, k_1 with an increase in the fractal dimension, D_{f1}. For the data presented in Figure 4.6a the binding rate coefficient, k_1 is given by:

$$k_1 = (0.6087 \pm 0.3854) D_{f1}^{8.658 \pm 1.512} \tag{4.5a}$$

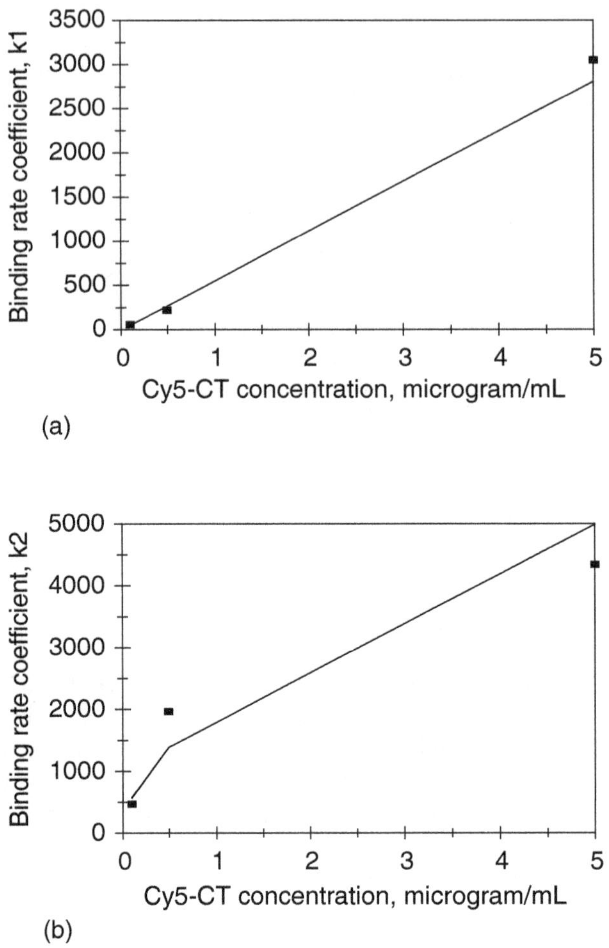

Figure 4.5 (a) Increase in the binding rate coefficient, k_1 with an increase in the Cy5–CT concentration (in μg/ml) in solution. (b) Increase in the binding rate coefficient, k_2 with an increase in the Cy5–CT concentration (in μg/ml) in solution.

The fit is good. Only three data points are available. The availability of more data points would lead to a more reliable fit. The binding rate coefficient, k_1 is extremely sensitive to the degree of heterogeneity or the fractal dimension, D_{f1} on the waveguide surface as noted by the order of dependence between 8.5 and 9 (equal to 8.658) exhibited.

Figure 4.6b and Table 4.3 show for a dual-fractal analysis the increase in the binding rate coefficient, k_2 with an increase in the fractal dimension, D_{f2}. For the data presented in Figure 4.6b the binding rate coefficient, k_2 is given by:

$$k_2 = (0.001244 \pm 0.000043) D_{f1}^{14.60 \pm 0.3115} \tag{4.5b}$$

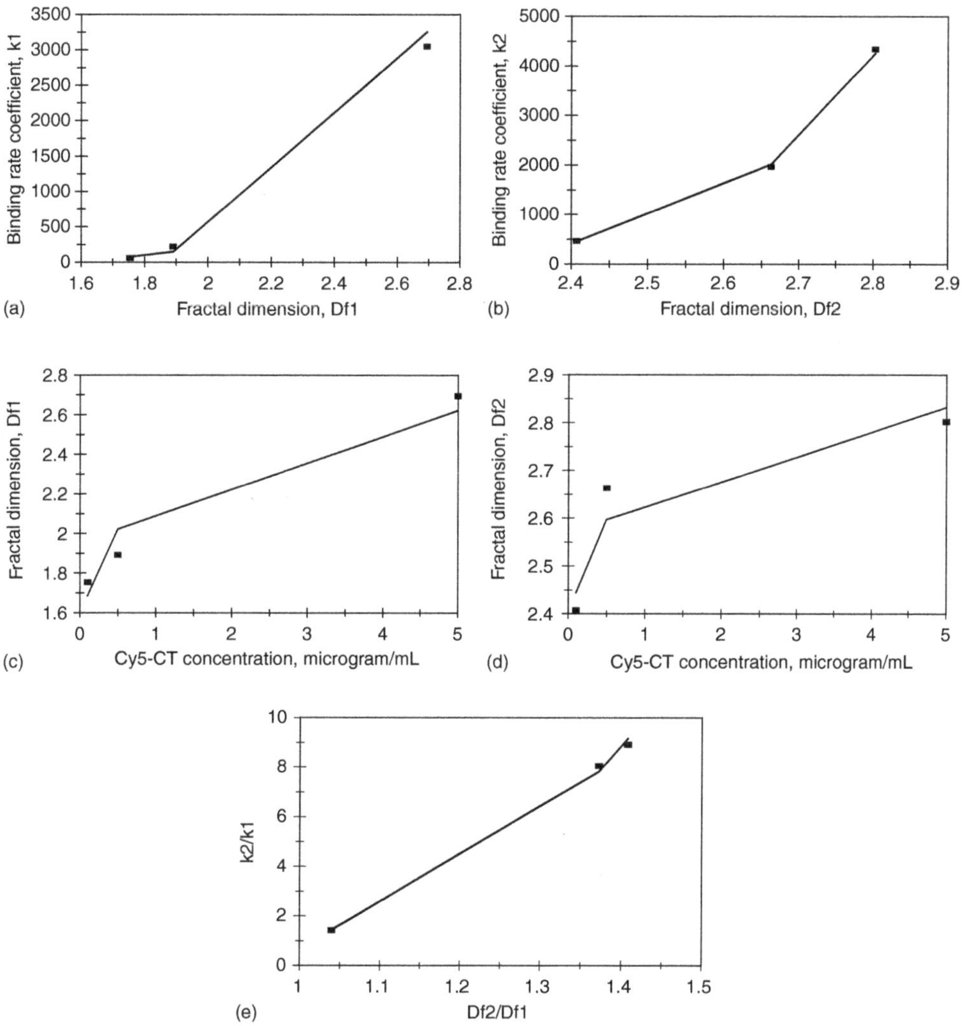

Figure 4.6 (a) Increase in the binding rate coefficient, k_1 with an increase in the fractal dimension, D_{f1}. (b) Increase in the binding rate coefficient, k_2 with an increase in the fractal dimension, D_{f2}. (c) Increase in the fractal dimension, D_{f1} with an increase in the Cy5–CT concentration (in μg/ml) in solution. (d) Increase in the fractal dimension, D_{f2} with an increase in the Cy5–CT concentration (in μg/ml) in solution. (e) Increase in the ratio of the binding rate coefficients, k_2/k_1 with an increase in the fractal dimension ratio, D_{f2}/D_{f1}.

The fit is good. Only three data points are available. The availability of more data points would lead to a more reliable fit. The binding rate coefficient, k_2 is extremely sensitive to the degree of heterogeneity or the fractal dimension, D_{f2} on the waveguide surface as noted by the order of dependence between 14 and 15 (equal to 14.60) exhibited.

Figure 4.6c and Table 4.3 show for a dual-fractal analysis the increase in the fractal dimension, D_{f1} with an increase in the Cy5–CT concentration in solution in the 0.1–5.0 μg/ml range. For the data shown in Figure 4.6c the fractal dimension, D_{f1} is given by:

$$D_{f1} = (2.187 \pm 0.188)[\text{Cy5} - \text{CT}]^{0.113 \pm 0.029} \tag{4.6a}$$

The fit is good. Only three data points are available. The availability of more data points would lead to a more reliable fit. The fractal dimension, D_{f1} exhibits only a very slight dependence on the Cy5–CT concentration in solution as noted by the 0.113 order of dependence exhibited.

Figure 4.6d and Table 4.3 show for a dual-fractal analysis the increase in the fractal dimension, D_{f2} with an increase in the Cy5–CT concentration in solution in the 0.1–5.0 μg/ml range. For the data shown in Figure 4.6d the fractal dimension, D_{f2} is given by:

$$D_{f2} = (2.665 \pm 0.084)[\text{Cy5} - \text{CT}]^{0.0377 \pm 0.0112} \tag{4.6b}$$

The fit is good. Only three data points are available. The availability of more data points would lead to a more reliable fit. The fractal dimension, D_{f2} exhibits only a very slight dependence on the Cy5–CT concentration in solution as noted by the 0.0377 order of dependence exhibited.

Figure 4.6e and Table 4.3 show for a dual-fractal analysis, the increase in the binding rate coefficient ratio, k_2/k_1 as the ratio of the fractal dimensions, D_{f2}/D_{f1} increases. For the data presented in Figure 4.6e and in Table 4.3, the binding rate coefficient ratio, k_2/k_1 is given by:

$$\frac{k_2}{k_1} = (1.123 \pm 0.459)\left(\frac{D_{f2}}{D_{f1}}\right)^{6.134 \pm 0.1683} \tag{4.7}$$

The fit is very good. Only three data points are available. The availability of more data points would lead to a more reliable fit. The binding rate coefficient ratio, k_2/k_1, is very sensitive to the fractal dimension ratio, D_{f2}/D_{f1} as noted by the higher than sixth (equal to 6.134) order of dependence exhibited.

Figure 4.7a shows the binding of 0.1 μg/ml Cy5–CT in solution to Neu5Ac sialic acid covalently immobilized on a maleimide-activated planar waveguide using a thiol-terminated linker (Ngundi *et al.*, 2006). Cy5 is a bisfunctional dye. A dual-fractal analysis is required to adequately describe the binding kinetics. The values of (a) the binding rate coefficient, k and the fractal dimension, D_f for a single-fractal analysis, and (b) the binding rate coefficients, k_1 and k_2, and the fractal dimensions, D_{f1} and D_{f2} for a dual-fractal analysis are given in Table 4.3. It is of interest to note that for a dual-fractal analysis an increase in the fractal dimension by a factor of 1.37 from a value of D_{f1} equal to 1.7544 to D_{f2} equal to 2.4076 leads to an increase in the binding rate coefficient by a factor of 8.06 from a value of k_1 equal to 58.046 to k_2 equal to 467.74. Note that changes in the binding rate coefficient and in the degree of heterogeneity on the surface or the fractal dimension are in the same direction.

Figure 4.7b shows the binding of 0.5 μg/ml Cy5–CT in solution to Neu5Ac sialic acid covalently immobilized on a maleimide-activated planar waveguide using a thiol-terminated linker (Ngundi *et al.*, 2006). A dual-fractal analysis is once again required to adequately

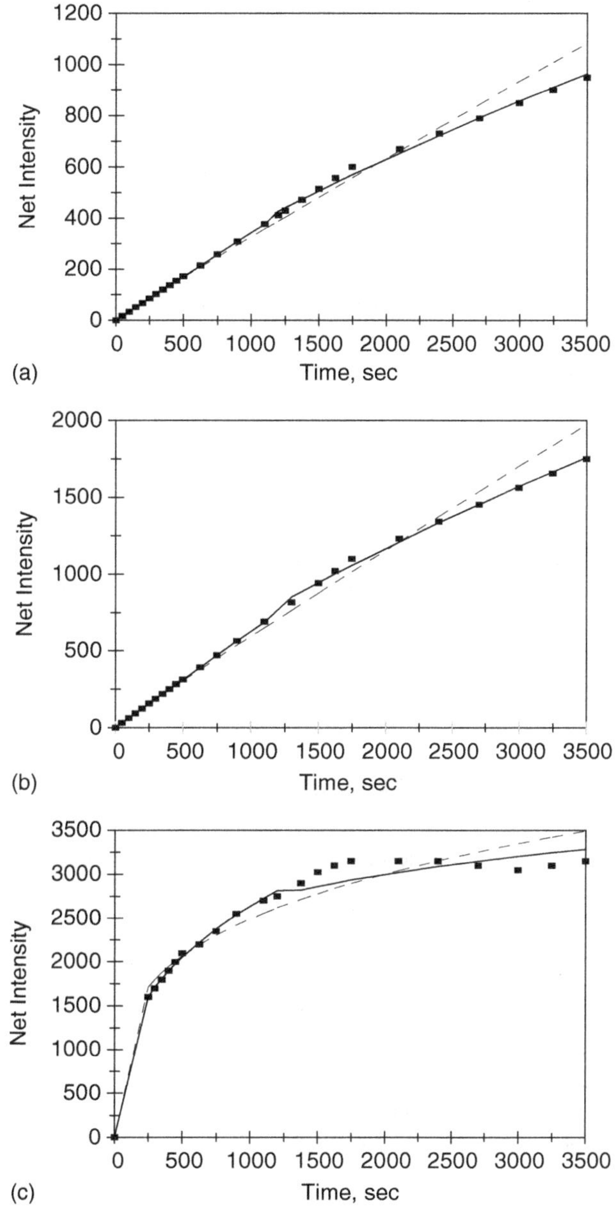

Figure 4.7 Binding of different concentrations (in μg/ml) of Cy5–CT in solution to Neu5Ac sialic acid covalently immobilized on a malemide-activated planar waveguide using a thiol-terminated linker (Ngundi *et al.*, 2006): (a) 0.1; (b) 0.5; (c) 5.0. When both a dashed (- - -) and a solid (——) line are used, then the solid line represents a dual-fractal analysis, and the dashed line represents a single-fractal analysis. In this case, the solid line is the best fit line.

describe the binding kinetics. The values of (a) the binding rate coefficient, k and the fractal dimension, D_f for a single-fractal analysis, and (b) the binding rate coefficients, k_1 and k_2, and the fractal dimensions, D_{f1} and D_{f2} for a dual-fractal analysis are given in Table 4.3. It is of interest to note that for a dual-fractal analysis an increase in the fractal dimension by a factor of 1.408 from a value of D_{f1} equal to 1.8910 to D_{f2} equal to 2.6630 leads to an increase in the binding rate coefficient by a factor of 8.92 from a value of k_1 equal to 220.046 to k_2 equal to 1965.33. Note that changes in the binding rate coefficient and in the degree of heterogeneity on the surface or the fractal dimension are once again in the same direction.

Figure 4.7c shows the binding of 5.0 µg/ml Cy5–CT in solution to Neu5Ac sialic acid covalently immobilized on a maleimide-activated planar waveguide using a thiol-terminated linker (Ngundi *et al.*, 2006). A dual-fractal analysis is ocne again required to adequately describe the binding kinetics. The values of (a) the binding rate coefficient, k and the fractal dimension, D_f for a single-fractal analysis, and (b) the binding rate coefficients, k_1 and k_2, and the fractal dimensions, D_{f1} and D_{f2} for a dual-fractal analysis are given in Table 4.3. It is of interest to note that for a dual-fractal analysis an increase in the fractal dimension by a factor of 1.040 from a value of D_{f1} equal to 2.6952 to D_{f2} equal to 2.8027 leads to an increase in the binding rate coefficient by a factor of 1.423 from a value of k_1 equal to 3049.07 to k_2 equal to 4338.93.

Figure 4.8a and Table 4.3 show for a dual-fractal analysis the increase in the binding rate coefficient, k_1 with an increase in the Neu5Ac sialic acid concentration in the 0.1–5.0 µg/ml range in solution. For the data presented in Figure 4.8a the binding rate coefficient, k_1 is given by:

$$k_1 = (8.0 + 34)[\text{Neu5Ac}\ \mu\text{g/mL}]^{1.715 \pm 0.599} \tag{4.8a}$$

The fit is not good. There is considerable scatter in the data. This is also reflected in the error in the rate coefficient. Only three data points are available. The availability of more data points would lead to a more reliable fit. The binding rate coefficient, k_1 exhibits an order of dependence between one and one-half and second (equal to 1.715) on the Neu5Ac sialic acid concentration in solution in the 0.1–5.0 µg/ml range. The non-integer order of dependence exhibited on the Cy5–CT concentration in solution lends support to the fractal nature of the system.

Figure 4.8b and Table 4.3 show for a dual-fractal analysis the increase in the binding rate coefficient, k_2 with an increase in the Neu5Ac sialic acid in the 0.1–5.0 µg/ml range in solution. For the data presented in Figure 4.8b the binding rate coefficient, k_2 is given by:

$$k_2 = (41.18 + 121.01)[\text{Neu5Ac}\ \mu\text{g/mL}]^{1.6095 \pm 0.4946} \tag{4.8b}$$

The fit is not good once again. There is considerable scatter in the data. This is also reflected in the error in the rate coefficient. Only three data points are available. The availability of more data points would lead to a more reliable fit. The binding rate coefficient, k_2 exhibits an order of dependence between one and one-half and second (equal to 1.609) on the Neu5Ac concentration in solution in the 0.1–5.0 µg/ml range. Note that the order of dependence on the Neu5Ac concentration in solution decreases by 6.15% as one goes from

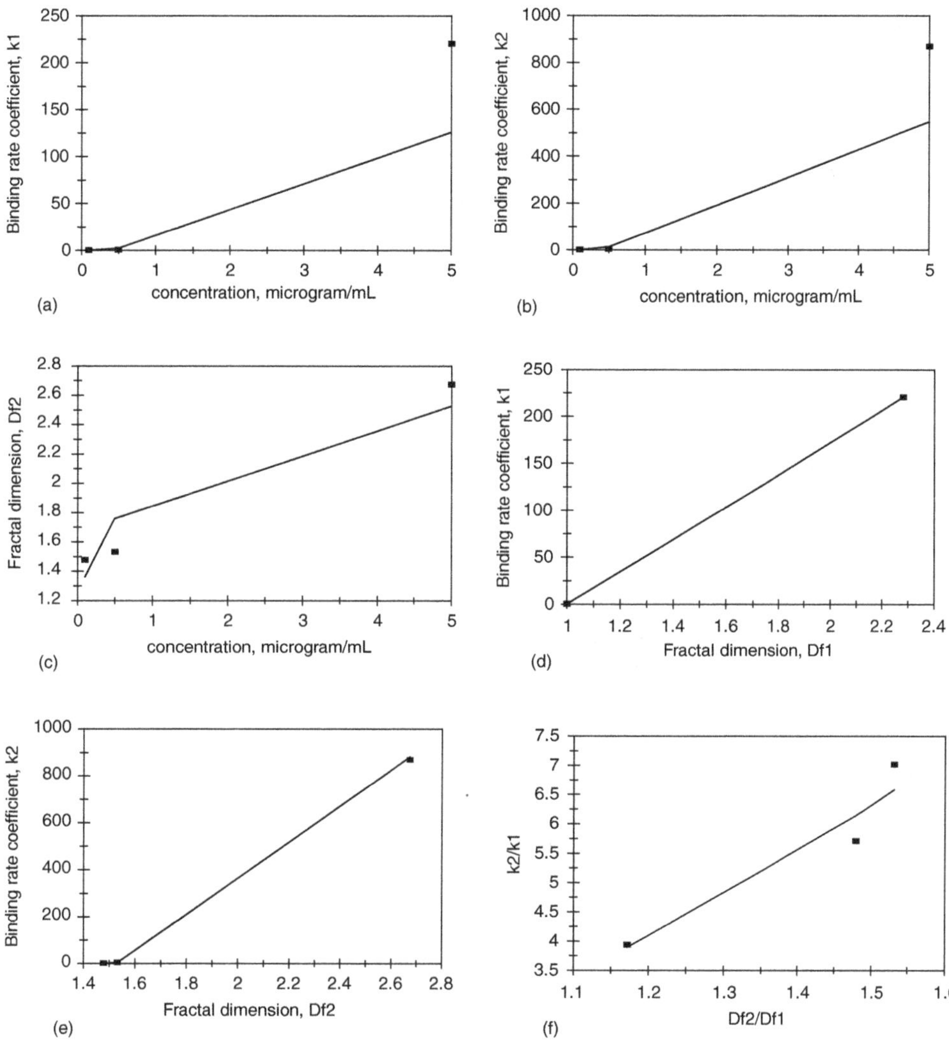

Figure 4.8 (a) Increase in the binding rate coefficient, k_1 with an increase in the Neu5Ac sialic acid concentration in solution. (b) Increase in the binding rate coefficient, k_2 with an increase in the Neu5Ac sialic acid concentration in solution. (c) Increase in the fractal dimension, D_{f2} with an increase in the Neu5Ac sialic acid in solution. (d) Increase in the binding rate coefficient, k_1 with an increase in the fractal dimension, D_{f1}. (e) Increase in the binding rate coefficient, k_2 with an increase in the fractal dimension, D_{f2}. (f) Increase in the ratio of the binding rate coefficients, k_2/k_1 with an increase in the fractal dimension ratio, D_{f2}/D_{f1}.

k_1 (equal to 1.715) to k_2 (equal to 0.1.6095). In other words, the binding rate coefficient, k_1 is slightly more sensitive than k_2 on the Neu5Ac concentration in solution in the 0.1–5.0 μg/ml in solution.

Figure 4.8c and Table 4.3 show for a dual-fractal analysis the increase in the fractal dimension, D_{f2} with an increase in the Neu5Ac sialic acid concentration in solution in the 0.1–5.0 μg/ml range. For the data shown in Figure 4.8c the fractal dimension, D_{f2} is given by:

$$D_{f2} = (1.961 \pm 0.361)[\text{Neu5Ac}]^{0.1576\pm0.0609} \tag{4.8c}$$

The fit is good. Only three data points are available. The availability of more data points would lead to a more reliable fit. The fractal dimension, D_{f2} exhibits only a very slight dependence on the Neu5Ac sialic acid concentration in solution as noted by the 0.1576 order of dependence exhibited.

Figure 4.8d and Table 4.3 show for a dual-fractal analysis the increase in the binding rate coefficient, k_1 with an increase in the fractal dimension, D_{f1}. For the data presented in Figure 4.8d the binding the binding rate coefficient, k_1 is given by:

$$k_1 = (0.4638 \pm 0.2499)D_{f1}^{7.473\pm0.6397} \tag{4.8d}$$

The fit is good. Only three data points are available. The availability of more data points would lead to a more reliable fit. The binding rate coefficient, k_1 is extremely sensitive to the degree of heterogeneity or the fractal dimension, D_{f1} on the waveguide surface as noted by the order of dependence between 7 and 7.5 (equal to 7.473) exhibited.

Figure 4.8e and Table 4.3 show for a dual-fractal analysis the increase in the binding rate coefficient, k_2 with an increase in the fractal dimension, D_{f2}. For the data presented in Figure 4.8e the binding the binding rate coefficient, k_2 is given by:

$$k_2 = (0.04988 \pm 0.01954)D_{f2}^{9.943\pm0.7040} \tag{4.8e}$$

The fit is good. Only three data points are available. The availability of more data points would lead to a more reliable fit. The binding rate coefficient, k_2 is extremely sensitive to the degree of heterogeneity or the fractal dimension, D_{f2} on the waveguide surface as noted by the order of dependence close to 10 (equal to 9.943) exhibited.

Figure 4.8f and Table 4.3 show for a dual-fractal analysis, the increase in the binding rate coefficient ratio, k_2/k_1 as the ratio of the fractal dimensions, D_{f2}/D_{f1} increases. For the data presented in Figure 4.8f and in Table 4.3, the binding rate coefficient ratio, k_2/k_1 is given by:

$$\frac{k_2}{k_1} = (2.862 \pm 0.295)\left(\frac{D_{f2}}{D_{f1}}\right)^{1.952 \pm 0.4759} \tag{4.8f}$$

The fit is good. Only three data points are available. The availability of more data points would lead to a more reliable fit. The binding rate coefficient ratio, k_2/k_1, is sensitive to the fractal dimension ratio, D_{f2}/D_{f1} as noted by the close to second (equal to 1.952) order of dependence exhibited.

Leonard *et al.* (2004) indicate that *L. monocytogenes* is an important food-borne pathogen, and has been linked to recent food poisoning outbreaks (Donelly, 2001; Schlech, 2000). *L. monocytogenes* is a gram-positive facultative anaerobic rod-shaped bacterium (Leonard *et al.*, 2004). This bacterium grows between 1 and 45 °C (Jones and Seeliger, 1992). Persons with weak immune defenses such as people with AIDS and diabetes, pregnant women, and newborns are at risk for listeriosis (Leonard *et al.*, 2004). These authors have developed a new 'real time' subtractive inhibition assay for the detection of low numbers of *L. monocytogenes* cells. A Biacore 3000 biosensor was used.

Figure 4.9a shows the binding of 1×10^9 cells/ml of *L. monocytogenes* cells with affinity-purified anti-*L. monocytogenes* in solution to the anti-Fab antibody immobilized on a

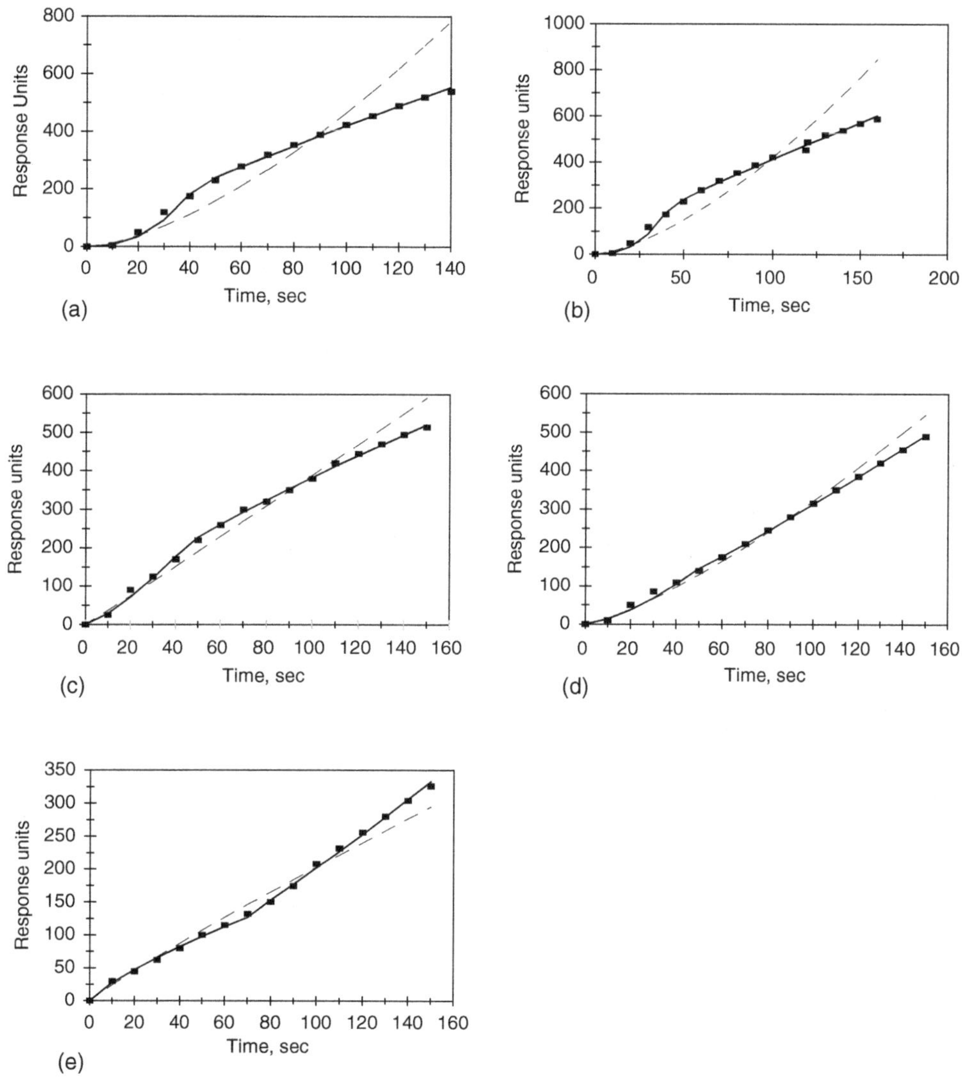

Figure 4.9 Binding of different concentrations (in cells/ml) of *L. monocytogenes* in solution to anti-Fab antibody immobilized on a Biacore 3000 biosensor chip surface (Leonard *et al.*, 2004): (a) 1×10^9; (b) 1×10^8; (c) 1×10^7; (d) 1×10^6; (e) 1×10^5. When both a dashed (---) and a solid (—) line are used, then the solid line represents a dual-fractal analysis, and the dashed line represents a single-fractal analysis. In this case, the solid line is the best fit line.

Biacore 3000 biosensor chip surface. A dual-fractal analysis is required to adequately describe the binding kinetics. The values of (a) the binding rate coefficient, k and the fractal dimension, D_f for a single-fractal analysis, and (b) the binding rate coefficients, k_1 and k_2 and the fractal dimensions, D_{f1} and D_{f2} for a dual-fractal analysis are given in Table 4.4. Note that an increase in the fractal dimension from D_{f1} equal to approximately zero to D_{f2} equal to 1.3620 leads to an increase in the binding rate coefficient by a factor of 357.2 from a value of k_1 equal to 0.02711 to k_2 equal to 9.6836.

Figure 4.9b shows the binding of 1×10^8 cells/ml of *L. monocytogenes* cells with affinity-purified anti-*L. monocytogenes* in solution to the anti-Fab antibody immobilized on a Biacore 3000 biosensor chip surface. A dual-fractal analysis is required to adequately describe the binding kinetics. The values of (a) the binding rate coefficient, k and the fractal dimension, D_f for a single-fractal analysis, and (b) the binding rate coefficients, k_1 and k_2 and the fractal dimensions, D_{f1} and D_{f2} for a dual-fractal analysis are given in Table 4.4. Note that an increase in the fractal dimension from D_{f1} equal to approximately zero to D_{f2} equal to 1.4118 leads to an increase in the binding rate coefficient by a factor of 497.6 from a value of k_1 equal to 0.02146 to k_2 equal to 10.6786.

Figure 4.9c shows the binding of 1×10^7 cells/ml of *L. monocytogenes* cells with affinity-purified anti-*L. monocytogenes* in solution to the anti-Fab antibody immobilized on a Biacore 3000 biosensor chip surface. A dual-fractal analysis is required to adequately describe the binding kinetics. The values of (a) the binding rate coefficient, k and the fractal dimension,

Table 4.4

Binding rate coefficients and fractal dimensions for purified anti-*L. monocytogenes* polyclonal antibody in solution to immobilized anti-Fab antibody immobilized on a sensor chip (Leonard *et al.*, 2004)

Analyte in solution	k	k_1	k_2	D_f	D_{f1}	D_{f2}
Anti-*L. monocytogenes* polyclonal antibody + 1×10^9 monocytogenes cells/ml	0.3641 ± 0.1850	0.02711 ± 0.01153	9.6836 ± 0.1744	0 + 0.2946	0 + 0.5576	1.3620 ± 0.03680
Anti-*L. monocytogenes* polyclonal antibody + 1×10^8 monocytogenes cells/ml	0.4357 ± 0.2313	0.02146 ± 0.00953	10.6786 ± 0.2903	0 + 0.5782	0 + 0.5782	1.4118 ± 0.04296
Anti-*L. monocytogenes* polyclonal antibody + 1×10^7 monocytogenes cells/ml	3.4032 ± 0.5484	1.3777 ± 0.2470	11.9950 ± 0.1694	0.9412 ± 0.1021	0.3652 ± 0.2596	1.4946 ± 0.0307
Anti-*L. monocytogenes* polyclonal antibody + 1×10^6 monocytogenes cells/ml	0.7677 ± 0.1449	0.4280 ± 0.1063	1.7981 ± 0.0103	0.3788 ± 0.1021	0.0172+ 0.2646	0.7590 ± 0.10255
Anti-*L. monocytogenes* polyclonal antibody + 1×10^5 monocytogenes cells/ml	2.9145 ± 0.3031	4.6752 ± 0.2675	0.6731 ± 0.0147	1.1568 ± 0.06762	1.4454 ± 0.06634	0.5232 ± 0.07466

D_f for a single-fractal analysis, and (b) the binding rate coefficients, k_1 and k_2 and the fractal dimensions, D_{f1} and D_{f2} for a dual-fractal analysis are given in Table 4.4. Note that an increase in the fractal dimension by a factor of 4.1 from D_{f1} equal to 0.3652 to D_{f2} equal to 1.4964 leads to an increase in the binding rate coefficient by a factor of 8.71 from a value of k_1 equal to 1.3777 to k_2 equal to 11.9950.

Figure 4.9d shows the binding of 1×10^6 cells/ml of *L. monocytogenes* cells with affinity-purified anti-*L. monocytogenes* in solution to the anti-Fab antibody immobilized on a Biacore 3000 biosensor chip surface. A dual-fractal analysis is required to adequately describe the binding kinetics. The values of (a) the binding rate coefficient, k and the fractal dimension, D_f for a single-fractal analysis, and (b) the binding rate coefficients, k_1 and k_2 and the fractal dimensions, D_{f1} and D_{f2} for a dual-fractal analysis are given in Table 4.4. Note that an increase in the fractal dimension by a factor of 44.13 from D_{f1} equal to 0.0172 to D_{f2} equal to 0.7590 leads to an increase in the binding rate coefficient by a factor of 4.20 from a value of k_1 equal to 0.4280 to k_2 equal to 1.7981.

Figure 4.9e shows the binding of 1×10^5 cells/ml of *L. monocytogenes* cells with affinity-purified anti-*L. monocytogenes* in solution to the anti-Fab antibody immobilized on a Biacore 3000 biosensor chip surface. A dual-fractal analysis is required to adequately describe the binding kinetics. The values of (a) the binding rate coefficient, k and the fractal dimension, D_f for a single-fractal analysis, and (b) the binding rate coefficients, k_1 and k_2 and the fractal dimensions, D_{f1} and D_{f2} for a dual-fractal analysis are given in Table 4.4. Note that a decrease in the fractal dimension by a factor of 2.763 from D_{f1} equal to 1.4454 to D_{f2} equal to 0.5232 leads to a decrease in the binding rate coefficient by a factor of 6.95 from a value of k_1 equal to 4.6752 to k_2 equal to 0.6731.

Figure 4.10a and Table 4.4 indicate for a dual-fractal analysis the increase in the binding rate coefficient, k_1 with an increase in the fractal dimension, D_{f1}. For the data shown in Figure 4.10a, the binding rate coefficient, k_1 is given by:

$$k_1 = (3.135 \pm 1.461) D_{f1}^{0.5135 \pm 0.1193} \tag{4.9a}$$

The fit is reasonable. Only three data points are available. The availability of more data points would lead to a more reliable fit. The binding rate coefficient, k_1 is only mildly sensitive to the fractal dimension, D_{f1} as noted by the slightly greater than one-half (equal to 0.5135) order of dependence exhibited.

Figure 4.10b and Table 4.4 indicate for a dual-fractal analysis the increase in the binding rate coefficient, k_2 with an increase in the fractal dimension, D_{f2}. For the data shown in Figure 4.10b, the binding rate coefficient, k_2 is given by:

$$k_2 = (4.011 \pm 0.127) D_{f2}^{2.785 \pm 00334} \tag{4.9b}$$

The fit is very good. Five data points are available. The availability of more data points would lead to a more reliable fit. The binding rate coefficient, k_2 is sensitive to the fractal dimension, D_{f2} as noted by the order of dependence between 2.5 and 3 (equal to 2.785) exhibited.

Balasubramanian *et al.* (2006) have recently used a lytic phage as a specific and selective probe for the detection of *S. aureus*. These authors emphasize that community acquired as well as hospital acquired *S. aureus* infections are a constant threat to the human

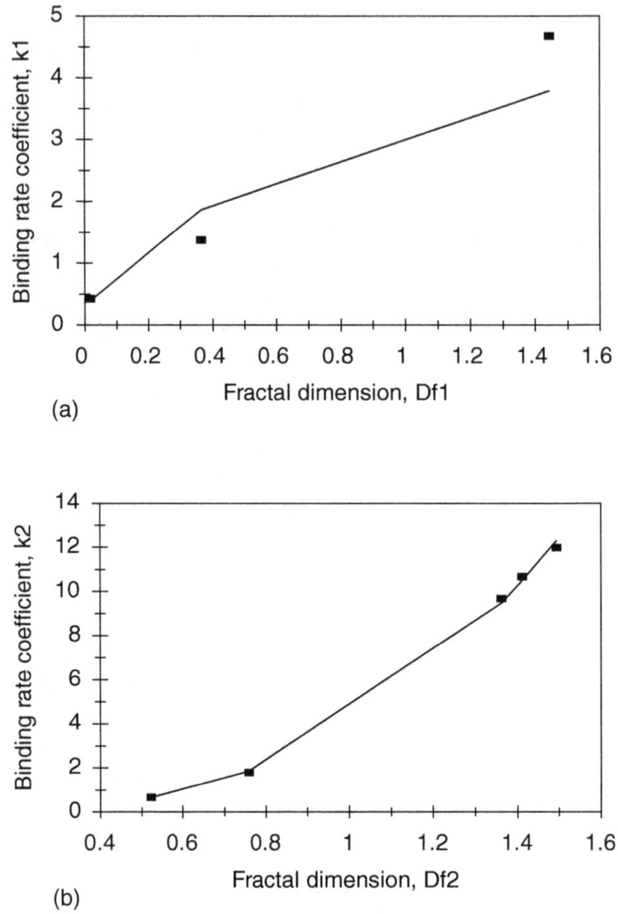

Figure 4.10 (a) Increase in the binding rate coefficient, k_1 with an increase in the fractal dimension, D_{f1}. (b) Increase in the binding rate coefficient, k_2 with an increase in the fractal dimension, D_{f2}.

population (Aucken *et al.*, 2002; Jay, 2000; Lowy, 1998). Balasubramanian *et al.* (2006) emphasize that *S. aureus* is responsible for multiple illnesses such as abdominal cramps, diarrhea, urinary tract infections, pneumonia, respiratory diseases, gastroenteritis, and bronchial diseases. They emphasize the need for a rapid and reliable detection device for *S. aureus* and other harmful pathogens which may be detected at low levels. These authors have used surface plasmon resonance spectroscopy using a lytic phage for the label-free detection of *S. aureus*. A surface plasmon resonance-based SPREETA^{TM} sensor was used as a detection platform.

Figure 4.11a shows the binding of 10^5cfu/ml of *S. aureus* in solution by the lytic phage immobilized on the sensing channel which was blocked by BSA (bovine serum albumin) to minimize the nonspecific adsorption of cells (Balasubramanian *et al.*, 2006). A single-fractal

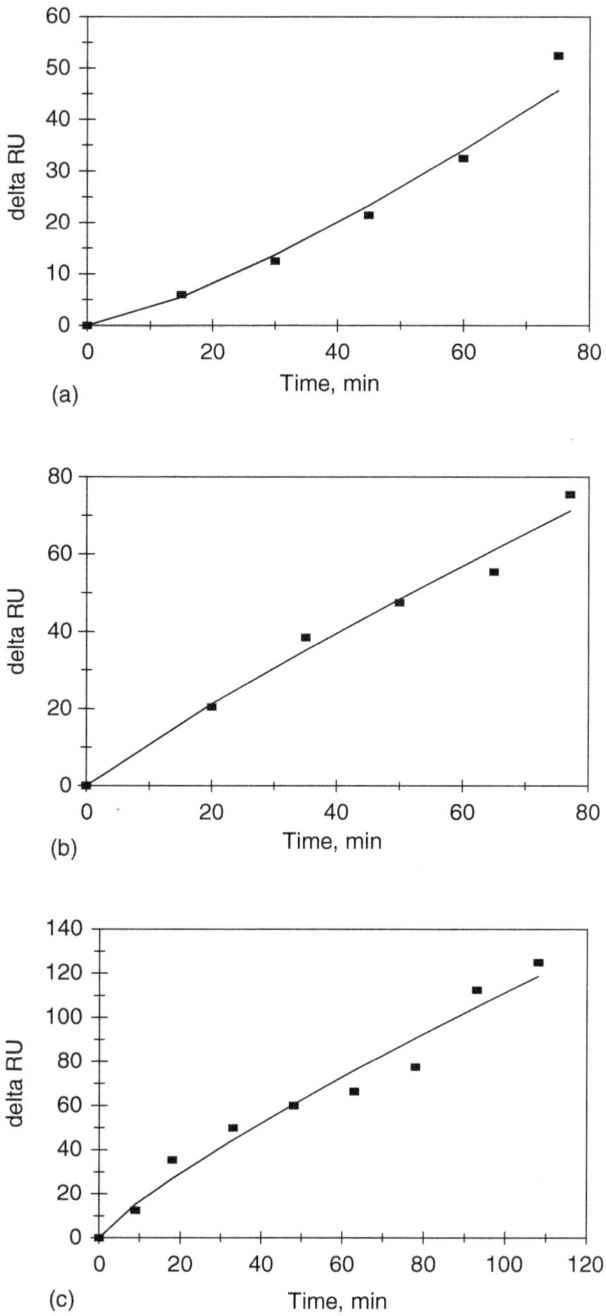

Figure 4.11 Binding of different concentrations (in cfu/ml) of *S. aureus* in solution to the lytic phage immobilized on the sensing channel (Balasubramnian *et al.*, 2006) (a) 10^5; (b) 10^6; (c) 10^7. When both a dashed (- - -) and a solid (——) line are used, then the solid line represents a dual-fractal analysis, and the dashed line represents a single-fractal analysis. In this case, the solid line is the best fit line.

Table 4.5

Binding rate coefficients and fractal dimensions for the different concentrations of *S. aureus* in solution to the lytic phage immobilized on the sensing channel blocked by BSA in a SPREETA™ sensor (Balasubramanian *et al.*, 2006)

S. aureus concentration (cfu/ml)	k	D_f
10^5	0.1564 ± 0.0201	0.3698 ± 0.1906
10^6	1.476 ± 0.1324	1.2064 ± 0.1630
10^7	2.4894 ± 0.4672	1.3490 ± 0.1504

analysis is adequate to describe the binding kinetics. The values of the binding rate coefficient, k and the fractal dimension, D_f for a single-fractal analysis are given in Table 4.5.

Figure 4.11b shows the binding of 10^6 cfu/ml of *S. aureus* in solution by the lytic phage immobilized on the sensing channel which was blocked by BSA to minimize the non-specific adsorption of cells (Balasubramanian *et al.*, 2006). A single-fractal analysis is adequate to describe the binding kinetics. The values of the binding rate coefficient, k and the fractal dimension, D_f for a single-fractal analysis are given in Table 4.5.

It is of interest to note that as the *S. aureus* concentration in solution increases by an order of magnitude from 10^5 to 10^6 cfu/ml the fractal dimension, D_f increases by a factor of 3.26, and the binding rate coefficient, k increases by a factor of 9.26. The fractal dimension increases from a value of D_f equal to 0.3698 to 1.2064. Similarly, the binding rate coefficient, k increases from a value of 0.1564 to 1.476.

Figure 4.11c shows the binding of 10^7 cfu/ml of *S. aureus* in solution by the lytic phage immobilized on the sensing channel which was blocked by BSA to minimize the non-specific adsorption of cells (Balasubramanian *et al.*, 2006). A single-fractal analysis is adequate to describe the binding kinetics. The values of the binding rate coefficient, k and the fractal dimension, D_f for a single-fractal analysis are given in Table 4.5.

Figure 4.12a and Table 4.5 indicate for a single-fractal analysis the increase in the binding rate coefficient, k with an increase in the *S. aureus* concentration in solution in the 10^5–10^7 cfu/ml concentration range. For the data presented in Figure 4.12a the binding rate coefficient, k is given by:

$$k = (0.000205 \pm 0.0000202)[Staphylococcus\ aureus]^{0.6009 \pm 0.2110} \tag{4.10a}$$

The fit is reasonable. Only three data points are available. The availability of more data points would lead to a more reliable fit. The binding rate coefficient, k is only mildly sensitive to the *S. aureus* concentration in solution as noted by the 0.6009 order of dependence exhibited. The non-integer order of dependence exhibited by the binding rate coefficient, k on the *S. aureus* concentration in solution lends support to the fractal nature of the system.

Figure 4.12b and Table 4.5 indicate for a single-fractal analysis the increase in the binding rate coefficient, k with an increase in the fractal dimension, D_f. For the data presented in Figure 4.12b the binding rate coefficient, k is given by:

$$k = (1.165 \pm 0.292) D_f^{2.033 \pm 0.2203} \tag{4.10b}$$

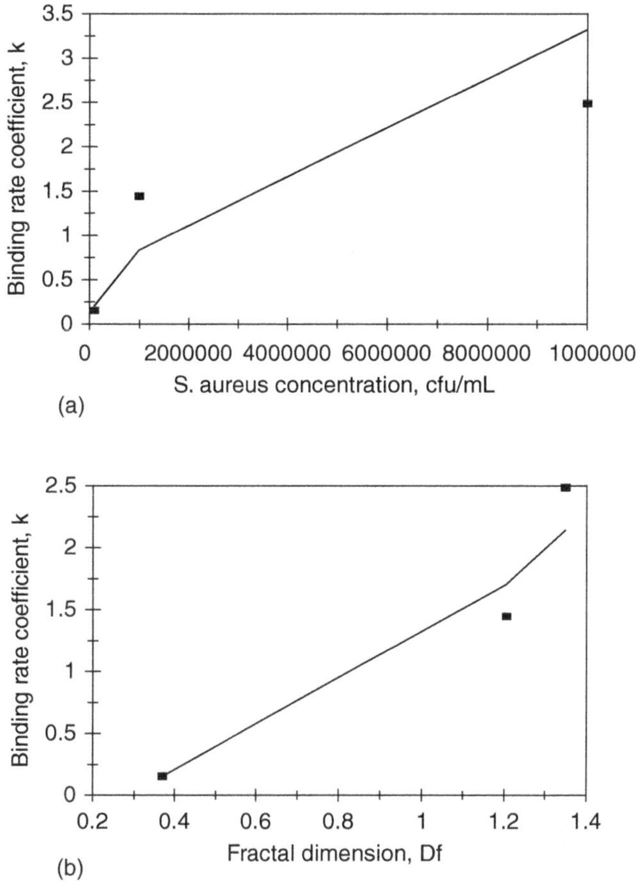

Figure 4.12 (a) Increase in the binding rate coefficient, k with an increase in the *S. aureus* concentration (in cfu/ml) in solution. (b) Increase in the binding rate coefficient, k with an increase in the fractal dimension, D_f.

The fit is reasonable. Only three data points are available. The availability of more data points would lead to a more reliable fit. The binding rate coefficient, k is quite sensitive to the fractal dimension, D_f or the degree of heterogeneity that exists on the SPREETA^{TM} biosensor surface as noted by the close to second order of dependence exhibited.

Zourob *et al.* (2005) have developed a novel disposable absorbing material clad leaky waveguide sensor device (LWD) for the detection of bacteria. These authors indicate that there has been a substantial increase (600%) in food borne illnesses in the year 2001 when compared with those occurring in the year 1982 (Deisingh, 2003). Buzby *et al.* (1996) and Mead *et al.* (1999) have also indicated that there were about 4500 deaths resulting from food borne illnesses and up to about 67 million food borne cases. These references are from the 1990s, and the real numbers could be very different nowadays. However, there is an urgent need for devices capable of rapid, specific, accurate, sensitive, and cost-effective detection of bacteria that cause illnesses (Zourob *et al.*, 2005). These authors indicate that

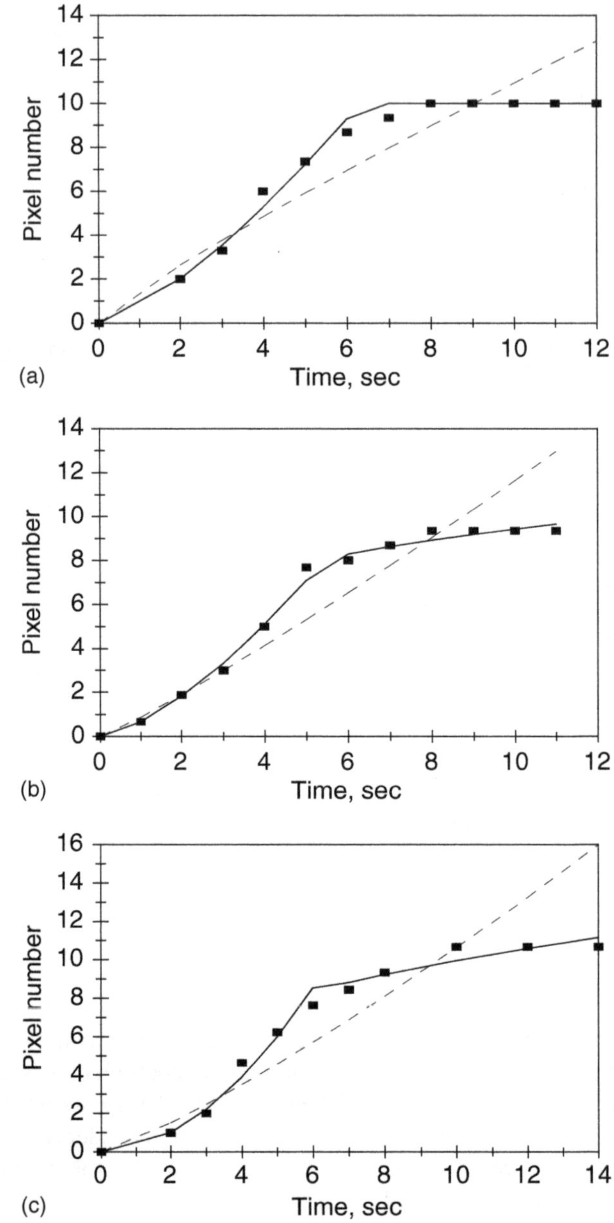

Figure 4.13 Binding of different concentrations of the bacteria *Bacillus subtilis var niger* BG to the FITC-labeled anti-BG immobilized on a leaky waveguide sensor chip surface (Zourob *et al.*, 2005) (a) 1×10^5; (b) 3×10^5; (c) 7×10^5. When both a dashed (---) and a solid (—) line are used, then the solid line represents a dual-fractal analysis, and the dashed line represents a single-fractal analysis. In this case, the solid line is the best fit line.

their intent is to develop a device to detect *Bacillus anthracis*. They used *Bacillus subtilis* var. *niger* spores as a non-pathogenic bacterial stimulant.

Figure 4.13a shows the binding of 1×10^5 spores/ml of bacteria, *B. subtilis var. niger* (BG) to the FITC (fluorescein isothiocyanate)-labeled anti-*B. subtilis var. niger* immobilized on the leaky waveguide sensor chip (Zourob *et al.*, 2005). A dual-fractal analysis is required to adequately describe the binding kinetics. The values of (a) the binding rate coefficient, k and the fractal dimension, D_f for a single-fractal analysis, and (b) the binding rate coefficients, k_1 and k_2 and the fractal dimensions, D_{f1} and D_{f2} for a dual-fractal analysis are given in Table 4.6. It is of interest to note that as the fractal dimension increases by a factor of 14.27 from a value of D_{f1} equal to 0.2102 to D_{f2} equal to 3.0 the binding rate coefficient increases by a factor of 13.1 from a value of k_1 equal to 0.7646 to k_2 equal to 10. Changes in the binding rate coefficient and in the fractal dimension or the degree of heterogeneity on the LWD sensor chip surface are in the same direction.

Figure 4.13b shows the binding of 3×10^5 spores/ml of bacteria, BG to the FITC-labeled anti-BG immobilized on the leaky waveguide sensor chip (Zourob *et al.*, 2005). A dual-fractal analysis is once again required to adequately describe the binding kinetics. The values of (a) the binding rate coefficient, k and the fractal dimension, D_f for a single-fractal analysis, and (b) the binding rate coefficients, k_1 and k_2 and the fractal dimensions, D_{f1} and D_{f2} for a dual-fractal analysis are given in Table 4.6. It is of interest to note that as the fractal dimension increases by a factor of 63.25 from a value of D_{f1} equal to 0.0396 to D_{f2} equal to 2.5048 the binding rate coefficient increases by a factor of 8.127 from a value of k_1 equal to 0.6559 to k_2 equal to 5.3310. Changes in the binding rate coefficient and in the fractal dimension or the degree of heterogeneity on the LWD sensor chip surface are once again in the same direction.

Figure 4.13c shows the binding of 7×10^5 spores/ml of bacteria, BG to the FITC-labeled anti-BG immobilized on the leaky waveguide sensor chip (Zourob *et al.*, 2005). A dual-fractal analysis is required to adequately describe the binding kinetics. The values of (a) the binding rate coefficient, k and the fractal dimension, D_f for a single-fractal analysis, and (b) the binding rate coefficients, k_1 and k_2 and the fractal dimensions, D_{f1} and D_{f2} for a dual-fractal analysis are given in Table 4.6. It is of interest to note that as the fractal dimension increases from a value of D_{f1} equal to ~0 to D_{f2} equal to 2.320 the binding rate

Table 4.6

Binding rate coefficients and fractal dimensions for the detection of different concentrations of bacterial spores (in spores/ml) in solution using a PT (polythiophene)–CLW (clad leaky waveguide) sensor (Zourob *et al.*, 2005)

Analyte in solution, bacteria (spores/ml)/ receptor on surface	k	k_1	k_2	D_f	D_{f1}	D_{f2}
1×10^5 BG/anti-BG on CLW surface	1.4242 ± 0.3210	0.7646 ± 0.0746	10 ± 0	1.2294 ± 0.2252	0.2102+ 0.2146	3.0 + 1.0E–14
3×10^5 BG/anti-BG on CLW surface	0.8660 ± 0.2183	0.6559 ± 0.0544	5.3310 ± 0.1910	0.7432 ± 0.1907	0.0396 + 0.1253	2.5048 ± 0.1390
7×10^5 BG/anti-BG on CLW surface	0.6489 ± 0.2308	0.2628 ± 0.0384	4.5705 ± 0.2563	0.5718 ± 0.3278	~0	2.3240 ± 0.1921

coefficient increases by a factor of 17.39 from a value of k_1 equal to 0.2628 to k_2 equal to 4.5705. Changes in the binding rate coefficient and in the fractal dimension or the degree of heterogeneity on the LWD sensor chip surface as noted above are in the same direction.

Table 4.6 and Figure 4.14a show for a dual-fractal analysis the decrease in the binding rate coefficient, k_1 with an increase in the concentration of the BG bacterial spores in solution. For the data shown in Figure 4.14a the binding rate coefficient, k_1 is given by:

$$k_1 = (384.59 \pm 170.02)[\text{BG concentration spores/ml}]^{-0.5290 \pm 0.2653} \tag{4.11a}$$

The fit is reasonable. Only three data points are available. The availability of more data points would lead to a more reliable fit. The binding rate coefficient, k_1 exhibits close to a negative one-half (equal to –0.5290) order of dependence on the BG bacteria concentration in solution. The non-integer order of dependence exhibited lends support to the fractal nature of the system.

Table 4.6 and Figure 4.14b show for a dual-fractal analysis the decrease in the binding rate coefficient, k_2 with an increase in the concentration of the BG bacterial spores in solution. For the data shown in Figure 4.14b the binding rate coefficient, k_2 is given by:

$$k_2 = (1070.32 \pm 176.04)[\text{BG concentration spores/ml}]^{-0.411 \pm 0.110} \tag{4.11b}$$

The fit is reasonable. Only three data points are available. The availability of more data points would lead to a more reliable fit. The binding rate coefficient, k_2 exhibits close to a negative one-half (equal to –0.411) order of dependence on the BG bacteria concentration in solution. The non-integer order of dependence exhibited lends support to the fractal nature of the system.

Table 4.6 and Figure 4.14c show for a dual-fractal analysis the decrease in the fractal dimension, D_{f1} with an increase in the concentration of the BG bacterial spores in solution. For the data shown in Figure 4.14c the fractal dimension, D_{f1} is given by:

$$D_{f1} = (1.4\text{E}+07 \pm 0)[\text{BG concentration spores/ml}]^{-1.563 \pm 0.0296} \tag{4.11c}$$

The fit is reasonable. Only three data points are available. The availability of more data points would lead to a more reliable fit. The fractal dimension, D_{f1} exhibits close to a negative one and one-half (equal to –1.563) order of dependence on the BG bacteria concentration in solution.

Table 4.6 and Figure 4.14d show for a dual-fractal analysis the decrease in the fractal dimension, D_{f2} with an increase in the concentration of the BG bacterial spores in solution. For the data shown in Figure 4.14d the fractal dimension, D_{f2} is given by:

$$D_{f2} = (13.696 \pm 0.411)[\text{BG concentration spores/ml}]^{-0.133 \pm 0.0214} \tag{4.11d}$$

The fit is reasonable. Only three data points are available. The availability of more data points would lead to a more reliable fit. The fractal dimension, D_{f2} exhibits less than negative one-half (actually closer to zero; equal to –0.133) order of dependence on the BG bacteria concentration in solution.

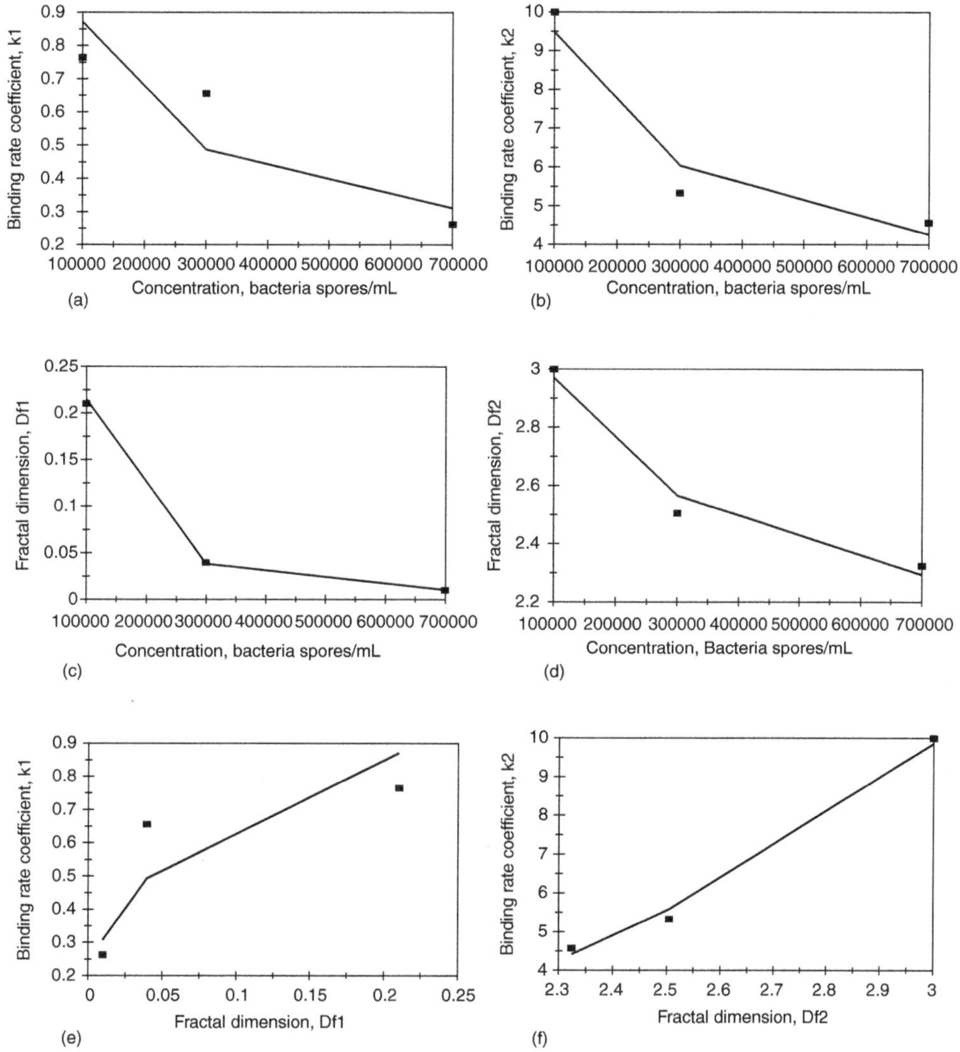

Figure 4.14 (a) Decrease in the binding rate coefficient, k_1 with an increase in the BG bacterial spores concentration in solution. (b) Decrease in the binding rate coefficient, k_2 with an increase in the BG bacterial spores concentration in solution. (c) Decrease in the fractal dimension, D_{f1} with an increase in the BG bacterial spores concentration in solution. (d) Decrease in the fractal dimension, D_{f2} with an increase in the BG bacterial spores concentration in solution. (e) Increase in the binding rate coefficient, k_1 with an increase in the fractal dimension, D_{f1}. (f) Increase in the binding rate coefficient, k_2 with an increase in the fractal dimension, D_{f2}.

Table 4.6 and Figure 4.14e show for a dual-fractal analysis the increase in the binding rate coefficient, k_1 with an increase in the fractal dimension, D_{f1}. For the data shown in Figure 4.14e the binding rate coefficient, k_1 is given by:

$$k_1 = (1.483 \pm 0.626) D_{f1}^{0.342 \pm 0.163} \tag{4.11e}$$

The fit is reasonable. Only three data points are available. The availability of more data points would lead to a more reliable fit. The binding rate coefficient, k_1 is only mildly sensitive to the fractal dimension, D_{f1} or the degree of heterogeneity that exists on the sensor chip surface as noted by the 0.342 order of dependence exhibited.

Table 4.6 and Figure 4.14f show for a dual-fractal analysis the increase in the binding rate coefficient, k_2 with an increase in the fractal dimension, D_{f2}. For the data shown in Figure 4.14f the binding rate coefficient, k_2 is given by:

$$k_2 = (0.3118 \pm 0.0193)D_{f2}^{3.144 \pm 0.3245} \tag{4.11f}$$

The fit is reasonable. Only three data points are available. The availability of more data points would lead to a more reliable fit. The binding rate coefficient, k_2 is very sensitive to the fractal dimension, D_{f2} or the degree of heterogeneity that exists on the sensor chip surface as noted by the greater than third (equal to 3.144) order of dependence exhibited.

4.4 CONCLUSIONS

A fractal analysis is presented for the binding of different pathogens such as Dengue virus (Zaytseva *et al.*, 2005) and Cy5-labeled CT (Ngundi *et al.*, 2004), and bacteria such as *E. coli* (Subramanian *et al.*, 2006), *B. globiggi* 9372 (Karasinski *et al.*, 2005), anti-*L. monocytogenes* (Leonard *et al.*, 2004), *S. aureus* (Balasubramanian *et al.*, 2006), and bacterial spores (Zourob *et al.*, 2005) in solution to appropriate receptors on biosensor surfaces. The binding kinetics may be adequately described by either a single- or a dual-fractal analysis. A dual-fractal analysis was used only if a single-fractal analysis did not provide an adequate fit. This was determined by a regression analysis provided by Sigmaplot (1993).

An increase in the fractal dimension value or the degree of heterogeneity on the biosensor surface, leads, in general, to an increase in the binding rate coefficient. For example, for the binding of Cy5–CT in solution to anti-CT mAb covalently immobilized on a maleimide-activated planar waveguide using a thiol-terminated linker (Ngundi *et al.*, 2004) the binding rate coefficients, k_1 and k_2 are very sensitive to the fractal dimension or the degree of heterogeneity on the sensor chip surface as noted by the 8.658 and 14.16 order of dependence exhibited, respectively. Predictive equations are also developed for the binding rate coefficients, k_1 and k_2 as a function of the Cy5-labeled CT concentration in solution (Ngundi *et al.*, 2004), and k_1 and k_2 as a function of BG in solution (Zourob *et al.*, 2005).

The predictive relationships developed for the binding rate coefficients, k_1 and k_2 as a function of the fractal dimensions, D_{f1} and D_{f2}, respectively, are of considerable value because they directly link the binding rate coefficient to the degree of heterogeneity that exists on the biosensor chip surface, and provide a means by which the binding rate coefficient may be manipulated by changing the degree of heterogeneity on the sensor chip surface. The binding rate coefficient is, in general, rather sensitive to the fractal dimension that exists on the biosensor chip surface. This may be noted by the high orders of dependence exhibited, and as indicated above.

It is suggested that the fractal surface (roughness) leads to turbulence, which enhances mixing, decreases diffusional limitations, and leads to an increase in the binding rate coefficient

(Martin *et al.*, 1991). For this to occur the characteristic length of the turbulent boundary layer may have to extend a few monolayers above the sensor chip surface to affect bulk diffusion to and from the surface. However, given the extremely laminar flow regimes in most biosensors this may not actually take place. The sensor chip surface is characterized by grooves and ridges, and this surface morphology may lead to eddy diffusion. This eddy diffusion can then help to enhance the mixing and extend the length of the boundary layer to affect the bulk diffusion to and from the surface.

More such studies are required to determine whether the binding rate coefficient(s) are sensitive to their respective fractal dimensions or the degree of heterogeneity that exists on the sensor chip surface. If this is correct, then experimentalists may find it worth their effort to pay a little more attention to the nature of the surface, and how it may be manipulated to control the relevant parameters and biosensor performance in desired directions.

Finally in a general sense, fractal models are fascinating. Newer avenues are required to analyze and to help detect pathogens and harmful bacteria at very dilute concentrations. The analysis of the studies of the boundaries (scale) over which the fractal behavior occurs should prove useful. The real interesting test of the fractal model would be if it can make a prediction that turns out to be correct. This would be extremely valuable, especially in the detection of pathogens and harmful bacteria. For example, if the fractal analysis enhances (or predicts) a biosensor performance parameter(s) (such as stability, sensitivity, response time, etc.) as an experimental variable is changed, then the value of the analysis will be substantially increased. Any increase in time that is made available to help in the evacuation process (for example, by making better biosensors) after the establishment of, for example, a pathogenic threat is invaluable.

REFERENCES

Aucken, HM, M Ganner, S Murchan, BD Cookson, and AP Johnson, A new UK strain of epidemic methicillin-resistant *Staphylococcus aureus* (EMSRA-17) resistant to multiple antibiotics. *Journal of Antimicrobial Chemotherapy*, 2002, **50**, 171–175.

Balasubramanian, S, IB Sorokulova, VJ Vodyanoy, and AL Simonian, Lytic phage as a specific and selective probe for detection of *Stapphylococcus aureus* – a surface plasmon resonance spectroscopic study. *Biosensors & Bioelectronics*, 2006, in press.

Buzby, JC, T Roberts, J Lin, and JM McDonald, Bacterial food-borne disease: medical costs and productivity losses, *USDA Economic Report*, 741, Washington, DC, 1996.

Campbell, GA and R Mutharasan, Detection of pathogen *Escherichia coli* O157:H7 using self-excited PZT-glass microcantilevers. *Biosensors & Bioelectronics*, 2005, **21**, 462–473.

Corel Corporation, Corel Quattro Pro 8.0, Ottawa, Canada, 1997.

Deisingh, A, Biosensors for microbial detection. *Microbiologist*, 2003, 30–33.

Donelly, CW, *Listeria monocytogenes*: a continual challenge, *Nutrition Reviews*, 2001, **59**, 183–194.

Emsley, P, C Fotinou, I Black, NF Fairweathers, IG Charles, C Watts, E Hewitt, and NW Isaacs, *Journal of Biological Chemistry*, 2000, **275**, 8889–8894.

ERS, Economic of foodborne disease. http://ers.usda.gov/Briefing/Foodborne Disease, 2003.

Fotinou, G, P Emsley, I Blacj, H Ando, H Ishida, M Kiso, KA Sinha, NF Fairweathers, and NW Isaacs. *Journal of Biological Chemistry*, 2001, **276**, 32274–32281.

Gubler, DJ. Dengue and dengue hemorrhagic fever: its history and resurgence as a global health problem, in *Dengue and Dengue Hemorrhagic Fever* (eds. DJ Gubler and G Kuno), CAB International, Cambridge, UK, 1997, pp. 1–22.

Havlin, S, Molecular diffusion and reaction, in *The Fractal Approach to Heterogenous Chemistry: Surfaces, Colloids, Polymers* (ed. D. Avnir), Wiley, New York, 1989, pp. 251–269.

Horner, SR, CR Mace, LJ Rothberg, and BL Miller, A proteomic biosensor for enteropathogenic *E. coli*. *Biosensors & Bioelectronics*, 2006, **21**, 1283–1290.

Jay, JM, *Modern Food Microbiology*, Aspen Publication, Colorado, USA, 2000.

Jones, D and H Seeliger, The genus *Listeria*, in *The Parokaryites* (eds. CA Balows, HG Truper, M Dworkin, W, Harder, KH Schleur), 2nd Edn., Springer-Verlag, Heidelberg, 1992, pp. 1595–1616.

Karasinski, J, S Andreescu, OA Sadik, B Lavine, and MN Vora, Multiarray sensors with pattern recognition for the detection, classification, and differentiation of bacteria at subspecies and strain levels. *Analytical Chemistry*, 2005, **77**, 7941–7949.

Ko, S and SA Grant, A novel FRET-based optical fiber biosensor for rapid detection of *Salmonella typhimurium*. *Biosensors & Bioelectronics*, 2006, **21**, 1283–1290.

Leonard, P, S Hearty, J Quinn, and R O'Kennedy, A generic approach for the detection of whole *Listeria monocytogenes* cells in contaminated samples using surface plasmon resonance. *Biosensors & Bioelectronics*, 2004, **19**, 1331–1335.

Lowy, FD, *Staphylococcus aureus* infections. *New England Journal of Medicine*, 1998, **339**, 520–532.

Martin, SJ, VE Granstaff, and GC Frye, Effect of surface roughness on the response of thickness-shear mode resonators in liquids. *Analytical Chemistry*, 1991, **65**, 2910–2922.

Mead, PS, L Slutsker, V Dietz, LF McCaige, JS Bresse, C Shapiro, PM Griffin, and RV Tauxe, Food related illness and death in the United States. *Emerging Infectious Diseases*, 1999, **5**, 520–532.

Moats, RK and BM Sullivan, Combinatorial augmentation for a multi-pathogen biosensor: signal analysis and design. *Biosensors & Bioelectronics*, 2004, **19**, 1673–1683.

Ngundi, MM, CR Taitt, SA Mc Murry, D Kahne, and FS Ligler, Detection of bacterial toxins with monosaccharide arrays. *Biosensors & Bioelectronics*, 2006, **21**, 1195–1201.

Ramakrishnan, A and A Sadana, A single-fractal analysis of cellular analyte-receptor binding kinetics utilizing biosensors. *Biosystems*, 2001, **59**(1), 35–45.

Rijal, K, A Leung, P Mohana Shankar, and R Mutharasan, Detection of pathogen *Escherichia coli* O157:H7 at 70 cells/ml using antibody-immobilized biconal tapered fiber sensors. *Biosensors & Bioelectronics*, 2005, **21**, 871–880.

Sadana, A, A fractal analysis approach for the evaluation of hybridization kinetics in biosensors. *Journal of Colloid and Interface Science*, 2001, **234**, 9–18.

Sadana, A, Fractal binding and dissociation kinetics for different biosensor applications. Elsevier, Amsterdam, 2005.

Schlech III, WF, Foodborne listeriosis. *Clinical and Infectious Diseases*, 2000, **31**, 770–775.

Sigmaplot, Scientific Graphic Software, *User's Manual*. Jandel Scientific, San Rafael, CA, 1993.

Subramanian, A, J Irudayaraj, and T Ryan, A mixed self-assembled monolayer-based surface plasmon resonance immunosensor for detection of *E. coli* O157:H7. *Biosensors & Bioelectronics*, 2006, **21**, 998–1006.

World Health Organization, Strengthening the implementation of the global strategy for dengue fever/dengue hemorrhagic fever prevention and control, Geneva, Switzerland, 2000.

Zang, RG, ML Westbrook, EM Westbrrok, DL Scott, Z Otwinowski, PR Maulik, RA Reed, and GG Shipley. *Journal of Molecular Biology*, 1995, **251**, 550–562.

Zaytseva, NV, RA Montagna, and AJ Baeumner, Microfluidic biosensor for the serotype-specific detection of Dengue virus RNA. *Analytical Chemistry*, 2005, **77**, 7520–7527.

Zhu, P, DR Shelton, JS Karns, A Sundaram, S Li, P Amstutz, and CM Tang, Detection of water-borne *E. coli* O157:H7 using the integrating waveguide biosensor. *Biosensors & Bioelectronics*, 2005, **21**, 687–683.

Zourob, M, S Mohr, BJT Brown, PR Fielden, MB McDonell, and NJ Goddard, Bacteria detection using disposable optical leaky waveguide sensors. *Biosensors & Bioelectronics*, 2005, **21**, 293–302.

– 5 –

Fractal Binding and Dissociation Kinetics of Disease-Related Compounds on Biosensor Surfaces

5.1 INTRODUCTION

The first known use of the biosensor was the detection of glucose levels in blood for the better management of diabetes. Over the years due to the ease of usage, the reliability, and the precision of results, biosensors are being used increasingly in different areas of application. Nevertheless, the detection of different compounds for the onset or management of diseases still remains a predominant area of application. McCord *et al.* (2001) emphasize the early detection of serum markers that would help medical personnel to make accurate and reliable diagnoses that helps improve prognosis for patients, especially with life-threatening diseases.

Some of the current applications of biosensor use for the 'early' detection of analytes that indicate the onset of diseases include:

(a) Detection of C-reactive protein (CRP) and other cardiac markers in human plasma for acute myocardial infarction (AMI) (Wolf *et al.*, 2004).
(b) Detection of hepatitis B surface antigen by amperometric and potentiometric immunosensors-based gold nanoparticles (Tang *et al.*, 2005).
(c) Detection of integrin $\alpha_3\beta_3$ human umbilical endothelial cell (HUVEC) using a novel optical biosensor (Worsfold *et al.*, 2004). Hynes (1992) indicates that $\alpha_3\beta_3$ is a cell surface bound adhesion receptor. It plays an important role in cell–cell and cell–extracellular matrix interactions. Furthermore, Brooks *et al.* (2004) indicate that integrin $\alpha_3\beta_3$ is one of the targets for the inhibition of tumor growth.
(d) Detection of estrogenic endocrine disrupting chemicals (EDC) by luminescent yeast cells entrapped in hydrogels (Fine *et al.*, 2006). These authors indicate that the exposure to these chemicals may cause reproductive abnormalities and feminization of wildlife (Facemire *et al.*, 1995; Fry and Toone, 1981; Jobling *et al.*, 1996), and also possible reproductive disorders in humans (Carlsen *et al.*, 1992; Giwercman *et al.*, 1993; Raloff, 1993).

(e) Detection of tyrosine in biological fluids by an infrared optical sensor (Huang and Yang, 2005). These authors state that tyrosine is a precursor of compounds that are essential to the human body such as dopa, dopamine, catecholamine, melanin, liothyronine, and epinephrine (Greenstein and Winitz, 1961; Scriver *et al.*, 1995). Furthermore, the metabolism of tyrosine is involved in atherosclerosis (Heitzer *et al.*, 2001), Parkinson's disease (Offen *et al.*, 1999), lung diseases (Ischiropoulos *et al.*, 1995), and liver diseases (Sherlock, 1989).

(f) Detection of pentamer and modified CRP using a surface plasmon resonance (SPR) biosensor (Hu *et al.*, 2006). These authors indicate that CRP exists in two distinct forms: pentamer (or native) CRP (pCRP) and modified CRP (mCRP).

(g) Detection of antiglutamic acid decarboxylase (anti-GAD) antibody using SPR (Lee *et al.*, 2005). Baekkeskov *et al.* (1990) indicate that GAD (an auto-antigen) is one of the major markers for the detection of insulin-dependent diabetes mellitus (IDDM).

In this chapter we analyze the influence of the degree of heterogeneity on the biosensor surface on the binding and dissociation (if applicable) kinetics for the above-mentioned reactions. The intent is to obtain valuable insights into these reactions especially as far as biosensor performance parameters are concerned. For example, we analyze the influence of the degree of heterogeneity or the fractal dimension, D_f on the binding and dissociation rate coefficients, and affinity values. It is hoped that one may be able to manipulate these values by changing the degree of heterogeneity on the biosensor surface. The analysis to be presented in this chapter and in other chapters in this book should help complement the experimental results and the theoretical procedures for evaluating the binding and dissociation rate coefficient constants obtained, for example by the Biacore (2002) software that comes along with their SPR biosensor.

5.2 THEORY

Havlin (1989) has reviewed and analyzed the diffusion of reactants towards fractal surfaces. The details of the theory and the equations involved for the binding and the dissociation phases for analyte–receptor binding are available (Sadana, 2001). The details are not repeated here except that just the equations are given to permit an easier reading. These equations have been applied to other biosensor systems (Sadana, 2001; Ramakrishnan and Sadana, 2001; Sadana, 2005). For most applications, a single- or a dual-fractal analysis is often adequate to describe the binding and the dissociation kinetics. Peculiarities in the values of the binding and the dissociation rate coefficients, as well as in the values of the fractal dimensions with regard to the dilute analyte systems being analyzed will be carefully noted, if applicable.

5.2.1 Single-fractal analysis

Binding rate coefficient

Havlin (1989) indicates that the diffusion of a particle (analyte [Ag]) from a homogeneous solution to a solid surface (e.g., receptor [Ab]-coated surface) on which it reacts to form a product (analyte–receptor complex; (Ab·Ag)) is given by

$$(\text{Ab}\cdot\text{Ag}) \approx \begin{cases} t^{(3-D_{\text{f,bind}})/2} = t^p & t < t_c \\ t^{1/2} & t < t_c \end{cases} \tag{5.1}$$

Here $D_{\text{f,bind}}$ or D_f (used later on in the chapter) is the fractal dimension of the surface during the binding step. t_c is the cross-over value. Havlin (1989) indicates that the cross-over value may be determined by $r_c^2 \sim t_c$. Above the characteristic length, r_c, the self-similarity of the surface is lost and the surface may be considered homogeneous. Above time, t_c the surface may be considered homogeneous, since the self-similarity property disappears, and 'regular' diffusion is now present. For a homogeneous surface where D_f is equal to 2, and when only diffusional limitations are present, $p = 1/2$ as it should be. Another way of looking at the $p = 1/2$ case (where $D_{\text{f,bind}}$ is equal to two) is that the analyte in solution views the fractal object, in our case, the receptor-coated biosensor surface, from a 'large distance.' In essence, in the association process, the diffusion of the analyte from the solution to the receptor surface creates a depletion layer of width $(Đt)^{1/2}$ where $Đ$ is the diffusion constant. This gives rise to the fractal power law, $(\text{Analyte} - \text{Receptor}) \sim t^{(3-D_{\text{f,bind}})/2}$. For the present analysis, t_c is arbitrarily chosen and we assume that the value of the t_c is not reached. One may consider the approach as an intermediate 'heuristic' approach that may be used in the future to develop an autonomous (and not time-dependent) model for diffusion-controlled kinetics.

Dissociation rate coefficient

The diffusion of the dissociated particle (receptor [Ab] or analyte [Ag]) from the solid surface (e.g., analyte [Ag]-receptor [Ab]) complex coated surface) into solution may be given, as a first approximation by

$$(\text{Ab}\cdot\text{Ag}) \approx -t^{(3-D_{\text{f,diss}})/2} = -t^p \quad (t > t_{\text{diss}}) \tag{5.2}$$

Here $D_{\text{f,diss}}$ is the fractal dimension of the surface for the dissociation step. This corresponds to the highest concentration of the analyte–receptor complex on the surface. Henceforth, its concentration only decreases. The dissociation kinetics may be analyzed in a manner 'similar' to the binding kinetics.

5.2.2 Dual-fractal analysis

Binding rate coefficient

Sometimes, the binding curve exhibits complexities and two parameters (k, D_f) are not sufficient to adequately describe the binding kinetics. This is further corroborated by low values of r^2 factor (goodness of fit). In that case, one resorts to a dual-fractal analysis (four parameters; k_1, k_2, D_{f1}, and D_{f2}) to adequately describe the binding kinetics. The single-fractal analysis presented above is thus extended to include two fractal dimensions. At present, the time ($t = t_1$) at which the 'first' fractal dimension 'changes' to the 'second' fractal dimension is arbitrary and empirical. For the most part, it is dictated by the data analyzed and experience gained by handling a single-fractal analysis. A smoother curve is obtained in the 'transition' region, if care is taken to select the correct number of points for the two regions. In this case, the product (antibody–antigen; or analyte–receptor complex, Ab·Ag or analyte–receptor) is given by

$$(\mathrm{Ab}\cdot\mathrm{Ag}) \approx \begin{cases} t^{(3-D_{\mathrm{f1,bind}})/2} = t^{p1} & (t < t_1) \\ t^{(3-D_{\mathrm{f2,bind}})/2} = t^{p2} & (t_1 < t < t_2 = t_\mathrm{c}) \\ t^{1/2} & (t > t_\mathrm{c}) \end{cases} \tag{5.3}$$

In some cases, as mentioned above, a triple-fractal analysis with six parameters (k_1, k_2, k_3, D_{f1}, D_{f2}, and D_{f3}) may be required to adequately model the binding kinetics. This is when the binding curve exhibits convolutions and complexities in its shape due to perhaps to the very dilute nature of the analyte (in some of the cases to be presented) or for some other reasons. Also, in some cases, a dual-fractal analysis may be required to describe the dissociation kinetics.

5.3 RESULTS

The fractal analysis will be applied to different analyte–receptor reactions occurring on biosensor chip surfaces with the specific purpose of trying to better understand the binding and dissociation kinetics of disease-related compounds occurring on biosensor surfaces. The intent is to provide a better and more accurate diagnosis, which should eventually lead to a better prognosis for at least a few diseases. Needless to say any assistance obtained, no matter how small, for the management of life-threatening diseases should prove invaluable.

Wolf *et al.* (2004) have recently shown a proof-of-concept rapidly screen markers for AMI. Using micromosaic immunoassays and self-regulating microfluidic networks these authors have detected CRP and other cardiac markers in human plasma. These authors were interested in simultaneously detecting cardiac markers such as myoglobin (Mb), cardiac Troponin I (cTnI), B-type natriuretic peptide (BNP) and S100α (S100A1), and CRP in plasma. Pleabani and Zaninotto (1998) indicate that Mb is found after skeletal muscle or myocardial cells are damaged. Adams *et al.* (1993) indicate that cardiac TnI is a highly specific marker for myocardium damages. It has become a standard marker for damages of the myocardium (ESC/ACC, 2000). Maisel *et al.* (2002) indicate that BNP is useful for the emergency diagnosis of heart failure, and for prognosis in patients with acute coronary syndrome (ACS) (Brett *et al.*, 2001).

Figure 5.1a shows the binding of 0.1 μg/ml of αCRP in PBS in solution to capture antibody of αCRP immobilized on a silica surface in a self-regulating microfluidic network (μFN). A single-fractal analysis is adequate to describe the binding kinetics. The values of the binding rate coefficient, k and the fractal dimension, D_f are given in Table 5.1. The values of the binding rate coefficients and the fractal dimensions for the binding phase were obtained using Corel Quattro Pro 8.0 (1997). The binding rate coefficients obtained are

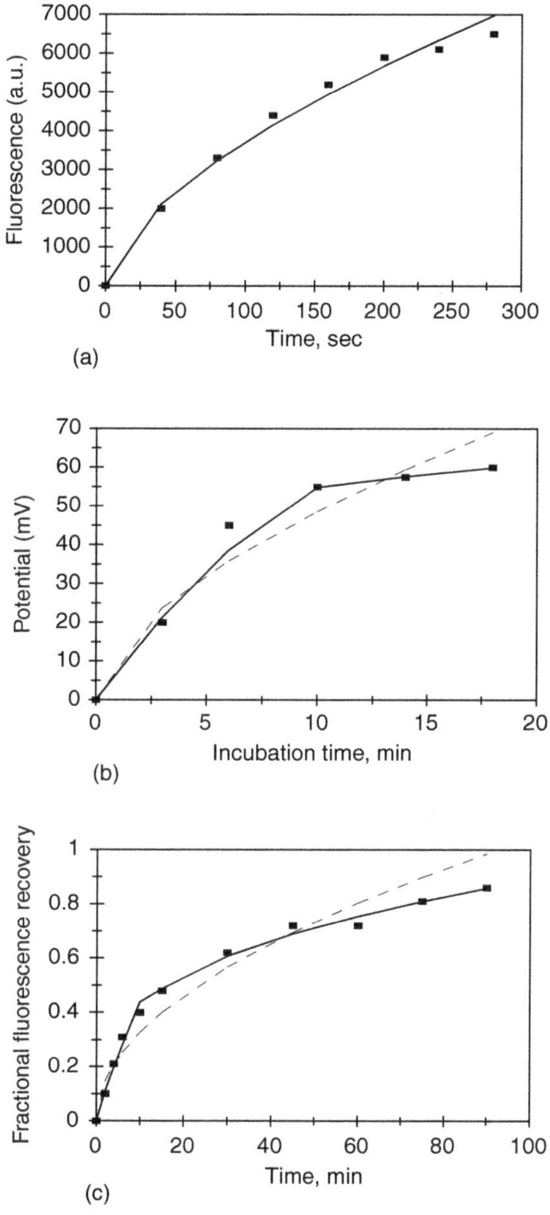

Figure 5.1 Binding of (a) C-reactive protein (CRP) in solution to capture antibody of CRP immobilized on a silica surface in a self-regulating microfluidic network (Wolf *et al.*, 2004). (b) Hepatitis B surface antigen in solution to the hepatitis B surface antibody electrostatically adsorbed on gold nanoparticles (tris(2,2′-bipyridyl) cobalt multilayer film (Tang *et al.*, 2005). (c) Integrin $\alpha_v\beta_3$ loaded HUVEC in solution to arginine–glycine–aspartate (RGD)-containing peptide attached to BODIPY (acceptor nanoparticles lipid dye Worsfold *et al.*, 2004).

Table 5.1

Binding rate coefficients and fractal dimensions for the (a) binding of C-reactive protein (CRP) to capture antibody of CRP immobilized on a silica surface in a self-regulating microfluidic network (Wolf *et al.*, 2004); (b) binding of hepatitis B surface antigen to hepatitis B surface antibody electrostatically adsorbed on gold nanoparticles/tris (2,2′-bipyridyl) cobalt multilayer film (Tang *et al.*, 2005); and (c) binding of integrin $\alpha_v\beta_3$ loaded HUVEC to arginine–glycine–aspartate (RGD)-containing peptide covalently attached to BODIPY (acceptor lipid dye) (Worsfold *et al.*, 2004)

Analyte in solution/receptor on surface	k	k_1	k_2	D_f	D_{f1}	D_{f2}	Reference
CRP/capture antibody of CRP	220.18 ± 13.41	na	na	1.7734 ± 0.0704	na	na	Wolf *et al.* (2004)
Hepatitis B surface antigen/ hepatitis B surface antibody electrostatically adsorbed on nanoparticles	12.377 ± 2.698	8.297 ± 1.695	39.133 ± 0.171	1.8106 ± 0.2758	1.1816 ± 0.4350	2.7056 ± 0.0210	Tang *et al.* (2005)
Integrin $\alpha_v\beta_3$ loaded HUVEC/ RGD containing peptide covalently attached to BODIPY (acceptor lipid dye)	0.1028 ± 0.0223	0.05825 ± 0.00727	0.2089 ± 0.0075	1.9956 ± 0.09842	1.2438 ± 0.1997	2.3724 ± 0.04694	Worsfold *et al.* (2004)

within 95% confidence limits. For example, for the binding of 0.1 μg/ml of αCRP in PBS in solution to capture antibody of αCRP immobilized on a silica surface, the binding rate coefficient, k is 220.18 ± 13.41. The 95% confidence limit indicates that the binding rate coefficient, k value lies between 206.77 and 233.59. This indicates hat the value is precise and significant.

Tang *et al.* (2005) have recently developed a fast, sensitive, and novel amperometric and potentiometric immunosensor based on gold nanoparticles/tris (2,2′-bipyridyl) cobalt (II) multilayer films for the detection of hepatitis B surface antigen. These authors indicate that their layer-by-layer (LBL)-modified immunosensor was easily constructed and effective in detecting HBsAg (hepatitis B surface antigen). They emphasize that the LBL technique is versatile, and has been used to assemble mono- or multilayers of proteins onto oppositely charged substrates (Yuri *et al.*, 1995; Wang *et al.*, 2002a, 2002b).

Figure 5.1b shows the binding of hepatitis B surface antigen in solution to the hepatitis B surface antibody electrostatically adsorbed on gold nanoparticles/tris (2,2′-bipyridyl) cobalt (II) multilayer film. A dual-fractal analysis is required to adequately describe the binding kinetics. The values of (a) the binding rate coefficient, k and the fractal dimension, D_f for a single-fractal analysis, and (b) the binding rate coefficients, k_1 and k_2, and the fractal dimensions, D_{f1} and D_{f2} for a dual-fractal analysis are given in Table 5.1. Note that for a dual-fractal analysis an increase in the fractal dimension by a factor of 2.29 from a value of D_{f1} equal to 1.1816 to D_{f2} equal to 2.7056 leads to an increase in the binding rate coefficient by a factor of 4.72 from a value of k_1 equal to 8.297 to k_2 equal to 39.13. Note that changes in the degree of heterogeneity on the sensor surface or the fractal dimension on the sensor surface and in the binding rate coefficient are in the same direction.

Worsfold *et al.* (2004) have developed a novel optical bionanosensor for the detection of $\alpha_v\beta_3$ loaded HUVEC to argine–glycine—aspartate (RGD)-containing peptide covalently attached to a BODIPY (donor) lipid dye. The authors indicate that the novel optical bionanosensor platform used a supported lipid membrane (SBLM). This membrane has generic multi-analyte sensing properties. BODIPY 530/550-donor is 2-(4,4-difluoro-5, 7-diphenyl-4-bora-3a, 4a-diaza-s-indacene-3-dodedecanoyl)-1-hexadecanoyl-*sn*-glycero-3-phosphoethanolamine. A second BODIPY (acceptor) lipid dye was integrated into the SBLM. This permitted signal amplification via a forster resonance energy transfer (FRET) mechanism. Worsfold *et al.* (2004) indicate that integrins are heterodimers comprising of α and β subunits.

Figure 5.1c shows the binding of integrin $\alpha_v\beta_3$ loaded HUVEC to RGD-containing peptide covalently attached to a BODIPY (donor) lipid dye used as an optical sensor. The BODIPY dye-RGD peptide labeled lipid bilayer was immobilized on a HMDS substrate. A dual-fractal analysis is required to adequately describe the binding kinetics. The values of (a) the binding rate coefficient, k and the fractal dimension, D_f for a single-fractal analysis, and (b) the binding rate coefficients, k_1 and k_2, and the fractal dimensions, D_{f1} and D_{f2} for a dual-fractal analysis are given in Table 5.1. Note that for a dual-fractal analysis an increase in the fractal dimension by a factor of 1.907 from a value of D_{f1} equal to 1.2438 to D_{f2} equal to 2.3724 leads to an increase in the binding rate coefficient by a factor of 3.59 from a value of k_1 equal to 0.05825 to k_2 equal to 0.2089. Note that changes in the degree of heterogeneity on the sensor surface or the fractal dimension on the sensor surface and in the binding rate coefficient are in the same direction.

Hu *et al.* (2006) have recently used SPR biosensing to detect pentamer and modified CRP. These authors indicate that structural modification of pentamer C-reactive protein (pCRP) produces modified CRP (mCRP). These authors indicate that mCRP exhibits different biological activities in the body. mCRP is now regarded as a more powerful inducer than pCRP for the assessment of risk of developing cardiovascular disease (CVD). These authors immobilized the monoclonal antibodies MabC8, Mab8D8, and Mab9C9 on a protein G layer to detect both pCRP and mCRP.

Figure 5.2a shows the binding and dissociation of pCRP in solution to MabC8 immobilized on a SPR biosensor chip surface. A single-fractal analysis is used to describe the binding kinetics. A dual-fractal analysis is required to describe the dissociation kinetics. The values of (a) the binding rate coefficient, k and the fractal dimension, D_f for a single-fractal analysis, and (b) the dissociation rate coefficient, k_d and the fractal dimension for the dissociation phase, D_{fd} for a single-fractal analysis, and (c) the dissociation rate coefficients, $k_{d1}$and k_{d2}, and the fractal dimensions, D_{fd1} and D_{fd2} for a dual-fractal analysis are given in Table 5.2.

Figure 5.2b shows the binding and dissociation of mCRP in solution to MabC8 immobilized on a SPR biosensor chip surface. A single-fractal analysis is used to describe the binding kinetics. A dual-fractal analysis is required to describe the dissociation kinetics. The values of (a) the binding rate coefficient, k and the fractal dimension, D_f for a single-fractal analysis, and (b) the dissociation rate coefficient, k_d and the fractal dimension for the dissociation phase, D_{fd}, for a single-fractal analysis, and (c) the dissociation rate coefficients, $k_{d1}$and k_{d2}, and the fractal dimensions, D_{fd1} and D_{fd2} for a dual-fractal analysis are given in Table 5.2.

Figure 5.2c shows the binding and dissociation of pCRP in solution to MabD8 immobilized on a SPR biosensor chip surface. A dual-fractal analysis is used to adequately describe the binding and dissociation kinetics. The values of (a) the binding rate coefficient, k and the fractal dimension, D_f for a single-fractal analysis, (b) the binding rate coefficients, k_1 and k_2 and the fractal dimensions D_{f1} and D_{f2} for a dual-fractal analysis, (c) the dissociation rate coefficient, k_d and the fractal dimension for the dissociation phase, D_{fd}, for a single-fractal analysis, and (d) the dissociation rate coefficients, $k_{d1}$and k_{d2}, and the fractal dimensions, D_{fd1} and D_{fd2} for a dual-fractal analysis are given in Table 5.2.

Note that for a dual-fractal analysis an increase in the fractal dimension by a factor of 1.466 from a value of D_{f1} equal to 2.0466 to D_{f2} equal to 3.0 leads to a decrease in the binding rate coefficient by a factor of 2.25 from a value of k_1 equal to 6.3238 to k_2 equal to 2.8156. Note that changes in the degree of heterogeneity on the sensor surface or the fractal dimension on the sensor surface and in the binding rate coefficient in this case are in opposite directions.

Figure 5.2d shows the binding and dissociation of mCRP in solution to Mab8D8 immobilized on a SPR biosensor chip surface. A single-fractal analysis is used to describe the binding kinetics. A dual-fractal analysis is required to describe the dissociation kinetics. The values of (a) the binding rate coefficient, k and the fractal dimension, D_f for a single-fractal analysis, and (b) the dissociation rate coefficient, k_d and the fractal dimension for the dissociation phase, D_{fd}, for a single-fractal analysis, and (c) the dissociation rate coefficients, k_{d1} and k_{d2}, and the fractal dimensions, D_{fd1} and D_{fd2} for a dual-fractal analysis are given in Table 5.2.

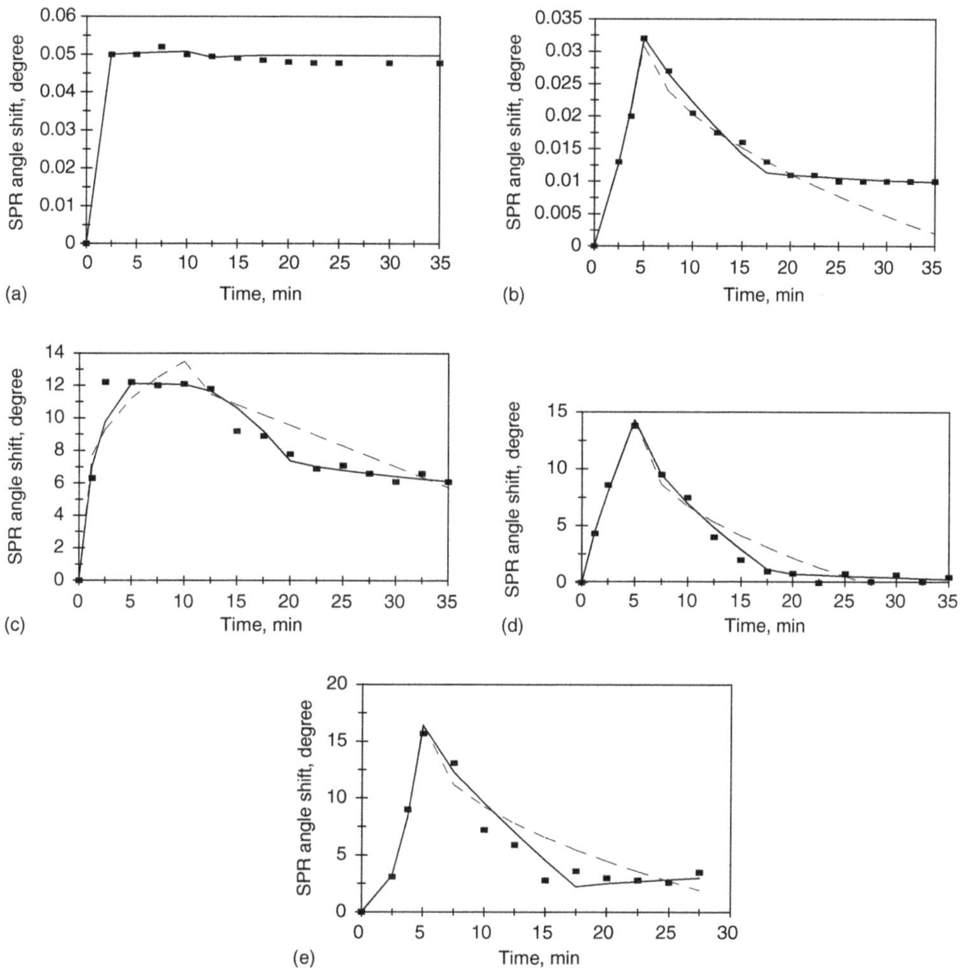

Figure 5.2 Binding of (a) pentamer C-reactive protein in solution to monoclonal antibody, C8 (MabC8) immobilized on a sensor chip surface; (b) modified C-reactive protein in solution to monoclonal antibody, C8 (MabC8) immobilized on a sensor chip surface; (c) pentamer C-reactive protein in solution to monoclonal antibody, D8 (MabD8) immobilized on a sensor chip surface; (d) modified C-reactive protein in solution to monoclonal antibody, D8 (MabD8) immobilized on a sensor chip surface; and (e) modified C-reactive protein in solution to monoclonal antibody, C9 (MabC9) immobilized on a sensor chip surface.

Figure 5.2e shows the binding and dissociation of mCRP in solution to Mab9C9 immobilized on a SPR biosensor chip surface. A single-fractal analysis is used to describe the binding kinetics. A dual-fractal analysis is required to describe the dissociation kinetics. The values of (a) the binding rate coefficient, k and the fractal dimension, D_f for a single-fractal

Table 5.2a

Binding and dissociation rate coefficients for the detection of pentamer C-reactive protein (pCRP) and modified C-reactive protein by monoclonal antibodies (mAbs) C8, 8D8, and CD9 immobilized on a protein G layer on a sensor chip surface (Hu *et al.*, 2006)

Analyte in solution/receptor on surface	k	k_1	k_2	k_d	k_{d1}	k_{d2}
pCRP/MabC8	0.04955 ± 0.0013	na	na	0.000843 ± 0.000490	0.00141 ± 0	0.00022 ± 0
mCRP/MabC8	0.003904 ± 0.000318	na	na	0.005032 ± 0.004102	0.002886 ± 0.00040	0.01717 ± 0.00024
pCRP/Mab8D8	7.2678 ± 1.7231	6.3238 ± 1.9585	2.8156 ± 0.1379	0.2539 ± 0.1468	0.09566 ± 0.06130	2.8156 ± 0.1379
mCRP/Mab8D8	3.6974 ± 0.3478	na	na	3.4148 ± 0.5448	2.3337 ± 0.2111	11.421 ± 0.312
mCRP/Mab9C9	0.3684 ± 0.0368	na	na	2.8445 ± 0.8698	1.5492 ± 0.4597	16.788 ± 0.386

Table 5.2b

Fractal dimensions for the binding and the dissociation phase during the detection of pentamer C-reactive protein (pCRP) and modified C-reactive protein by monoclonal antibodies (mAbs) C8, 8D8, and CD9 immobilized on a protein G layer on a sensor chip surface (Hu *et al.*, 2006)

Analyte in solution/receptor on surface	D_f	D_{f1}	D_{f2}	D_{fd}	D_{fd1}	D_{fd2}
pCRP/MabC8	2.9780 ± 0.04344	na	na	0 + 0.4162	0 + 1.0812	0 + 2E-13
mCRP/MabC8	0.4320 ± 0.3172	na	na	1.9450 + 0.4756	1.2888 ± 0.2862	2.8482 ± 0.04538
pCRP/Mab8D8	2.4610 ± 0.02518	2.0466 ± 0.5504	3.0 − 0.0134	0.8178 ± 0.4150	0 + 0.7790	2.5322 ± 0.1652
mCRP/Mab8D8	1.3178 ± 0.1834	na	na	2.0956 ± 0.1181	1.6622 ± 0.1169	2.024 ± 0.1679
mCRP/Mab9C9	0 ± 0.3856	na	na	1.9862 ± 0.2128	1.2870 ± 0.3508	0 + 0.2558

analysis, and (b) the dissociation rate coefficient, k_d and the fractal dimension for the dissociation phase, D_{fd}, for a single-fractal analysis, and (c) the dissociation rate coefficients, k_{d1} and k_{d2}, and the fractal dimensions, D_{fd1} and D_{fd2} for a dual-fractal analysis are given in Table 5.2.

Zeng *et al.* (2006) have recently developed a quartz crystal microbalance (QCM) immunosensor array for the clinical immunotyping of acute leukemias. These authors fabricated each QCM with a plasma-polymerized film (PPF) of *n*-butylamine, nanogold particles, and protein A (PA) to immobilize the antibodies in oriented form. Leukemic lineage-associated monoclonal antibodies were immobilized onto the nanogold-PA-modified surface of the QCM. These authors indicate that immunotyping with monoclonal (usually) antibodies is an important means to recognize the various differentiated antigens of leukocytes. Zeng *et al.* (2006) used their QCM array technique for the immunotyping of 120 human bone marrow (BM) samples. These included 96 samples of untreated acute leukemic patients and 24 normal human samples. These authors emphasize that their QCM array technique demonstrates increased sensing characteristics with regard to the response signal. Also, their technique is biocompatible and exhibits reduced nonspecific interference with different types of samples.

Figure 5.3a shows the binding of CD7 antigen expressed in nucleated cells from 96 acute leukemic patients to leukemic-associated monoclonal antibodies immobilized onto the nanogold-PA (protein A)-modified surface of a QCM immunosensor array. A dual-fractal analysis is required to adequately describe the binding kinetics. The values of (a) the binding rate coefficient, k and the fractal dimension, D_f for a single-fractal analysis, and (b) the binding rate coefficients, k_1 and k_2 and the fractal dimensions, D_{f1} and D_{f2} for a dual-fractal analysis are given in Table 5.3. Once again note that for a dual-fractal analysis an increase in the fractal dimension by a factor of 2.894 from a value of D_{f1} equal to 0.9048 to D_{f2} equal to 2.6188, there is an increase in the binding rate coefficient by a factor of 5.79 from a value of k_1 equal to 19.161 to k_2 equal to 110.988. Changes in the degree of heterogeneity or the fractal dimension on the QCM surface and in the binding rate coefficient are in the same direction.

Figure 5.3b shows the binding of CD3 antigen expressed in nucleated cells from 96 acute leukemic patients to leukemic-associated monoclonal antibodies immobilized onto the nanogold-PA (protein A)-modified surface of a QCM immunosensor array. A dual-fractal analysis is required to adequately describe the binding kinetics. The values of (a) the binding rate coefficient, k and the fractal dimension, D_f for a single-fractal analysis, and (b) the binding rate coefficients, k_1 and k_2 and the fractal dimensions, D_{f1} and D_{f2} for a dual-fractal analysis are given in Table 5.3. It is of interest to note that for a dual-fractal analysis an increase in the fractal dimension by a factor of 3.22 from a value of D_{f1} equal to 0.7904 to D_{f2} equal to 2.5466, there is an increase in the binding rate coefficient by a factor of 6.80 from a value of k_1 equal to 12.625 to k_2 equal to 85.837. Once again, changes in the degree of heterogeneity or the fractal dimension on the QCM surface and in the binding rate coefficient are in the same direction.

Figure 5.3c shows the binding of CD5 antigen expressed in nucleated cells from 96 acute leukemic patients to leukemic-associated monoclonal antibodies immobilized onto the nanogold-PA (protein A)-modified surface of a QCM immunosensor array. A dual-fractal analysis is required to adequately describe the binding kinetics. The values of (a) the binding rate coefficient, k and the fractal dimension, D_f for a single-fractal analysis, and (b) the binding rate coefficients, k_1 and k_2 and the fractal dimensions, D_{f1} and D_{f2} for a dual-fractal analysis are given in Table 5.3. Once again note that for a dual-fractal analysis an increase in the fractal dimension by a factor of 3.81 from a value of D_{f1} equal to 0.7874 to D_{f2} equal to 3.0, there is an increase in the binding rate coefficient by a factor of 6.09 from a value of k_1 equal to 7.5588 to k_2 equal to 46. Changes in the degree of heterogeneity or the fractal dimension on the QCM surface and in the binding rate coefficient are, once again, in the same direction.

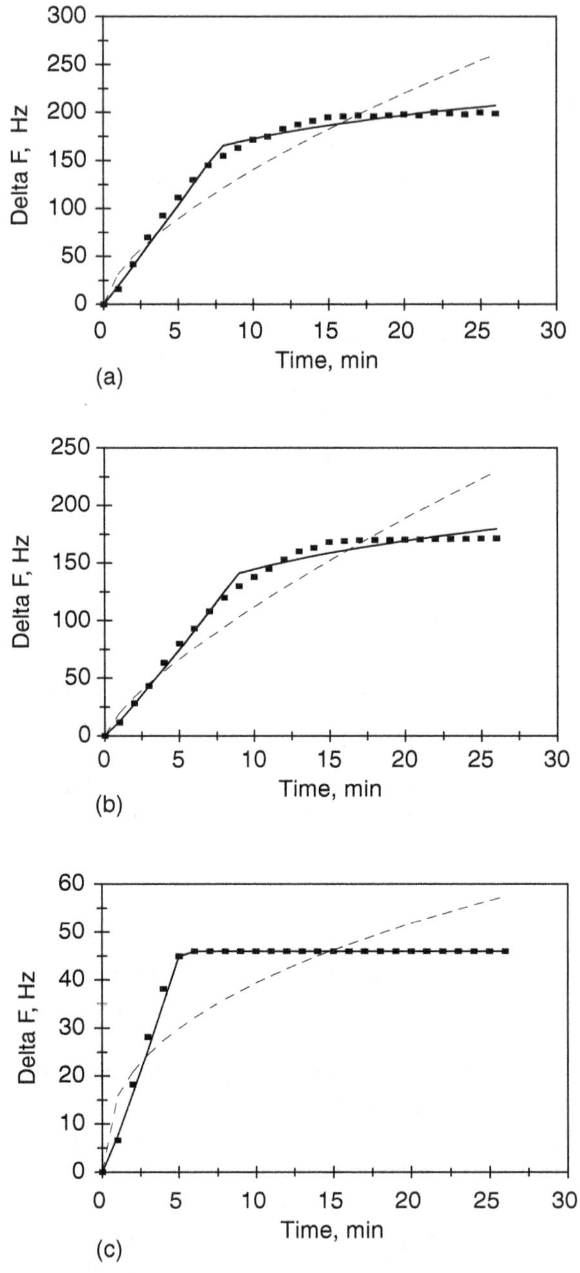

Figure 5.3 Binding of different antigens expressed in nucleated cells from 96 acute leukemic patients to lineage-associated monoclonal antibodies onto the nanogold protein A (PA)-modified surface of a QCM biosensor (Zeng *et al.*, 2006): (a) CD7, (b) CD3, (c) CD5.

Table 5.3

Binding rate coefficients and fractal dimensions of different antigens expressed in nucleated cells from 96 acute leukemic patients to lineage-associated monoclonal antibodies immobilized onto the nanogold protein A (PA)-modified surface of a QCM biosensor (Zeng *et al.*, 2006)

Analyte in solution	k	k_1	k_2	D_f	D_{f1}	D_{f2}
CD7	31.382 ± 7.916	19.161 ± 2.61	110.988 ± 3.586	1.6996 ± 0.1074	0.9048 ± 0.1255	2.6188 ± 0.0417
CD3	19.783 ± 4.598	12.625 ± 0.88	85.837 ± 3.614	1.4944 ± 0.0997	0.7904 ± 0.0662	2.5466 ± 0.0612
CD5	15.861 ± 4.79	7.5588 ± 1.1396	46 ± 0	2.2094 ± 0.1259	0.7874 ± 0.1895	3.0 – 4E-14

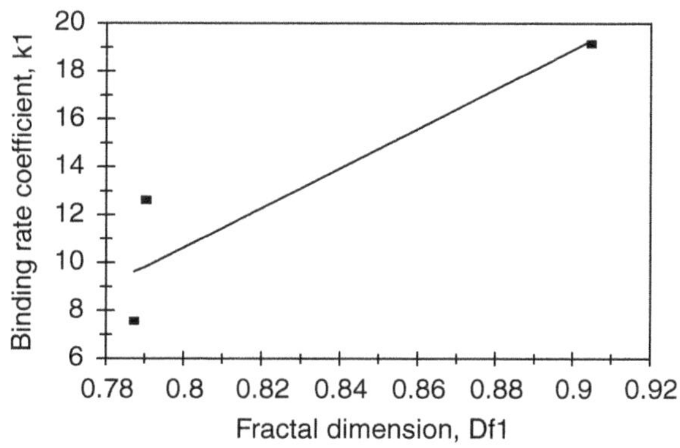

Figure 5.4 Increase in the binding rate coefficient, k_1 with an increase in the fractal dimension, D_{f1} for a dual-fractal analysis.

Figure 5.4 and Table 5.3 show for a dual-fractal analysis the increase in the binding rate coefficient, k_1 with an increase in the fractal dimension, D_{f1} or the degree of heterogeneity that exists on the QCM surface. For the data presented in Figure 5.4, the binding rate coefficient, k_1 is given by

$$k_1 = (31.782 \pm 13.292) D_{f1}^{4.989 \pm 3.121} \tag{5.4}$$

The fit is reasonable. There is scatter in the data. The availability of more data points would lead to a more reliable fit. The binding rate coefficient, k_1 is very sensitive to the degree of heterogeneity or the fractal dimension on the QCM surface as noted by the close to fifth (equal to 4.989) order of dependence exhibited.

Fine *et al.* (2006) have recently used luminescent yeast cells entrapped in hydrogels to detect EDCs. These authors entrapped genetically modified *Saccharomyces cerevisiae* cells containing the estrogen receptor alpha-mediated expression of the *luc* reporter gene in hydrogel matrices. They state that the organization for economic cooperation and development (OCED) defines an EDC as "an exogeneous substance that causes adverse health effects in an intact organism, or its progeny, consequent to changes in endocrine function." Witorsch (2002) indicates that the chemicals that produce estrogenic effects in some organisms include phthalates and alkylphenols, pharmaceutical agents such as diethylstilbestrol (DES) and pesticides (for example, kepone and methoxychlor).

Figure 5.5a shows the binding and dissociation of 17-β-estradiol (β –E_2) in solution to genetically modified *S. cerevisiae* cells in a calcium alginate hydrogel matrix. A single-fractal analysis is adequate to describe the binding and dissociation kinetics. The values of (a) the binding rate coefficient, k and the fractal dimension, D_f for the binding phase, and (b) the dissociation rate coefficient, k_d and the fractal dimension, D_{fd} for the dissociation phase are given in Tables 5.4a and 5.4b.

Figure 5.5b shows the binding of 10 nM 17-β-estradiol (β –E_2) in solution to genetically modified *S. cerevisiae* cells in a polyvinyl alcohol (PVA)-based hydrogel matrix.

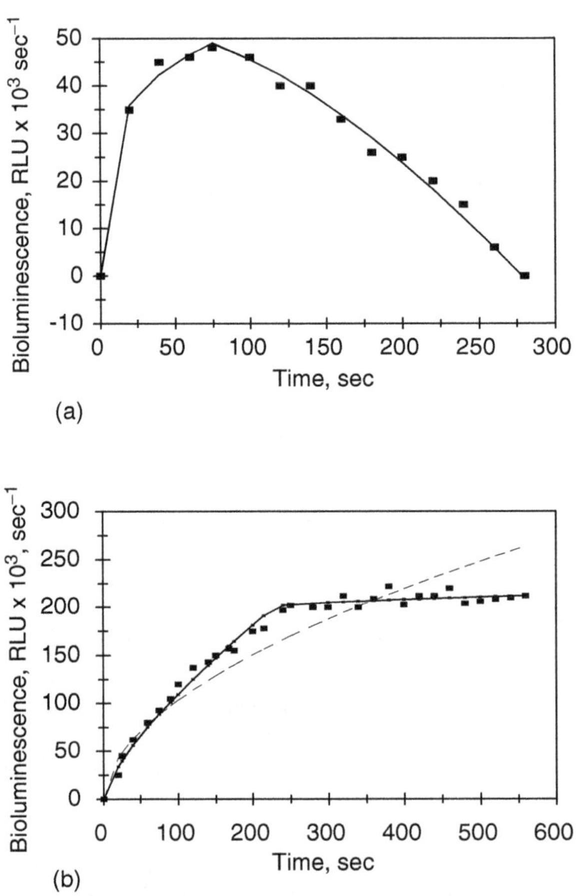

Figure 5.5 Binding and dissociation of 17-β estradiol (β-E2) to genetically modified *Saccharomyces cerevisiae* cells in (a) calcium alginate beads (b) polyvinyl (PVA) hydrogels (Fine *et al.*, 2006).

Table 5.4a

Binding and dissociation rate coefficients for (a) 17-β estradiol (β-E2) to genetically modified *Saccharomyces cerevisiae* cells in calcium alginate beads and in polyvinyl (PVA) hydrogels (Fine *et al.*, 2006), and (b) binding and dissociation rate coefficients for β-E2 induced alginate beads after 4 weeks of storage in slow freeze conditions at −80 °C and in fast freezing conditions (liquid nitrogen treatment) at −80 °C (Fine *et al.*, 2006)

Experimental conditions	k	k_1	k_2	k_d	k_{d1}	k_{d2}
Calcium alginate beads	17.778 ± 0.923	na	na	0.02561 ± 0.00478	na	na
Polyvinyl alcohol (PVA)	8.4402 ± 1.4991	3.8192 ± 0.4271	148.12 ± 4.50	na	na	na
Slow freeze conditions	7.326 ± 1.093	1.1052 ± 0.2288	24.171 ± 0.874	0.5793 ± 0.1468	0.3481 ± 0.0875	67.763 ± 0.456
Fast freeze conditions	3.2605 ± 0.4145	2.2202 ± 0.3040	92.880 ± 0.463	0.01403 ± 0.0080	0.004372 ± 0.003506	1.3002 ± 0.0324

Table 5.4b

Fractal dimensions for the binding and the dissociation phase for the binding and dissociation of (a) 17-β estradiol (β-E2) to genetically modified *Saccharomyces cerevisiae* cells in calcium alginate beads and in polyvinyl (PVA) hydrogels (Fine *et al.*, 2006), and (b) β-E2 induced alginate beads after 4 weeks of storage in slow freeze conditions at −80 °C and in fast freezing conditions (liquid nitrogen treatment) at −80 °C (Fine *et al.*, 2006)

Experimental conditions	D_f	D_{f1}	D_{f2}	D_{fd}	D_{fd1}	D_{fd2}
Calcium alginate beads	2.5298 ± 0.1006	na	na	0.02561 ± 0.00478	na	na
Polyvinyl alcohol (PVA)	1.9122 ± 0.0657	1.5430 ± 0.06968	2.8863 ± 0.06290	na	na	na
Slow freeze conditions	1.7582 ± 0.1476	0.8968 ± 0.5858	2.1956 ± 0.07486	0.5793 ± 0.1468	1.0164 ± 0.1721	2.7128 ± 0.0427
Fast freeze conditions	1.6176 ± 0.1270	1.4470 ± 0.1895	2.8042 ± 0.04024	0.01403 ± 0.0080	0. + 0.5632	1.6858 ± 0.09190

A dual-fractal analysis is required to adequately describe the binding kinetics. The values of (a) the binding rate coefficient, k and the fractal dimension, D_f for the binding phase for a single-fractal analysis, and (b) the binding rate coefficients, k_1 and k_2 and the fractal dimensions, D_{f1} and D_{f2}, for a dual-fractal analysis are given in Tables 5.4a and 5.4b. It is of interest to note that as the fractal dimension in the binding phase for a dual-fractal analysis increases by a factor of 1.87 from a value of D_{f1} equal to 1.5430 to D_{f2} equal to 2.8863, the binding rate coefficient increases by factor of 38.72 from a value of k_1 equal to 3.8192 to k_2 equal to 148.12. Changes in the degree of heterogeneity or the fractal dimension and in the binding rate coefficient are in the same direction.

Figure 5.6a shows the binding of 10 nM 17-β-estradiol (β –E_2) to genetically modified *Saccharomyces cerevisiae* cells in solution to alginate beads after 4 weeks of storage in –80 °C in slow freeze conditions. A dual-fractal analysis is required to adequately describe the binding kinetics. The values of (a) the binding rate coefficient, k and the fractal dimension, D_f for the binding phase for a single-fractal analysis, and (b) the binding rate coefficients, k_1 and k_2 and the fractal dimensions, D_{f1} and D_{f2}, for a dual-fractal analysis are given in Tables 5.4a and 5.4b. It is of interest to note that as the fractal dimension for a dual-fractal analysis increases by a factor of 2.45 from a value of D_{f1} equal to 0.8968 to D_{f2} equal to 2.1956, the binding rate coefficient increases by a factor of 21.87 from a value of k_1 equal to 1.1052 to k_2 equal to 24.17. Changes in the degree of heterogeneity or the fractal dimension and in the binding rate coefficient are once again in the same direction.

Figure 5.6b shows the binding of 10 nM 17-β-estradiol (β–E_2) to genetically modified *Saccharomyces cerevisiae* cells in solution to alginate beads after 4 weeks of storage in

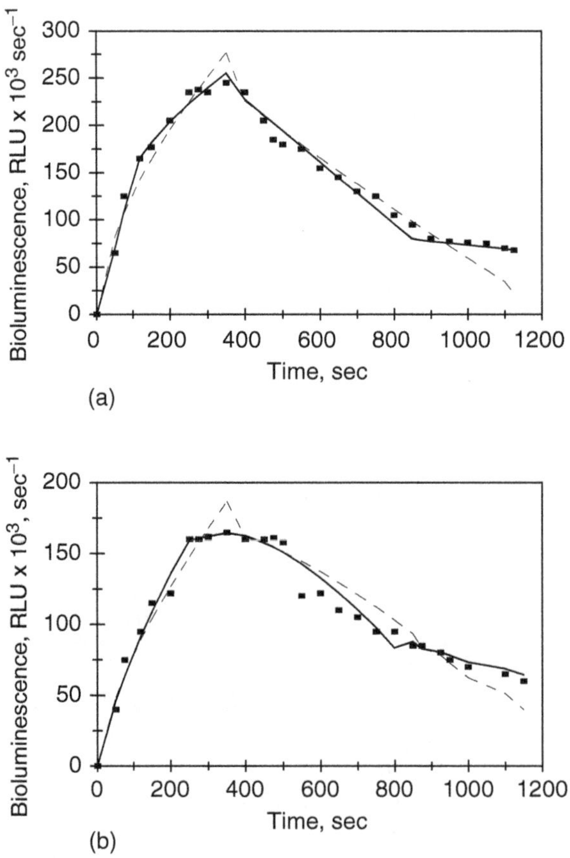

Figure 5.6 Binding and dissociation of 17-β estradiol (β-E2) to genetically modified *Saccharomyces cerevisiae* cells in alginate beads after 4 weeks of storage in: (a) slow freeze conditions (−80 °C); (b) fast freezing conditions (liquid nitrogen treatment) at −80 °C (Fine *et al.*, 2006).

$-80\,°C$ at fast freeze conditions (liquid nitrogen treatment). A dual-fractal analysis is required to adequately describe the binding kinetics. The values of (a) the binding rate coefficient, k and the fractal dimension, D_f for the binding phase for a single-fractal analysis, and (b) the binding rate coefficients, k_1 and k_2 and the fractal dimensions, D_{f1} and D_{f2}, for a dual-fractal analysis are given in Tables 5.4a and 5.4b. It is of interest to note that as the fractal dimension for a dual-fractal analysis increases by a factor of 1.94 from a value of D_{f1} equal to 1.4470 to D_{f2} equal to 2.8042, the binding rate coefficient increases by factor of 41.83 from a value of k_1 equal to 2.2202 to k_2 equal to 92.880. Changes in the degree of heterogeneity or the fractal dimension and in the binding rate coefficient are once again, as above, in the same direction.

Figure 5.7a and Tables 5.4a and 5.4b show for a dual-fractal analysis the increase in the binding rate coefficient, k_1 with an increase in the fractal dimension, D_{f1}. For the data plotted in Figure 5.7a the binding rate coefficient, k_1 is given by

$$k_1 = (1.335 \pm 0.359) D_{f1}^{1.9744 \pm 0.705} \tag{5.5a}$$

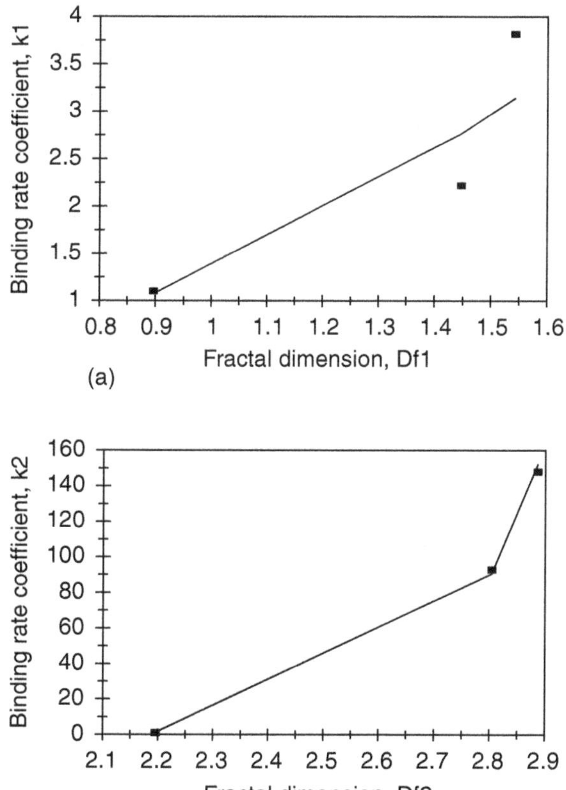

Figure 5.7 (a) Increase in the binding rate coefficient, k_1 with an increase in the fractal dimension, D_{f1}. (b) Increase in the binding rate coefficient, k_2 with an increase in the fractal dimension, D_{f2}.

The fit is reasonable. Only three data points are available. The availability of more data points would lead to a more reliable fit. The binding rate coefficient, k_1 exhibits close to a second (equal to 1.9744) order of dependence on the fractal dimension or the degree of heterogeneity that exists on the sensor surface.

Figure 5.7b and Tables 5.4a and 5.4b show for a dual-fractal analysis the increase in the binding rate coefficient, k_2 with an increase in the fractal dimension, D_{f2}. For the data plotted in Figure 5.7a the binding rate coefficient, k_2 is given by

$$k_2 = (8\text{E-}07 \pm 0.3\text{E-}07)\, D_{f2}^{17.984 \pm 0.1699} \tag{5.5b}$$

The fit is good. Only three data points are available. The availability of more data points would lead to a more reliable fit. The binding rate coefficient, k_2 exhibits close to an eighteenth (equal to 17.984) order of dependence on the fractal dimension or the degree of heterogeneity that exists on the sensor surface.

Huang and Yang (2005) have developed an infrared optical sensor for the detection of tyrosine in biological fluids. They indicate that tyrosine is a semi-essential amino acid that can be biosynthesized through the hydroxylation of phenylalanine. These authors have proposed a new IT spectroscopic sensing method that uses evanescent waves. Their method uses a proline-modified phase. The first step is the pre-equilibrium of the proline phase with copper ions to form a complex, which is unstable due to steric hindrance. In step two this phase is used to form a more stable proline-Cu^{2+} tyrosine complex wherein the less stable proline ligands are replaced by tyrosine units. The IR radiation detects this tyrosine that is attached to the surface of the sensing phase. Note that the common IR spectroscopic method to detect an analyte in solution uses the attenuated total reflection (ATR) technique (Harrick, 1967).

Figure 5.8a shows the binding of 1 mM tyrosine in 75 mM ammonia buffer and in the presence of 0.5 mM copper ions to a proline-modified sensing phase on the surface of the internal reflection element of the sensor (Huang and Yang, 2005). A dual-fractal analysis is required to adequately describe the binding kinetics. The values of (a) the binding rate coefficient, k and the fractal dimension, D_f for a single-fractal analysis, and (b) the binding rate coefficients, k_1 and k_2, and the fractal dimensions, D_{f1} and D_{f2}, for a dual-fractal analysis are given in Table 5.5. It is of interest to note for a dual-fractal analysis that as the fractal dimension increases by a factor of 1.27 from a value of D_{f1} equal to 2.192 to D_{f2} equal to 2.7856, the binding rate coefficient increases by a factor of 1.73 from a value of k_1 equal to 28.051 to k_2 equal to 48.692. Changes in the degree of heterogeneity or the fractal dimension on the sensing surface and in the binding rate coefficient are in the same direction.

Figure 5.8b shows the binding of 1 mM tyrosine in 75 mM ammonia buffer and in the presence of 0.5 mM nickel ions to a proline-modified sensing phase on the surface of an internal reflection element of the sensor (Huang and Yang, 2005). A single-fractal analysis is adequate to describe the binding kinetics. The values of the binding rate coefficient, k and the fractal dimension, D_f are given in Table 5.5. In this case, the binding mechanism is not as complex as is the case for copper ions, since here a single-fractal analysis is adequate to describe the binding kinetics, whereas for copper ions a dual-fractal analysis is required to adequately describe the binding kinetics.

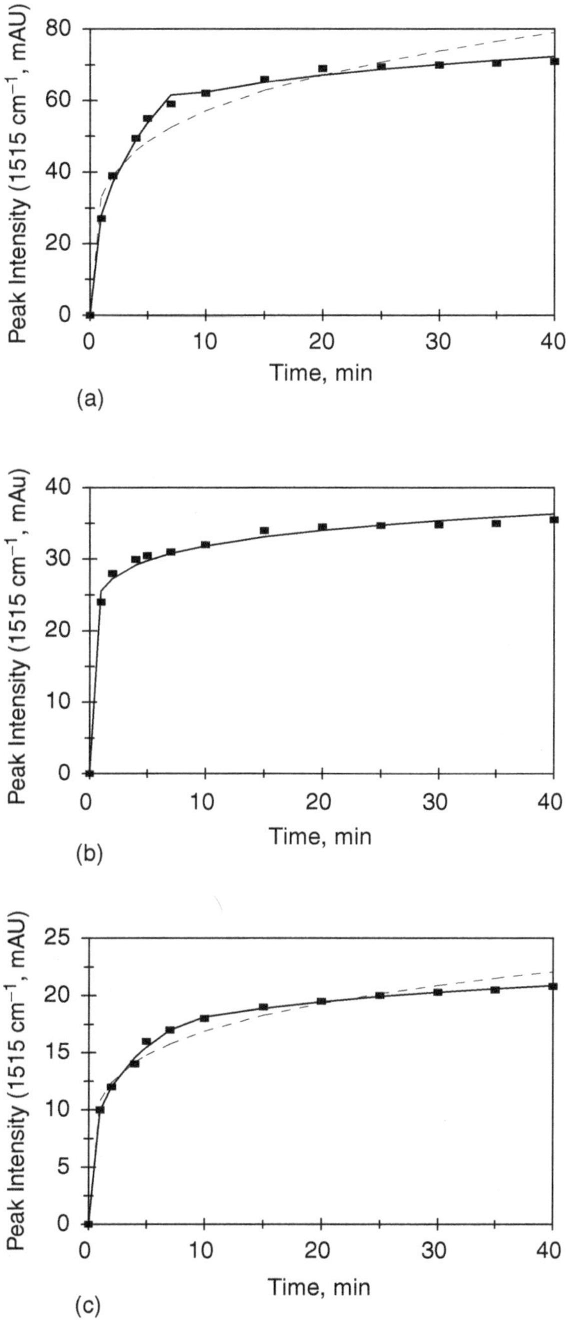

Figure 5.8 Binding of 1 mM tyrosine in 75 mM ammonia to a proline-modified sensing phase on the surface of an internal reflection element (Huang and Yang, 2005). Influence of the addition of 0.5 mM of (a) copper (b) nickel, and (c) zinc.

Table 5.5

Binding rate coefficients and fractal dimensions for the binding of 1 mM tyrosine in 75 mM ammonia to a proline-modified sensing phase on the surface of an internal reflection element (Huang and Yang, 2005)

Metal ion	k	k_1	k_2	D_f	D_{f1}	D_{f2}
Copper	33.252 ± 3.671	28.051 ± 1.318	48.692 ± 0.856	2.531 ± 0.0532	2.192 ± 0.05892	2.7856 ± 0.0322
Nickel	25.566 ± 0.749	na	na	2.8093 ± 0.01466	na	na
Zinc	10.826 ± 0.629	9.933 ± 0.300	14.339 ± 0.079	2.6134 ± 0.02866	2.488 ± 0.03822	2.7964 ± 0.009054

Note: Influence of the addition of 0.5 mM copper, nickel, and zinc ions.

Figure 5.8c shows the binding of 1 mM tyrosine in 75 mM ammonia buffer and in the presence of 0.5 mM zinc ions to a proline-modified sensing phase on the surface of the internal reflection element of the sensor (Huang and Yang, 2005). A dual-fractal analysis is required to adequately describe the binding kinetics. The values of (a) the binding rate coefficient, k and the fractal dimension, D_f for a single-fractal analysis, and (b) the binding rate coefficients, k_1 and k_2, and the fractal dimensions, D_{f1} and D_{f2}, for a dual-fractal analysis are given in Table 5.5. It is of interest to note for a dual-fractal analysis that as the fractal dimension increases by a factor of 1.27 from a value of D_{f1} equal to 2.192 to D_{f2} equal to 2.7856, the binding rate coefficient increases by a factor of 1.73 from a value of k_1 equal to 28.051 to k_2 equal to 48.692. Changes in the degree of heterogeneity or the fractal dimension on the sensing surface and in the binding rate coefficient are in the same direction.

Finally, it is of interest to compare the fractal dimension and the binding rate coefficient values when copper and zinc are used. In both of these cases, a dual-fractal analysis is required to adequately describe the binding kinetics. However, the fractal dimensions (D_{f1} and D_{f2}) are both lower for copper ions when compared with zinc; however, the binding rate coefficients (k_1 and k_2) are both higher for copper than for zinc. This indicates that not only is the heterogeneity of the sensing surface or the fractal dimension involved in the binding rate coefficient value, but also the transition metal (in this case copper or zinc) must be taken into account. It would be of interest to see if other transition metal ions such as cobalt (Co^{2+}) or Manganese (Mn^{++}) could also be used.

Figure 5.9a shows the binding of 1000 μM tyrosine at pH 9.1 ± 0.1 using 75 mM ammonia buffer and 0.5 mM copper ions to a proline-modified sensing phase on the surface of an internal reflection element of the sensor (Huang and Yang, 2005). A dual-fractal analysis is required to adequately describe the binding kinetics. The values of (a) the binding rate coefficient, k and the fractal dimension, D_f for a single-fractal analysis, and (b) the binding rate coefficients, k_1 and k_2, and the fractal dimensions, D_{f1} and D_{f2} for a dual-fractal analysis are given in Table 5.6. It is of interest to note that as the fractal dimension increases by a factor of 1.34 from a value of D_{f1} equal to 2.0724 to D_{f2} equal to 2.7724, the binding rate coefficient increases by a factor of 1.8 from a value of k_1 equal to 26.76 to k_2 equal to 48.171. Changes in the binding rate coefficient

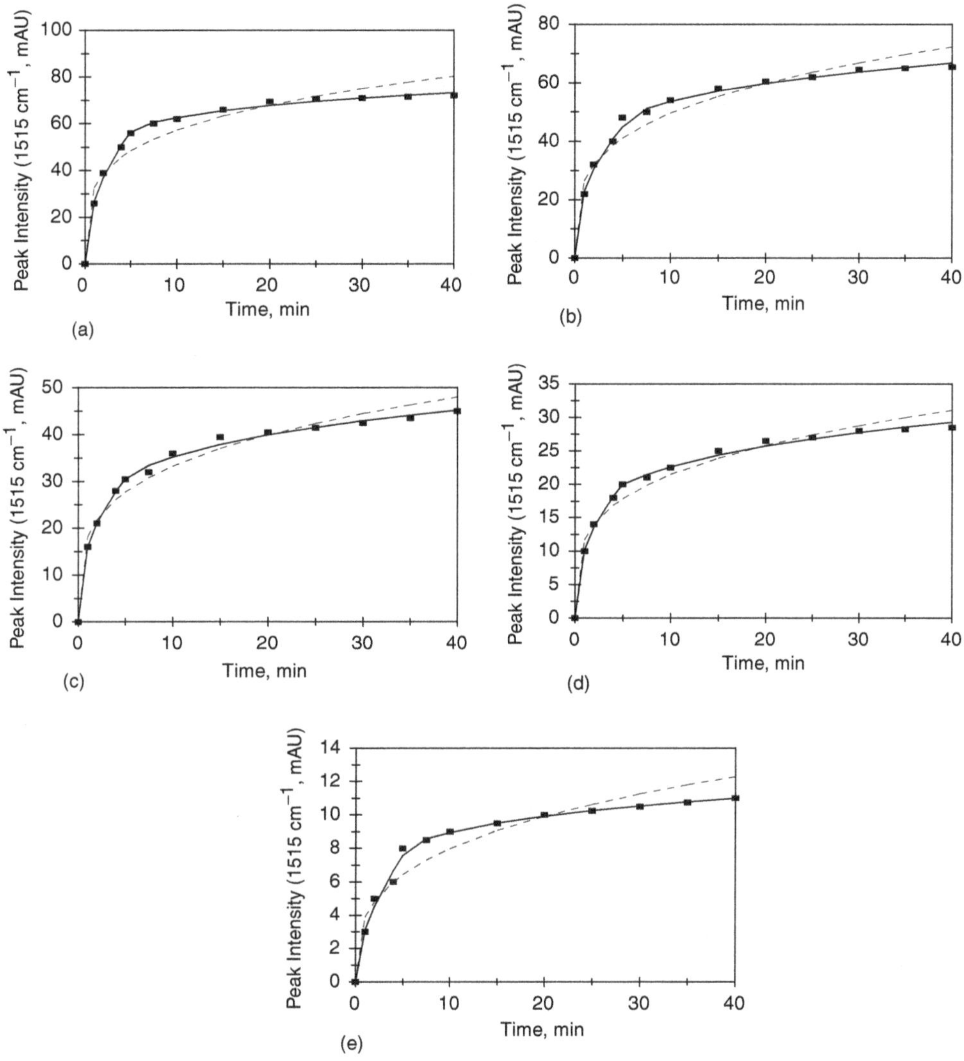

Figure 5.9 Binding of 1 mM tyrosine in 75 mM ammonia to a proline-modified sensing phase on the surface of an internal reflection element (Huang and Yang, 2005): (a) 1000, (b) 800, (c) 500, (d) 300, (e) 100.

are in the same direction as the degree of heterogeneity or fractal dimension on the sensing surface.

Figure 5.9b shows the binding of 800 μM tyrosine at pH 9.1 ± 0.1 using 75 mM ammonia buffer and 0.5 mM copper ions to a proline-modified sensing phase on the surface of an internal reflection element of the sensor (Huang and Yang, 2005). A dual-fractal analysis is required to adequately describe the binding kinetics. The values of (a) the binding

Table 5.6

Binding rate coefficients and fractal dimensions for the different concentrations of tyrosine (in μM) in solution to the proline-modified sensing phase synthesized on the surface of an internal reflection element (Huang and Yang, 2005)

Tyrosine concentration in solution (μM)	D_f	D_{f1}	D_{f2}	k	k_1	k_2
1000	2.5120 ± 0.05698	2.0724 ± 0.07316	2.7724 ± 0.01998	32.697 ± 3.878	26.761 ± 1.264	48.171 ± 0.772
800	2.4556 ± 0.07316	2.1654 ± 0.0805	2.6814 ± 0.01878	26.498 ± 2.697	22.868 ± 1.516	37.100 ± 0.588
500	2.4718 ± 0.03706	2.1944 ± 0.007356	2.6394 ± 0.03520	18.123 ± 1.370	15.963 ± 0.074	23.262 ± 0.661
300	2.4688 ± 0.03934	2.1550 ± 0.03908	2.6298 ± 0.02754	11.664 ± 0.938	10.148 ± 0.253	14.792 ± 0.328
100	2.3768 ± 0.06892	1.8952 ± 0.1361	2.7012 ± 0.007096	3.8994 ± 0.5662	3.1123 ± 0.3654	6.3399 ± 0.0359

Note: Sensor is based on the formation of copper complexes between the sensing phase and tyrosine.

rate coefficient, k and the fractal dimension, D_f for a single-fractal analysis, and (b) the binding rate coefficients, k_1 and k_2, and the fractal dimensions, D_{f1} and D_{f2} for a dual-fractal analysis are given in Table 5.6. In this case too, note that as the fractal dimension increases by a factor of 1.24 from a value of D_{f1} equal to 2.1654 to D_{f2} equal to 2.6814, the binding rate coefficient increases by a factor of 1.622 from a value of k_1 equal to 22.868 to k_2 equal to 37.1. Changes in the binding rate coefficient are, once again, in the same direction as the degree of heterogeneity or fractal dimension on the sensing surface.

Figure 5.9c shows the binding of 500 μM tyrosine at pH 9.1 ± 0.1 using 75 mM ammonia buffer and 0.5 mM copper ions to a proline-modified sensing phase on the surface of an internal reflection element of the sensor (Huang and Yang, 2005). A dual-fractal analysis is required to adequately describe the binding kinetics. The values of (a) the binding rate coefficient, k and the fractal dimension, D_f for a single-fractal analysis, and (b) the binding rate coefficients, k_1 and k_2, and the fractal dimensions, D_{f1} and D_{f2} for a dual-fractal analysis are given in Table 5.6. In this case too, note that as the fractal dimension increases by a factor of 1.202 from a value of D_{f1} equal to 2.1944 to D_{f2} equal to 2.6394, the binding rate coefficient increases by a factor of 1.46 from a value of k_1 equal to 15.963 to k_2 equal to 23.262. Changes in the binding rate coefficient are, once again, in the same direction as the degree of heterogeneity or fractal dimension on the sensing surface.

Figure 5.9d shows the binding of 300 μM tyrosine at pH 9.1 ± 0.1 using 75 mM ammonia buffer and 0.5 mM copper ions to a proline-modified sensing phase on the surface of an internal reflection element of the sensor (Huang and Yang, 2005). A dual-fractal analysis is required to adequately describe the binding kinetics. The values of (a) the binding rate coefficient, k and the fractal dimension, D_f for a single-fractal analysis, and

(b) the binding rate coefficients, k_1 and k_2, and the fractal dimensions, D_{f1} and D_{f2} for a dual-fractal analysis are given in Table 5.6. In this case too, note that as the fractal dimension increases by a factor of 1.22 from a value of D_{f1} equal to 2.1550 to D_{f2} equal to 2.6298, the binding rate coefficient increases by a factor of 1.46 from a value of k_1 equal to 10.148 to k_2 equal to 14.792. Changes in the binding rate coefficient are, once again, in the same direction as the degree of heterogeneity or fractal dimension on the sensing surface.

Figure 5.9e shows the binding of 100 μM tyrosine at pH 9.1 ± 0.1 using 75 mM ammonia buffer and 0.5 mM copper ions to a proline-modified sensing phase on the surface of an internal reflection element of the sensor (Huang and Yang, 2005). A dual-fractal analysis is required to adequately describe the binding kinetics. The values of (a) the binding rate coefficient, k and the fractal dimension, D_f for a single-fractal analysis, and (b) the binding rate coefficients, k_1 and k_2, and the fractal dimensions, D_{f1} and D_{f2} for a dual-fractal analysis are given in Table 5.6. In this case too, note that as the fractal dimension increases by a factor of 1.425 from a value of D_{f1} equal to 1.8952 to D_{f2} equal to 2.7012, the binding rate coefficient increases by a factor of 2.04 from a value of k_1 equal to 3.1123 to k_2 equal to 6.3399. Changes in the binding rate coefficient are, once again, in the same direction as the degree of heterogeneity or fractal dimension on the sensing surface.

Figure 5.10a and Table 5.6 show the increase in the binding rate coefficient, k_1 with an increase in the tyrosine concentration in the range 100–1000 μM in solution. For the data given in Figure 5.10a, the binding rate coefficient, k_1 is given by

$$k_1 = (0.04429 \pm 0.00387)[\text{tyrosine concentration, }\mu\text{M}]^{0.9372 \pm 0.0457} \tag{5.6a}$$

The fit is very good. Only five data points are available. The availability of more data points would lead to a more reliable fit. The binding rate coefficient, k_1 exhibits close to a first (equal to 0.9372) order of dependence on the tyrosine concentration (100–1000 μM) in solution.

Figure 5.10b and Table 5.6 show the increase in the binding rate coefficient, k_2 with an increase in the tyrosine concentration in the range 100–1000 μM in solution. For the data given in Figure 5.10b, the binding rate coefficient, k_2 is given by

$$k_2 = (0.1084 \pm 0.0077)[\text{tyrosine concentration, }\mu\text{M}]^{0.873 \pm 0.0372} \tag{5.6b}$$

The fit is very good. Only five data points are available. The availability of more data points would lead to a more reliable fit. The binding rate coefficient, k_2 exhibits a dependence between one-half and first (equal to 0.873) order of dependence on the tyrosine concentration (100–1000 μM) in solution.

Figure 5.10c and Table 5.6 show the increase in the fractal dimension, D_{f2} with an increase in the tyrosine concentration in the range 100–1000 μM in solution. For the data given in Figure 5.10c the fractal dimension, D_{f2} is given by

$$D_{f2} = (2.0902 \pm 0.0308)[\text{tyrosine concentration, }\mu\text{M}]^{0.0389 \pm 0.0160} \tag{5.6c}$$

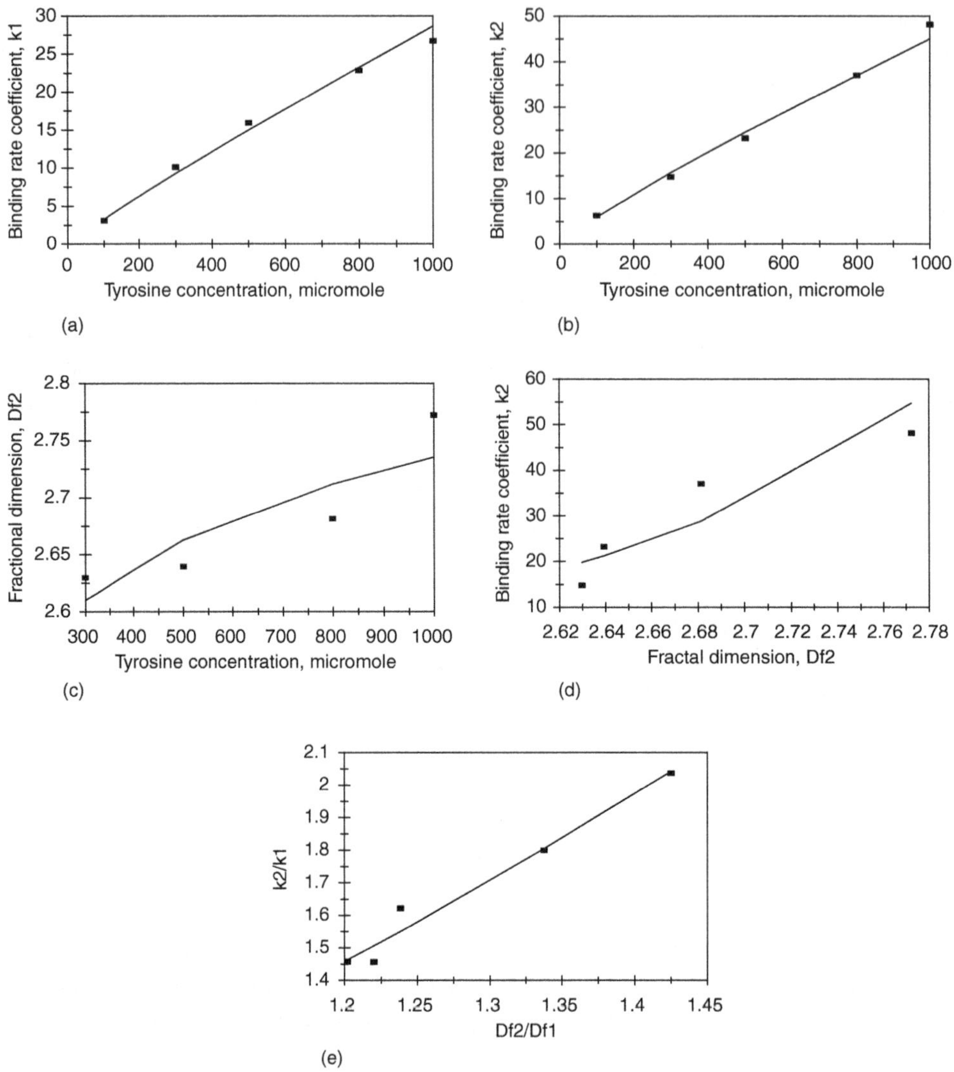

Figure 5.10 Increase in the (a) binding rate coefficient, k_1; (b) binding rate coefficient, k_2; and (c) fractal dimension, D_{f2} with an increase in the tyrosine concentration (in μM) in solution. (d) Increase in the binding rate coefficient, k_2 with an increase in the fractal dimension, D_{f2}. (e) Increase in the binding rate coefficient ratio, k_2/k_1 with an increase in the fractal dimension ratio, D_{f2}/D_{f1}.

The fit is reasonable. Only four data points are available. The availability of more data points would lead to a more reliable fit. The fractal dimension, D_{f2}, exhibits only a slight dependence on the tyrosine concentration (100–1000 μM) in solution as noted by the close to zero (equal to 0.0389) order of dependence exhibited.

Figure 5.10d and Table 5.6 show the increase in the binding rate coefficient, k_2 with an increase in the fractal dimension, D_{f2}. For the data given in Figure 5.10d, the binding rate coefficient, k_2 is given by

$$k_2 = (1.7\text{E-}07 \pm 0.5\text{E-}07) D_{f2}^{19.22 \pm 7.043} \tag{5.6d}$$

The fit is good. Only five data points are available. The availability of more data points would lead to a more reliable fit. The binding rate coefficient, k_2 exhibits a dependence between one-half and first (equal to 0.873) order on the tyrosine concentration (100–1000 μM) in solution.

Figure 5.10e and Table 5.6 show the increase in the binding rate coefficient ratio, k_2/k_1 with an increase in the fractal dimension ratio, D_{f2}/D_{f1}. For the data given in Figure 5.10e, the binding rate coefficient ratio, k_2/k_1 is given by

$$\frac{k_2}{k_1} = (1.0213 \pm 0.0399) \left(\frac{D_{f2}}{D_{f1}} \right)^{1.957 \pm 0.2258} \tag{5.6e}$$

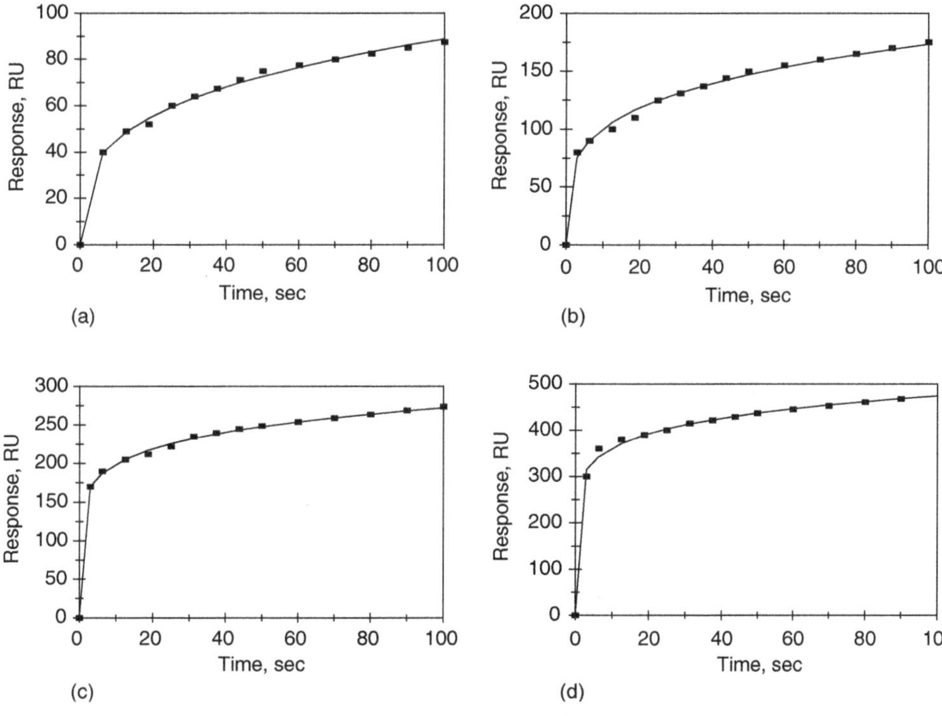

Figure 5.11 Binding of different concentrations of anti-GAD antibody (in μM) in solution to GAD immobilized on Fc2 of SAM2 on a SPR biosensor chip surface (Lee *et al.*, 2005): (a) 0.5, (b) 1.0, (c) 2.0, (d) 4.0.

The fit is very good. Only five data points are available. The availability of more data points would lead to a more reliable fit. The binding rate coefficient ratio, k_2/k_1 exhibits close to a second order of dependence (equal to 1.957) on the fractal dimension ratio, D_{f2}/D_{f1}.

Lee *et al.* (2005) have recently used an SPR biosensor to detect anti-GAD antibody. This antibody according to these authors is an indicator for the presence of type I diabetes mellitus (DM). These authors used a self-assembled monolayer (SAM) of thiol on a gold surface of a biosensor chip. They indicate that GAD (EC 4.1.1.15) is an enzyme that regulates the synthesis of γ-aminobutyric acid in human islets and in the brain. It exists in two homologous forms, GAD65 and GAD67. The auto-antibody (anti-GAD antibody) recognizes GAD65 in the sera of IDDM (insulin-dependent diabetes mellitus) patients (Figures 5.11–5.15; Table 5.7).

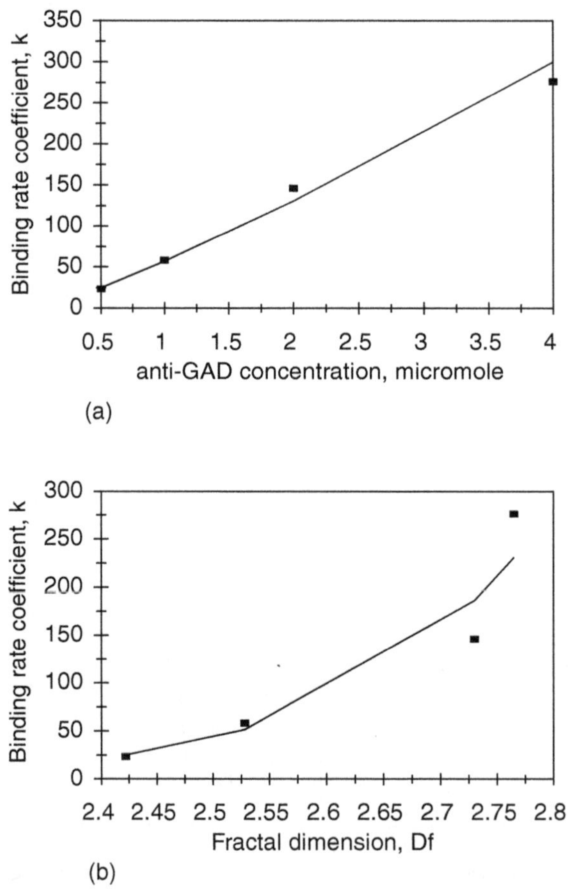

Figure 5.12 (a) Increase in the binding rate coefficient, *k* with an increase in the anti-GAD concentration (in μM) in solution. (b) Increase in the binding rate coefficient, *k* with an increase in the fractal dimension, D_f.

Table 5.7

Binding rate coefficients and fractal dimensions for (a) different concentrations of anti-GAD antibody (in μM) in solution to GAD immobilized on Fc2 of SAM2 on a SPR biosensor chip surface and (b) specific binding of different concentrations of anti-GAD antibody (in μM) in solution to GAD immobilized on Fc2 wherein the response of Fc1 (where BSA immobilized) is subtracted from that of Fc2 on a SPR biosensor chip surface (Lee *et al.*, 2005)

Anti-GAD concentration (in μM) in solution/GAD immobilized on Fc2 of SAM2	k	D_f
(a)		
0.5	23.476 ± 0.500	2.4220 ± 0.01477
1.0	58.405 ± 1.774	2.5282 ± 0.5976
2.0	146.18 ± 1.552	2.7296 ± 0.00563
4.0	276.30 ± 5.83	2.7650 ± 0.0116
(b)		
0.5	4.9024 ± 0.0780	2.0018 ± 0.01697
1.0	5.7413 ± 0.1292	1.8296 ± 0.02392
2.0	7.5119 ± 0.1730	1.9220 ± 0.02446
4.0	8.8466 ± 0.2337	1.8686 ± 0.02356

Figure 5.11a shows the binding of 0.5 μM anti-GAD antibody in solution to immobilized GAD on Fc2 of SAM2 on a SPR biosensor chip surface (Lee *et al.*, 2005). SAM2 is an alkanethiol monolayer self-assembled on the gold surface with a 10:1 ratio of 3-MPA to 11-UA. 3-MPA is 3-mercaptopropionic acid and 11-MUA is 11-mercaptoundecanoic acid. A single-fractal analysis is adequate to describe the binding kinetics. The values of (a) the binding rate coefficient, k and the fractal dimension, D_f for a single-fractal analysis are given in Table 5.7a.

Figure 5.11b shows the binding of 1.0 μM anti-GAD antibody in solution to immobilized GAD on Fc2 of SAM2 on a SPR biosensor chip surface (Lee *et al.*, 2005). A single-fractal analysis is adequate to describe the binding kinetics. The values of (a) the binding rate coefficient, k and the fractal dimension, D_f for a single-fractal analysis are given in Table 5.7a.

Figure 5.11c shows the binding of 2.0 μM anti-GAD antibody in solution to immobilized GAD on Fc2 of SAM2 on a SPR biosensor chip surface (Lee *et al.*, 2005). A single-fractal analysis is adequate to describe the binding kinetics. The values of (a) the binding rate coefficient, k and the fractal dimension, D_f for a single-fractal analysis are given in Table 5.7a.

Figure 5.11d shows the binding of 4.0 μM anti-GAD antibody in solution to immobilized GAD on Fc2 of SAM2 on a SPR biosensor chip surface (Lee *et al.*, 2005). A single-fractal analysis is adequate to describe the binding kinetics. The values of (a) the binding rate coefficient, k and the fractal dimension, D_f for a single-fractal analysis are given in Table 5.7a.

Figure 5.12a shows the increase in the binding rate coefficient, k with an increase in the anti-GAD concentration in solution in the 0.5–4.0 μM range. For the data presented in Table 5.7a and in Figure 5.12a, the binding rate coefficient, k is given by

$$k = (56.92 \pm 6.45)[\text{anti-GAD concentration, in } \mu\text{M}]^{1.199 \pm 0.0692} \qquad (5.7a)$$

The fit is very good. Only four data points are available. The availability of more data points would lead to a more reliable fit. The binding rate coefficient, k exhibits close to a first- (equal to 1.199) order of dependence on the anti-GAD concentration in solution. The non-integer order of dependence exhibited by the binding rate coefficient, k on the anti-GAD concentration in solution lends support to the fractal nature of the system.

Figure 5.12b shows the increase in the binding rate coefficient, k with an increase in the fractal dimension, D_f. For the data presented in Figure 5.12b and in Table 5.7a the binding rate coefficient, k is given by

$$k = (8.92 \pm 2.22)D_f^{16.785 \pm 2.167} \tag{5.7b}$$

The fit is reasonable. Only four data points are available. The availability of more data points would lead to a more reliable fit. The binding rate coefficient, k is extremely sensitive to the fractal dimension, D_f or the degree of heterogeneity that exists on the biosensor surface as noted by the order of dependence between sixteen and one-half and seventeen (equal to 16.785) exhibited.

Figure 5.13a shows the binding of 0.5 μM anti-GAD in solution to the immobilized GAD surface (Lee *et al.*, 2005). The response of Fcl (where BSA immobilized) is subtracted from

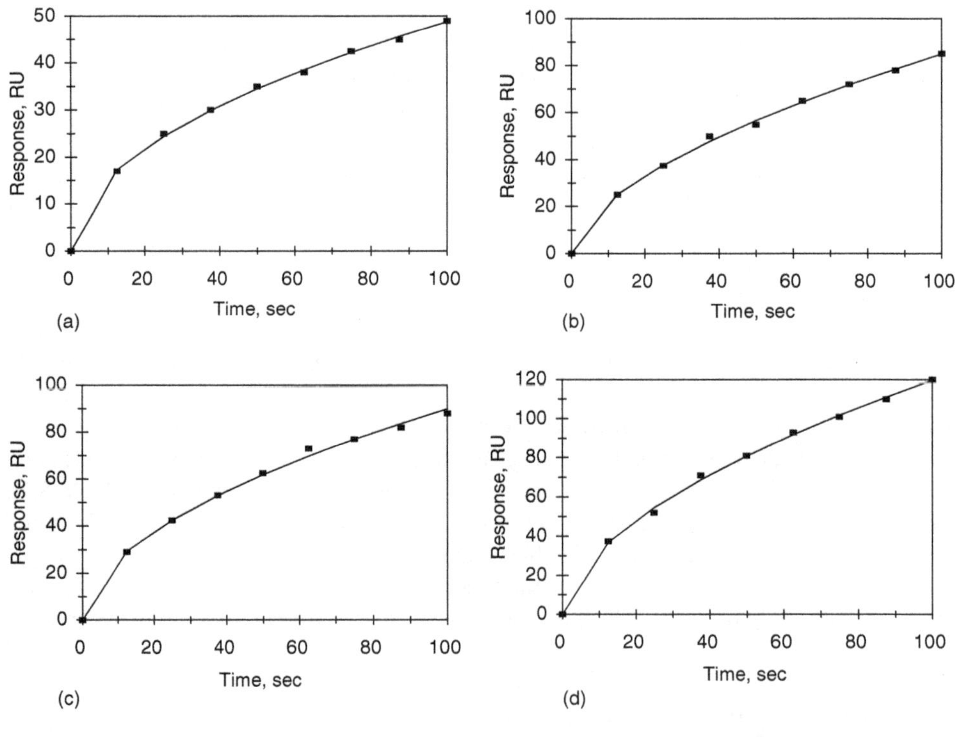

Figure 5.13 Specific binding of different concentration of anti-GAD antibody (in μM) in solution to GAD immobilized on Fc2 wherein the response of Fc1 (where BSA immobilized) is subtracted from that of Fc2 (Lee *et al.*, 2005): (a) 0.5, (b) 1.0, (c) 2.0, (d) 4.0.

that of Fc2 when the analyte is simultaneously injected over SAM2 on a SPR biosensor chip surface. A single-fractal analysis is adequate to describe the binding kinetics. The values of the binding rate coefficient, *k* and the fractal dimension, D_f are given in Table 5.7b.

Figure 5.13b shows the binding of 1.0 μM anti-GAD in solution to the immobilized GAD surface (Lee *et al.*, 2005). The response of Fcl (where BSA immobilized) is subtracted from that of Fc2 when the analyte is simultaneously injected over SAM2 on a SPR biosensor chip surface. A single-fractal analysis is adequate to describe the binding kinetics. The values of the binding rate coefficient, *k* and the fractal dimension, D_f are given in Table 5.7b.

Figure 5.13c shows the binding of 2.0 μM anti-GAD in solution to the immobilized GAD surface (Lee *et al.*, 2005). The response of Fcl (where BSA immobilized) is subtracted from that of Fc2 when the analyte is simultaneously injected over SAM2 on a SPR biosensor chip surface. A single-fractal analysis is adequate to describe the binding kinetics. The values of the binding rate coefficient, *k* and the fractal dimension, D_f are given in Table 5.7b.

Figure 5.13d shows the binding of 4.0 μM anti-GAD in solution to the immobilized GAD surface (Lee *et al.*, 2005). The response of Fcl (where BSA immobilized) is subtracted from that of Fc2 when the analyte is simultaneously injected over SAM2 on a SPR biosensor chip surface. A single-fractal analysis is adequate to describe the binding kinetics. The values of the binding rate coefficient, *k* and the fractal dimension, D_f are given in Table 5.7b.

Figure 5.14 shows the increase in the binding rate coefficient, *k* with an increase in the anti-GAD concentration in solution in the 0.5–4.0 μM range. For the data presented in Table 5.7a and in Figure 5.14, the binding rate coefficient, k is given by

$$k = (5.938 \pm 0.207)[\text{anti-GAD concentration, in } \mu\text{M}]^{0.294 \pm 0.0221} \tag{5.8}$$

The fit is very good. Only four data points are available. The availability of more data points would lead to a more reliable fit. The binding rate coefficient, *k* is only mildly sensitive to the anti-GAD concentration in solution as noted by the 0.294 order of dependence

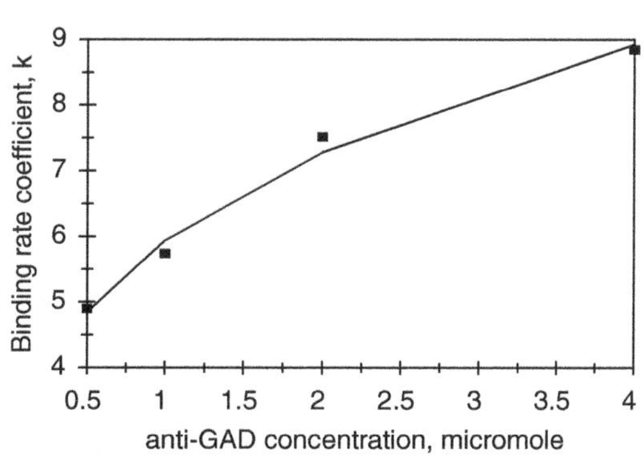

Figure 5.14 Increase in the binding rate coefficient, *k* with an increase in the anti-GAD concentration (in μM) in solution.

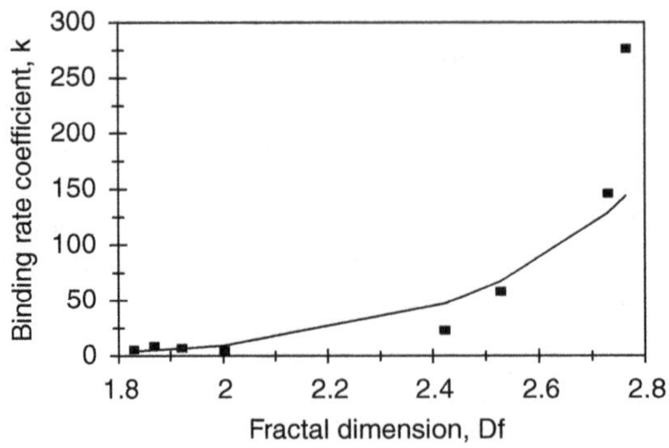

Figure 5.15 Increase in the binding rate coefficient, k with an increase in the fractal dimension, D_f (data in Figures 5.12b and 5.14 are combined).

exhibited. Once again, the non-integer order of dependence exhibited by the binding rate coefficient, k on the anti-GAD concentration in solution lends support to the fractal nature of the system.

Figure 5.15 shows the increase in the binding rate coefficient, k with an increase in the fractal dimension, D_f. For the data presented in Figure 5.15 and in Table 5.7a and b (note both data sets are plotted together) the binding rate coefficient, k is given by

$$k = (0.0276 \pm 0.0196)D_f^{8.436\pm1.158} \tag{5.9}$$

The fit is reasonable considering that two data sets are plotted together. The binding rate coefficient, k is extremely sensitive to the fractal dimension, D_f or the degree of heterogeneity that exists on the biosensor surface as noted by the close to eight and one-half (equal to 8.436) order of dependence exhibited.

5.4 CONCLUSIONS

A fractal analysis is presented for the binding of different disease-related compounds on biosensor surfaces. The binding and the dissociation kinetics may be described by a single- or a dual-fractal analysis. A dual-fractal analysis is only used when a single-fractal analysis does not provide an adequate fit. This was done using Corel Quattro Pro 8.0 (1989), and only when the regression coefficient for the sum of least squares was less than 0.97. The fractal dimension values provide a quantitative indication of the degree of heterogeneity that exists on the sensing surface. Binding and dissociation rate coefficient values are provided. The fractal dimension for the binding and dissociation phase, D_f and D_{fd}, respectively is not a typical independent variable, such as analyte concentration that may be directly manipulated. It is estimated from Eqs. (5.1)–(5.3), and one may consider it as a derived variable.

An increase in the fractal dimension value or the degree of heterogeneity on the sensing surface leads, in general, to an increase in the binding rate coefficient. One notes an increase in (a) the binding rate coefficient, k with an increase on the fractal dimension, D_f for the binding of CD7 antigen expressed in nucleated cells from 96 acute leukemic patients to leukemic-associated monoclonal antibodies immobilized onto the nanogold-PA (protein A)-modified surface of a QCM immunosensor array (Zeng *et al.*, 2006), (b) the binding and dissociation of 17-β estradiol (β -E2) to genetically modified *Saccharomyces cerevisiae* cells in solution in a calcium alginate hydrogel matrix (Fine *et al.*, 2006), (c) an infrared optical sensor for the detection of tyrosine in biological fluids (Huang and Yang, 2005), and (d) for the detection of anti-glutamic acid decarboxylase (GAD) antibody on a SPR biosensor (Lee *et al.*, 2005).

The predictive relationships developed for the binding rate coefficients as a function of their respective fractal dimensions are of considerable value since they directly link the binding rate coefficient to the degree of heterogeneity that exists on the sensor chip surface, and provide a means by which the binding rate coefficients may be manipulated by changing the degree of heterogeneity on the sensor chip surface.

More such studies are required to determine whether the binding rate coefficient(s) are sensitive to their respective fractal dimensions or the degree of heterogeneity that exists on the sensor chip surface. If this is correct, then experimentalists may find it worth their effort to pay a little more attention to the nature of the surface, and how it may be manipulated by changing the degree of heterogeneity on the sensor chip surface to enhance biosensor performance parameters. This, as one clearly recognizes, is of particular importance for the early detection of the onset of diseases.

Some of the precursors or the early disease markers (biomarkers) presented in this chapter include the detection of (a) markers for acute myocardial infarction (AMI), such as C-reactive protein (CRP) (Wolf *et al.*, 2004), (b) hepatitis B surface antigen by hepatitis B surface antibody (Tang *et al.*, 2005), (c) integrin $\alpha_3\beta_3$ (Worsfold *et al.*, 2004), which is a target for the inhibition of tumor growth (Brooks *et al.*, 1994), (d) estrogenic endocrine disrupting chemicals (EDCs) by luminescent yeast cells entrapped in hydrogels (Fine *et al.*, 2006), (e) tyrosine in biological fluids (Huang and Yang, 2005), whose metabolism is involved in atherosclerosis (Heitzer *et al.*, 2001), lung diseases (Ischiropoulos *et al.*, 1995), and liver diseases (Sherlock, 1989), and anti-GAD (anti-glutamic acid decarboxylase) antibody using a SPR biosensor (Lee *et al.*, 2005), which is a major marker for the detection of IDDM (insulin-dependent diabetes mellitus).

Needless to say, the early detection of the above-mentioned markers (by biosensors or otherwise) for the onset of the diseases, as well as the detection of other biomarkers for other diseases, will go a long way for the diagnosis and prognosis of especially intractable diseases. This chapter is written specifically with this purpose in mind. Any assistance in this matter is invaluable to the patient involved and his/her family as far as pain and suffering and, in general, economic hardships that are involved with these diseases. Hopefully, more better biosensor parameters developed in the future can help mitigate some of this pain and suffering and the economic hardships involved.

REFERENCES

Adams JE, GS Bodor, and VG Davila-Roman, Cardiac troponin I: a marker with high specificity for cardiac injury, *Circulation*, 1993, **88**, 101–106.

Baekkeskov S, HJ Aanstoot, S Christgau, A Reetz, M Solimena, M Cascalho, F Folli, H Richter-Olesen, PDe Camilli, and PD Camilli, Identification of the 64K autoantigen insulin-dependent diabetes as the GABA-synthesizing enzyme glutamic acid decarboxylase. *Nature*, 1990, **347**, 151–156.

Biacore AB, *BIAEvaluation 3.2 software*, Uppsala, Sweden, 2002.

Brett W, A Mandinova, and A Remppis, Translocation of S100A1(1) calcium binding protein during heart surgery. *Biochemical and Biophysical Research Communication*, 2001, **284**, 698–703.

Brooks PC, RF Clark, and DA Cheresh, Requirement of vascular integrin $\alpha_3\beta_3$ for angiogenesis. *Science*, 2004, **264**, 569–571.

Carlsen E, A Giwercman, N Keiding, and NE Skakkebaeck, Evidence for decreasing quality of semen during the past 50 years. *British Medical Journal*, 1992, **305**, 609–613.

Corel Quattro Pro 8.0, Corel Corporation Limited, Ottawa, Canada, 1997.

ESC/ACC, 2000. Consensus document, Myocardial infarction redefined: A consensus document of the Joint European Society of Cardiology/American College of Cardiology Committee for the redefinition of myocardial infarction. *European Heart Journal*, 2000, **21**, 1502–1513.

Facemire CF, TS Gross, and JJL Guillette, Reproductive impairment in the Florida panther: nature or nuture. *Environment Health Perspective*, 1995, **103**(4), 79–86.

Fine T, P Leskinen, T Isobe, H Shiraishi, M Morita, RS marks, and M Virta, Luminescent yeast cells entrapped in hydrogels for estrogenic endocrine disrupting chemical biodetection, *Biosensors and Bioelectronics*, 2006, **21**, 2263–2269.

Fry DM, and CK Toone, DDT-induced feminization of gull embryos. *Science*, 1981, **213**, 922–924.

Givercam A, E Carlsen, N Keiding, and NE Skakkebaek, Evidence for increasing incidence of abnormalities of the human testis: a review. *Environmental Health Perspective*, 1993, **101**, 65–71.

Greenstein JP, and M Winitz, *Chemistry of the Amino Acids*, Vols. 1–3, Wiley, New York, 1961.

Harrick NJ, Internal Reflection Spectroscopy, Wiley, New York, 1967.

Havlin S, Molecular diffusion and reaction, in *The Fractal Approach to Heterogeneous Chemistry: Surfaces, Colloids, Polymers* (ed. D Avnir), Wiley, New York, 1989, pp. 251–269.

Hietzer T, T Schlinzig, K Krohn, T Meinertz, and T Munzel, Endothelial disfunction, oxidative stress, and risk of cardiovascular events in patients with coronary artery disease. *Circulation*, 2001, **104**, 2673–2678.

Hu WP, HY Hsu, A Chiou, KY Tseng, HY Lin, GL Chang, and SJ Chen, Immunodetection of pentamer and modified C-reactive protein using surface plasmon resonance biosensing. *Biosensors and Bioelectronics*, 2006, **21**, 1631–1637.

Huang GG, and J Yang, Development of infrared optical sensor for selective detection of tyrosine in biological fluids. *Biosensors and Bioelectronics*, 2005, **21**, 408–418.

Hynes RO, Integrins: versatility, modulation, and signaling in cell adhesion. *Cell*, 1992, **69**(1), 11–25.

Ichiropoulos H, AB Al-Mehdi, and AB Fisher, Reactive species in ischemic rat lung injury: contribution of peroxynitrite. *American Journal of Physiology*, 1995, **269**, L158–L164.

Jobling S, D Sheahan, JA Osborne, P Matthiessen, and JP Sumpter, Inhibition of testicular growth in rainbow trout (*Oncorhynchus myksis*) exposed to estrogenic alkylphenolic chemicals. *Environmental and Toxicological Chemistry*, 1996, **15**, 194–202.

Lee JW, SJ Sim, SM Cho, and J lee, Characterization of a self-assembled monolayer of thiol on a gold surface and the fabrication of a biosensor chip based on surface plasmon resonance for detecting anti-GAD antibody. *Biosensors and Bioelectronics*, 2005, **20**, 1422–1427.

Maisel AS, P Krishnaswamy, and RM Nowak, Rapid measurement of a B-type natriuretic peptide in the emergency diagnosis of heart failure. *New England Journal of Medicine*, 2002, **347**, 161–167.

McCord J, RM Nowak, PA McCullough, C Foreback, S Borzak, G Tokarski, MC Tomlanovich, G Jacobsen, and WD Weaver, Ninety-minute exclusion of acute myocardial infarction by use of quantitative point-of-care testing of myoglobin and troponin I. *Circulation*, 2001, **104**, 1483–1488.

Offen D, A Hochman, S Gorodin I Ziv, A Shirvan, A Barzilai, and E Melamed, Oxidative stress and neuroprotection in Parkinson's disease: implications form studies on dopamine-induced apoptosis. *Advances in Neurology*, 1999, **80**, 265–269.

Pleabani M, and M Zaninotto, Diagnostic strategies in myocardial infarction using myoglobin measurement. *European Heart Journal*, 1998, **19**, N12–N15.

Raloff J, Ecocancers: do environmental factors underlie a breast cancer epidemic? *Science News*, 1993, **144**, 10–13.

Ramakrishnan A, and A Sadana, A single-fractal analysis of cellular analyte-receptor binding kinetics using biosensors. *BioSystems*, 2001, **59**, 35–51.

Sadana A, A fractal analysis for the evaluation of hybridization kinetics in biosensors. *Journal of Colloid and Interface Science*, 2001, **151**(1), 166–177.

Sadana A, *Fractal Binding and Dissociation Kinetics for Different Biosensor Applications*, Elsevier, Amsterdam, 2005.

Scriver CR, AL Beaudet, WS Sly, and D Valle, *The Metabolic and Molecular Bases of Inherited Disease*, Vols. 1–3, McGraw-Hill, New York, 1955.

Sherlock S, 1989. *Diseases of the Liver and Bilary System.* Blackwell Scientific Publications, Oxford, UK.

Tang D, R Yuan, Y Chai, Y Fu, J Dai, Y Liu, and X Zhong, New amperometric and potentioimetric immunosensors based on gold nanoparticles/tris (2,2′-bipyridyl) cobalt (III) multilayer films for hepatitis B surface antigen determinations. *Biosensors and Bioelectronics*, 2005, **21**, 534–548.

Wang CL, H Wang, ZY Wu, GL Shen, and RQ Yu, A piezoelectric immunoassay based on self-assembled monolayers of cystamine and polystyrene sulfonate for determination of *Schistosoma japonicum* antibodies. *Analytical and Bioanalytical Chemistry*, 2002a, **373**, 803–809.

Wang HY, GB Xu, and SJ Dong, Electrochemiluminescence of tris (2,2′-bipyridyl)ruthenium (II) ion-exchanged in polyelectrolyte-silica composite thin-films. Electroanalysis, 2002b, **14**, 853–857.

Witorsch RJ, Endocrine disruptors: can effects and environmental risks be predicted? *Regulatory Toxicology and Pharmacology*, 2002, **36**, 118–130.

Wolf M, D Juncker, B Michel, P Hunziker, and E Delamarche, Simultaneous detection of C-reactive protein and other cardiac markers in human plasma using micromosaic immunoassays and self-regulating microfluidic networks. *Biosensors and Bioelectronics*, 2004, **19**, 1193–1202.

Worsfold O, C Toma, and T Nishiya, Development of a novel optical bionanosensor. *Biosensors and Bioelectronics*, 2004, **19**, 15105–1511.

Yuri L, A Katsuhiko, I Izumi, and K Toyoki, Assembly of multi-component protein films by means of electrostatic layer-by-layer adsorption. *Journal of the American Chemical Society*, 1995, **117**, 6117–6123.

Zeng H, H Wang, F Chen, H Xin, G Wang, L Xiao, K Song, D Wu, Q He, and G Shen, Development of quartz-crystal-microbalance-based immunosensor array for clinical immunophenotyping of acute leukemias. *Analytical Biochemistry*, 2006, **351**, 69–76.

– 6 –

Fractal Analysis of Binding and Dissociation of Analytes that Help Control Diseases on Biosensor Surfaces

6.1 INTRODUCTION

Biosensors are being used increasingly nowadays for the management, control, and diagnosis for a wide variety of diseases. Historically, biosensors were initially developed for the control of diabetes by quantitatively determining glucose levels in the blood of humans. The ease of biosensor use has prompted and permitted the application of this device for the detection of analytes that are involved in other diseases. Some of the more recent applications of the detection by biosensors of disease-related analytes that have recently appeared in the literature include:

(a) Control of blood pressure drugs by an endothelial biosensing system (Kamei *et al.*, 2004). These authors indicate that if blood pressure becomes abnormal due to hypertension (Olschewski *et al.*, 2001), or various arteriosclerotic diseases (Lusis, 2000; Berk *et al.*, 2001), then drugs are required to control and regulate blood pressure.
(b) The use of a fluorescent coagulation assay using a fiber optic evanescent wave sensor for the detection of thrombin (Garden *et al.*, 2004). Thrombin and other factors are involved in the blood coagulation system.
(c) The use of *in vitro* and *in vivo* dithiol probes for the detection of thiols (Pullela *et al.*, 2006). These authors indicate that thiols are important components of protein structures; they act as metabolic intermediates and play a central role in combating stress and in maintaining redox homeostasis (Rahman and MacNee, 2000).
(d) The application of cell structures on a biosensor chip for the detection of an allergic response (Matsubara *et al.*, 2004).
(e) The detection of kinases using surface plasmon resonance (SPR) biosensor technology (Nordin *et al.*, 2005). These authors indicate that one-third of all human proteins are thought to be substrates for protein kinases (Davies *et al.*, 2000; Cohen, 1999). Nordin *et al.* (2005) further indicate that a malfunction of cellular signaling pathways

that rely heavily on the phosphorylation status of proteins involved in these pathways may lead to pathological diseases such as cancer and inflammatory ailments. Thus, these protein kinases are frequently targeted groups for drug discovery.

(f) The detection of endothelin-1 using a SPR biosensor and the quartz crystal microbalance (QCM) method (Laricchia-Robio and Revoltella, 2004). These authors indicate that endothelin (ET) is a potent vasoconstrictor peptide, and three isoforms of human ET (ET-1, ET-2, and ET-3) have been identified (Hirata *et al.*, 1989).

(g) The detection of choline (ChO) and acetylcholine (ACh) using a polyvinylferrocenium modified Pt enzyme electrode (Sen *et al.*, 2004). These authors indicate that ACh is a neurotransmitter, and the determination of ACh in the brain is important for the understanding of the mechanisms of neurotransmission and neuroregulation. This should assist in the early detection and treatment of Alzheimer's disease (Campanella *et al.*, 1985; Huang *et al.*, 1993; Hale *et al.*, 1991).

(h) The detection of phosphatase using a SPR biosensor (Stenlund *et al.*, 2006). Phosphatase controls many cellular functions, and since their disturbances lead to a growing number of diseases, they are potential targets for therapeutic intervention.

Binding and dissociation (if applicable) kinetic curves for the above mentioned examples will be taken from the literature and analyzed using fractal methods to help determine binding and dissociation (if applicable) rate coefficient(s) values, as well as values of the fractal dimensions present in the binding and dissociation (if applicable) phases. Affinity values are also of interest to practicing biosensorists, and these values are also presented.

6.2 THEORY

Havlin (1989) has reviewed and analyzed the diffusion of reactants toward fractal surfaces. The details of the theory and the equations involved for the binding and the dissociation phases for analyte–receptor binding are available (Sadana, 2001). The details are not repeated here, except that just the equations are given to permit an easier reading. These equations have been applied to other biosensor systems (Sadana, 2001, 2005; Ramakrishnan and Sadana, 2001). For most applications, a single- or a dual-fractal analysis is often adequate to describe the binding and the dissociation kinetics. Peculiarities in the values of the binding and the dissociation rate coefficients, as well as in the values of the fractal dimensions with regard to the dilute analyte systems being analyzed will be carefully noted, if applicable.

6.2.1 Single-fractal analysis

Binding rate coefficient

Havlin (1989) indicates that the diffusion of a particle (analyte [Ag]) from a homogeneous solution to a solid surface (e.g., receptor [Ab]-coated surface) on which it reacts to form a product (analyte–receptor complex; Ab·Ag) is given by:

$$(\mathrm{Ab}\cdot\mathrm{Ag})\approx\begin{cases} t^{(3-D_{\mathrm{f,bind}})/2}=t^{\mathrm{p}} & t<t_{\mathrm{c}} \\ t^{1/2} & t>t_{\mathrm{c}} \end{cases} \tag{6.1}$$

Here $D_{\mathrm{f,bind}}$ or D_{f} is the fractal dimension of the surface during the binding step. t_{c} is the cross-over value. Havlin (1989) indicates that the crossover value may be determined by $r_{\mathrm{c}}^2 \sim t_{\mathrm{c}}$. Above the characteristic length, r_{c} the self-similarity of the surface is lost and the surface may be considered homogeneous. Above time, t_{c} the surface may be considered homogeneous, since the self-similarity property disappears, and 'regular' diffusion is now present. For a homogeneous surface where D_{f} is equal to 2, and when only diffusional limitations are present, $p = 1/2$ as it should be. Another way of looking at the $p = 1/2$ case (where $D_{\mathrm{f,bind}}$ is equal to two) is that the analyte in solution views the fractal object, in our case, the receptor-coated biosensor surface, from a 'large distance.' In essence, in the association process, the diffusion of the analyte from the solution to the receptor surface creates a depletion layer of width $(Ɖt)^{1/2}$ where $Ɖ$ is the diffusion constant. This gives rise to the fractal power law, $(\mathrm{Analyte}-\mathrm{Receptor})\sim t^{(3-D_{\mathrm{f,bind}})/2}$. For the present analysis, t_{c} is arbitrarily chosen and we assume that the value of the t_{c} is not reached. One may consider the approach as an intermediate 'heuristic' approach that may be used in the future to develop an autonomous (and not time-dependent) model for diffusion-controlled kinetics.

Dissociation rate coefficient

The diffusion of the dissociated particle (receptor [Ab] or analyte [Ag]) from the solid surface (e.g., analyte [Ag]–receptor [Ab]) complex coated surface) into solution may be given, as a first approximation by:

$$(\mathrm{Ab}\cdot\mathrm{Ag})\approx -t^{(3-D_{\mathrm{f,diss}})/2} = -t^{\mathrm{p}} \quad (t>t_{\mathrm{diss}}) \tag{6.2}$$

Here $D_{\mathrm{f,diss}}$ is the fractal dimension of the surface for the dissociation step. t_{diss} is the start of the dissociation step. This corresponds to the highest concentration of the analyte–receptor complex on the surface. Henceforth, its concentration only decreases. The dissociation kinetics may be analyzed in a manner 'similar' to the binding kinetics.

6.2.2 Dual-fractal analysis

Binding rate coefficient

Sometimes, the binding curve exhibits complexities and two parameters (k, D_{f}) are not sufficient to adequately describe the binding kinetics. This is further corroborated by low values of r^2 factor (goodness-of-fit). In that case, one resorts to a dual-fractal analysis (four parameters; k_1, k_2, D_{f1}, and D_{f2}) to adequately describe the binding kinetics. The single-fractal analysis presented above is thus extended to include two fractal dimensions. At present, the time ($t = t_1$) at which the 'first' fractal dimension 'changes' to the 'second' fractal dimension is arbitrary and empirical. For the most part, it is dictated by the data analyzed and experience gained by handling a single-fractal analysis. A smoother curve is obtained in the 'transition' region, if care is taken to select the correct number of points for the two regions. In this case, the product (antibody–antigen; or analyte–receptor complex, Ab · Ag or analyte–receptor) is given by:

$$(\text{Ab}\cdot\text{Ag}) \approx \begin{cases} t^{(3-D_{\text{f1,bind}})/2} = t^{\text{p1}} & (t < t_1) \\ t^{(3-D_{\text{f2,bind}})/2} = t^{\text{p2}} & (t_1 < t < t_2 = t_c) \\ t^{1/2} & (t > t_c) \end{cases} \quad (6.3)$$

In some cases, as mentioned above, a triple-fractal analysis with six parameters (k_1, k_2, k_3, D_{f1}, D_{f2}, and D_{f3}) may be required to adequately model the binding kinetics. This is when the binding curve exhibits convolutions and complexities in its shape due to perhaps the very dilute nature of the analyte or for some other reasons. Also, in some cases, a dual-fractal analysis may be required to describe the dissociation kinetics.

6.3 RESULTS

The fractal analysis will be applied for the detection of analytes that help control diseases on biosensor surfaces. Some of the analytes that will be analyzed include blood pressure drugs (Kamei *et al.*, 2004), thrombin (Garden *et al.*, 2004), thiols (Pullela *et al.*, 2006), analytes involved in an allergic response (Matsubara *et al.*, 2004), kinases (Nordin *et al.*, 2005), endothelin-1 (Laricchia-Robio and Revoltella, 2004), choline and acetylcholine (Sen *et al.*, 2004), and phosphatases (Stenlund *et al.*, 2006).

Alternate expressions for fitting the data are available that include saturation, first-order reaction, and no diffusion limitations, but these expressions are apparently deficient in describing the heterogeneity that inherently exists on the surface. One might justifiably argue that the appropriate modeling may be achieved by using a Langmuirian or other approach. The Langmuirian approach may be used to model the data presented if one assumes the presence of discrete classes of sites (for example, double exponential analysis as compared with a single-fractal analysis). Lee and Lee (1995) indicate that the fractal approach has been applied to surface science, for example, adsorption and reaction processes. These authors emphasize that the fractal approach provides a convenient means to represent the different structures and morphology at the reaction surface. These authors also emphasize using the fractal approach to develop optimal structures and as a predictive approach. Another advantage of the fractal technique is that the analyte–receptor association (as well as the dissociation reaction) is a complex reaction, and the fractal analysis via the fractal dimension and the rate coefficient provides a useful lumped parameter(s) analysis of the diffusion-limited reaction occurring on a heterogeneous surface.

In a classical situation, to demonstrate fractality, one should make a log–log plot, and one should definitely have a large amount of data. It may be useful to compare the fit to some other forms, such as exponential, or one involving saturation, etc. At present, we do not present any independent proof or physical evidence of fractals in the examples presented. It is a convenient means (since it is a lumped parameter) to make the degree of heterogeneity that exists on the surface more quantitative. Thus, there is some arbitrariness in the fractal model to be presented. The fractal approach provides additional information about interactions that may not be obtained by conventional analysis of biosensor data.

There is no nonselective adsorption of the analyte. The present system (environmental pollutants in the aqueous or the gas phase) being analyzed may be typically very dilute. Nonselective adsorption would skew the results obtained very significantly. In these types of systems, it is imperative to minimize this nonselective adsorption. We also do recognize that, in some cases, this nonselective adsorption may not be a significant component of the adsorbed material and that this rate of association, which is of a temporal nature, would depend on surface availability. If we were to accommodate the nonselective adsorption into the model, there would be an increase in the heterogeneity on the surface, since, by its very nature, nonspecific adsorption is more heterogeneous than specific adsorption. This would lead to higher fractal dimension values since the fractal dimension is a direct measure of the degree of heterogeneity that exists on the surface.

Matsubara *et al.* (2004) have recently used cell cultures immobilized on a biosensor chip for the detection of an allergic response. Figure 6.1a shows the binding of 5 μM quinacrine in solution to rat basophic leukemic cells (RBL-2H3), a tumor analog of rat

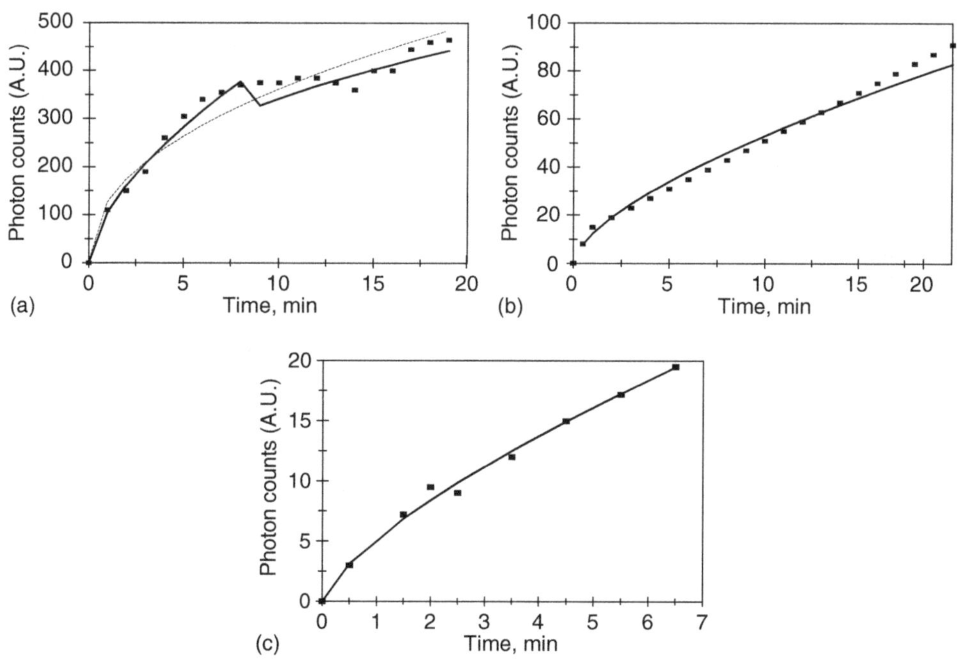

Figure 6.1 Binding of different concentrations (in μM) of quinacrine in solution to rat basophic leukemic cells (RBL-2H3), a tumor analog of rat mucosal mast cells, on a PDMS chip (Matsubara *et al.*, 2004): (a) 5; (b) 1. (c) Allergic response of mast cells/quinacrine release after exocytosis in a microfluidic chip (Matsubara *et al.*, 2004). (When only a solid line (___) is used, then a single-fractal analysis applies. When a (---) dashed and a solid (___) line is used, then the dashed line represents a single-fractal analysis and the solid line represents a dual-fractal analysis. In this case the solid line provides the better fit).

mucosal mast cells, immobilized on a poly(dimethylsiloxane) (PDMS) chip. A dual-fractal analysis is required to adequately describe the binding kinetics. The values of (a) the binding rate coefficient, k and the fractal dimension, D_f for a single-fractal analysis, and (b) the binding rate coefficients, k_1 and k_2 and the fractal dimensions, D_{f1} and D_{f2} for a dual-fractal analysis are given in Table 6.1.

The values of the binding rate coefficient presented in Table 6.1 were obtained from a regression analysis using Corel Quattro Pro 8.0 (1989) to model the data using eqs. (6.1a) and (6.1b), wherein $(\text{Analyte}-\text{Receptor}) = kt^{(3-D_f)/2}$ for a single-fractal analysis for the binding phase, and $(\text{Analyte}-\text{Receptor}) = kt^{(3-D_{f1})/2}$ and $kt^{(3-D_{f2})/2}$ for a dual-fractal analysis. The binding rate coefficient values presented in Table 6.1a are within 95% confidence values. For example, for the binding of 5.0 μM quinacrine in solution to RBL-2H3 immobilized on a PDMS chip the binding rate coefficient, k_1 value for a dual-fractal analysis is 124.89 ± 14.23. The 95% confidence limit indicates that the k_1 value lies between 110.66 and 139.12, and the values are precise and significant. It is of interest to note that as the fractal dimension or the degree of heterogeneity on the sensor chip surface increases by a factor of 1.272 from a value of D_{f1} equal to 1.7756 to D_{f2} equal to 2.2602, the binding rate coefficient increases by a factor of 1.435 from a value of k_1 equal to 105.67 to k_2 equal to 151.68. Note that changes in the fractal dimension or the degree of heterogeneity on the surface and in the binding rate coefficient are in the same direction.

Figure 6.1b shows the binding of 1 μM quinacrine in solution to RBL-2H3 immobilized on a PDMS chip. A single-fractal analysis is adequate to describe the binding kinetics. The values of the binding rate coefficient, k and the fractal dimension, D_f for a single-fractal analysis are given in Table 6.1.

Figure 6.1c shows the binding of quinacrine released after exocytosis of mast cells. Once again, a single-fractal analysis is adequate to describe the binding kinetics. The values of the binding rate coefficient, k and the fractal dimension, D_f for a single-fractal analysis are given in Table 6.1.

Garden *et al.* (2004) have recently developed a fiber optic sensor to detect thrombin. Figure 6.2a shows the binding of thrombin (denoted by NIH standards in NIH/ml) plus fluorescein 5-isothiocyanate (FITC)-labeled fibrinogen in solution to unlabeled fibrinogen

Table 6.1

Binding rate coefficients and fractal dimensions for (a) 5 and 1 μM quinacrine in solution to RBL-2H3, a tumor analog of rat mucosal mast cells, on a PDMS chip (Matsubara *et al.*, 2004), and (b) allergic response of mast cells/quinacrine release after exocytosis in a microfluidic chip (Matsubara *et al.*, 2004)

Analyte in solution/ receptor on surface	k	k_1	k_2	D_f	D_{f1}	D_{f2}
(a) 5 μM quinacrine/RBL-2H3 cells on PDMS chip	124.89 ± 14.23	105.67 ± 7.73	151.68 ± 8.71	2.0838 ± 0.0610	1.7756 ± 0.0694	2.2602 ± 0.1546
1 μM quinacrine/RBL-2H3 cells on PDMS chip	12.18 ± 1.00	na	na	1.7204 ± 0.0352	na	na
(b) Quinacrine release after exocytosis of mast cells	5.126 ± 0.372	na	na	1.5750 ± 0.6334	na	na

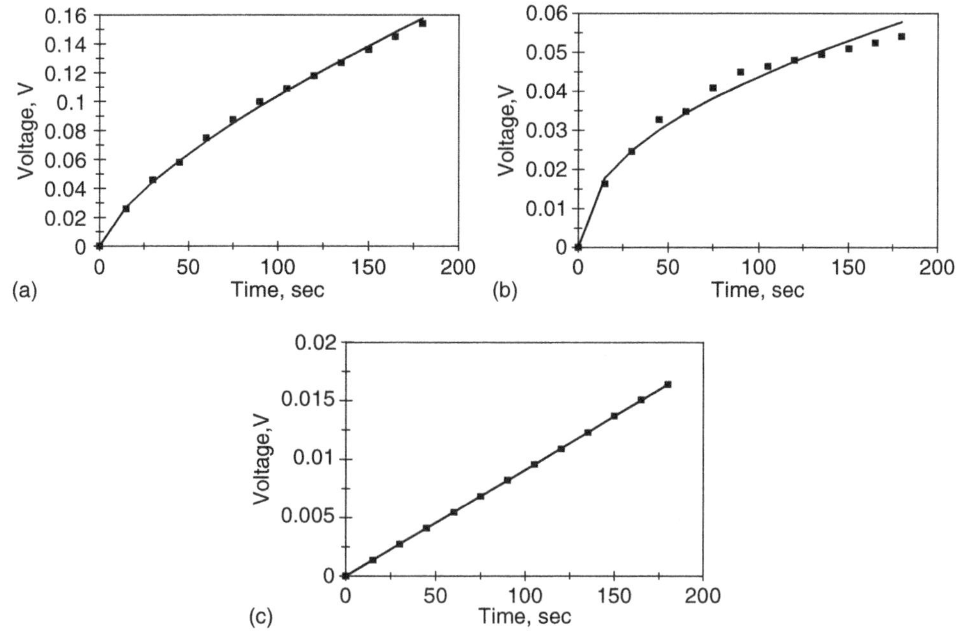

Figure 6.2 Binding of different concentrations of thrombin (denoted by NIH standards in NIH/ml) plus FITC-labeled fibrinogen to unlabeled fibrinogen coated on an optic fiber (Garden *et al.*, 2004): (a) 0.1 NIH/ml (b) 1 NIH/ml (c) 0.01 NIH/ml.

coated on an optic fiber. In this case the thrombin concentration is 0.1 NIH/ml in solution. A single-fractal analysis is adequate to describe the binding kinetics. The values of the binding rate coefficient, k and the fractal dimension, D_f are given in Table 6.2.

Figure 6.2b shows the binding of 1 NIH/ml thrombin plus FITC-labeled fibrinogen in solution to unlabeled fibrinogen coated on an optic fiber. In this case too a single-fractal analysis is adequate to describe the binding kinetics. The values of the binding rate coefficient, k and the fractal dimension, D_f are given in Table 6.3.

Figure 6.2c shows the binding of 0.01 NIH/ml thrombin plus FITC-labeled fibrinogen in solution to unlabeled fibrinogen coated on an optic fiber. Here too a single-fractal analysis is adequate to describe the binding kinetics. The values of the binding rate coefficient, k and the fractal dimension, D_f are given in Table 6.3.

Pullela *et al.* (2006) very recently indicated that thiols play a critical role in maintaining biological homeostasis. These authors have developed dithiol probes using a fluorescent reagent (DSSA) based on a dithiol linker. Figure 6.3a shows the binding of 1 mM glutathione in solution to the $DSSA_{Al}$ probe. The Al subscript denotes the aliphatic linker. A single-fractal analysis is adequate to describe the binding kinetics. The values of the binding rate coefficient, k and the fractal dimension, D_f for a single-fractal analysis are given in Table 6.2b.

Figure 6.3b shows the binding of 2 mM glutathione in solution to the $DSSA_{Al}$ probe. Here too, a single-fractal analysis is adequate to describe the binding kinetics. The values of the binding rate coefficient, k and the fractal dimension, D_f for a single-fractal analysis are given in Table 6.2b.

Table 6.2

Binding rate coefficients and fractal dimensions for (a) the binding of thrombin (denoted by NIH standards in NIH/ml) plus FITC-labeled fibrinogen to unlabeled fibrinogen coated on a optic fiber (Garden *et al.*, 2004), and (b) binding of different concentrations (in mM) of glutathione in solution to the $DSSA_{Al}$ probe (Pullela *et al.*, 2006)

Analyte in solution/receptor on surface	k	D_f	Reference
(a) 0.1 NIH/ml/unlabeled fibrinogen	0.004028 ± 0.000119	1.5878 ± 0.02330	Garden *et al.* (2004)
1 NIH/ml/unlabeled fibrinogen	0.005029 ± 0.000315	2.0598 ± 0.04848	Garden *et al.* (2004)
0.01 NIH/ml/unlabeled fibrinogen	9.1E-05 ± 0.0	2.0 ± 0.001742	Garden *et al.* (2004)
(b) 1 mM glutathione/$DSSA_{Al}$ probe	56.067 ± 0.860	2.2974 ± 0.01636	Pullela *et al.* (2006)
2 mM glutathione/$DSSA_{Al}$ probe	91.340 ± 1.311	2.3128 ± 0.01532	Pullela *et al.* (2006)
4 mM glutathione/$DSSA_{Al}$ probe	86.187 ± 2.875	2.2318 ± 0.02466	Pullela *et al.* (2006)
8 mM glutathione/$DSSA_{Al}$ probe	71.478 ± 4.916	2.1556 ± 0.04998	Pullela *et al.* (2006)
16 mM glutathione/$DSSA_{Al}$ probe	86.348 ± 5.028	2.1862 ± 0.0346	Pullela *et al.* (2006)

Table 6.3

Binding rate coefficients and fractal dimensions for the binding of different concentrations of TCEP hydrochloride to the $DSSA_{Al}$ probe (Pullela *et al.*, 2006)

Analyte in solution/ receptor on probe surface	k	k_1	k_2	D_f	D_{f1}	D_{f2}
11.5 mM TCEP/$DSSA_{Al}$	50.881 ± 4.478	16.297 ± 0.993	185.33 ± 2.46	2.1360 ± 0.0499	1.8750 ± 0.0700	2.3836 ± 0.0251
23 mM TCEP/$DSSA_{Al}$	313.70 ± 6.99	na	na	2.3786 ± 0.0130	na	na
46 mM TCEP/$DSSA_{Al}$	1232.33 ± 76.31	613.93 ± 33.17	4788.24 ± 23.03	2.6096 ± 0.0356	2.4512 ± 0.0623	2.8674 ± 0.0119
92 mM TCEP/$DSSA_{Al}$	3519.55 ± 145.02	2102.1 ± 59.89	8001.12 ± 6.50	2.8033 ± 0.0239	2.6856 ± 0.03326	2.9597 ± 0.00202

Figures 6.3c–e show the binding of 4–16 mM glutathione in solution to the $DSSA_{Al}$ probe. In each of these cases too, a single-fractal analysis is adequate to describe the binding kinetics. The values of the binding rate coefficient, k and the fractal dimension, D_f for a single-fractal analysis for each of these cases are given in Table 6.2b.

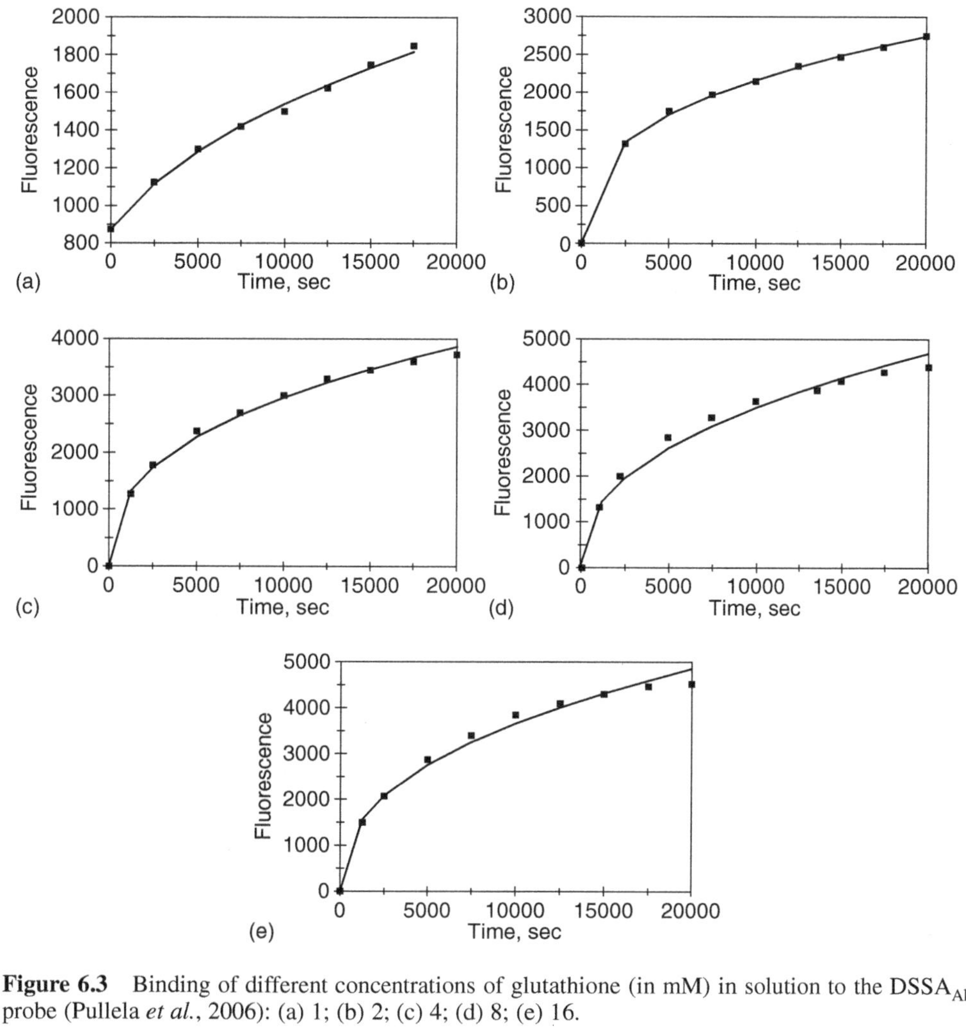

Figure 6.3 Binding of different concentrations of glutathione (in mM) in solution to the $DSSA_{Al}$ probe (Pullela *et al.*, 2006): (a) 1; (b) 2; (c) 4; (d) 8; (e) 16.

Figure 6.4a shows the binding of 11.5 mM tris(2-carboxyethyl)phosphine (TCEP) in solution to the $DSSA_{Al}$ probe (Pullela *et al.*, 2006). A dual-fractal analysis is required to adequately describe the binding kinetics. The values of (a) the binding rate coefficient, k and the fractal dimension, D_f for a single-fractal analysis, and (b) the binding rate coefficients, k_1 and k_2 and the fractal dimensions, D_{f1} and D_{f2} for a dual-fractal analysis are given in Table 6.3. It is of interest to note that for a dual-fractal analysis, as the fractal dimensions increases by 27.1% from a value of D_{f1} equal to 1.8750 to D_{f2} equal to 2.3836, the binding rate coefficient increases by a factor of 11.37 from a value of k_1 equal to 16.297 to k_2 equal to 185.33. Changes in the binding rate coefficient and in the fractal dimension or the degree of heterogeneity on the $DSSA_{Al}$ probe surface are in the same direction.

Figure 6.4b shows the binding of 23 mM TCEP in solution to the $DSSA_{Al}$ probe (Pullela *et al.*, 2006). A single-fractal analysis is adequate to describe the binding kinetics. The

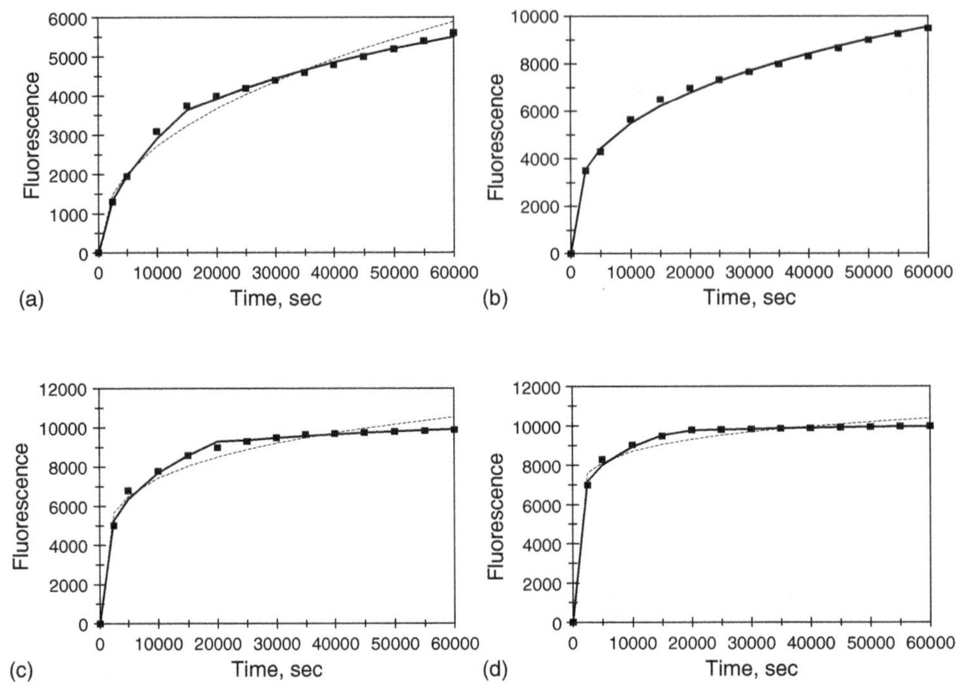

Figure 6.4 Binding of different concentrations of TCEP hydrochloride (in mM) in solution to the $DSSA_{Al}$ probe (Pullela *et al.*, 2006): (a) 11.5; (b) 23; (c) 46; (d) 92. (When only a solid line (__) is used, then a single-fractal analysis applies. When a (---) dashed and a solid (____) line is used, then the dashed line represents a single-fractal analysis and the solid line represents a dual-fractal analysis. In this case the solid line provides the better fit).

values of the binding rate coefficient, k and the fractal dimension, D_f for a single-fractal analysis are given in Table 6.3.

Figure 6.4c shows the binding of 46 mM TCEP in solution to the $DSSA_{Al}$ probe (Pullela *et al.*, 2006). A dual-fractal analysis is required to adequately describe the binding kinetics. The values of (a) the binding rate coefficient, k and the fractal dimension, D_f for a single-fractal analysis, and (b) the binding rate coefficients, k_1 and k_2 and the fractal dimensions, D_{f1} and D_{f2} for a dual-fractal analysis are given in Table 6.3. It is of interest to note that for a dual-fractal analysis, as the fractal dimension increases by 17% from a value of D_{f1} equal to 2.4512 to D_{f2} equal to 2.8674, the binding rate coefficient increases by a factor of 7.8 from a value of k_1 equal to 613.93 to k_2 equal to 4788.24. Changes in the binding rate coefficient and in the fractal dimension or the degree of heterogeneity on the $DSSA_{Al}$ probe surface are in the same direction.

Figure 6.4d shows the binding of 92 mM TCEP in solution to the $DSSA_{Al}$ probe (Pullela *et al.*, 2006). A dual-fractal analysis is required to adequately describe the binding kinetics. The values of (a) the binding rate coefficient, k and the fractal dimension, D_f for a single-fractal analysis, and (b) the binding rate coefficients, k_1 and k_2 and the fractal dimensions, D_{f1} and D_{f2} for a dual-fractal analysis are given in Table 6.3. It is of interest to note that for a dual-fractal analysis, as the fractal dimension increases by 10.2% from a value of D_{f1} equal to 2.6856 to D_{f2} equal to 2.9597, the binding rate coefficient increases by a factor of 3.806 from a value of k_1 equal to 2102.1 to k_2 equal to 8001.12. Changes in the binding rate coefficient and in the fractal dimension or the degree of heterogeneity on the $DSSA_{Al}$ probe surface are in the same direction.

Figure 6.5a and Table 6.3 show for a dual-fractal analysis the increase in the binding rate coefficient, k_1 with an increase in the TCEP concentration in solution. For the 11.5–92 mM TCEP concentration range, the binding rate coefficient, k_1 is given by:

$$k_1 = (0.05331 \pm 0.01953)\,[\mathrm{TCEP}]^{2.377 \pm 0.2084} \tag{6.4a}$$

The fit is good. Only three data points are available. The availability of more data points would lead to a more reliable fit. The binding rate coefficient, k_1 exhibits an order of dependence between second and two and one-half (equal to 2.377) on the TCEP concentration in the 11.5–92 mM range in solution. The non-integer order of dependence exhibited lends support to the fractal nature of the system.

Figure 6.5b and Table 6.3 show for a dual-fractal analysis the increase in the binding rate coefficient, k_2 with an increase in the TCEP concentration in solution. For the 11.5 to 92 mM TCEP concentration range, the binding rate coefficient, k_2 is given by:

$$k_2 = (2.164 \pm 1.758)\,[\mathrm{TCEP}]^{1.887 \pm 0.397} \tag{6.4b}$$

The fit is good. Only three data points are available. The availability of more data points would lead to a more reliable fit. The binding rate coefficient, k_2 is very sensitive to the TCEP concentration in solution exhibiting close to a second (equal to 1.887) order of dependence on the TCEP concentration in the 11.5–92 mM range in solution. The non-integer order of dependence exhibited once again lends support to the fractal nature of the system.

Figure 6.5c and Table 6.3 show the increase in the binding rate coefficient, k_1 with an increase in the fractal dimension, D_{f1}. For the 11.5 to 92 mM TCEP concentration in solution, the binding rate coefficient, k_1 is given by:

$$k_1 = (0.003302 \pm 0.000012) D_{f1}^{13.53 \pm 0.013} \tag{6.4c}$$

The fit is good. Only three data points are available. The availability of more data points would lead to a more reliable fit. The binding rate coefficient, k_1 is very sensitive to the fractal dimension, D_{f1} or the degree of heterogeneity that exists on the surface as noted by the slightly greater than 13 and one-half (equal to 13.53) order of dependence exhibited.

Figure 6.5d and Table 6.3 show the increase in the binding rate coefficient, k_2 with an increase in the fractal dimension, D_{f2}. For the 11.5 to 92 mM TCEP concentration in solution, the binding rate coefficient, k_2 is given by:

$$k_2 = (4.8\mathrm{E}{-05} \pm 0.4\mathrm{E}{-05}) D_{f2}^{17.46 \pm 0.172} \tag{6.4d}$$

The fit is good. Only three data points are available. The availability of more data points would lead to a more reliable fit. The binding rate coefficient, k_2 is very sensitive to the fractal dimension, D_{f2} or the degree of heterogeneity that exists on the surface as noted by the close to 17 and one-half (equal to 17.46) order of dependence exhibited.

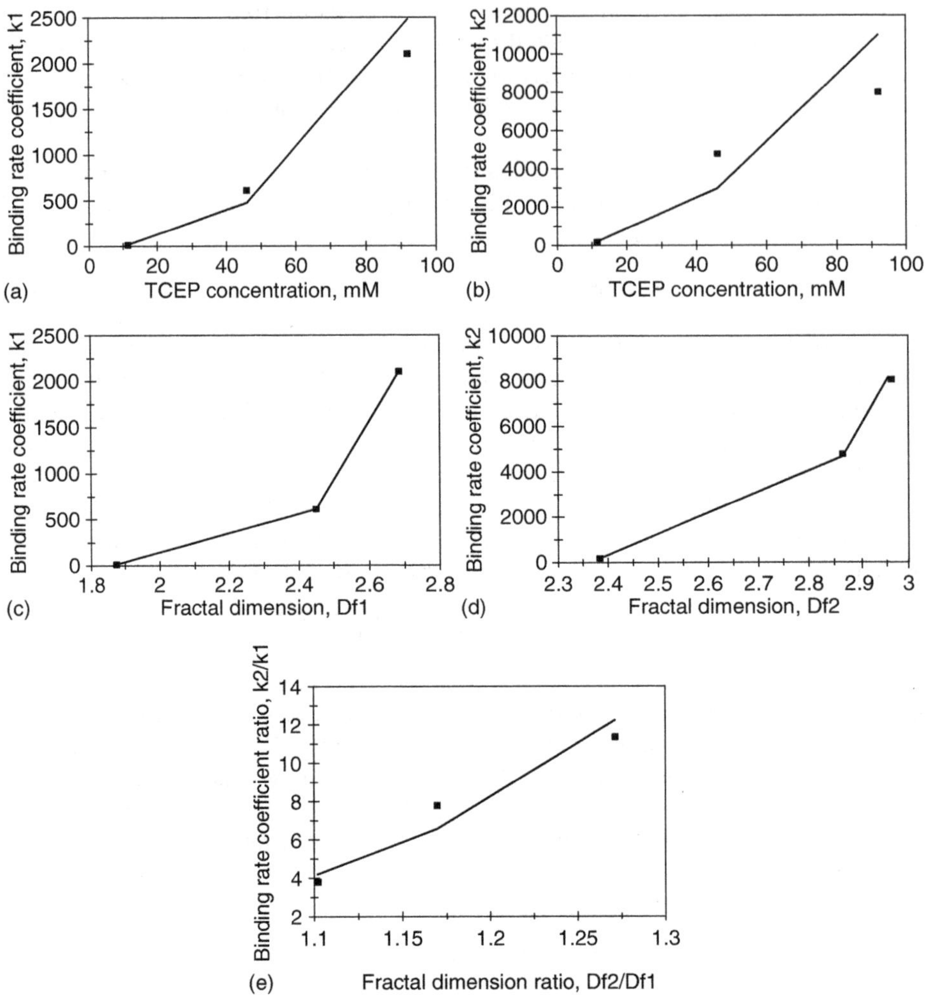

Figure 6.5 (a) Increase in the binding rate coefficient, k_1 with an increase in the TCEP concentration in solution. (b) Increase in the binding rate coefficient, k_2 with an increase in the TCEP concentration in solution. (c) Increase in the binding rate coefficient, k_1 with an increase in the fractal dimension, D_{f1}. (d) Increase in the binding rate coefficient, k_2 with an increase in the fractal dimension, D_{f2}. (e) Increase in the ratio of the binding rate coefficients, k_2/k_1 with an increase in the fractal dimension ratio, D_{f2}/D_{f1}.

Figure 6.5e and Table 6.3 show the increase in the ratio of the binding rate coefficients, k_2/k_1 with an increase in the ratio of the fractal dimensions, D_{f2}/D_{f1}. For the 11.5 to 92 mM TCEP concentration in solution, the ratio of the binding rate coefficients, k_2/k_1 is given by:

$$\frac{k_2}{k_1} = (2.038 \pm 0.481)\left(\frac{D_{f2}}{D_{f1}}\right)^{7.46 \pm 0.0172} \tag{6.4e}$$

The fit is good. Only three data points are available. The availability of more data points would lead to a more reliable fit. The ratio of the binding rate coefficients, k_2/k_1 is very sensitive to the ratio of fractal dimensions, D_{f2}/D_{f1} as noted by the close to seven and one-half order of dependence exhibited.

Figure 6.6a shows the binding of thiols in solution to the $DSSA_{Ar}$ probe with an aromatic linker (Pullela *et al.*, 2006). A single-fractal analysis is adequate to describe the binding kinetics. The values of the binding rate coefficient, k and the fractal dimension, D_f for a single-fractal analysis are 42.40 ± 2.41 and 2.3106 ± 0.04526, respectively. The authors indicate that all of the results presented were averages of five replicates, and the standard deviation is less than 10%.

Figure 6.6b shows the binding of thiols in solution to the $DSSA_{Al}$ probe with an aliphatic linker (Pullela *et al.*, 2006). A dual-fractal analysis is required to adequately describe the binding kinetics. The values of the binding rate coefficient, k and the fractal dimension, D_f for a single fractal analysis are 42.21 ± 4.7, and 2.664 ± 0.07902, respectively. The values of the binding rate coefficients, k_1 and k_2 for a dual-fractal analysis are 56.54 ± 0.83 and 24.62 ± 1.03, respectively. Similarly, the values of the fractal dimensions, D_{f1} and D_{f2} are 2.8309 ± 0.0206 and 2.368 ± 0.0911, respectively.

Figure 6.7a shows the binding of 1.5 mM TCEP hydrochloride in solution to the $DSSA_{Al}$ probe. A dual-fractal analysis is required to adequately describe the binding kinetics. The values of (a) the binding rate coefficient, k and the fractal dimension, D_f for a single- fractal analysis, and (b) the binding rate coefficients, k_1 and k_2, and the fractal dimensions, D_{f1} and D_{f2} for a dual-fractal analysis are given in Table 6.4. It is of interest to note that as the fractal dimension increases by a factor of 1.344 from a value of D_{f1} equal to 1.9922 to D_{f2} equal to 2.6782, the binding rate coefficient increases by a factor of 16.06 from a value of k_1 equal to 88.13 to k_2 equal to 1415.13. Changes in the degree of heterogeneity or the fractal dimension on the $DSSA_{Al}$ probe surface and in the binding rate coefficient are in the same direction.

Figure 6.7b shows the binding of 1.5 mM dithiothreitol (DTT) in solution to the $DSSA_{Al}$ probe. A dual-fractal analysis is required to adequately describe the binding kinetics. The values of the binding rate coefficient, k and the fractal dimension, D_f for a single-fractal analysis are given in Table 6.4.

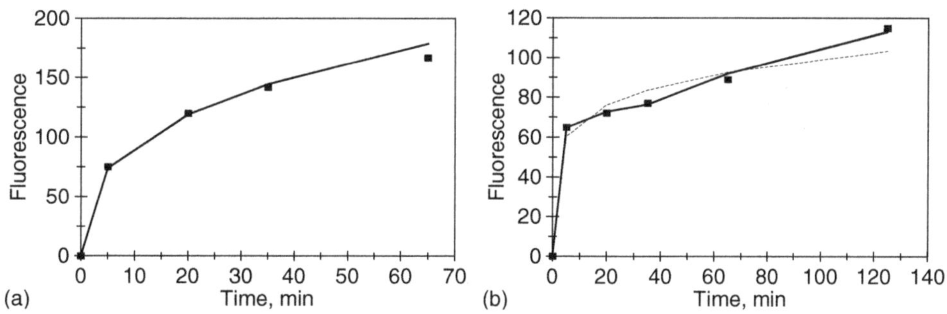

Figure 6.6 (a) Binding of thiols in *E. coli* cells to the $DSSA_{Ar}$ probe (aromatic linker) (Pullela *et al.*, 2006). (b) Binding of thiols in *E. coli* cells to the $DSSA_{Al}$ probe (aliphatic linker) (Pullela *et al.*, 2006). (When only a solid line (__) is used, then a single-fractal analysis applies. When a (---) dashed and a solid (____) line is used, then the dashed line represents a single-fractal analysis and the solid line represents a dual-fractal analysis. In this case the solid line provides the better fit).

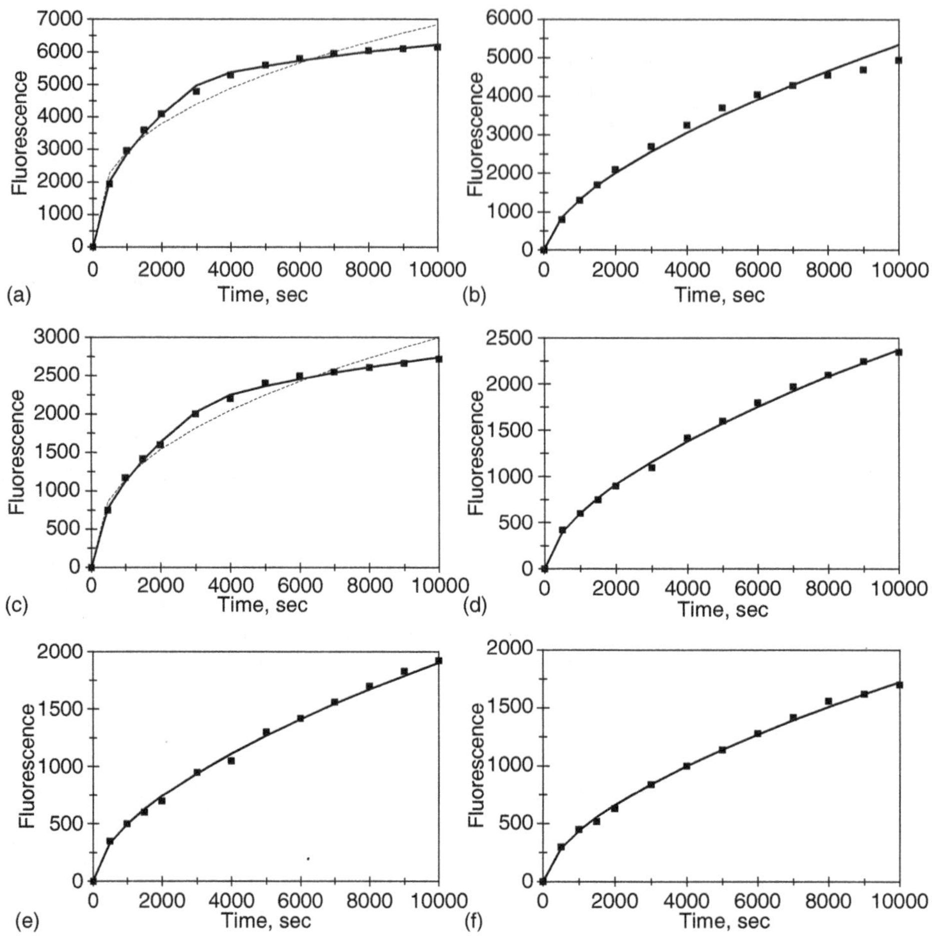

Figure 6.7 Binding of TCEP and thiols at 1.5 mM–5 μM of $DSSA_{Al}$ probe (Pullela *et al.*, 2006): (a) TCEP hydrochloride; (b) dithiothreitol (DTT); (c) homocysteine (Hcy); (d) *N*-methyl (2-thiopropionyl) glycine (NMPG); (e) 2-mercaptoethanol (βME); (f) reduced glutathione (GSH). (When only a solid line (__) is used, then a single-fractal analysis applies. When a (---) dashed and a solid (___) line is used, then the dashed line represents a single-fractal analysis and the solid line represents a dual-fractal analysis. In this case the solid line provides the better fit).

Figure 6.7c shows the binding of 1.5 mM homocysteine (Hcy) in solution to the $DSSA_{Al}$ probe. A dual-fractal analysis is required to adequately describe the binding kinetics. The values of (a) the binding rate coefficient, k and the fractal dimension, D_f for a single-fractal analysis, and (b) the binding rate coefficients, k_1 and k_2 and the fractal dimensions, D_{f1} and D_{f2} for a dual-fractal analysis are given in Table 6.4. It is of interest to note that as the fractal dimension increases by a factor of 1.335 from a value of D_{f1} equal to 1.9238 to D_{f2} equal to 2.5684, the binding rate coefficient increases by a factor of 13.78 from a value of k_1 equal to 27.285 to k_2 equal to 375.93. Changes in the degree of heterogeneity or the fractal dimension on the $DSSA_{Al}$ probe surface and in the binding rate coefficient are in the same direction.

Table 6.4

Binding rate coefficients and fractal dimensions for different thiols and TCEP at 5 μM–1.5 mM of the $DSSA_{Al}$ in Tris buffer (100 mM, pH 8.2) at 25 °C (Pullela *et al.*, 2006)

Analyte in solution	k	k_1	k_2	D_f	D_{f1}	D_{f2}
tris(2-carboxyethyl) phosphine chloride (TCEP)	232.40 ± 20.24	88.13 ± 2.58	1415.13 ± 16.82	2.2654 ± 0.0529	1.9922 ± 0.0580	2.6782 ± 0.02942
Dithiothreitol (DTT)	19.537 ± 1.077	na	na	1.7814 ± 0.0339	na	na
Homocysteine (Hcy)	64.83 ± 5.23	27.285 ± 0.991	375.93 ± 5.94	2.1674 ± 0.049	1.9238 ± 0.0520	2.5684 ± 0.0391
N-methyl(2-thiopropionyl)glycine (NMPG)	9.910 ± 0.41	na	na	1.8100 ± 0.0254	na	na
2-mercaptoethanol (βME)	8.725 ± 0.334	na	na	1.8306 ± 0.0238	na	na
Reduced glutathione (GSH)	7.086 ± 0.236	na	na	1.8070 ± 0.021	na	na

Figure 6.7d shows the binding of 1.5 mM *N*-methyl(2-thiopropionyl)glycine (NMPG) in solution to the $DSSA_{Al}$ probe. A dual-fractal analysis is required to adequately describe the binding kinetics. The values of the binding rate coefficient, k and the fractal dimension, D_f for a single-fractal analysis are given in Table 6.4.

Figure 6.7e shows the binding of 1.5 mM 2-mercaptoethanol (βME) in solution to the $DSSA_{Al}$ probe. A dual-fractal analysis is required to adequately describe the binding kinetics. The values of the binding rate coefficient, k and the fractal dimension, D_f for a single-fractal analysis are given in Table 6.4.

Figure 6.7f shows the binding of 1.5 mM reduced glutathione in solution to the $DSSA_{Al}$ probe. A dual-fractal analysis is required to adequately describe the binding kinetics. The values of the binding rate coefficient, k and the fractal dimension, D_f for a single-fractal analysis are given in Table 6.4.

Figure 6.8 and Table 6.4 show for a single-fractal analysis the decrease in the binding rate coefficient, k with an increase in the fractal dimension, D_f. For the data presented in Figure 6.8, the binding rate coefficient, k is given by:

$$k = (6.4 \pm 2.6)D_f^{-30.29 \pm 17.65} \tag{6.5}$$

The fit is reasonable. Only four data points are available. The availability of more data points would lead to a more reliable fit. The binding rate coefficient, k is extremely sensitive to the fractal dimension, D_f or the degree of heterogeneity that exists on the $DSSA_{Al}$ probe surface as noted by the slightly greater than 13th order of dependence exhibited.

Kamei *et al.* (2004) indicate that blood pressure control is essential for homeostasis of our bodies. These authors wanted to develop an assessment system for blood pressure control drugs. They indicate that nitric oxide (NO) is an indicator for assessing drugs (chemicals) for

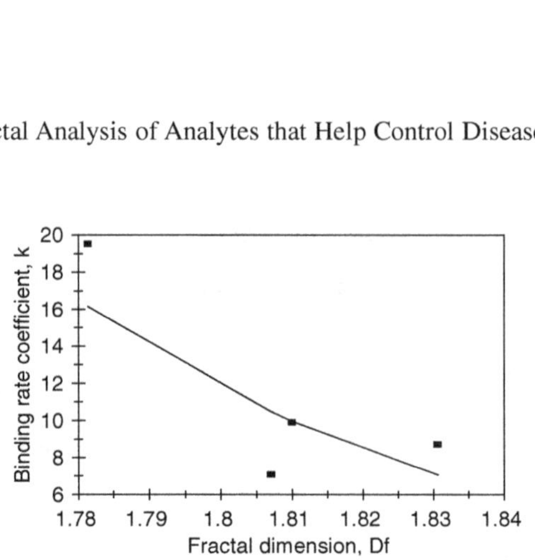

Figure 6.8 Decrease in the binding rate coefficient, k with an increase in the fractal dimension, D_f.

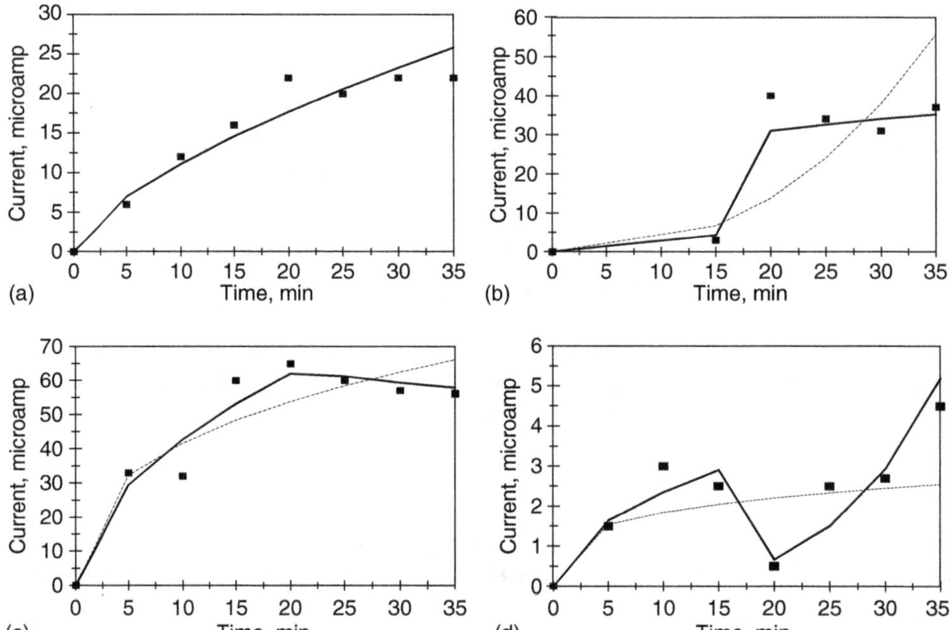

Figure 6.9 Binding of blood pressure drugs to the endothelial biosensing system (Kamei *et al.*, 2004): (a) no stimulation; (b) 1 mM acetyl choline chloride (AcChCl); (c) 1mM NOC 7 (NO, nitric oxide donor); (d) 1 mM AcChCl + 5 mM *N*-monomethyl-L-arginine (L-NMMA). (When only a solid line (__) is used, then a single-fractal analysis applies. When a (---) dashed and a solid (____) line is used, then the dashed line represents a single-fractal analysis and the solid line represents a dual-fractal analysis. In this case the solid line is the best fit line).

blood pressure control. These authors developed the endothelial biosensing system to assess blood pressure control drugs.

Figures 6.9a–d shows the quantitative analysis of NO released using differential pulse voltametry (DPV) with the gold electrode coated with a polyion complex layer. Kamei *et al.* (2004) use ACh (Furchgott, 1983), NOC 7 (Keefer *et al.*, 1996), and

N-monomethyl-L-arginine (L-NMMA) (Aisaka *et al.*, 1989) as standard samples of blood pressure control drugs.

Figure 6.9a shows the release of nitric oxide without stimulation. A single-fractal analysis is adequate to describe the binding kinetics of nitric oxide to the gold electrode coated with a polyion complex layer in the endothelial cellular biosensing system. The values of the binding rate coefficient, k and the fractal dimension, D_f for a single-fractal analysis is given in Table 6.5.

Figure 6.9b shows the release of nitric oxide stimulated by 1 mM AcChCl (acetylcholine chloride), and its binding to the endothelial cellular biosensing system. A dual-fractal analysis is required to adequately describe the binding kinetics. The values of (a) the binding rate coefficient, k and the fractal dimension, D_f for a single-fractal analysis, and (b) the binding rate coefficients, k_1 and k_2 and the fractal dimensions, D_{f1} and D_{f2} for a dual-fractal analysis are given in Table 6.5. Note that due to the convex nature of the shape of the curve at time, t close to zero, the estimated fractal dimension value (D_f or D_{f1}) is equal to zero. This indicates that the surface acts as a 'Cantor' like dust (Viscek, 1989).

Figure 6.9c shows the release of nitric oxide stimulated by 1 mM NOC 7, and its binding to the endothelial cellular biosensing system. A dual-fractal analysis is required to adequately describe the binding kinetics. The values of (a) the binding rate coefficient, k and the fractal dimension, D_f for a single-fractal analysis, and (b) the binding rate coefficients, k_1 and k_2 and the fractal dimensions, D_{f1} and D_{f2} for a dual-fractal analysis are given in Table 6.5. It is of interest to note that as the fractal dimension on the surface increases by a factor of 1.563 from a value of D_{f1} equal to 1.9188 to D_{f2} equal to 3.0, and the binding rate coefficient increases by a factor of 8.56 from a value of k_1 equal to 12.284 to k_2 equal to 105.16. Changes in the binding rate coefficient and in the fractal dimension or the degree of heterogeneity on the endothelial biosensing system (gold electrode coated with a polyion complex layer) surface are in the same direction.

Figure 6.9d shows the release of nitric oxide stimulated by 1 mM AcChCl + 5 mM L-NMMA, and its binding to the endothelial cellular biosensing system. A dual-fractal analysis is required to adequately describe the binding kinetics. The values of (a) the binding

Table 6.5

Binding rate coefficients and fractal dimensions for the binding of blood pressure drugs to the endothelial biosensing system. Time course of NO release by HUVEC stimulated by 1 mM AcChCl, 1 mM NOC 7, and 1 mM AcChCl + 5 mM L-NMMA (Kamei *et al.*, 2004)

Analyte in solution	k	k_1	k_2	D_f	D_{f1}	D_{f2}
No stimulation	2.3386 ± 0.3797	na	na	1.6488 ± 0.1794	na	na
1 mM AcChCl	0.00781 ± 0.010	6.4E-06 + 10.6E-06	15.626 ± 1.864	0 + 2.508	0	2.5432 ± 0.9462
1 mM NOC 7	17.748 ± 3.603	12.284 ± 3.305	105.16 ± 4.74	2.2594 ± 0.2204	1.9188 ± 0.4578	3.0 − 0.2112
1 mM AcChCl + 5 mM L-NMMA	1.003 + 1.122	0.7072 ± 0.2469	1.1E-05 ± 0.6E-05	1.4830 ± 0.8948	1.9458 ± 0.7622	0 + 2.0438

rate coefficient, k and the fractal dimension, D_f for a single-fractal analysis, and (b) the binding rate coefficients, k_1 and k_2 and the fractal dimensions, D_{f1} and D_{f2} for a dual-fractal analysis are given in Table 6.5. Note that in this case, the estimated value of the fractal dimension during the second phase of binding, D_{f2} is equal to zero.

Table 6.5 and Figure 6.10 show for a dual-fractal analysis the increase in the binding rate coefficient, k_2 with an increase in the fractal dimension, D_{f2}. For the data shown in Figure 6.10, the binding rate coefficient, k_2 is given by:

$$k_2 = (0.396 \pm 0.499)D_{f2}^{4.569 \pm 0.301} \tag{6.6}$$

The fit is reasonable. There is considerable scatter in the data as may be noted by the large positive error in the binding rate coefficient. The availability of more data points would lead to a more reliable fit. The binding rate coefficient, k_2 is very sensitive to the fractal dimension, D_{f2} or the degree of heterogeneity that exists on the surface as noted by the close to four and one-half order of dependence exhibited.

Sen *et al.* (2004) indicate that enzyme electrodes are important for the determination of specific substrates in clinical analysis. Potter *et al.* (1983) indicate that the combination of immobilized enzyme technology and electrochemical sensors has led to the development of low-cost devices that are not only rapid but also sensitive. Sen *et al.* (2004) have developed a polyvinylferrocenium modified Pt electrode to be used as amperometric choline and acetylcholine electrodes. These authors co-immobilized choline oxidase and acetylcholoinesterase in a polyvinylferrocenium perchlorate matrix coated on a Pt electrode surface. They analyzed the influence of the polymeric film thickness, temperature, pH, enzyme, and substrate concentration on the response of the enzyme electrode.

Figure 6.11a shows the binding and dissociation of 0.15 mM acetylcholine in solution to the acetylcholine enzyme electrode. A single-fractal analysis is adequate to describe the binding and dissociation kinetics. The values of (a) the binding rate coefficient, k and the fractal dimension, D_f for a single-fractal analysis, and (b) the dissociation rate coefficient, k_d and the fractal dimension in the dissociation phase, D_{fd} for a single-fractal analysis are given in Table 6.6. The value of the affinity, $K = k/k_d$ is equal to 360.

Figure 6.11b shows the binding and dissociation of 0.15 mM choline in solution to the acetylcholine enzyme electrode. A dual-fractal analysis is required to adequately describe the

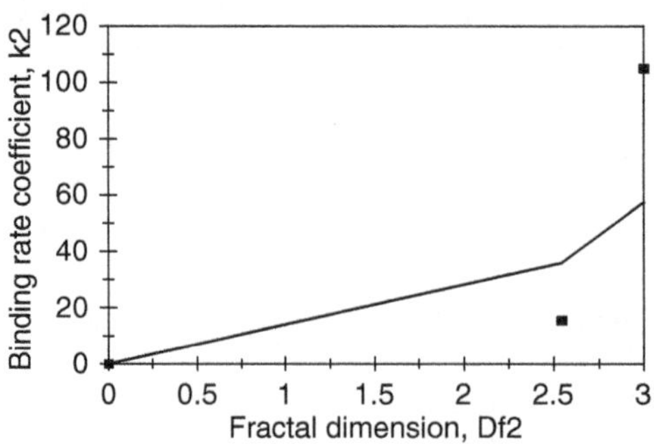

Figure 6.10 Increase in the binding rate coefficient, k_2 with an increase in the fractal dimension, D_{f2}.

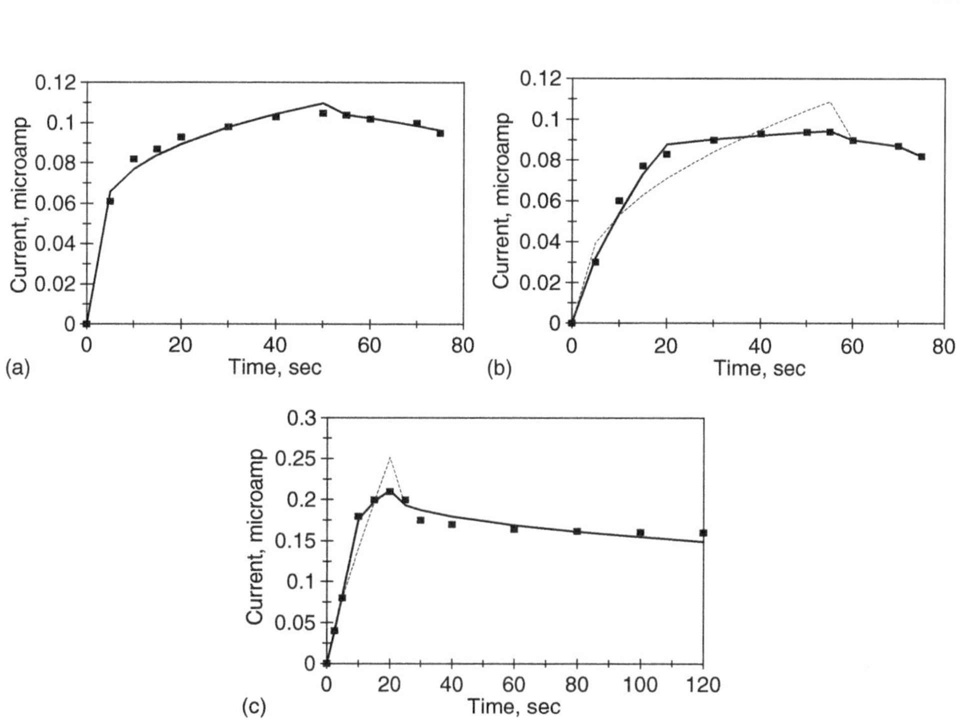

Figure 6.11 Binding of (Sen *et al.*, 2004): (a) acetylcholine to an acetylcholine electrode; (b) choline to an acetylcholine electrode; (c) choline to a choline electrode. (When only a solid line (__) is used, then a single-fractal analysis applies. When a (---) dashed and a solid (____) line is used, then the dashed line represents a single-fractal analysis and the solid line represents a dual-fractal analysis. In this case the solid line provides the best fit).

binding kinetics. A single-fractal analysis is adequate to describe the dissociation kinetics. The values of (a) the binding rate coefficient, k and the fractal dimension, D_f for a single-fractal analysis, (b) the binding rate coefficients, k_1 and k_2 and the fractal dimensions, D_{f1} and D_{f2} for a dual-fractal analysis, and (c) the dissociation rate coefficient, k_d and the fractal dimension in the dissociation phase, D_{fd} for a single-fractal analysis are given in Table 6.6. It is of interest to note that as the fractal dimension increases by a factor of 1.92 from a value of D_{f1} equal to 1.49 to D_{f2} equal to 2.8567, the binding rate coefficient increases by a factor of 7.47 from a value of k_1 equal to 0.009483 to k_2 equal to 0.07084. Changes in the fractal dimension or the degree of heterogeneity on the surface and in the binding rate coefficient are in the same direction. Affinities are defined by, $K_1 = k_1/k_d$ and $K_2 = k_2/k_d$. Then, the K_1 and K_2 values are 7.83 and 58.5, respectively.

Figure 6.11c shows the binding and dissociation of 0.15 mM choline in solution to the choline enzyme electrode. A dual-fractal analysis is required to adequately describe the binding kinetics. A single-fractal analysis is adequate to describe the dissociation kinetics. The values of (a) the binding rate coefficient, k and the fractal dimension, D_f for a single-fractal analysis, (b) the binding rate coefficients, k_1 and k_2 and the fractal dimensions, D_{f1} and D_{f2} for a dual-fractal analysis, and (c) the dissociation rate coefficient, k_d and the fractal dimension for the dissociation phase, D_{fd} for a single-fractal analysis are given in Table 6.6. It is of interest to note that as the fractal dimension increases by a factor of 3.0722

Table 6.6

Binding and dissociation rate coefficients and fractal dimensions for (a) choline and acetylcholine to an enzyme electrode (Sen *et al.*, 2004), (b) different concentrations (1.5–25 μM) of cantharidin in solution to PP1 (protein phosphatase1) immobilized on a sensor surface by amine coupling (Stenlund *et al.*, 2006)

Analyte in solution	k	k_1	k_2	k_d	D_f	D_{f1}	D_{f2}	D_{fd}	References
(a) Acetylcholine/ acetylcholine electrode	0.04608 ± 0.00259	na	na	0.000128 ± 0.000034	2.5562 ± 0.05498	na	na	0.3876 ± 0.3698	Sen *et al.* (2004)
Choline/acetylcholine electrode	0.0200 ± 0.00395	0.009483 ± 0.00115	0.07084 ± 0.00050	0.001211 ± 0.000010	2.1546 ± 0.1623	1.4900 ± 0.2206	2.8567 ± 0.0301	1.4150 ± 0.01717	Sen *et al.* (2004)
Choline/choline electrode	0.0204 ± 0.0041	0.01451 ± 0.00072	0.1077 ± 0.0013	0.008331 ± 0.003440	1.3230 ± 0.2192	0.8302 ± 0.0981	2.5504 ± 0.0501	2.1368 ± 0.2516	Sen *et al.* (2004)
(b) 25 μM cantharidin/ PP1 on a sensor chip	0.7817 ± 0.0686	na	na	0.1324 ± 0.1031	2.0372 ± 0.0504	na	na	1.8080 ± 0.7867	Stenlund *et al.* (2006)
12 μM cantharidin/ PP1 on a sensor chip	0.3336 ± 0.0124	na	na	0.1446 ± 0.0091	1.7866 ± 0.0436	na	na	1.9176 ± 0.0513	Stenlund *et al.* (2006)
6 μM cantharidin/PP1 on a sensor chip	0.1207 ± 0.0030	na	na	0.09288 ± 0.00456	1.5338 ± 0.2962	na	na	1.8862 ± 0.0400	Stenlund *et al.* (2006)
3 μM cantharidin/PP1 on a sensor chip	0.0607 ± 0.00597	na	na	0.01412 ± 0.00116	1.5430 ± 0.1124	na	na	1.4218 ± 0.0662	Stenlund *et al.* (2006)
1.5 μM cantharidin/PP1 on a sensor chip	0.00674 ± 0.00093	na	na	0.003598 ± 0.000021	0.9548 ± 0.0149	na	na	0.9969 ± 0.00589	Stenlund *et al.* (2006)

from a value of D_{f1} equal to 0.8302 to D_{f2} equal to 2.5504, the binding rate coefficient increases by a factor of 7.42 from a value of k_1 equal to 0.014513 to k_2 equal to 0.1077. Changes in the fractal dimension or the degree of heterogeneity on the surface and in the binding rate coefficient are once again in the same direction. The K_1 and K_2 values are 1.74 and 12.93, respectively.

Note that the affinity values for choline for the acetylcholine electrode are higher than those for the choline electrode. The K_1 value is higher by a factor of 4.5, and the K_2 value is higher by a factor of 4.52. Both of the affinity values increase by almost the same amount.

Stenlund *et al.* (2006) indicate that the analysis of protein interactions using the SPR biosensor provides useful information in drug discovery applications (Myszka and Rich, 2000). Stenlund *et al.* (2006) have presented kinetic data for the interaction of three different phosphatases: PP1, PP2B, and PTP1B. The first two are serine/threonine phosphatases, and the third one is a tyrosine phosphatase. Phosphatases, these authors emphasize, are involved in cellular signal transduction, and are thus potential targets for therapeutic intervention. They presented assays for the kinetic characterization of inhibitor–phosphatase interactions.

Figure 6.12a shows the binding and dissociation of 25 μM canthardin in solution to protein phosphatase1 (PP1) immobilized on a SPR biosensor chip surface by amine coupling in the presence of the detergent P20. Stenlund *et al.* (2006) indicate that P20 enhances the data quality of Biacore assays (Brogan *et al.*, 2004), and did improve the data quality for the phosphatases used in the present study. A single-fractal analysis is adequate to describe the binding and dissociation kinetics. The values of (a) the binding rate coefficient, k and the fractal dimension, D_f for a single-fractal analysis, and (b) the dissociation rate coefficient, k_d and the fractal dimension, D_{fd} for a single-fractal analysis in the dissociation phase are given in Table 6.6.

Figure 6.12b shows the binding and dissociation of 12 μM canthardin in solution to PP1 immobilized on a SPR biosensor chip surface by amine coupling in the presence P20. A single-fractal analysis is once again adequate to describe the binding and dissociation kinetics. The values of (a) the binding rate coefficient, k and the fractal dimension, D_f for a single-fractal analysis, and (b) the dissociation rate coefficient, k_d and the fractal dimension, D_{fd} for a single-fractal analysis in the dissociation phase are given in Table 6.6.

Figure 6.12c shows the binding and dissociation of 6 μM canthardin in solution to PP1 immobilized on a SPR biosensor chip surface by amine coupling in the presence of P20. A single-fractal analysis is adequate to describe the binding and dissociation kinetics. The values of (a) the binding rate coefficient, k and the fractal dimension, D_f for a single-fractal analysis, and (b) the dissociation rate coefficient, k_d and the fractal dimension, D_{fd} for a sinlge-fractal analysis in the dissociation phase are given in Table 6.6.

Figure 6.12d shows the binding and dissociation of 3 μM canthardin in solution to PP1 immobilized on a SPR biosensor chip surface by amine coupling in the presence of P20. A single-fractal analysis is adequate to describe the binding and dissociation kinetics. The values of (a) the binding rate coefficient, k and the fractal dimension, D_f for a single-fractal analysis, and (b) the dissociation rate coefficient, k_d and the fractal dimension, D_{fd} for a single-fractal analysis in the dissociation phase are given in Table 6.6.

Figure 6.12e shows the binding and dissociation of 3 μM canthardin in solution to PP1 immobilized on a SPR biosensor chip surface by amine coupling in the presence of P20. A single-fractal analysis is adequate to describe the binding and dissociation kinetics. The

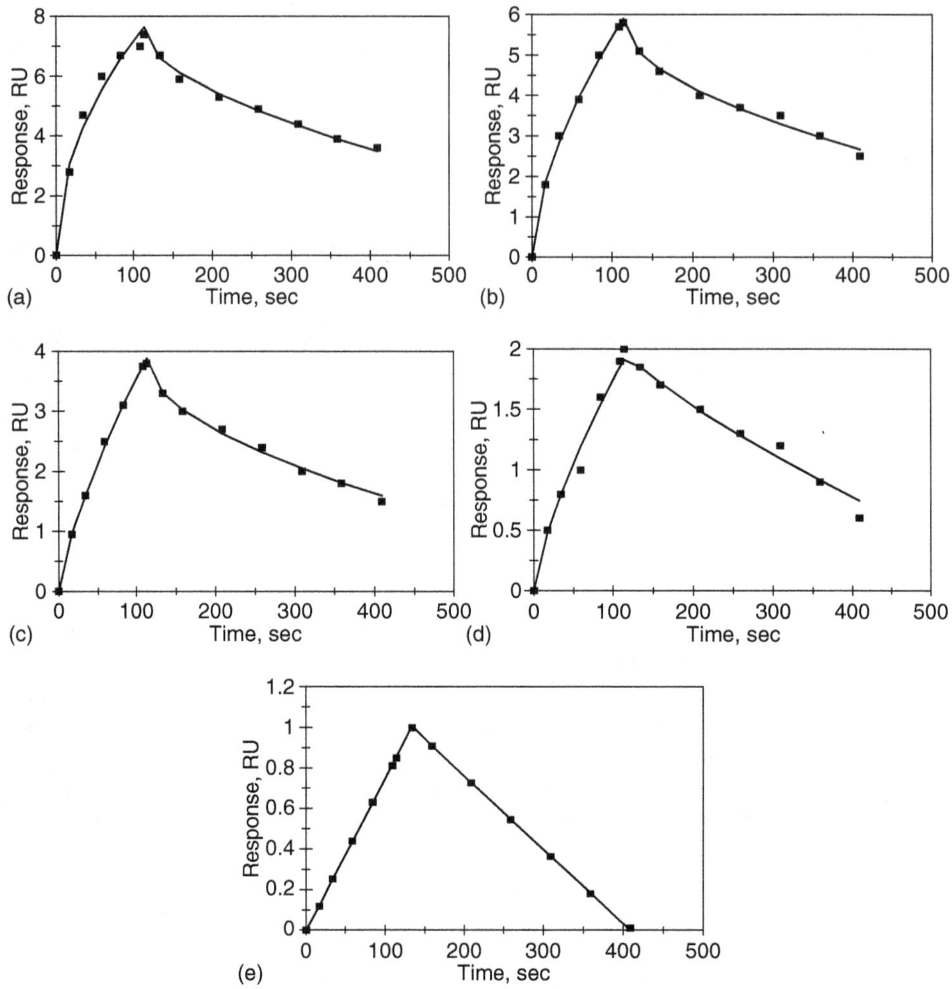

Figure 6.12 Binding of different concentrations of cantharidin (in μM) in solution to protein phosphate 1 (PP1) immobilized on a sensor chip surface by amine coupling (Stenlund *et al.*, 2006): (a) 25; (b) 12; (c) 6; (d) 3; (e) 1.5.

values of (a) the binding rate coefficient, k and the fractal dimension, D_f for a single-fractal analysis, and (b) the dissociation rate coefficient, k_d and the fractal dimension, D_{fd} for a single-fractal analysis in the dissociation phase are given in Table 6.6.

Figure 6.13a and Table 6.6 show for a single-fractal analysis the increase in the binding rate coefficient, k with an increase in the cantharidin concentration in solution in the 1.5–25 μM range. For the data presented in Figure 6.13a, the binding rate coefficient, k is given by:

$$k = (0.005945 \pm 0.003671)[\text{cantharidin } \mu\text{M}]^{1.596 \pm 0.2164} \tag{6.7a}$$

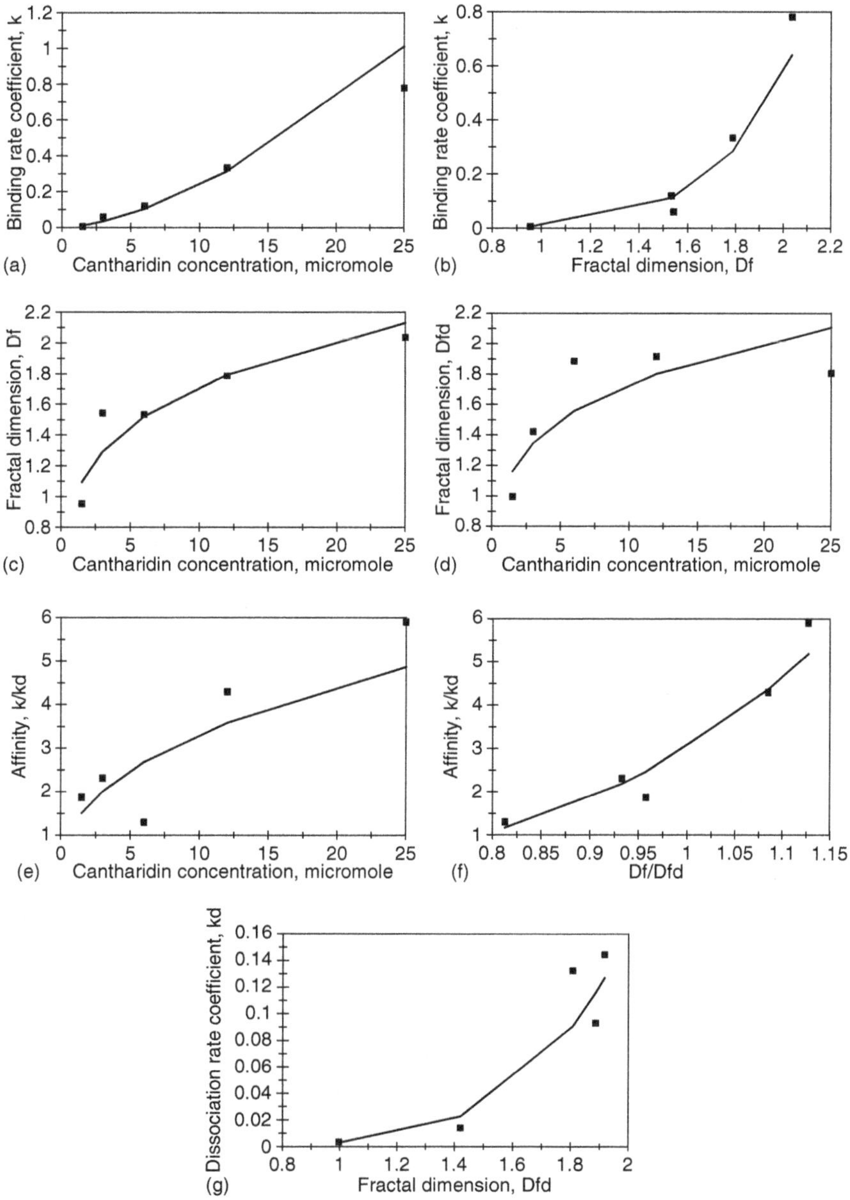

Figure 6.13 (a) Increase in the binding rate coefficient, k with an increase in the cantharidin concentration (in μM) in solution. (b) Increase in the binding rate coefficient, k with an increase in the fractal dimension, D_f. (c) Increase in the fractal dimension, D_f with an increase in the cantharidin concentration (in μM) in solution. (d) Increase in the fractal dimension in the dissociation phase, D_{fd} with an increase in the cantharidin concentration (in μM) in solution. (e) Increase in the affinity, K ($=k/k_d$) with an increase in the cantharidin concentration (in μM) in solution. (f) Increase in the affinity, K ($=k/k_d$) with an increase in the ratio of fractal dimensions, D_f/D_{fd}. (g) Increase in the dissociation rate coefficient, k_d with an increase in the fractal dimension, D_{fd}.

The fit is very good. Only five data points are available. The availability of more data points would lead to a more reliable fit. The binding rate coefficient, k exhibits close to one and one-half (equal to 1.596) order of dependence on the cantharidin concentration in solution in the 1.5–25μM range. The non-integer order of dependence exhibited by the binding rate coefficient, k on the cantharidin concentration in solution lends support to the fractal nature of the system.

Figure 6.13b and Table 6.6 show for a single-fractal analysis the increase in the binding rate coefficient, k with an increase in the fractal dimension, D_f. For the data presented in Figure 6.13b, the binding rate coefficient, k is given by:

$$k = (0.007627 \pm 0.003828)D_f^{6.227 \pm 0.7098} \tag{6.7b}$$

The fit is good. Only five data points are available. The availability of more data points would lead to a more reliable fit. The binding rate coefficient, k is very sensitive to the fractal dimension, D_f or the degree of heterogeneity on the sensor chip surface as noted by a higher than sixth (equal to 6.227) order of dependence exhibited.

Figure 6.13c and Table 6.6 show the increase in the fractal dimension, D_f with an increase in the cantharidin concentration in solution in the 1.5–25 μM range. For the data presented in Figure 6.13c, the fractal dimension, D_f is given by:

$$D_f = (0.9955 \pm 0.1413)[\text{cantharidin}]^{0.2366 \pm 0.0598} \tag{6.7c}$$

The fit is very good. Only five data points are available. The availability of more data points would lead to a more reliable fit. The fractal dimension, D_f is only mildly sensitive to the cantharidin concentration in solution as noted by the low (equal to 0.2366) order of dependence exhibited. Note that the fractal dimension is based on a log scale, and even small changes in the fractal dimension value will lead to significant changes in the degree of heterogeneity on the sensor chip surface.

Figure 6.13d and Table 6.6 show the increase in the fractal dimension, D_{fd} with an increase in the cantharidin concentration in solution in the 1.5–25 μM range. For the data presented in Figure 6.13d, the fractal dimension, D_{fd} is given by:

$$D_{fd} = (1.067 \pm 0.2021)[\text{cantharidin}]^{0.2111 \pm 0.0782} \tag{6.7d}$$

The fit is reasonable. Only five data points are available. The availability of more data points would lead to a more reliable fit. Just like the fractal dimension, D_f, the fractal dimension in the dissociation phase, D_{fd} is only mildly sensitive to the cantharidin concentration in solution as noted by the low (equal to 0.2111) order of dependence exhibited.

Figure 6.13e and Table 6.6 show the increase in the affinity, K $(= k/k_d)$ with an increase in the cantharidin concentration in solution in the 1.5–25 μM range. For the data presented in Figure 6.13e, the affinity, K is given by:

$$K = \frac{k}{k_d} = (1.272 \pm 0.764)[\text{cantharidin}]^{0.417 \pm 0.212} \tag{6.7e}$$

The fit is good. Only five data points are available. The availability of more data points would lead to a more reliable fit. The affinity, K exhibits a 0.417 order of dependence on the cantharidin concentration in solution in the 1.5–25 μM range. The non-integer order of dependence exhibited by the affinity on the cantharidin concentration in solution lends support to the fractal nature of the system.

Figure 6.13f and Table 6.6 show the increase in the affinity, K $(=k/k_d)$ with an increase in the ratio of the fractal dimensions, D_f/D_{fd}. For the data presented in Figure 6.13f, the affinity, K is given by:

$$K = \frac{k}{k_d} = (3.005 \pm 0.0625)\left(\frac{D_f}{D_{fd}}\right)^{4.517 \pm 0.727} \tag{6.7f}$$

The fit is good. Only five data points are available. The availability of more data points would lead to a more reliable fit. The affinity, K is very sensitive to the ratio of the fractal dimensions as noted by the slightly higher than four and one-half (equal to 4.517) order of dependence exhibited.

Figure 6.13g and Table 6.6 show for a single-fractal analysis the increase in the dissociation rate coefficient, k_d with an increase in the fractal dimension in the dissociation phase, D_{fd}. For the data presented in Figure 6.13g, the dissociation rate coefficient, k_d is given by:

$$k_d = (0.003026 \pm 0.001479) D_{fd}^{5.740 \pm 0.7150} \tag{6.7g}$$

The fit is good. Only five data points are available. The availability of more data points would lead to a more reliable fit. The dissociation rate coefficient, k_d is very sensitive to the fractal dimension in the dissociation phase or the degree of heterogeneity that exists on the sensor chip surface in the dissociation phase as noted by the very high order of dependence between five and one-half and six (equal to 5.740) exhibited.

6.4 CONCLUSIONS

A fractal analysis is presented for the binding of different analytes in solution that help control diseases on biosensor surfaces. The fractal analysis provides values of (a) the binding rate coefficient between the analyte (in solution)–receptor (on the surface) interactions occurring on biosensor surfaces, and (b) the fractal dimension, D_f which provides an indication of the degree of heterogeneity on the biosensor surface. The fractal analysis is offered as an alternate way to improve the understanding of the kinetics in the heterogeneous case with the diffusion-limited reactions occurring on structured surfaces.

Data is taken from the literature for blood pressure drugs (Kamei *et al.*, 2004), thrombin (Garden *et al.*, 2004), probes for the detection of thiols (Pullela *et al.*, 2006), cell structures for the detection of an allergic response (Matsubara *et al.*, 2004), kinases (Nordin *et al.*, 2005), endothelin-1 (Laricchia-Robio and Revoltella, 2004), ChO and ACh, and phosphatases (Stenlund *et al.*, 2006). The analysis of both the binding as well as the dissociation (wherever applicable) steps provide a more quantitative and complete picture of the reaction occurring on the sensor chip surface, besides providing a value of the affinity, K which is the ratio of the binding rate coefficient, k and the dissociation rate coefficient, k_d.

The fractal dimension value provides a quantitative measure of the degree of heterogeneity that exists on the surface for the analyte–receptor systems analyzed. The degree of heterogeneity for the binding and the dissociation phases is, in general, different. Both types of examples are presented wherein either a single- or a dual-fractal analysis is required to adequately describe the binding and/or dissociation kinetics. The dual-fractal analysis is used only when the single-fractal analysis does not provide an adequate fit. This was done by regression analysis provided by Corel Quattro Pro 8.0 (1989).

In accordance with the prefactor analysis for fractal aggregates (Sorenson and Roberts, 1997), quantitative (predictive) expressions are developed for: (a) k_1 and k_2 as a function of the TCEP concentration in solution (Pullela *et al.*, 2006), and k as a function of the cantharidin concentration in solution (Stenlund *et al.*, 2006); (b) k_1 and k_2, as a function of the fractal dimensions, D_{f1} and D_{f2}, respectively; (c) the ratio k_2/k_1 as a function of the fractal dimension ratio, D_{f2}/D_{f1}; (d) the fractal dimension, D_f as a function of the cantharidin concentration in solution (Stenlund *et al.*, 2006); (e) the fractal dimension in the dissociation phase, D_{fd} as a function of the cantharidin concentration in solution (Stenlund *et al.*, 2006); (f) the affinity, K ($=k/k_d$) as a function of cantharidin concentration in solution and the ratio of the fractal dimensions in the binding and in the dissociation phase (D_f/D_{fd}); and (g) the dissociation rate coefficient, k_d as a function of the fractal dimension in the dissociation phase, D_{fd}.

The analysis is presented for the binding and dissociation of analytes that help control diseases on biosensor surfaces. The analysis provides much needed insights into these reactions occurring on biosensor surfaces. More such studies are required to determine if the binding and dissociation rate coefficients is sensitive to the degree of heterogeneity that exists on biosensor surfaces.

The early detection of analytes that help control disease cannot be overemphasized. Biosensors can be of very significant use in this area. It is apparently critical to identify such analytes, most probably in rather very dilute solution, and to be able to manipulate and control them, leading eventually to a better understanding of their mechanisms of action *in vivo*. Such understanding should significantly assist in arresting or, at least slow down, the progress of diseases, especially debilitating ones that cause so much harm emotionally and financially. Besides, they generally, decrease substantially the quality of life of the individuals and families afflicted especially by pathogenic diseases.

REFERENCES

Aisaka, K, SS Gross, DW Griffith, and R Levi, *N*-methylarginine, and inhibitor of endothelium-derived nitric oxide synthesis, is a potent pressor agent in the guinea pig: does nitric oxide regulate blood pressure *in vivo*? *Biochemical Biophysical Research Communication*, 1989, **160**(2), 881–886.

Berk, BC, JI Abe, J Surapisitchat, and C Yan, Endothelial atheroprotective and anti-inflammatory mechanisms. *Annals of the New York Academy of Sciences*, 2001, **947**, 93–111.

Brogan, KL, JH Shin, and MH Schoenfish, Influence of surfactants and antibody immobilization strategy on reducing nonspecific protein interactions for molecular recognition force microscopy. *Langmuir*, 2004, **20**, 9729–9735.

Campanella, L, M Mascini, G Palleschi, and M Tomassetti, Determination of choline-containing phospholipids in amniotic fluid by an enzyme sensor. *Clinica Chimica Acta*, 1985, **151**, 71–83.

Cohen, P, The development and therapeutic potential of protein kinase inhibitors. *Current Opinion in Chemical Biology*, 1999, **3**, 459–465.

Corel Quattro Pro 8.0, Corel Corporation, Ottawa, Canada, 1989.

Davies, SP, H Reddy, M Caivano, and P Cohen, Specificity and mechanism of action of some commonly used protein kinase inhibitors. *Biochemical Journal*, 2000, **351**, 95–105.

Furchgott, RF, Role of endothelium in responses of vascular smooth muscle. *Circulation Research*, 1983, **53**(5), 557–573.

Garden, SR, GJ Doellgast, KS Killham, and NJC Strachan, A fluorescent coagulation assay for thrombin using a fiber optic evanescent sensor. *Biosensors and Bioelectronics*, 2004, **19**, 737–740.

Hale, PD, LF Liu, and TA Skotheim, Enzyme-modified carbon paste: tetrathiafulvalene electrodes for the determination of acetylcholine. *Electroanalysis*, 1991, **3**, 751–756.

Havlin, S, Molecular diffusion and reactions, in *The Fractal Approach to Heterogeneous Chemistry: Surfaces, Colloids, Polymers* (ed. D Avnir), Wiley, New York, 1989.

Hirata, Y, H Yoshimi, T Emori, M Shichiri, F Marumo, TX Watanabe, S Kumagaye, K Nakajima, T Kimura, and S Sakakibara, Receptor binding activity and cytosolic free calcium response by synthetic endothelin analogs in cultured rat vascular smooth muscle cells. *Biochemical and Biophysical Research Communications*, 1989, **160**(1), 228–234.

Huang, Z, R Villarta-Snow, GJ Lubrano, and GG Guilbault, Development of choline and acetylcholine microelectrodes. *Analytical Biochemistry*, 1993, **215**, 31–37.

Kamei, KI, T Haruyama, M Masayasu, Y Yanagida, M Aizawa, and E Kobatake, The construction of endothelial cellular biosensing system for the control of blood pressure drugs. *Biosensors and Bioelectronics*, 2004, **19**, 1121–1124.

Keefer, LK, RW Nims, KM Davies, and DA Wink, "NONOates" 1-substituted diagen-1-ium-1,2-diolates as nitric oxide donors: convenient nitric oxide dosage forms. *Methods in Enzymology*, 1996, **268**, 281–293.

Laricchia-Robio, L and RP Revoltella, Comparison between the surface plasmon resonance (SPR) biosensor and the quartz crystal microbalance (QCM) method in a structural analysis of human endothelin-1. *Biosensors and Bioelectronics*, 2004, **19**, 1753–1758.

Lee, CK and SL Lee, Multi-fractal scaling analysis of reactions over fractal surfaces. *Surface Science*, 1995, **325**, 298–308.

Lusis, AJ, Atherosclerosis, *Nature*, 2000, **407**(6801), 233–241.

Martin, SJ, VE Granstaff, and GC Frye, Effect of surface roughness on the response of thickness-shear mode resonators in liquids. *Analytical Chemistry*, 1991, **65**, 2910–2922.

Matsubara, Y, Y Murakami, M Kobayashi, Y Morita, and E Tamiya, Application of on-chip cell cultures for the detection of allergic response. *Biosensors and Bioelectronics*, 2004, **19**, 741–747.

Myszka, DG and RL Rich, Implementing surface plasmon resonance biosensors in drug discovery. *Pharmaceutical Science and Technology Today*, 2000, **3**, 130–137.

Nordin, H, M Jungnelius, R Karlsson, and O Karlsson, Kinetic studies of small molecule interactions with protein kinases using biosensor technology. *Analytical Biochemistry*, 2005, **340**, 359–368.

Olschewski, H, F Rose, E Gruning, HA Ghofani, D Walmrath, R Schulz, R Schermuly, F Grimminger, and W Seeger, Cellular pathophysiology and therapy of pulmonary hypertension. *Journal of Laboratory Clinical Medicine*, 2001, **138**(6), 367–377.

Potter, PE, JL Meek, and NH Neff, Acetylcholine and choline in neuronal tissue measured by HPLC with electrochemical detection. *Journal of Neurochemistry*, 1983, **41**, 188.

Pullela, PK, T Chiku, MJ Carvan III, and DS Sem, Fluorescence-based detection of thiols in vitro and in vivo using dithiol probes. *Analytical Biochemistry*, 2006.

Rahman, I and W MacNee, Regulation of reduction glutathione levels and gene transcription in lung inflammation: therapeutic approaches. *Free Radicals in Biology and Medicine*, 2000, **28**, 1405–1420.

Ramakrishnan, A and A Sadana, A single fractal analysis of cellular analyte–receptor binding kinetics using biosensors. *Biosystems*, 2001, **59**, 35–51

Sadana, A, A fractal analysis approach for the evaluation of hybridization kinetics in biosensors. *Journal of Colloid and Interface Science*, 2001, **234**, 9–18.

Sadana, A, A fractal analysis for the evaluation of hybridization kinetics using biosensors. *Journal of Colloid and Interface Science*, 2001, **151**(1), 166–177.

Sadana, A, *Fractal Binding and Dissociation Kinetics for Different Biosensor Applications*, Elsevier, Amsterdam, 2005.

Sen, S, A Gulce, and H Gulce, Polyvinylferrocenium modified Pt electrode for the design of amperometric choline an acetylcholine enzyme electrodes. *Biosensors and Bioelectronics*, 2004, **19**, 1261–1268.

Stenlund, P, A Frostell-Karlsson, and OP Karlsson, Studies of small molecule interactions with protein phosphatases using biosensor technology. *Analytical Biochemistry*, 2006, **353**(2), 217–225.

Viscek, T, *Fractal Growth Phenomena*, World Scientific, Singapore, 1989.

– 7 –

Fractal Analysis of Binding and Dissociation of Small Molecules Involved in Drug Discovery on Biosensor Surfaces

7.1 INTRODUCTION

Biosensors may be used very effectively in drug discovery. Recently, Myszka (2004) has used the Biacore S51 surface plasmon resonance (SPR) biosensor for drug discovery by analyzing the binding of a number of small molecule inhibitors that interact with the enzyme carbonic anhydrase II. The compounds analyzed by Myszka (2004) varied in molecular mass from 95 to 340 Daltons (Da), and exhibited an approximately four orders of magnitude change in affinity values. This author suggests that the mechanistic data obtained by such analysis along with the kinetic constants for binding and dissociation assist considerably in drug discovery. Wear *et al.* (2005) have used a SPR-based assay for small molecule inhibitors of human cyclophilin A (CypA). These authors indicate that their sensor surfaces permitted them to assess and rank the equilibrium dissociation rate coefficient, K_d values for quite a few new small-molecule (~300 to 500 Da) inhibitors of CypA. Other studies where the SPR biosensor has been used for drug discovery are also available (Abdiche and Myszka, 2004; Baird *et al.*, 2002; Cannon *et al.*, 2004a,b; Lofas, 2004; Myszka, 2004; Rich and Myszka, 2003; Zhukov *et al.*, 2004).

Gopalakrishanan *et al.* (2005) have recently used a cell-based microarrayed compound screening (μARCS) format for identifying agonists of G-protein-coupled receptors (GPCRs). These authors indicate that these GPCRs initiate the primary signal transduction process in response to hormones and neurotransmitters.

In this chapter we will use fractal analysis to analyze the binding and dissociation (if applicable) kinetics of (a) cyclosporine A (CsA) in solution to covalently immobilized hexahistidine cyclophilin A (His-CypA) on a sensor chip surface (Wear *et al.*, 2005), (b) different concentrations of dopamine in solution to the microarray compound (μARCS) cells expressing $D_{4.4}$ and $G_{\alpha q05}$ protein into a agarose cell (Gopalakrishnan *et al.*, 2003), and (c) different concentrations of beta amyloid in solution to a fibril sensor surface (Cannon *et al.*, 2004a). Though the last example (beta amyloid) is not strictly a small molecule compound, it is involved very significantly in the progression of Alzheimers disease.

The number of studies that have appeared in the recent literature involving Alzheimers disease has increased substantially. Thus, the analysis of the binding and dissociation of beta-amyloid (Aβ) in solution to fibril surfaces is presented here. Binding and dissociation (if applicable) rate coefficient values for the above mentioned cases are presented here. The fractal dimension values presented provide an indication of the degree of heterogeneity present on the biosensor chip surface.

7.2 THEORY

Havlin (1989) has reviewed and analyzed the diffusion of reactants towards fractal surfaces. The details of the theory and the equations involved for the binding and the dissociation phases for analyte–receptor binding are available (Sadana, 2001). The details are not repeated here except that just the equations are given to permit an easier reading. These equations have been applied to other biosensor systems (Sadana, 2001, 2005; Ramakrishnan and Sadana, 2001). For most applications, a single- or a dual-fractal analysis is often adequate to describe the binding and the dissociation kinetics. Peculiarities in the values of the binding and the dissociation rate coefficients, as well as in the values of the fractal dimensions with regard to the dilute analyte systems being analyzed will be carefully noted, if applicable.

7.2.1 Single-fractal analysis

Binding rate coefficient

Havlin (1989) indicates that the diffusion of a particle (analyte [Ag]) from a homogeneous solution to a solid surface (e.g. receptor [Ab]-coated surface) on which it reacts to form a product (analyte–receptor complex; (Ab · Ag)) is given by

$$(\mathrm{Ab}\cdot\mathrm{Ag}) \approx \begin{cases} t^{(3-D_{\mathrm{f,bind}})/2} = t^{p} & t < t_{c} \\ t^{1/2} & t > t_{c} \end{cases} \tag{7.1}$$

Here $D_{\mathrm{f,bind}}$ or D_{f} (used later on in the chapter) is the fractal dimension of the surface during the binding step. t_c is the cross-over value. Havlin (1989) indicates that the cross-over value may be determined by $r_c^2 \sim t_c$. Above the characteristic length, r_c, the self-similarity of the surface is lost and the surface may be considered homogeneous. Above time, t_c the surface may be considered homogeneous, since the self-similarity property disappears, and 'regular' diffusion is now present. For a homogeneous surface where D_f is equal to 2, and when only diffusional limitations are present, $p = 1/2$ as it should be. Another way of looking at the $p = 1/2$ case (where $D_{\mathrm{f,bind}}$ is equal to two) is that the analyte in solution views the fractal object, in our case, the receptor-coated biosensor surface, from a 'large distance.' In essence, in the association process, the diffusion of the analyte from the solution to the receptor surface creates a depletion layer of width $(Đt)^{1/2}$ where $Đ$ is the diffusion constant. This gives rise to the fractal power law, $(\mathrm{Analyte}-\mathrm{Receptor}) \sim t^{(3-D_{\mathrm{f,bind}})/2}$. For the present analysis, t_c is arbitrarily chosen and we assume that the value of the t_c is not reached. One may consider the approach as an intermediate 'heuristic' approach that may be used in the future to develop an autonomous (and not time-dependent) model for diffusion-controlled kinetics.

Dissociation rate coefficient

The diffusion of the dissociated particle (receptor [Ab] or analyte [Ag]) from the solid surface (e.g., analyte [Ag]–receptor [Ab]) complex coated surface) into solution may be given, as a first approximation by

$$(\mathrm{Ab}\cdot\mathrm{Ag}) \approx -t^{(3-D_{\mathrm{f,diss}})/2} = -t^{p}\,(t > t_{\mathrm{diss}}) \tag{7.2}$$

Here $D_{\mathrm{f,diss}}$ is the fractal dimension of the surface for the dissociation step. This corresponds to the highest concentration of the analyte–receptor complex on the surface.

Henceforth, its concentration only decreases. The dissociation kinetics may be analyzed in a manner 'similar' to the binding kinetics.

7.2.2 Dual-fractal analysis

Binding rate coefficient

Sometimes, the binding curve exhibits complexities and two parameters (k, D_f) are not sufficient to adequately describe the binding kinetics. This is further corroborated by low values of r^2 factor (goodness of fit). In that case, one resorts to a dual-fractal analysis (four parameters; k_1, k_2, D_{f1}, and D_{f2}) to adequately describe the binding kinetics. The single-fractal analysis presented above is thus extended to include two fractal dimensions. At present, the time ($t = t_1$) at which the 'first' fractal dimension 'changes' to the 'second' fractal dimension is arbitrary and empirical. For the most part, it is dictated by the data analyzed and experience gained by handling a single-fractal analysis. A smoother curve is obtained in the 'transition' region, if care is taken to select the correct number of points for the two regions. In this case, the product (antibody–antigen; or analyte–receptor complex, Ab · Ag or analyte–receptor) is given by

$$(\mathrm{Ab}\cdot\mathrm{Ag}) \approx \begin{cases} t^{(3-D_{\mathrm{f1,bind}})/2} = t^{p1} & (t < t_1) \\ t^{(3-D_{\mathrm{f2,bind}})/2} = t^{p2} & (t_1 < t < t_2 = t_c) \\ t^{1/2} & (t > t_c) \end{cases} \tag{7.3}$$

In some cases, as mentioned above, a triple-fractal analysis with six parameters (k_1, k_2, k_3, D_{f1}, D_{f2}, and D_{f3}) may be required to adequately model the binding kinetics. This is when the binding curve exhibits convolutions and complexities in its shape due to perhaps to the very dilute nature of the analyte (in some of the cases to be presented) or for some other reasons. Also, in some cases, a dual-fractal analysis may be required to describe the dissociation kinetics.

7.3 RESULTS

At the outset it is appropriate to indicate that a fractal analysis will be applied to data obtained from the literature for the binding and dissociation (if applicable) kinetics of (a) CsA in solution to covalently immobilized His-CypA on a sensor chip surface (Wear *et al.*, 2005), (b) different concentrations of dopamine in solution to the μARCS cells expressing $D_{4.4}$ and $G_{\alpha q05}$ protein into a agarose cell (Gopalakrishnan *et al.*, 2003), and (c) different concentrations of beta amyloid in solution to a fibril sensor surface (Cannon *et al.*, 2004a). This is one possible explanation to analyze diffusion-limited binding kinetics assumed to be present in the system to be analyzed. The parameters thus obtained would provide a useful comparison of different situations. Alternate explanations involving saturation, first-order reaction, and no diffusion limitations are possible, but they are apparently deficient in describing the heterogeneity that inherently exists on the surface. The binding (and dissociation) on the biosensor surface (SPR or other biosensor) is a complex reaction, and the fractal analysis via the fractal dimension and the binding and the dissociation rate coefficient provide a useful lumped parameter(s) analysis of the diffusion-limited situation.

Note that the SPR, as indicated elsewhere in this book, (for example, the BIAcore instrumentation manufactured by Biacore, Uppsala, Sweden) utilizes a carboxymethylated dextran surface which, according to them, under appropriate and careful usage leads to diffusion-free binding and dissociation kinetics. There are references to this effect available in the literature whereby first-order kinetics without heterogeneity on the surface describes the diffusion-free binding kinetics (Karlsson *et al.*, 1991; Lundstrom, 1994). Furthermore, good performance is also demonstrated for small molecules (Karlsson *et al.*, 1995). This is a widely used and expensive biosensor, and as indicated elsewhere in this book, what we are offering or presenting here is an alternate explanation to describe the binding (and dissociation) kinetics that includes both diffusional limitations and heterogeneity on the surface. This would especially be true if the SPR is not carefully utilized. Finally, the analysis to be presented here would be of more value if we could offer an analysis of two different sets of experiments on the same sensing surface, one clearly diffusion-limited and one kinetically limited to see if a fractal analysis is really required for the second case. However, since we are analyzing the data available in the literature, we are unable to judge or estimate if the data have been obtained under diffusion-free conditions. Thus, to be conservative, we have assumed that diffusion limitations are present and heterogeneity exists in all of the cases analyzed.

Wear *et al.* (2005) have recently developed a SPR-based binding assay that helps screen for novel small molecule cyclophilin inhibitors of potential therapeutic interest. These authors emphasize that though the *in vivo* role of CypA is not properly understood, its inherent ability to assist in the protein folding process will assist the cell when it is stressed (Galat, 2003; Dornan *et al.*, 2003). Wear *et al.* (2005) captured His-CypA and oriented it via its N-terminal hexahistidine tag on a NTA (Ni^{2+} -nitiriloacetic acid) sensor surface. This permitted these authors to rank the equilibrium dissociation rate constants for several small-molecule (~300–500 Da) inhibitors of CypA.

Figure 7.1 shows the binding of 20 nM CsA in solution to the covalently immobilized His-CypA on a sensor chip surface. A dual-fractal analysis is required to adequately describe the binding and the dissociation kinetics. The values of (a) the binding rate coefficient, k and the

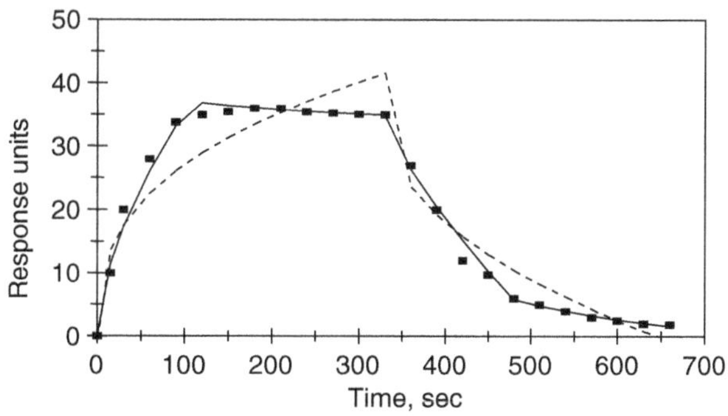

Figure 7.1 Binding of 20 μM CsA in solution to covalently stabilized His-CypA on a sensor chip surface (Wear *et al.*, 2005).

Table 7.1a

Binding and dissociation rate coefficients for Cyclosporin A (CsA) in the presence and in the absence of a small molecule inhibitors KM19 in solution to hexahistidine cyclophilin A (His-CypA) covalently stabilized on a nitrilotriacetic acid (NTA) surface (Wear *et al.*, 2005)

CSA concentration (nM) + KM19 concentration (μM)	k	k_1	k_2	k_d	k_{d1}	k_{d2}
20 nM CsA	5.513 ± 1.116	2.257 ± 0.339	47.434 ± 0.154	2.084 ± 0.357	0.6601 ± 0.071	12.662 ± 0.105
25 nM CsA	4.228 ± 0.820	2.812 ± 0.493	23.094 ± 0.042	1.1065 ± 0.204	0.6723 ± 0.107	11.189 ± 0.010
25 nM CsA + 62.5 μM KM19	3.556 ± 0.662	1.877 ± 0.253	21.151 ± 0.014	3.2733 ± 0.2755	2.5962± 0.0963	15.283 ± 0.082
25 nM CsA + 25 μM KM19	2.797 ± 0.420	1.523 ± 0.071	12.603 ± 0.269	1.498 ± 0.222	0.9859 ± 0.0795	12.629 ± 0.078
25 nM CsA + 12 μM KM19	2.1905 ± 0.347	1.247 ± 0.088	16.259 ± 0.135	0.7611 ± 0.0541	na	na
25 nM CsA + 6 μM KM19	2.0699 ± 0.378	1.060 ± 0.0727	15.906 ± 0.019	0.545 ± 0.105	0.3376 ± 0.0436	4.9587 ± 0.0401

fractal dimension, D_f for a single-fractal analysis, (b) the binding rate coefficients, k_1 and k_2, and the fractal dimensions, D_{f1} and D_{f2} for a dual-fractal analysis, (c) the dissociation rate coefficient, k_d and the fractal dimension, D_{fd} for the dissociation phase for a single-fractal analysis, and (d) the dissociation rate coefficients, k_{d1} and k_{d2} and the fractal dimensions, D_{fd1} and D_{fd2} for a dual-fractal analysis are given in Table 7.1. Corel Quattro Pro 8.0 (1997) was used to fit the data given in Figure 7.1 using eqs. (7.1a)–(7.1d). These equations were used to obtain the fractal dimension and the binding and dissociation rate coefficient values.The values of the parameters presented in Table 7.1 are within 95% confidence limits.

Table 7.1b

Fractal dimensions for the binding and the dissociation phase for Cyclosporin A (CsA) in the presence and in the absence of a small molecule inhibitors KM19 in solution to hexahistidine cyclophilin A (His-CypA) covalently stabilized on an nitrilotriacetic acid (NTA) surface (Wear *et al.*, 2005)

CSA concentration (nM) + KM19 concentration (μM)	D_f	D_{f1}	D_{f2}	D_{fd}	D_{fd1}	D_{fd2}
20 nM CsA	2.2988 ± 0.1147	1.8060 ± 0.166	3.0 − 0.10285	2.0244 ± 0.1133	1.4874 ± 0.137	1.665 ± 0.0327
25 nM CsA	2.2922 ± 0.1169	2.051 ± 0.160	2.945 ± 0.0073	1.5574 ± 0.102	1.6256 ± 0.117	2.719 ± 0.0097
25 nM CsA + 62.5 μM KM19	2.2720 ± 0.1126	1.8840 ± 0.163	2.956 ± 0.00511	2.3302 ± 0.0452	2.199 ± 0.0335	2.8792 ± 0.0203
25 nM CsA + 25 μM KM19	2.2032 ± 0.102	1.8376 ± 0.0666	2.8046 ± 0.0616	2.0962 ± 0.0773	1.88580 ± 0.0713	2.9305 ± 0.0234
25 nM CsA + 12 μM KM19	2.1632 ± 0.0971	1.823 ± 0.0878	2.9250 ± 0.0332	1.8950 ± 0.0384	na	na
25 nM CsA + 6 μM KM19	2.1720 ± 0.1105	1.766 ± 0.0855	2.951 ± 0.0048	1.7492 ± 0.1053	1.471 ± 0.112	2.564 ± 0.0404

For example, for the binding of 20 nM CsA in solution to His-CypA immobilized on a sensor chip surface the binding rate coefficient, k_1 value is given by 2.257 ± 0.339. The 95% confidence limit indicates that 95% of the k_1 values will lie between 1.918 and 2.596. This indicates that the values are precise and significant.

It is of interest to note that as the fractal dimension increases by a factor of 1.66 from a value of D_{f1} equal to 1.8060 to D_{f2} equal to 3.0, the binding rate coefficient increases by a factor of 21.01 from a value of k_1 equal to 3.257 to k_2 equal to 47.434. Similarly, as the fractal dimension in the dissociation phase increases by a factor of 1.12 from a value of D_{fd1} equal to 1.4874 to D_{fd2} equal to 1.662, the dissociation rate coefficient increases by a factor of 19.2 from a value of k_{d1} equal to 0.6601 to k_{d2} equal to 12.662.

Figure 7.2a shows the binding of 20 nM CsA in solution to the covalently immobilized His-CypA on a sensor chip surface. A dual-fractal analysis is required to adequately describe the binding and the dissociation kinetics. The values of (a) the binding rate coefficient, k and the fractal dimension, D_f for a single-fractal analysis, (b) the binding rate coefficients, k_1 and k_2, and the fractal dimensions, D_{f1} and D_{f2} for a dual-fractal analysis, (c) the dissociation rate coefficient, k_d and the fractal dimension, D_{fd} for the dissociation phase for a single-fractal analysis, and (d) the dissociation rate coefficients, k_{d1} and k_{d2} and the fractal dimensions, D_{fd1} and D_{fd2} for a dual-fractal analysis are given in Table 7.1.

Note that as the fractal dimension increases by a factor of 1.436 from a value of D_{f1} equal to 2.051 to D_{f2} equal to 2.945, the binding rate coefficient increases by a factor of 8.213 from a value of k_1 equal to 2.812 to k_2 equal to 2.719. Similarly, as the fractal dimension in the dissociation phase increases by a factor of 1.673 from a value of D_{fd1} equal to 1.6256 to D_{fd2} equal to 2.719, the dissociation rate coefficient increases by a factor of 16.64 from a value of k_{d1} equal to 0.6723 to k_{d2} equal to 11.189.

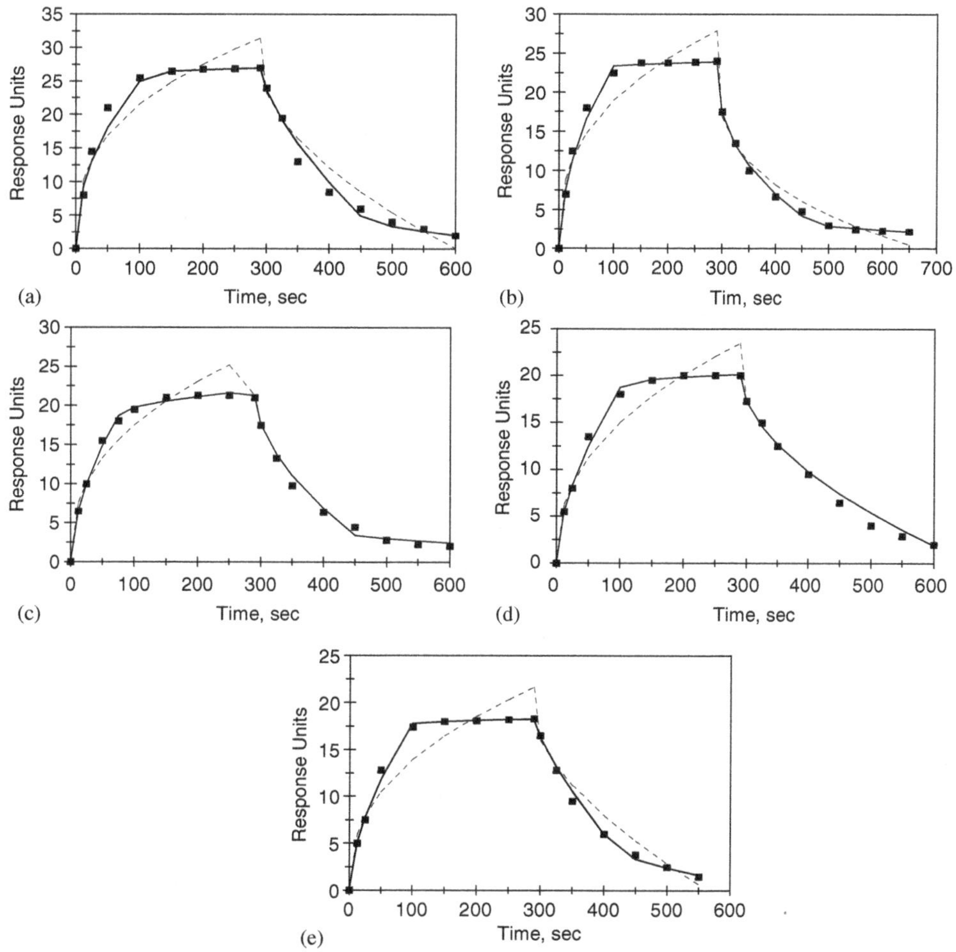

Figure 7.2 Binding of different concentrations of CsA and KM19 in solution to an NTA surface with 925 RU of covalently stabilized His-CypA on a sensor chip surface (Wear *et al.*, 2005): (a) 25 nM CsA, (b) 25 nM CsA + 62.5 μM KM19, (c) 25 nM CsA + 25 μM KM19, (d) 25 nM CsA + 12 μM KM19, (e) 25 nM CsA + 6.25 μM KM19.

Figure 7.2b shows the binding of 20 nM CsA + 62.5 μM KM19 in solution to the covalently immobilized His-CypA on a sensor chip surface. A dual-fractal analysis is once again required to adequately describe the binding and the dissociation kinetics. The values of (a) the binding rate coefficient, k and the fractal dimension, D_f for a single-fractal analysis, (b) the binding rate coefficients, k_1 and k_2, and the fractal dimensions, D_{f1} and D_{f2} for a dual-fractal analysis, (c) the dissociation rate coefficient, k_d and the fractal dimension, D_{fd} for the dissociation phase for a single-fractal analysis, and (d) the dissociation rate coefficients, k_{d1} and k_{d2} and the fractal dimensions, D_{fd1} and D_{fd2} for a dual-fractal analysis are given in Table 7.1.

Note that as the fractal dimension increases by a factor of 1.569 from a value of D_{f1} equal to 1.8840 to D_{f2} equal to 2.956, the binding rate coefficient increases by a factor of 11.27 from a value of k_1 equal to 1.877 to k_2 equal to 21.151. Similarly, as the fractal dimension in the dissociation phase increases by a factor of 1.309 from a value of D_{fd1} equal to 2.199 to D_{fd2} equal to 2.8792, the dissociation rate coefficient increases by a factor of 5.89 from a value of k_{d1} equal to 2.5962 to k_{d2} equal to 15.283.

Figure 7.2c shows the binding of 20 nM CsA + 25 μM KM19 in solution to the covalently immobilized His-CypA on a sensor chip surface. A dual-fractal analysis is required to adequately describe the binding and the dissociation kinetics. The values of (a) the binding rate coefficient, k and the fractal dimension, D_f for a single-fractal analysis, (b) the binding rate coefficients, k_1 and k_2, and the fractal dimensions, D_{f1} and D_{f2} for a dual-fractal analysis, (c) the dissociation rate coefficient, k_d and the fractal dimension, D_{fd} for the dissociation phase for a single-fractal analysis, and (d) the dissociation rate coefficients, k_{d1} and k_{d2} and the fractal dimensions, D_{fd1} and D_{fd2} for a dual-fractal analysis are given in Table 7.1.

Note that as the fractal dimension increases by a factor of 1.526 from a value of D_{f1} equal to 1.8376 to D_{f2} equal to 2.8046, the binding rate coefficient increases by a factor of 8.28 from a value of k_1 equal to 1.523 to k_2 equal to 12.603. Similarly, as the fractal dimension in the dissociation phase increases by a factor of 1.553 from a value of D_{fd1} equal to 1.8858 to D_{fd2} equal to 2.9305 the dissociation rate coefficient increases by a factor of 12.81 from a value of k_{d1} equal to 0.9859 to k_{d2} equal to 12.629.

Figure 7.2d shows the binding of 20 nM CsA + 12 μM KM19 in solution to the covalently immobilized His-CypA on a sensor chip surface. A dual-fractal analysis is required to adequately describe the binding kinetics. A single-fractal analysis is adequate to describe the dissociation kinetics. The values of (a) the binding rate coefficient, k and the fractal dimension, D_f for a single-fractal analysis, (b) the binding rate coefficients, k_1 and k_2, and the fractal dimensions, D_{f1} and D_{f2} for a dual-fractal analysis, and (c) the dissociation rate coefficient, k_d and the fractal dimension, D_{fd} for the dissociation phase for a single-fractal analysis, are given in Table 7.1. Note that as the fractal dimension increases by a factor of 1.604 from a value of D_{f1} equal to 1.823 to D_{f2} equal to 2.9250, the binding rate coefficient increases by a factor of 13.04 from a value of k_1 equal to 1.247 to k_2 equal to 16.259.

Figure 7.2e shows the binding of 20 nM CsA + 6 μM KM19 in solution to the covalently immobilized His-CypA on a sensor chip surface. A dual-fractal analysis is once again required to adequately describe the binding and the dissociation kinetics. The values of (a) the binding rate coefficient, k and the fractal dimension, D_f for a single-fractal analysis, (b) the binding rate coefficients, k_1 and k_2, and the fractal dimensions, D_{f1} and D_{f2} for a dual-fractal analysis, (c) the dissociation rate coefficient, k_d and the fractal dimension, D_{fd} for the dissociation phase for a single-fractal analysis, and (d) the dissociation rate coefficients, k_{d1} and k_{d2} and the fractal dimensions, D_{fd1} and D_{fd2} for a dual-fractal analysis are given in Table 7.1.

Note that as the fractal dimension increases by a factor of 1.671 from a value of D_{f1} equal to 1.766 to D_{f2} equal to 2.951, the binding rate coefficient increases by a factor of 115 from a value of k_1 equal to 1.060 to k_2 equal to 15.906. Similarly, as the fractal dimension in the dissociation phase increases by a factor of 1.74 from a value of D_{fd1} equal to 1.471 to D_{fd2} equal to 2.564, the dissociation rate coefficient increases by a factor of 14.7 from a value of k_{d1} equal to 0.3376 to k_{d2} equal to 4.9587.

Table 7.1 and Figure 7.3a show the increase in the binding rate coefficient, k_1 with an increase in the KM19 concentration (in μM) in solution. For the data presented in Figure 7.3a, the binding rate coefficient, k_1 is given by

$$k_1 = (0.6723 \pm 0.0075)[25\,\text{nm CsA} + \text{KM19}]^{0.250 \pm 0.0065} \quad (7.4a)$$

The fit is very good. Only four data points are available. The availability of more data points would lead to a more reliable fit. The binding rate coefficient, k_1 exhibits only a mild (equal to 0.25) order of dependence on the KM19 concentration in solution. The non-integer order of dependence exhibited lends support to the fractal nature of the system.

Table 7.1 and Figure 7.3b show for a dual-fractal analysis the increase in the binding rate coefficient, k_1 with an increase in the fractal dimension, D_{f1}. For the data presented in Figure 7.3b, the binding rate coefficient, k_1 is given by

$$k_1 = (0.02645 \pm 0.0026)D_{f1}^{6.559 \pm 0.8289} \quad (7.4b)$$

The fit is very good. Only five data points are available. The availability of more data points would lead to a more reliable fit. The binding rate coefficient, k_1 is very sensitive to the fractal dimension, D_{f1} or the degree of heterogeneity that exists on the sensor chip surface as noted by the close to six and one-half (equal to 6.559) order of dependence exhibited.

Table 7.1 and Figure 7.3c show the for a dual-fractal analysis the increase in the binding rate coefficient, k_2 with an increase in the fractal dimension, D_{f2}. For the data presented in Figure 7.3c, the binding rate coefficient, k_2 is given by

$$k_2 = (0.00723 \pm 0.00108)D_{f2}^{7.227 \pm 3.284} \quad (7.4c)$$

The fit is reasonable. There is scatter in the data. This is reflected in the error in the binding rate coefficient, k_2 value. Only four data points are available. The availability of more data points would lead to a more reliable fit. The binding rate coefficient, k_2 is very sensitive to the fractal dimension, D_{f2} or the degree of heterogeneity that exists on the sensor chip surface as noted by the greater than seventh (equal to 7.227) order of dependence exhibited.

Table 7.1 and Figure 7.3d show for a dual-fractal analysis the increase in the fractal dimension, D_{f1} with an increase in the KM19 concentration (in μM) in solution. For the data presented in Figure 7.3d, the fractal dimension, D_{f1} is given by

$$D_{f1} = (1.693 \pm 0.013)[25\,\text{nm CsA} + \text{KM19}]^{0.02612 \pm 0.00443} \quad (7.4d)$$

The fit is good. Only four data points are available. The availability of more data points would lead to a more reliable fit. The fractal dimension, D_{f1} or the degree of heterogeneity on the sensor chip surface exhibits only a very mild, almost negligible, order (equal to 0.02612) of dependence on the KM19 concentration in solution. Note that, and as mentioned earlier, the fractal dimension, D_{f1} is based on a log scale, and even small changes in the fractal dimension leads to significant changes in the degree of heterogeneity on the sensor chip surface.

Affinity, K_1 values are of interest to practicing biosensorists. Table 7.1 and Figure 7.3e show for a dual-fractal analysis the decrease in the affinity, K_1 value with an increase in

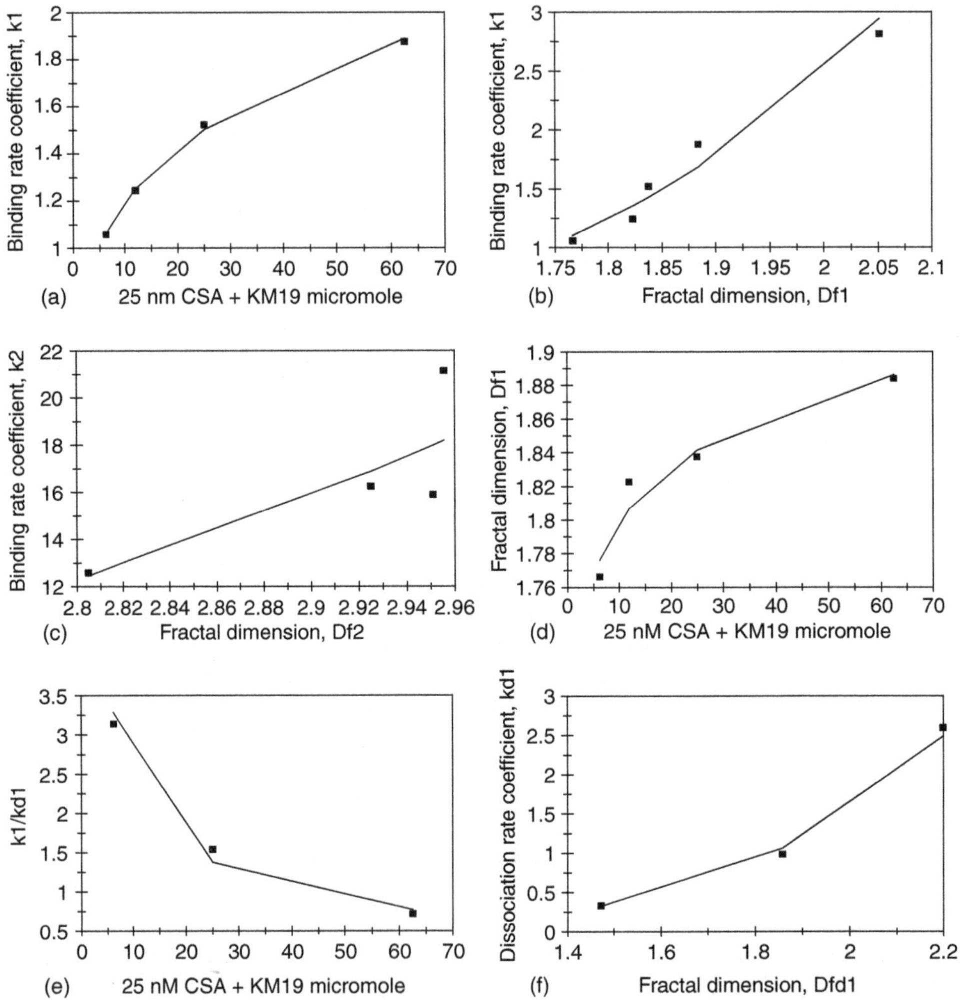

Figure 7.3 (a) Increase in the binding rate coefficient, k_1 with an increase in the KM19 concentration (in μM) in solution (CsA concentration 25 nm). (b) Increase in the binding rate coefficient, k_1 with an increase in the fractal dimension, D_{f1}. (c) Increase in the binding rate coefficient, k_2 with an increase in the fractal dimension, D_{f2}. (d) Increase in the fractal dimension for the dissociation phase, D_{fd} with an increase in the KM19 concentration (in μM) in solution (CsA concentration 25 nm). (e) Decrease in the affinity, K_1 $(= k_1/k_{d1})$ with an increase in the fractal dimension ratio, D_{f1}/D_{fd1}. (f) Increase in the dissociation rate coefficient, k_{d1} with an increase in the fractal dimension, D_{fd1}. (g) Increase in the affinity, K_1 $(= k_1/k_{d1})$ with an increase in the fractal dimension ratio, D_{f1}/D_{fd1}.(h) Increase in the affinity, K_2 $(= k_2/k_{d2})$ with an increase in the fractal dimension ratio, D_{f2}/D_{fd2}. (i) Increase in the binding rate coefficient ratio, k_2/k_1 with an increase in the fractal dimension ratio, D_{f2}/D_{f1}. (j) Increase in the dissociation rate coefficient, k_{d2} with an increase in the fractal dimension in the dissociation phase, $D_{fd2.}$ (k) Increase in the dissociation rate coefficient ratio, k_{d2}/k_{d1} with an increase in the fractal dimension ratio, D_{fd2}/D_{fd1}.

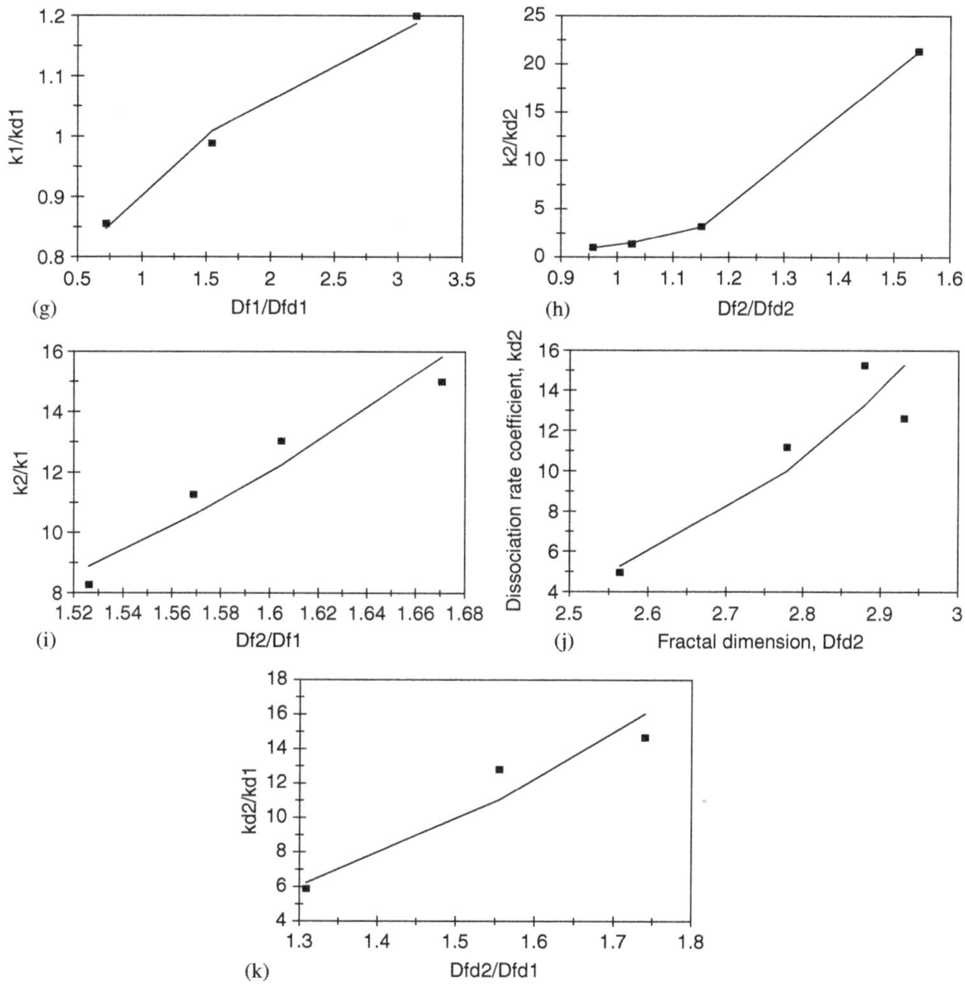

Figure 7.3

the KM19 concentration (in μM) in solution. For the data presented in Figure 7.3e, the affinity, K_1 is given by

$$K_1\left(=\frac{k_1}{k_d}\right) = (10.382 \pm 1.585)[25\,\mathrm{nm\,CsA} + \mathrm{KM19}]^{-0.6276\pm0.0867} \tag{7.4e}$$

The fit is good. Only three data points are available. The availability of more data points would lead to a more reliable fit. The affinity, K_1 exhibits a negative 0.6276 order of dependence on the KM19 concentration in solution. The non-integer order of dependence exhibited by the affinity, K_1 on the KM19 concentration in solution lends support to the fractal nature of the system.

Table 7.1 and Figure 7.3f show for a dual-fractal analysis the increase in the dissociation rate coefficient, k_{d1} with an increase in the fractal dimension in the dissociation phase, D_{fd1}. For the data presented in Figure 7.3f, the dissociation rate coefficient, k_{d1} is given by

$$k_{d1} = (0.04669 \pm 0.00447) D_{fd1}^{5.043 \pm 0.3202} \tag{7.4f}$$

The fit is good. Only three data points are available. The availability of more data points would lead to a more reliable fit. The dissociation rate coefficient, k_{d1} is very sensitive to the fractal dimension, D_{fd1} or the degree of heterogeneity that exists on the sensor chip surface in the dissociation phase as noted by the greater than fifth (equal to 5.043) order of dependence exhibited.

Table 7.1 and Figure 7.3g show the increase in the affinity, K_1 with an increase in the ratio of fractal dimensions, D_{f1}/D_{fd1}. For the data presented in Figure 7.3g, the affinity, K_1 is given by

$$\frac{k_1}{k_{d1}} = (0.9130 \pm 0.0227) \left(\frac{D_{f1}}{D_{fd1}} \right)^{0.2299 \pm 0.0236} \tag{7.4g}$$

The fit is very good. Only three data points are available. The availability of more data points would lead to a more reliable fit. The affinity, K_1 is only mildly sensitive to the ratio of the fractal dimensions, (D_{f1}/D_{fd1}) or the degree of heterogeneity that exists on the sensor chip surface in the dissociation phase as noted by the 0.2299 order of dependence exhibited.

Table 7.1 and Figure 7.3h show the increase in the affinity, K_2 with an increase in the ratio of fractal dimensions, D_{f2}/D_{fd2}. For the data presented in Figure 7.3h, the affinity, K_2 is given by

$$\frac{k_2}{k_{d2}} = (1.259 \pm 0.088) \left(\frac{D_{f2}}{D_{fd2}} \right)^{6.513 \pm 0.1852} \tag{7.4h}$$

The fit is very good. Only four data points are available. The availability of more data points would lead to a more reliable fit. The affinity, K_2 is very sensitive to the ratio of the fractal dimensions, (D_{f2}/D_{fd2}) or the degree of heterogeneity that exists on the sensor chip surface in the dissociation phase as noted by close to six and one-half (equal to 6.513) order of dependence exhibited.

Table 7.1 and Figure 7.3i show the increase in the ratio of the binding rate coefficients, k_2/k_1 with an increase in the ratio of fractal dimensions, D_{f2}/D_{f1}. For the data presented in Figure 7.3h, the ratio of the binding rate coefficients, k_2/k_1 is given by

$$\frac{k_2}{k_1} = (0.595 \pm 0.054) \left(\frac{D_{f2}}{D_{f1}} \right)^{6.395 \pm 1.342} \tag{7.4i}$$

The fit is good. Only four data points are available. The availability of more data points would lead to a more reliable fit. The ratio of the binding rate coefficients, k_2/k_1 is very sensitive to the ratio of the fractal dimensions, (D_{f2}/D_{f1}) or the degree of heterogeneity that exists on the sensor chip surface in the dissociation phase as noted by the greater than sixth (equal to 6.395) order of dependence exhibited.

Table 7.1 and Figure 7.3j show for a dual-fractal analysis the increase in the dissociation rate coefficient, k_{d2} with an increase in the fractal dimension in the dissociation phase, D_{fd2}. For the data presented in Figure 7.3j, the dissociation rate coefficient, k_{d2} is given by

$$k_{d2} = (0.00297 \pm 0.00062) D_{fd2}^{7.945 \pm 1.851} \tag{7.4j}$$

The fit is good. Only four data points are available. The availability of more data points would lead to a more reliable fit. The dissociation rate coefficient, k_{d2} is very sensitive to the fractal dimension, D_{fd2} or the degree of heterogeneity that exists on the sensor chip surface in the dissociation phase as noted by the close to eighth (equal to 7.945) order of dependence exhibited.

Table 7.1 and Figure 7.3k show the increase in the ratio of the dissociation rate coefficients, k_{d2}/k_{d1} with an increase in the ratio of fractal dimensions, D_{fd2}/D_{fd1}. For the data presented in Figure 7.3k, the ratio of the binding rate coefficients, k_{d2}/k_{d1} is given by

$$\frac{k_{d2}}{k_{d1}} = (2.547 \pm 0.508) \left(\frac{D_{fd2}}{D_{fd1}} \right)^{3.323 \pm 0.8968} \tag{7.4k}$$

The fit is good. Only three data points are available. The availability of more data points would lead to a more reliable fit. The ratio of the dissociation rate coefficients, k_{d2}/k_{d1} is sensitive to the ratio of the fractal dimensions, D_{fd2}/D_{fd1} or the degree of heterogeneity that exists on the sensor chip surface in the dissociation phases as noted by the greater than third (equal to 3.323) order of dependence exhibited.

Gopalakrishnan *et al.* (2003) have recently used a cell-based (μARCS) format for identifying agonists of GPCRs. These authors indicate that μARCS is a well-less, ultra high-throughput platform that permits the rapid screening of a large number of compounds (Schurdak *et al.*, 2001; Burns *et al.*, 2001). Marinissen and Gutkind (2001) indicate that approximately 30% of the currently approved drugs selectively target members of the GPCR family. The GPCRs are a well-known family of validated drug targets. Goapalkrishnan *et al.* (2003) demonstrated the use of μARCS to identify the D_4 receptor agonists with a throughput that was superior to the methodologies presently available.

Figure 7.4a shows the binding of 10 μM dopamine to the μARCS format (Gopalakrishnan *et al.*, 2003). In the μARCS format the cells expressing the $D_{4.4}$ receptor and $G_{\alpha q05}$ protein were preloaded with fluo-4, cast into a 1% agarose gel, and placed above the compound sheets. A dual-fractal analysis is required to adequately describe the binding and the dissociation kinetics. The values of (a) the binding rate coefficient, k and the fractal dimension, D_f for a single-fractal analysis, (b) the binding rate coefficients, k_1 and k_2, and the fractal dimensions, D_{f1} and D_{f2} for a dual-fractal analysis, (c) the dissociation rate coefficient, k_d and the fractal dimension for the dissociation phase, D_{fd} for a single-fractal analysis, and (d) the dissociation rate coefficients, k_{d1} and k_{d2} and the fractal dimensions, D_{fd1} and D_{fd2} for a dual-fractal analysis are given in Tables 7.2a and 7.2b.

It is of interest to note that for a dual-fractal analysis as the fractal dimension increases by a factor of 1.877 from a value of D_{f1} equal to 1.4066 to D_{f2} equal to 2.6402, the binding rate coefficient increases by a factor of 2.85 from a value of k_1 equal to 4095.9 to k_2 equal

Table 7.2a

Binding and dissociation rate coefficients for different concentrations (in μM) of dopamine to the microarray compound screening (μARCS) cells expressing $D_{4.4}$ receptor and $G_{\alpha q05}$ protein into a 1% agarose gel (Gopalakrishnan *et al.*, 2003)

Dopamine concentration (μM)	K	k_1	k_2	k_d	k_{d1}	k_{d2}
10	5153.26 ± 818.08	4095.90 ± 556.88	11683.19 ± 266.06	201.81 ± 73.66	96.545 ± 10.676	5112.79 ± 91.81
1	4590.41 ± 680.73	4089.58 ± 624.53	8704.10 ± 1.70	476.79 ± 181.52	226.56 ± 69.49	2039.58 ± 141.24
0.1	4860.74 ± 730.77	4313.97 ± 654.99	9291.77 ± 36.67	259.82 ± 79.15	152.83 ± 30.19	8313.90 ± 75.52
0.01	2963.05 ± 292.73	1745.18 ± 217.18	10758.79 ± 14.47	85.16 ± 12.39	41.33 ± 3.35	2909.02 ± 10.65
0.003	223.19 ± 38.13	154.63 ± 24.85	1802.33 ± 35.79	7.900 ± 7.398	0.08525 ± 0.06338	3043.56 ± 19.69

Table 7.2b

Fractal dimensions for the binding and dissociation of different concentrations (in μM) of dopamine to the microarray compound screening (μARCS) cells expressing $D_{4.4}$ receptor and $G_{\alpha q05}$ protein into a 1% agarose gel (Gopalakrishnan *et al.*, 2003)

Dopamine concentration (μM)	D_f	D_{f1}	D_{f2}	D_{fd}	D_{fd1}	D_{fd2}
10	1.8298 ± 0.2318	1.4066 ± 0.3246	2.6402 ± 0.1243	1.1102 ± 0.1282	0.4948 ± 0.0738	2.5664 ± 0.04624
1	1.6564 ± 0.2656	1.4320 ± 0.3618	2.3562 ± 0.00079	1.5210 ± 0.1192	0.8620 ± 0.2010	2.1978 ± 0.06318
0.1	1.8616 ± 0.2690	1.6298 ± 0.3598	2.5702 ± 0.01599	1.3748 ± 0.1173	0.9772 ± 0.1125	2.84898 ± 0.03750
0.01	2.1982 ± 0.1087	1.5192 ± 0.1711	2.9760 ± 0.00753	1.0150 ± 0.1461	0.5782 ± 0.1497	2.5638 ± 0.03590
0.003	1.0034 ± 0.1551	0.6534 ± 0.2346	2.2674 ± 0.130	0.2532 + 0.6010	0 + 1.0674	2.8509 ± 0.1491

to 11683.19. Similarly, for a dual-fractal analysis, as the fractal dimension for dissociation increases by a factor of 5.19 from a value of D_{fd1} equal to 0.4948 to D_{fd2} equal to 2.5644, the dissociation rate coefficient increases by a factor of 52.96 from a value of k_{d1} equal to 96.545 to k_{d2} equal to 5112.79.

Figure 7.4b shows the binding of 1 μM dopamine to the μARCS format (Gopalakrishnan *et al.*, 2003). A dual-fractal analysis is required to adequately describe the binding and the dissociation kinetics. The values of (a) the binding rate coefficient, k and the fractal

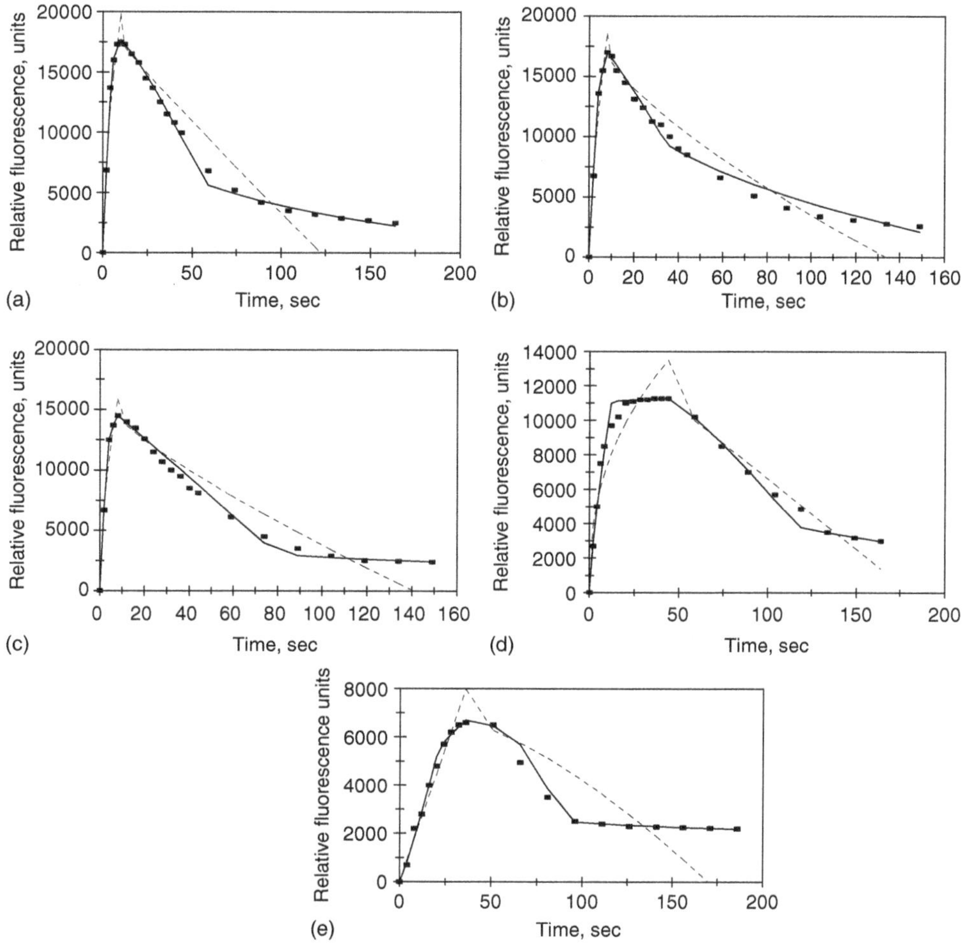

Figure 7.4 Binding of different concentrations (in μM) of dopamine in solution to the microarray compound (μARCS) cells expressing $D_{4.4}$ and $G_{\alpha q05}$ protein into an agarose cell (Gopalakrishnan *et al.*, 2003): (a) 10, (b) 1, (c) 0.1, (d) 0.010, (e) 0.003.

dimension, D_f for a single-fractal analysis, (b) the binding rate coefficients, k_1 and k_2, and the fractal dimensions, D_{f1} and D_{f2} for a dual-fractal analysis, (c) the dissociation rate coefficient, k_d and the fractal dimension for the dissociation phase, D_{fd} for a single-fractal analysis, and (d) the dissociation rate coefficients, k_{d1} and k_{d2} and the fractal dimensions, D_{fd1} and D_{fd2} for a dual-fractal analysis are given in Tables 7.2a and 7.2b.

It is of interest to note that for a dual-fractal analysis as the fractal dimension increases by a factor of 1.675 from a value of D_{f1} equal to 1.4320 to D_{f2} equal to 2.3562, the binding rate coefficient increases by a factor of 2.128 from a value of k_1 equal to 4089.58 to k_2 equal to 8704.1. Similarly, for a dual-fractal analysis, as the fractal dimension for dissociation increases by a factor of 2.55 from a value of D_{fd1} equal to 0.8620 to D_{fd2} equal to 2.1978, the dissociation rate coefficient increases by a factor of 9.0 from a value of k_{d1} equal to 226.56 to k_{d2} equal to 2039.58.

Figure 7.4c shows the binding of 0.1 μM dopamine to the μARCS format (Gopalakrishnan *et al.*, 2003). A dual-fractal analysis is required to adequately describe the binding and the dissociation kinetics. The values of (a) the binding rate coefficient, k and the fractal dimension, D_f for a single-fractal analysis, (b) the binding rate coefficients, k_1 and k_2, and the fractal dimensions, D_{f1} and D_{f2} for a dual-fractal analysis, (c) the dissociation rate coefficient, k_d and the fractal dimension for the dissociation phase, D_{fd} for a single-fractal analysis, and (d) the dissociation rate coefficients, k_{d1} and k_{d2} and the fractal dimensions, D_{fd1} and D_{fd2} for a dual-fractal analysis are given in Tables 7.2a and 7.2b.

It is of interest to note that for a dual-fractal analysis as the fractal dimension increases by a factor of 1.577 from a value of D_{f1} equal to 1.6298 to D_{f2} equal to 2.5702, the binding rate coefficient increases by a factor of 2.154 from a value of k_1 equal to 4313.97 to k_2 equal to 9291.77. Similarly, for a dual-fractal analysis, as the fractal dimension for dissociation increases by a factor of 2.91 from a value of D_{fd1} equal to 0.9772 to D_{fd2} equal to 2.8489, the dissociation rate coefficient increases by a factor of 54.4 from a value of k_{d1} equal to 152.83 to k_{d2} equal to 8313.9.

Figure 7.4d shows the binding of 0.01 μM dopamine to the μARCS format (Gopalakrishnan *et al.*, 2003). A dual-fractal analysis is required to adequately describe the binding and the dissociation kinetics. The values of (a) the binding rate coefficient, k and the fractal dimension, D_f for a single-fractal analysis, (b) the binding rate coefficients, k_1 and k_2, and the fractal dimensions, D_{f1} and D_{f2} for a dual-fractal analysis, (c) the dissociation rate coefficient, k_d and the fractal dimension for the dissociation phase, D_{fd} for a single-fractal analysis, and (d) the dissociation rate coefficients, k_{d1} and k_{d2} and the fractal dimensions, D_{fd1} and D_{fd2} for a dual-fractal analysis are given in Tables 7.2a and 7.2b.

It is of interest to note that for a dual-fractal analysis as the fractal dimension increases by a factor of 1.959 from a value of D_{f1} equal to 1.5192 to D_{f2} equal to 2.9760, the binding rate coefficient increases by a factor of 6.164 from a value of k_1 equal to 1745.18 to k_2 equal to 10758.79. Similarly, for a dual-fractal analysis, as the fractal dimension for dissociation increases by a factor of 4.434 from a value of D_{fd1} equal to 0.5782 to D_{fd2} equal to 2.5638, the dissociation rate coefficient increases by a factor of 70.39 from a value of k_{d1} equal to 41.33 to k_{d2} equal to 2909.02.

Figure 7.4e shows the binding of 0.003 μM dopamine to the μARCS format (Gopalakrishnan *et al.*, 2003). A dual-fractal analysis is required to adequately describe the binding and the dissociation kinetics. The values of (a) the binding rate coefficient, k and the fractal dimension, D_f for a single-fractal analysis, (b) the binding rate coefficients, k_1 and k_2, and the fractal dimensions, D_{f1} and D_{f2} for a dual-fractal analysis, (c) the dissociation rate coefficient, k_d and the fractal dimension for the dissociation phase, D_{fd} for a single-fractal analysis, and (d) the dissociation rate coefficients, k_{d1} and k_{d2} and the fractal dimensions, D_{fd1} and D_{fd2} for a dual-fractal analysis are given in Tables 7.2a and 7.2b.

It is of interest to note that for a dual-fractal analysis as the fractal dimension increases by a factor of 3.47 from a value of D_{f1} equal to 0.6534 to D_{f2} equal to 2.2674, the binding rate coefficient increases by a factor of 11.66 from a value of k_1 equal to 154.63 to k_2 equal to 1802.33. Similarly, for a dual-fractal analysis, as the fractal dimension for dissociation increases by a factor of 2.91 from a value of D_{fd1} equal to 0.9772 to D_{fd2} equal to 2.8489, the dissociation rate coefficient increases by a factor of 35701 from a value of k_{d1} equal to 0.08525 to k_{d2} equal to 3043.56.

Figure 7.5a and Table 7.2a show for a dual-fractal analysis the increase in the binding rate coefficient, k_1 with an increase in the fractal dimension, D_{f1}. For the data shown in Figure 7.5a, the binding rate coefficient, k_1 is given by

$$k_1 = (919.55 \pm 606) D_{f1}^{3.18 \pm 0.676} \tag{7.5a}$$

The fit is good. Only five data points are available. The availability of more data points would lead to a more reliable fit. The binding rate coefficient, k_1 is sensitive to the fractal dimension, D_{f1} or the degree of heterogeneity present on the biosensor surface as noted by the order greater than third (equal to 3.18) exhibited.

Figure 7.5b and Table 7.2a show for a dual-fractal analysis the increase in the dissociation rate coefficient, k_{d1} with an increase in the fractal dimension, D_{fd1}. For the data shown in Figure 7.5a, the dissociation rate coefficient, k_{d1} is given by:

$$k_{d1} = (196.73 \pm 136.05) D_{fd1}^{1.680 \pm 0.137} \tag{7.5b}$$

The fit is good. Only five data points are available. The availability of more data points would lead to a more reliable fit. The dissociation rate coefficient, k_{d1} is sensitive to the fractal dimension, D_{fd1} or the degree of heterogeneity present on the biosensor surface as noted by the order between one and one-half and second (equal to 1.680) exhibited.

Figure 7.5c and Table 7.2a show for a dual-fractal analysis the increase in the dissociation rate coefficient, k_{d2} with an increase in the fractal dimension, D_{fd2}. For the data shown in Figure 7.5c, the dissociation rate coefficient, k_{d2} is given by

$$k_{d2} = (29.455 \pm 10.803) D_{fd2}^{5.279 \pm 1.6875} \tag{7.5c}$$

The fit is good. Only four data points are available. The availability of more data points would lead to a more reliable fit. The dissociation rate coefficient, k_{d2} is very sensitive to the fractal dimension, D_{fd2} or the degree of heterogeneity present on the biosensor surface as noted by the order between five and five and one-half (equal to 5.279) exhibited.

Figure 7.5d and Table 7.2a show the increase in the affinity, K_1 (= k_1/k_{d1}) with an increase in the fractal dimension ratio D_{f1}/D_{fd1}. For the data shown in Figure 7.5d, the affinity, K_1 is given by

$$K_1 = (11.954 \pm 3.042) \left(\frac{D_{f1}}{D_{fd1}} \right)^{1.255 \pm 0.454} \tag{7.5d}$$

The fit is good. Only four data points are available. The availability of more data points would lead to a more reliable fit. The affinity, K_1 is quite sensitive to the fractal dimension

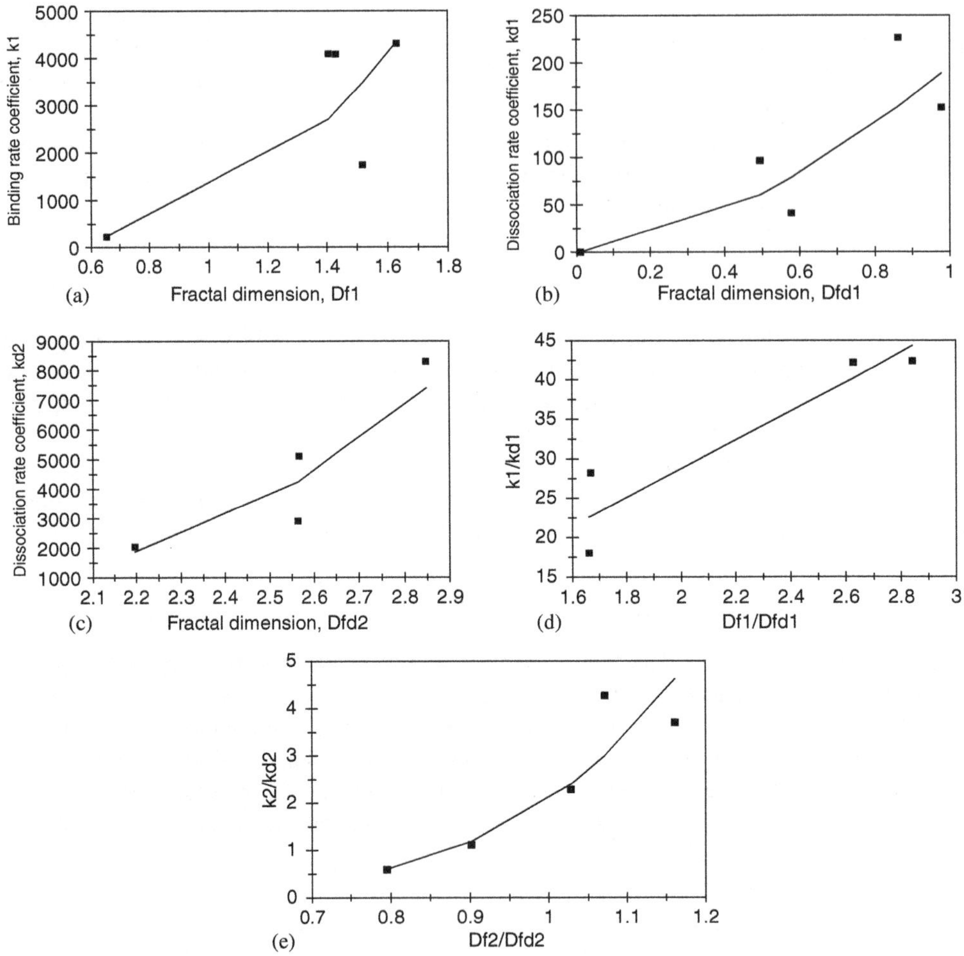

Figure 7.5 (a) Increase in the binding rate coefficient, k_1 with an increase in the fractal dimension, D_{f1}. (b) Increase in the dissociation rate coefficient, k_{d1} with an increase in the fractal dimension, D_{fd1}. (c) Increase in the dissociation rate coefficient, k_{d2} with an increase in the fractal dimension, D_{fd2}. (d) Increase in the affinity, K_1 (= k_1/k_{d1}) with an increase in the fractal dimension ratio, D_{f1}/D_{fd1}. (e) Increase in the affinity, K_2 (= k_2/k_{d2}) with an increase in the fractal dimension ratio, D_{f2}/D_{fd2}.

ratio D_{f1}/D_{fd2} as noted by the order between one and one and one-half (equal to 1.255) exhibited.

Figure 7.5e and Table 7.2a show the increase in the affinity, K_2 (= k_2/k_{d2}) with an increase in the fractal dimension ratio D_{f2}/D_{fd2}. For the data shown in Figure 7.5d, the affinity, K_2 is given by:

$$K_2 = (2.068 \pm 0.571)\left(\frac{D_{f2}}{D_{fd2}}\right)^{5.386 \pm 0.815} \tag{7.5e}$$

The fit is good. Only five data points are available. The availability of more data points would lead to a more reliable fit. The affinity, K_2 is very sensitive to the fractal dimension ratio D_{f2}/D_{fd2} as noted by the order between five and five and one-half (equal to 5.386) exhibited.

Cannon *et al.* (2004) have analyzed the kinetic interactions of Aβ peptide with amyloid fibrils using a SPR biosensor. These authors indicate that amyloid fibrils have been associated with quite a few neurodegenerative diseases including Alzheimer's (Falk *et al.*, 1997; Martin, 1999). Reproducible results using biosensor technology have been obtained for Aβ(1–40) peptides and Aβ(1–40) fibrils (Wood *et al.*, 1996; Tjernberg *et al.*, 1996; Myszka *et al.*, 1999; Cairo *et al.*, 2002; Hasegawa *et al.*, 2002). Cannon *et al.* (2004) have used the SPR biosensor to monitor the interactions of Aβ soluble peptide in solution to fibrils immobilized on a sensor chip surface.

Figure 7.6 shows the binding and dissociation of 3 μM Aβ(1–40) peptide in solution to fibrils immobilized on a sensor chip surface (Cannon *et al.*, 2004). These authors propose a peptide-fibril binding (elongation) model wherein the binding of the peptide to the fibril leads to two rearrangements that lead to an identical binding site on the fibril end for the polymerization to proceed. A single-fractal analysis is adequate to describe the binding and the dissociation kinetics. The values of (a) the binding rate coefficient, k and the fractal dimension, D_f for a single-fractal analysis, and (b) the dissociation rate coefficient, k_d and the fractal dimension, D_{fd} for a single-fractal analysis are given in Table 7.3a.

Figure 7.7a shows the binding of 3 μM Aβ(1–40) peptide in solution to a sonicated fibril surface (Cannon *et al.*, 2004). The pH of the solution is 7.4, phosphate buffered saline (PBS) is the buffer, and the solution contained 0.005% polysorbate 20 as the surfactant. The solution flow over the sensor chip is 100 μl/min. A single-fractal analysis is adequate to describe the binding and the dissociation kinetics. The values of (a) the binding rate coefficient, k and the fractal dimension, D_f for a single-fractal analysis, and (b) the dissociation rate coefficient, k_d and the fractal dimension in the dissociation phase, D_{fd} for a single-fractal analysis are given in Table 7.3b.

Figure 7.7b shows the binding of 3 μM Aβ(1–40) peptide in solution to an unsonicated fibril surface (Cannon *et al.*, 2004). The pH of the solution is 7.4, PBS is the buffer, and

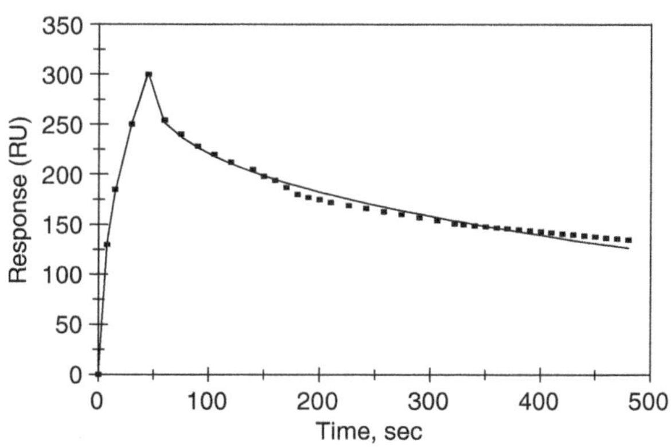

Figure 7.6 Binding of 3 μM of beta amyloid, Aβ(1–40) in solution to a fibril surface (Cannon *et al.*, 2004).

Table 7.3a

Binding and dissociation rate coefficients for (a) 3 μM Aβ(1–40) to a fibril immobilized surface, (b) 3 μM Aβ(1–40) to a fibril sonicated and unsonicated fibril surface at pH 7.4, (c) 10 μM soluble Aβ(1–40) to a low-(2500 RU beta-amyloid fibrils) and a high-(6000 RU beta-amyloid fibrils) density surface (Cannon *et al.*, 2004)

Analyte in solution/receptor on surface	k	k_d	k_{d1}	k_{d2}
(a) 3 μM Aβ(1–40); flow rate 100 μl/min/immobilized fibril surface	51.647 ± 0.876	17.439 ± 0.676	na	na
(b) 3 μM Aβ(1–40)/unsonicated surface(1200 RU fibrils)	20.907 ± 0.247	5.553 ± 0.776	na	na
3 μM Aβ(1–40)/sonicated surface (1200 RU fibrils)	11.145 ± 0.319	8.675 ± 0.646	na	na
(c) 10μ M Aβ(1–40); flow rate 3 μ l/min/high-density (6000 RU) beta-amyloid fibril surface	45.294 ± 2.027	60.574 ± 5.878	na	na
10 μM Aβ(1–40); flow rate 3 μl/min/low-density (2500 RU) beta-amyloid fibril surface	14.197 ± 0.484	31.935 ± 2.816	na	na
(d) 10 μM Aβ(1–40); flow rate 100 μl/min/high-density (6000 RU) beta-amyloid fibril surface	37.951 ± 0.648	192.56 ± 22.140	70.299 ± 1.034	34.724 ± 0.147
10 μM Aβ(1–40); flow rate 100 μl/min/low-density (2500 RU) beta-amyloid fibril surface	19.287 ± 0.977	35.413 ± 1.715	na	na

the solution contained 0.005% polysorbate 20 as the surfactant. The solution flow over the sensor chip is 100 μl/min. A single-fractal analysis is, once again, adequate to describe the binding and the dissociation kinetics. The values of (a) the binding rate coefficient, k and the fractal dimension, D_f for a single-fractal analysis, and (b) the dissociation rate coefficient, k_d and the fractal dimension in the dissociation phase, D_{fd} for a single-fractal analysis are given in Table 7.3b.

It is of interest to note that as one goes from the sonicated to the unsonicated fibril surface there is a decrease in the fractal dimension, D_f value by 4.955 from a value of D_f equal to 2.2101 to 2.101, and a corresponding decrease in the binding rate coefficient value by 46.7% from a value of k equal to 20.907 to 11.145. Note that changes in the degree of heterogeneity on the sensor chip surface and in the binding rate coefficient are in the same direction.

Figure 7.8a shows the binding of 10 μM Aβ(1–40) soluble peptide in solution to a high-density (6000 RU) fibril surface (Cannon *et al.*, 2004). The pH of the solution is 7.4, PBS is the buffer, and 0.005% polysorbate 20 surfactant was used as the running buffer. The solution flow over the sensor chip is 3 μl/min. A single-fractal analysis is, once again, adequate to describe the binding and the dissociation kinetics. The values of (a) the binding rate coefficient, k and the fractal dimension, D_f for a single-fractal analysis, and (b) the dissociation rate coefficient, k_d and the fractal dimension in the dissociation phase, D_{fd} for a single-fractal analysis are given in Table 7.3b.

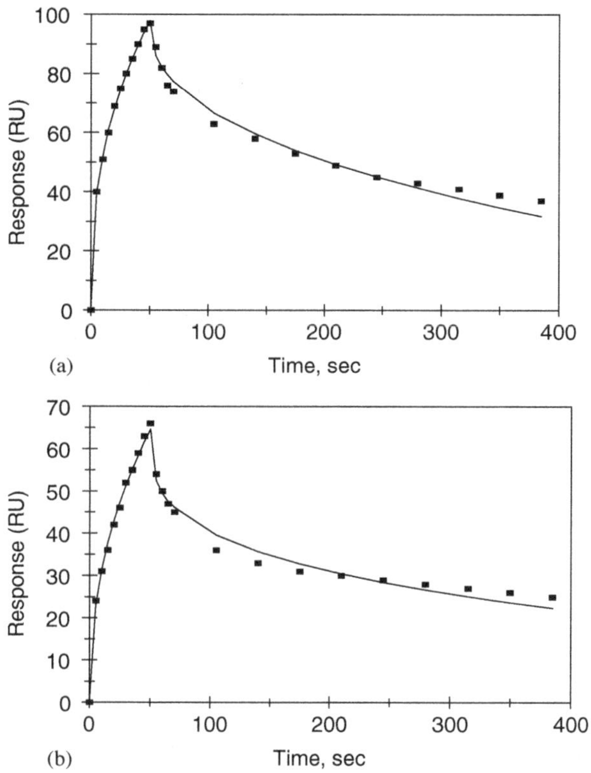

Figure 7.7 Binding of different concentrations (in μM) of beta amyloid, Aβ(1–40) in solution to a fibril surface; pH 7.4, PBS, and 0.005% polysorbate 20 surfactant (Cannon *et al.*, 2004): (a) 3 (sonicated surface), (b) 3 (unsonicated surface).

Figure 7.8b shows the binding of 10 μM Aβ(1–40) soluble peptide in solution to a low-density (2500 RU) fibril surface (Cannon *et al.*, 2004). The pH of the solution is 7.4, PBS is the buffer, and 0.005% polysorbate 20 surfactant was used as the running buffer. The solution flow over the sensor chip is 3 μl/min. A single-fractal analysis is, once again, adequate to describe the binding and the dissociation kinetics. The values of (a) the binding rate coefficient, k and the fractal dimension, D_f for a single-fractal analysis, and (b) the dissociation rate coefficient, k_d and the fractal dimension in the dissociation phase, D_{fd} are given in Table 7.3b.

It is of interest to note that as one goes from the high-density to the low-density fibril surface, the fractal dimension, D_f decreases by 18.54% from a value of D_f equal to 1.7796 to 1.7472, and the binding rate coefficient, k decreases by a factor of 3.19 from a value of k equal to 45.294 to 14.19. Note that, once again, changes in the degree of heterogeneity or the fractal dimension on the sensor chip surface are in the same direction.

Figure 7.9a shows the binding of 10 μM Aβ(1–40) soluble peptide in solution to a high-density (6000 RU) fibril surface (Cannon *et al.*, 2004). The pH of the solution is 7.4, PBS is the buffer, and 0.005% polysorbate 20 surfactant was used as the running buffer. The solution flow over the sensor chip is 100 μl/min. The flow rate is the only difference between Figures 7.8a,b and 7.9a,b. A single-fractal analysis is, once again, adequate to describe the binding

Table 7.3b

Fractal dimensions for the binding and the dissociation phases for (a) 3 μM Aβ(1–40) in solution to a fibril immobilized surface, (b) 3 μM Aβ(1–40) in solution to a fibril sonicated and unsonicated fibril surface at pH 7.4, (c) 10 μM soluble Aβ(1–40) in solution to a low-(2500 RU beta-amyloid fibrils) and a high-density (6000 RU beta-amyloid fibrils) surface (Cannon *et al.*, 2004)

Analyte in solution/ receptor on surface	D_f	D_{fd}	D_{fd1}	D_{fd2}
(a) μM Aβ(1–40); flow rate 100 μ l/min/immobilized fibril surface	2.0722 ± 0.0276	2.2536 ± 0.01598	na	na
(b) 3 μM Aβ(1–40)/sonicated surface (1200 RU fibrils)	2.2104 ± 0.0107	2.1530 ± 0.05274	na	na
3 μM Aβ(1–40)/unsonicated surface (1200 RU fibrils)	2.101 ± 0.0256	2.4444 ± 0.02894	na	na
(c) 10 μM Aβ(1–40); flow rate 3 μl/min/high-density (6000 RU) beta-amyloid fibril surface	1.7796 ± 0.02476	2.2000 ± 0.04916	na	na
10 μM Aβ(1–40); flow rate 3 μl/min/low-density (2500 RU) beta-amyloid fibril surface	1.7472 ± 0.01899	2.5146 ± 0.04340	na	na
(d) 10 μM Aβ(1–40); flow rate 100 μl/min/high-density (6000 RU) beta-amyloid fibril surface	1.7290 ± 0.0325	2.8006 ± 0.1056	2.6532 ± 0.02874	2.4466 ± 0.01462
10 μM Aβ(1–40); flow rate 100 μl/min/low-density (2500 RU) beta-amyloid fibril surface	1.8410 ± 0.1099	2.7628 ± 0.0161	na	na

kinetics. A dual-fractal analysis is required to adequately describe the dissociation kinetics. The values of (a) the binding rate coefficient, k and the fractal dimension, D_f for a single-fractal analysis, (b) the dissociation rate coefficient, k_d and the fractal dimension in the dissociation phase, D_{fd}, for a single-fractal analysis and (c) the dissociation rate coefficients, k_{d1} and k_{d2}, and the fractal dimensions in the dissociation phase, $D_{fd1}$and D_{fd2} for a dual-fractal analysis are given in Table 7.3b

Figure 7.9b shows the binding of 10 μM Aβ(1–40) soluble peptide in solution to a low-density (2500 RU) fibril surface (Cannon *et al.*, 2004). The pH of the solution is 7.4, PBS is the buffer, and 0.005% polysorbate 20 surfactant was used as the running buffer. The solution flow over the sensor chip is 100 μl/min. A single-fractal analysis is, once again, adequate to describe the binding and the dissociation kinetics. The values of (a) the binding rate coefficient, k and the fractal dimension, D_f for a single-fractal analysis, and (b) the dissociation rate coefficient, k_d and the fractal dimension in the dissociation phase, D_{fd} for a single-fractal analysis are given in Table 7.3b.

It is of interest to compare the results obtained in Figures 7.8a,b with 7.9a,b for the low (3 μl/min) with the high (100 μl/min) flow rates when the low- and high-density fibril surfaces are used. When the high-density (6000 RU) fibril surface is used, an increase in the flow rate from 3 to 100 μl/min leads to a decrease in the fractal dimension value by 2.93%

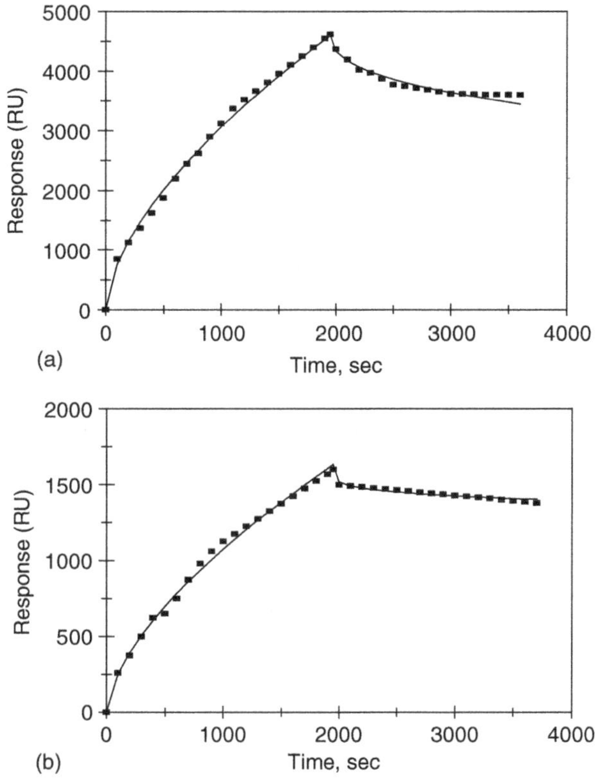

Figure 7.8 Binding and dissociation of soluble peptide to a (a) high-density and a (b) low-density fibril sensor surface (Cannon *et al.*, 2004).

from a value of 1.7796 to 1.7290, and to a corresponding decrease in the binding rate coefficient, k value by 19.35% from a value of k equal to 45.294 to 37.951.

Similarly, when the high-density (6000 RU) fibril surface is used, an increase in the flow rate from 3 to 100 μl/min leads to an increase in the fractal dimension value by 5.37% from a value of 1.7472 to 1.8410, and to a corresponding increase in the binding rate coefficient, k value by 35.85% from a value of k equal to 14.197 to 19.287.

Cannon *et al.* (2004) analyzed the influence of different concentrations (in μM) of the mutant peptide, F19P Aβ(1–40) in solution on its binding and dissociation kinetics to fibrils immobilized on a sensor chip surface. Figure 7.10a shows the binding and dissociation kinetics of 1 μM F19P Aβ(1–40) in solution to fibrils immobilized on a sensor chip surface (Cannon *et al.*, 2004). A dual-fractal analysis is adequate to describe the binding and dissociation kinetics. The values of (a) the binding rate coefficient, k and the fractal dimension, D_f for a single-fractal analysis, (b) the binding rate coefficients, k_1 and k_2 and the fractal dimensions, D_{f1} and D_{f2} for a dual-fractal analysis, (c) the dissociation rate coefficient, k_d and the fractal dimension, D_{fd} for a single-fractal analysis, and (d) the dissociation rate coefficients, k_{d1} and k_{d2}, and the fractal dimensions for the dissociation phase, D_{fd1} and D_{fd2} for a dual-fractal analysis are given in Tables 7.4a and 7.4b.

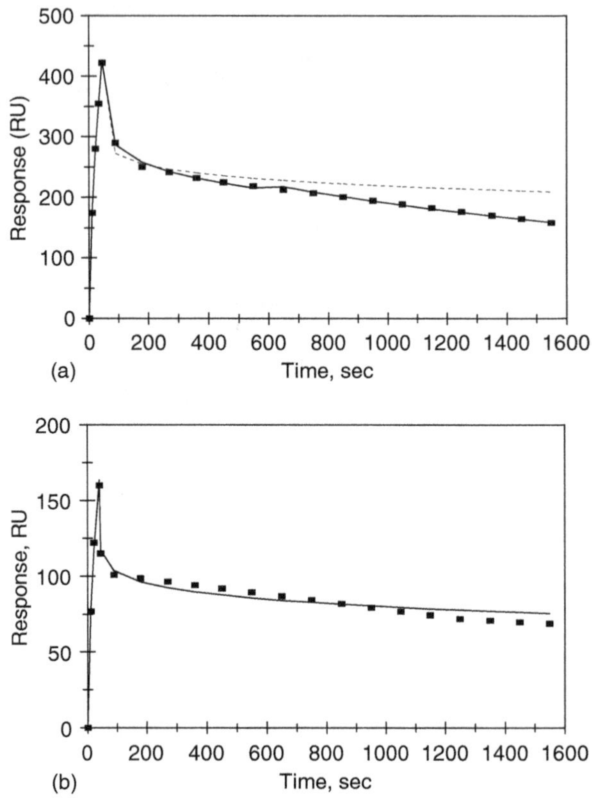

Figure 7.9 Binding and dissociation of soluble peptide to a (a) high-density and a (b) low-density fibril sensor surface (Cannon *et al.*, 2004).

It is of interest to note that for a dual-fractal analysis, as the fractal dimension increases by a factor of 1.33 from a value of D_{f1} equal to 2.0582 to 2.7666, the binding rate coefficient increases by a factor of 2.20 from a value of k_1 equal to 12.888 to k_2 equal to 28.378. Once again, an increase in the fractal dimension or the degree of heterogeneity on the sensor chip surface leads to an increase in the binding rate coefficient.

Similarly, an increase in the fractal dimension in the dissociation phase by a factor of 1.516 from a value of D_{fd1} equal to 1.8094 to D_{fd2} equal to 2.7440 leads to an increase in the dissociation rate coefficient value by a factor of 3.78 from k_{d1} equal to 5.788 to k_{d2} equal to 21.914.

Figure 7.10b shows the binding and dissociation kinetics of 0.5 μM F19P Aβ(1–40) in solution to fibrils immobilized on a sensor chip surface (Cannon *et al.*, 2004). A single-fractal analysis is adequate to describe the binding kinetics, and a dual-fractal analysis is required to describe the dissociation kinetics. The values of (a) the binding rate coefficient, k and the fractal dimension, D_f for a single-fractal analysis, (b) the dissociation rate coefficient, k_d and the fractal dimension, D_{fd} for a single-fractal analysis, and (c) the dissociation rate coefficients, k_{d1} and k_{d2}, and the fractal dimensions for the dissociation phase,

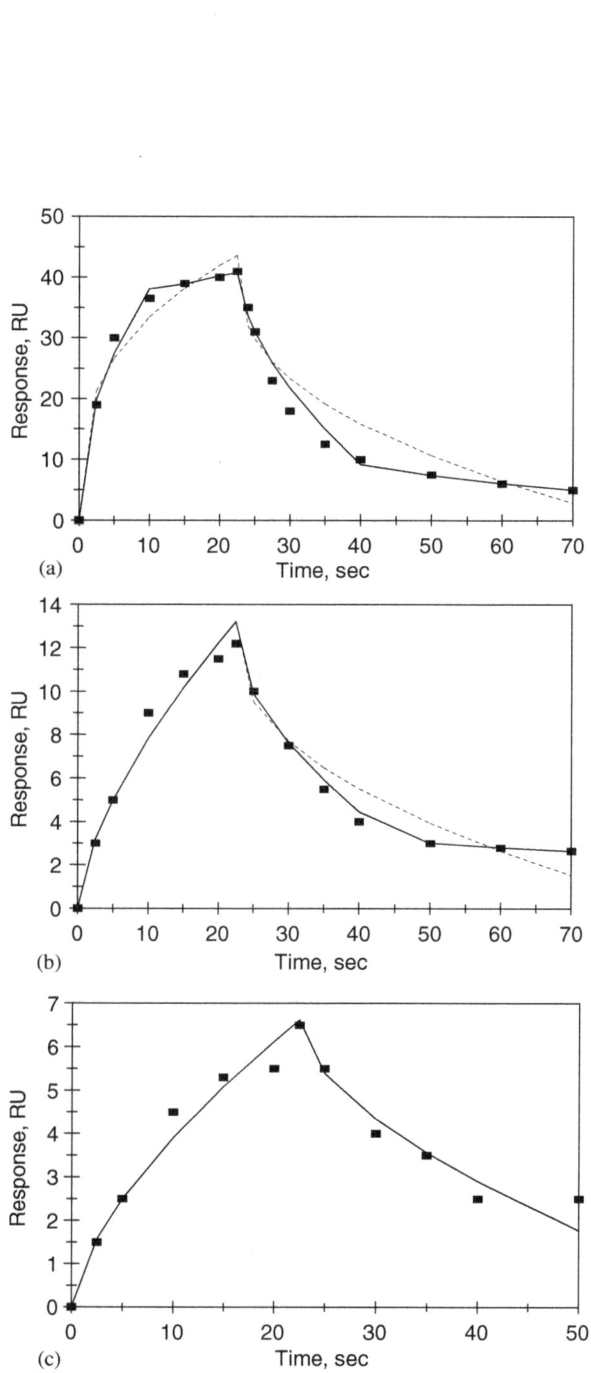

Figure 7.10 Binding of different concentrations (in μM) of F19P Aβ(1–40) (mutant) in solution to fibrils immobilized on a sensor chip surface; pH 7.4, PBS, and 0.005% polysorbate 20 surfactant (Cannon *et al.*, 2004): (a) 1.0, (b) 0.5, (c) 0.25.

D_{fd1} and D_{fd2} for a dual-fractal analysis are given in Tables 7.4a and 7.4b. An increase in the fractal dimension in the dissociation phase by a factor of 1.622 from a value of D_{fd1} equal to 1.7588 to D_{fd2} equal to 2.8535 leads to an increase in the dissociation rate coefficient value by a factor of 5.51 from k_{d1} equal to 1.3079 to k_{d2} equal to 7.2051.

Table 7.4a

Binding and dissociation rate coefficients for different concentrations (in μM) of F19P Aβ(1–40) mutant peptide in solution to fibrils immobilized on a sensor chip surface. Flow rate 100 μl/min; pH 7.4; PBS; and 0.005% polysorbate 20 surfactant used as a running buffer (Cannon *et al.*, 2004)

F19P Aβ (1–40) mutant peptide concentration in solution, μM	k	k_1	k_2	k_d	k_{d1}	k_{d2}
1.0	15.770 ± 1.673	12.888 ± 1.448	28.378 ± 0.230	7.592 ± 1.880	5.788 ± 1.141	21.914 ± 0.145
0.5	1.762 ± 0.179	na	na	0.4668 ± 0.0598	1.3079 ± 0.1178	7.2051 ± 0.009
0.25	0.9741 ± 0.0882	na	na	0.6398 ± 0.1076	na	na

Table 7.4b

Fractal dimensions for the binding and the dissociation phases for different concentrations (in μM) of F19P Aβ(1–40) mutant peptide in solution to fibrils immobilized on a sensor chip surface. Flow rate 100 μl/min; pH 7.4; PBS; and 0.005% polysorbate 20 surfactant used as a running buffer (Cannon *et al.*, 2004)

F19P Aβ (1–40) mutant peptide concentration in solution, μM	D_f	D_{f1}	D_{f2}	D_{fd}	D_{fd1}	D_{fd2}
1.0	2.3464 ± 0.1048	2.0582 ± 0.2172	2.7666 ± 0.0547	2.1636 ± 0.1004	1.8094 ± 0.1400	2.7440 ± 0.0152
0.5	1.7060± 0.1007	na	na	2.0624 ± 0.1196	1.7588 ± 0.0938	2.8535 ± 0.0083
0.25	1.6988 ± 0.0999	na	na	1.7932 ± 0.1688	na	na

Figure 7.10c shows the binding and dissociation kinetics of 0.25 μM F19P Aβ(1–40) in solution to fibrils immobilized on a sensor chip surface (Cannon *et al.*, 2004). A single-fractal analysis is adequate to describe the binding and the dissociation kinetics. The values of the binding rate coefficient, k and the fractal dimension, D_f for a single-fractal analysis are given in Tables 7.4a and 7.4b.

Tables 7.4a and 7.4b and Figure 7.11a show the increase in the binding rate coefficient (k, k_1, or k_2) with an increase in the fractal dimension (D_f, D_{f1}, and D_{f2}). All of these points are plotted together on the same graph, due to the scarcity of experimental data available. For the data shown in Figure 7.11a, the binding rate coefficient is given by

$$k, k_1, \text{or } k_2 = (0.0476 + 0.0720)(D_f, D_{f1}, \text{or } D_{f2})^{6.602 \pm 1.916} \tag{7.6a}$$

The fit is good. Only four data points are available. The availability of more data points would lead to a more reliable fit. There is scatter in the data. This is reflected in the error in the rate coefficient value. This is also due to the fact that the binding rate coefficients obtained for a single- and a dual-fractal analysis are plotted together on the same graph due

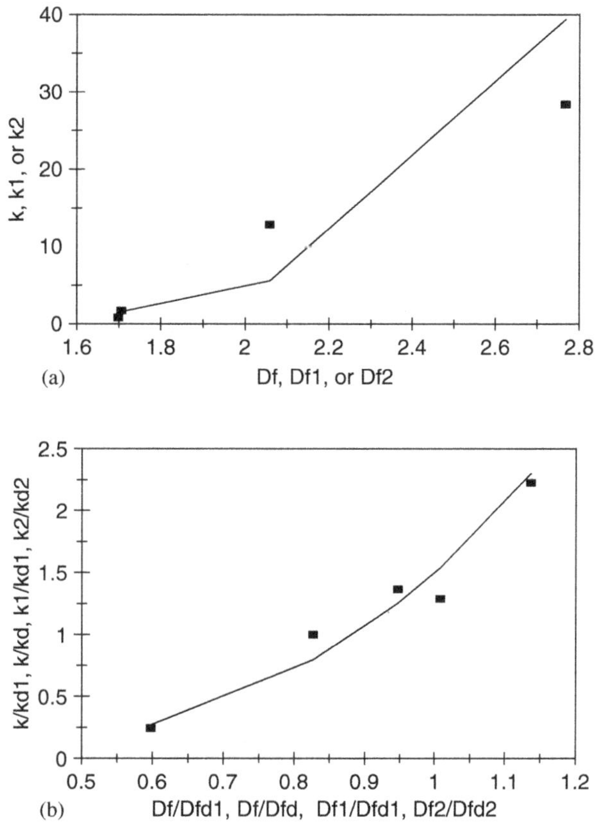

Figure 7.11 (a) Increase in the binding rate coefficient, k with an increase in the fractal dimension, D_f. (b) Increase in the affinity with an increase in the fractal dimension ratio.

to the scarcity of points available. The binding rate coefficient is very sensitive to the fractal dimension or the degree of heterogeneity present on the sensor chip surface as noted by the greater than six and one-half (equal to 6.602) order of dependence exhibited.

Table 7.4a and b and Figure 7.12b show the increase in the affinity (k/k_d, k/k_{d1}, k_1/k_{d1}, k_2/k_{d2}) with an increase in the ratio of fractal dimension values (D_f/D_{fd1}, D_f/D_{fd}, D_{f1}/D_{fd1}, D_{f2}/D_{fd2}). All of these points are plotted together on the same graph, once again due to the scarcity of data points available. For the data shown in Figure 7.11b, the affinity is given by

$$\left(\frac{k}{k_d}, \frac{k}{k_{d1}}, \frac{k_1}{k_{d1}}, \frac{k_2}{k_{d2}}\right) = \left(\frac{D_f}{D_{fd1}}, \frac{D_f}{D_{fd}}, \frac{D_{f1}}{D_{fd1}}, \frac{D_{f2}}{D_{fd2}}\right)^{3.3256\pm0.3739} \tag{7.6b}$$

The fit is very good. Only three data points are available. The availability of more data points would lead to a more reliable fit. The affinity is very sensitive to the fractal dimension or the degree of heterogeneity present on the sensor chip surface as noted by the greater than third (equal to 3.325) order of dependence exhibited.

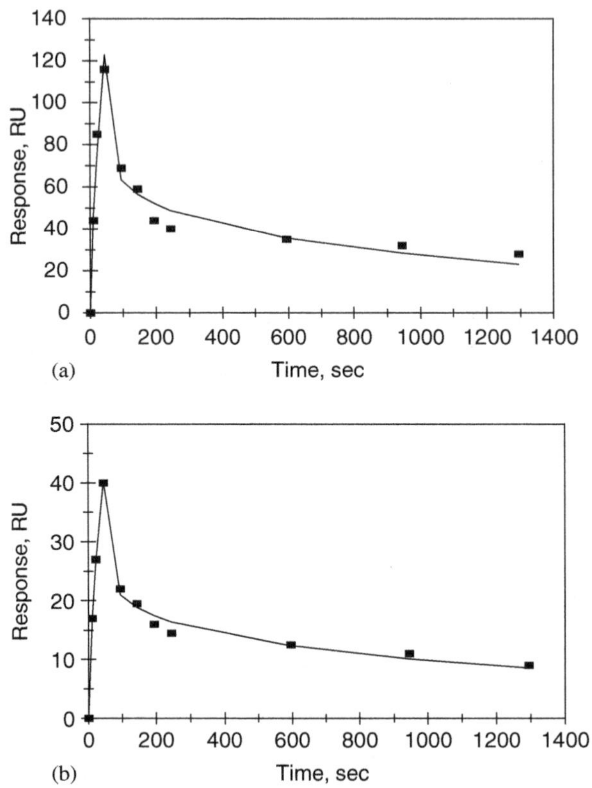

Figure 7.12 Binding of different concentrations (in μM) of beta amyloid, Aβ(1–40) in solution to an immobilized fibril surface (Cannon *et al.*, 2004). Influence of repeat runs: (a) 9, (b) 3.

Cannon *et al.* (2004) wanted to analyze the reproducibility in their studies. They performed three sets of reproducible experiments. We present now two out of three of these studies using fractals and attempt to analyze their kinetics of binding and dissociation of the analyte–receptor reactions. Figure 7.12a shows the binding and dissociation of 9 μM Aβ(1–40) peptide in solution to a sonicated fibril sensor surface (Cannon *et al.*, 2004). A single-fractal analysis is adequate to describe the binding and dissociation kinetics. The values of (a) the binding rate coefficient, k and the fractal dimension, D_f for a single-fractal analysis, and (b) the dissociation rate coefficient, k_d and the fractal dimension, D_{fd} for a single-fractal analysis are given in Tables 7.5a and 7.5b.

Figure 7.12b shows the binding and dissociation of 9 μM Aβ(1–40) peptide in solution to a sonicated fibril sensor surface (Cannon *et al.*, 2004). A single-fractal analysis is adequate to describe the binding and dissociation kinetics. The values of (a) the binding rate coefficient, k and the fractal dimension, D_f for a single-fractal analysis, and (b) the dissociation rate coefficient, k_d and the fractal dimension, D_{fd} for a single-fractal analysis are given in Tables 7.5a and 7.5b.

It is of interest to note that as the Aβ(1–40) peptide concentration in solution decreases by a factor of 3 from 9 to 3 μM, (a) the fractal dimension increases by a factor of 1.10 from

Table 7.5a

Binding and dissociation rate coefficients for different concentrations (in μM) of Aβ(1–40) in solution to immobilized sonicated fibril surfaces. Influence of repeated runs (Cannon *et al.*, 2004)

Aβ(1–40) concentration (μM)	K	k_d	k_{d1}	k_{d2}
Run #1, 9	8.581 ± 1.308	26.421 ± 2.743	na	na
Run #1, 3	3.608 ± 0.1113	10.286 ± 0.576	na	na
Run #2, 9	13.171 ± 1.614	6.225 ± 3.438	1.1927 ± 0.7180	30.035 ± 0.252
Run #2, 3	3.8666 ± 0.436	13.030 ± 0.380	na	na

Table 7.5b

Fractal dimensions for the binding and the dissociation phases for different concentrations (in μM) of Aβ(1–40) in solution to immobilized sonicated fibril surfaces. Influence of repeated runs (Cannon *et al.*, 2004)

Aβ(1–40) concentration (μM)	D_f	D_{fd}	D_{fd1}	D_{fd2}
Run #1, 9	1.6014 ± 0.2894	2.6476 ± 0.0675	na	na
Run #1, 3	1.7656 ± 0.5796	2.6864 ± 0.0372	na	na
Run #2, 9	1.6882 ±0.1738	2.2516 ± 0.1800	1.2514 ± 0.5450	2.7434 ± 0.0108
Run #2, 3	1.5150 ± 0.1606	2.7702 ± 0.118	na	na

a value of D_f equal to 1.6014 to 1.7656, and (b) the binding rate coefficient, k decreases by a factor of 2.22 from a value of k equal 8.581 to 3.608. In this case, changes in the fractal dimension and in the binding rate coefficient are in opposite directions.

Figure 7.13a shows the binding and dissociation of 9 μM Aβ(1–40) peptide concentration in solution to a sonicated fibril sensor surface (Cannon *et al.*, 2004). In this case, just as in Figure 7.12a, the binding kinetics is adequately described by a single-fractal analysis. A dual-fractal analysis is, however, required to adequately describe the dissociation kinetics. The values of (a) the binding rate coefficient, k and the fractal dimension, D_f for a single-fractal analysis, (b) the dissociation rate coefficient, k_d and the fractal dimension, D_{fd} for a single-fractal analysis, and (c) the dissociation rate coefficients, k_{d1} and k_{d2} and the fractal dimensions for the dissociation phase, D_{fd1} and D_{fd2} for a dual-fractal analysis are given in Tables 7.5a and 7.5b.

Figure 7.13b shows the binding and dissociation of 3 μM Aβ(1–40) peptide in solution to a sonicated fibril surface. Just as in Figure 7.12b, a single-fractal analysis is adequate to describe the binding and dissociation kinetics. The values of (a) the binding rate coefficient, k and the fractal dimension, D_f for a single-fractal analysis, and (b) the dissociation rate coefficient, k_d and the fractal dimension for the dissociation phase, D_{fd} for a single-fractal analysis is given in Tables 7.5a and 7.5b.

It is of interest to compare the kinetic parameters obtained in Figure 7.12b with 7.13b. In both of these cases, the binding and dissociation kinetics were adequately described by a single-fractal analysis for the 3 μM Aβ(1–40) peptide in solution. The average fractal

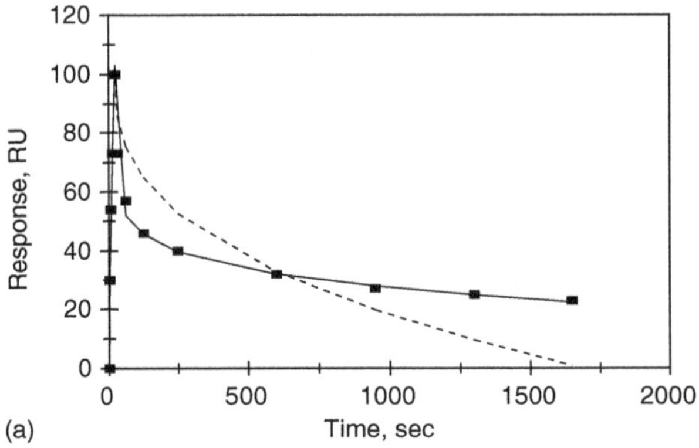

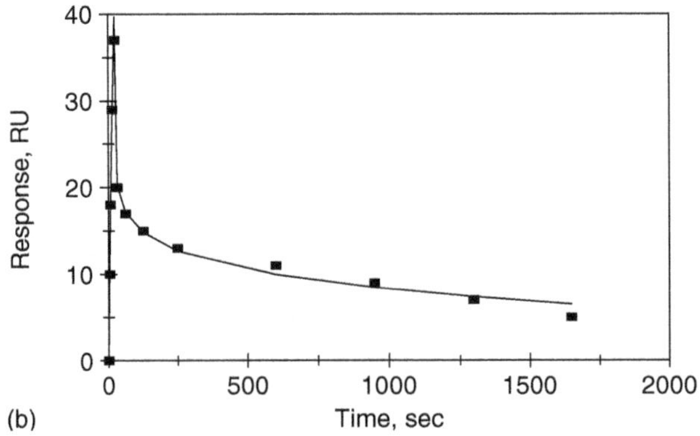

Figure 7.13 Binding and dissociation of different concentrations (in μM) of beta amyloid, Aβ(1–40) in solution to an immobilized fibril surface (Cannon *et al.*, 2004). Influence of repeat runs: (a) 9, (b) 3.

dimension for dissociation, $D_{fd,ave}$ is equal to 2.7283. The actual fractal dimension values in the dissociation phase, D_{fd} were 2.96% from the average fractal dimension value in the dissociation phase, $D_{fd,ave}$. This indicates that the fractal dimension values for the dissociation phase are reproducible, at least for these two sets of experimental data.

Similarly, the average binding rate coefficient, k_{ave} value for the 3 μM Aβ(1–40) peptide in solution case is 3.7373. The actual binding rate coefficient, k values obtained are 12.9% from the average value. The average dissociation rate coefficient, k_d are within 11.77% of the average binding rate coefficient value. Once again, this indicates that the rate coefficient values for the binding and the dissociation phases are reproducible, at least for these two sets of experimental results.

Figure 7.14 and Tables 7.5a and 7.5b show the increase in the affinity, $K(= k/k_d)$ with an increase in the fractal dimension ratio, D_f/D_{fd}. For the data shown in Figure 7.14, the affinity, K is given by

$$K\left(=\frac{k}{k_d}\right) = (0.6321 \pm 0.0288)\left(\frac{D_f}{D_{fd}}\right)^{1.275 \pm 0.226} \tag{7.7}$$

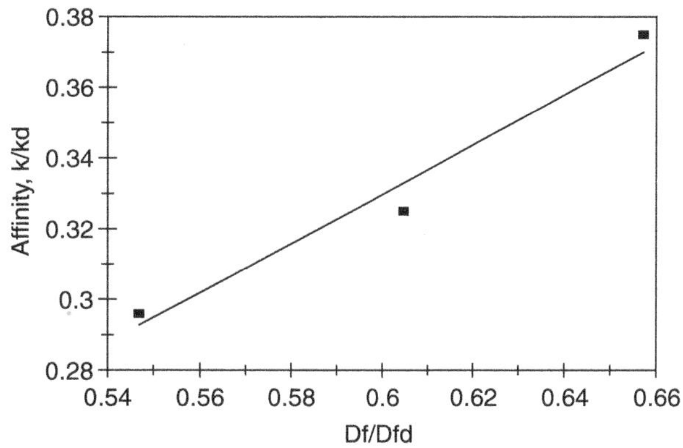

Figure 7.14 Increase in the affinity, $K(= k/k_d)$ with an increase in the fractal dimension ratio, D_f/D_{fd}.

The fit is good. Only three data points are available. The availability of more data points would lead to a more reliable fit. The affinity, $K(= k/k_d)$ exhibits an order of dependence between first and one and one-half (equal to 1.275) on the ratio of fractal dimensions present in the binding and in the dissociation phases.

Figure 7.15a and Table 7.5a show for a single-fractal analysis the increase in the binding rate coefficient, k with an increase in the Aβ(1–40) concentration (in μM) in solution. For the data shown in Figure 7.15a, the binding rate coefficient, k is given by

$$k = (1.312 \pm 0.318)[A\beta(1-40)]^{0.952 \pm 0.198} \tag{7.8a}$$

The fit is reasonable. There is scatter in the data. Only four data points are available. The availability of more data points would lead to a more reliable fit. The binding rate coefficient, k exhibits close to a first (equal to 0.952) order of dependence on the Aβ(1–40) concentration in solution. The non-integer order of dependence exhibited by the binding rate coefficient k on the Aβ(1–40) concentration in solution lends support to the fractal nature of the system.

Figure 7.15b and Table 7.5a show for a single-fractal analysis the decrease in the fractal dimension in the dissociation phase, D_{fd} with an increase in the Aβ(1–40) concentration (in μM) in solution. For the data shown in Figure 7.15b, the fractal dimension in the dissociation phase, D_{fd} is given by

$$k_d = (2.1808 \pm 0.0617)[A\beta(1-40)]^{-0.0272 \pm 0.0242} \tag{7.8b}$$

The fit is reasonable. There is scatter in the data. Only three data points are available. The availability of more data points would lead to a more reliable fit. The fractal dimension in the dissociation phase, D_{fd} exhibits a very mild (equal to −0.0272) negative order of dependence on the Aβ(1–40) concentration in solution.

7.4 CONCLUSIONS

A fractal analysis is presented for the binding and dissociation kinetics of small molecules involved in drug design on biosensor surfaces. The fractal analysis provides a quantitative

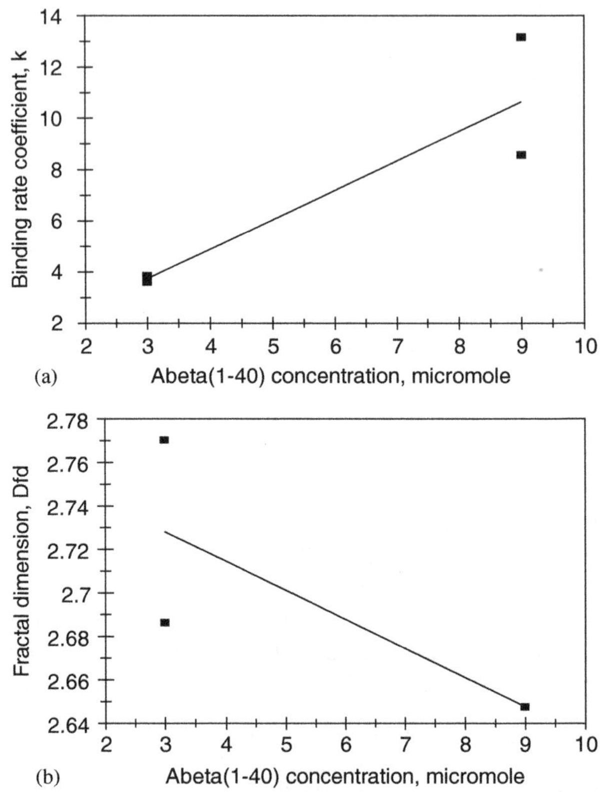

Figure 7.15 (a) Increase in the binding rate coefficient, k with an increase in the beta amyloid, Aβ(1–40) concentration (in μM) in solution. (b) Decrease in the fractal dimension for dissociation, D_{fd} with an increase in the beta amyloid, Aβ(1–40) concentration (in μM) in solution.

indication of the degree of heterogeneity present on the sensor chip surface (fractal dimension, D_f), and relates it to the binding rate coefficient, k. Both types of examples are given wherein either a single- or a dual-fractal analysis were used. The dual-fractal analysis was used only when the single-fractal analysis did not provide an adequate fit. This was done by the regression analysis provided by Corel Quattro Pro 8.0 (1997). In accord with the pre-factor analysis of fractal aggregates (Sorenson and Roberts, 1997) quantitative (predictive) expressions are developed for (a) the binding rate coefficient, k_1 as a function of the KM19 concentration in solution and for the binding of 20 nM CsA + KM19 concentration (in μM) in solution to the covalently immobilized His-CypA (Wear *et al.*, 2005), (b) the binding rate coefficients, k_1 and k_2 as a function of the fractal dimensions, D_{f1} and D_{f2}, respectively, for the above-mentioned reaction (Wear *et al.*, 2005), (c) the fractal dimension, D_{f1} as a function of the KM19 concentration (in μM) in solution (Wear *et al.*, 2005), (d) the affinity, K_1 as a function of the KM19 concentration (in μM) in solution (Wear *et al.*, 2005), (e) the dissociation rate coefficient, k_{d1} as a function of the fractal dimension, D_{fd1}, (f) the affinity, K_1 (= k_1/k_{d1}) as a function of the ratio of fractal dimensions in the binding and in the dissociation phases, D_{f1}/D_{fd1} (Wear *et al.*, 2005), (g) the binding rate coefficient, k_1 as a function of the fractal dimension, D_{f1} for the binding of different concentrations of dopamine in solution (in μM) to the μARCS format (Gopalakrishnan *et al.*, 2003), (h) the affinity, K_1 (= k_1/k_{d1}) as a function of the ratio of fractal dimensions present in the binding and in the dissociation phases, D_{f1}/D_{fd1} for the above mentioned reactions (Goapalkrishnan *et al.*, 2003), (i) the binding rate coefficient, k as a function of the fractal dimension for the binding (and dissociation) of different concentrations (in μM) of F19P Aβ(1–40) peptide in solution to fibrils immobilized on a sensor chip surface (Cannon *et al.*, 2004), (j) the affinity, K_1 as a function of the ratio of fractal dimensions present in the binding and in the dissociation phases, D_{f1}/D_{fd1} for the reaction mentioned above (Cannon *et al.*, 2004), and (k) the binding and the dissociation rate coefficients as a function of the Aβ(1–40) peptide in solution to the sonicated fibril surface (Cannon *et al.*, 2004).

The fractal dimension is not a classical independent variable such as analyte (antigen, antibody, or other biological molecule) concentration in solution. Nevertheless, the expressions obtained for the binding (and the dissociation) rate coefficients for a single- and a dual-fractal analysis as a function of the fractal dimension indicate a high sensitivity of these rate coefficients on their respective fractal dimensions on the SPR sensor chip surface. Note that the data analysis in itself does not provide any evidence for surface roughness or heterogeneity, and the existence of surface roughness or heterogeneity assumed may not be correct. Considering the complexity involved on the SPR chip surface, this is not an unreasonable assumption. Furthermore, there is deviation in the data that may be minimized by providing a correction for the depletion of the analyte.

It is hoped that the fractal analysis of the binding and dissociation of small molecules such as of CsA + KM19, dopamine, and of Aβ(1–40) peptide to their respective receptors on biosensor surfaces presented would not only provide a better understanding of the kinetics involved, but also a fresh perspective that should be of considerable use in drug discovery. It should be pointed out that the manipulation of heterogeneity on the biosensor surface, via the fractal dimension, provides an additional means by which these types of interactions may be manipulated in required directions, thereby facilitating in drug discovery.

REFERENCES

Abdiche YN and DG Myszka, Probing the mechanism of drug/lipid membrane interactions using Biacore. *Analytical Biochemistry*, 2004, **328**, 93–99.

Baird CL, ES Courtenay, and DG Myszka, Surface plasmon resonance characterization of drug/liposome interactions. *Analytical Biochemistry*, 2002, **310**, 93–99.

Burns DJ, JL Kofron, U Warrior, and BA Beutel, Well-less, gel-permeation formats for ultra-HTS. *Drug Discovery Today*, 2001, **6**, S40–S47.

Cairo CW, A Strzelec, RM Murphy, and LL Kiessling, Affinity-based inhibition of beta-amyloid toxicity. *Biochemistry*, 2002, **41**, 8620–8629.

Cannon MJ, GA Papalia, I Navratilova, RJ Fisher, LR Roberts, KM Worthy, AG Stephen, GR Marchesini, EJ Collins, and D Casper, Comparative analyses of a small molecule/enzyme interaction by multiple users of Biacore technology. *Analytical Biochemistry*, 2004a, **330**, 98–113.

Cannon MJ, AD Williams, R Wetzel, and DG Myszka, Kinetic analysis of beta-amyloid fibril elongation. *Analytical Biochemistry*, 2004b, **328**, 67–75.

Corel Quattro Pro 8.0, Corel Corporation Limited, Ottawa, Canada, 1997.

Dornan J, P Taylor, and MD Walkinsaw, Structures of immunophilins and their ligand complexes. *Current Topics in Medicinal Chemistry*, 2003, **3**, 1392–1409.

Falk RH, RL Comenzo, and M Skinner, The systemic amyloidoses. *New England Journal of Medicine*, 1997, **337**, 898–909.

Galat A, Peptodylprolyl *cis/trans*isomerases (immunophilins): biological diversity-targets-functions. *Current Topics in Medicinal Chemistry*, 2003, **3**, 1315–1347.

Gopalakrishnan SM, RB Moreland, JL Kofron, RJ Helfrich, E Gubbins, J McGowen, N Masters, D Donnelly Roberts, JD Brioni, DJ Burns, and U Warior, A cell-based microarrayed compound screening format for identifying agonists of G-protein-coupled receptors. *Analytical Biochemistry*, 2003, **321**, 192–201.

Hasegawa K, K Ono, M Yamada, and H Naiki, Kinetic modeling and determination of reaction constants of Alzheimer's beta-amyloid extension and dissociation using surface plasmon resonance. *Biochemistry*, 2002, **41**, 13489–13498.

Havlin S, The Fractal Approach to Heterogeneous Chemistry: Surfaces, Colloids, Polymers (ed. D. Avnir), Wiley, New York, 1989, pp. 251–269.

Lofas S, Optimizing the hit-to-lead process using SPR analysis. *Assay Drug Development Technology*, 2004, **2**, 407–415.

Marinissen MJ and JS Gutkind, G protein-coupled receptors and signaling networks: emerging paradigms. *Trends in Pharmacological Science*, 2001, **22**, 368–375.

Martin JB, Molecular basis of the neurodegenerative disorders. *New England Journal of Medicine*, 1999, **340**, 1970–1980.

Myszka DG, Analysis of small-molecule interactions using Biacore S51 technology. *Analytical Biochemistry*, 2004, **329**, 316–323.

Myszka DG and RL Rich, Implementing surface plasmon resonance biosensors in drug discovery. *Pharmaceutical Science and Technology*, 2000, **3**, 310–317.

Myszka DG, SJ Wood, and AL Biere, Analysis of fibril elongation using surface plasmon resonance biosensors. *Methods in Enzymology*, 1999, **309**, 386–402.

Rich RL and DG Myszka, A survey of the year 2002 commercial optical biosensor literature. *Journal of Molecular Recognition*, 2003, **16**, 351–382.

Schurdak ME, MJ Voorbach, L Gao, X Cheng, K Comess, S Rottinghaus, U Warrior, H Truong, DJ Burns, and BA Beutel, Complex gel permeation assays for screening combinatorial libraries. *Journal of Biomolecular Screening*, 2001, **6**, 313–322.

Sorenson CM and GC Roberts, The prefactor of fractal aggregates. *Journal of Colloid and Interface Science*, 1997, **186**, 447–452.

Tjernberg LO, J Naslund, F Lindqvist, J Johannson, AR Karlsson, J Thyberg, L Terenius, and C Norstedt, Arrest of beta-amyloid fibril formation by a pentapeptide ligand. *Journal of Biological Chemistry*, 1996, **271**, 8545–8548.

Wear MA, A Patterson, IC Malone, C Dunsmore, NJ Turner, and MD Walkinsaw, A surface plasmon resonance-based assay for small molecule inhibitors of human cyclophilin A. *Analytical Biochemistry*, 2005, **345**, 214–226.

Wood SJ, W Chan, and R Wetzel, An ApoE-Aβ inhibition complex in Aβ fibril extension. *Chemical Biology*, 1996, **3**, 949–956.

Zhukov A, M Schurenberg, O Jansson, D Areskoug, and J Buijs, Integration of surface plasmon resonance with mass spectroscopy: automated ligand fishing and sample preparation for MALDI MS using a Biacore 3000 biosensor. *Journal of Biomolecular Technology*, 2004, **15**, 112–119.

– 8 –

Fractal Binding and Dissociation Kinetics of Prion-Related Interactions on Biosensor Surfaces

8.1 INTRODUCTION

Prion-related diseases such as transmissible spongiform encephalopathies (TSEs) are slowly forming, insidious, and intractable. Maxson *et al.* (2003) indicate that though the precise composition of the infectious agent in these types of diseases is not yet been fully described, prions are involved in these types of diseases. Pruisner (1998) has proposed that protease resistant prion protein (PrP-res) may be the infectious agent, and is linked with the pathogenesis of these types of diseases. Maxson *et al.* (2003) have also indicated the heightened concern with regard to TSEs crossing species barriers to infect new taxa. Thus, there is an urgent need to develop therapeutic treatments and diagnostics for these presently incurable diseases. Biosensors offer this type of opportunity for the early detection of prions and prion-related interactions. For example, Maxson *et al.* (2003) have developed a solid-phase assay for the identification of modulators of prion protein interactions.

αB crystallin is another protein that is involved in the pathogenesis of insidious diseases such as Alzheimer's, Parkinson's, and Alexander's (Dabir *et al.*, 2004; Renkawek *et al.*, 1994; Iwaki *et al.*, 1989; Iwaki, 1997). Liu *et al.* (2006) have recently analyzed the subunit dynamics of αB crystallin using surface plasmon resonance (SPR). These authors indicate that this is a small heat shock protein (sHSP), and its three dimensional structure consists of a helical *N*-terminal domain, a α crystallin core domain, and a short flexible *C*-terminus extension (Kim *et al.*, 1998; Ghosh and Clark, 2005). α Crystallin is apparently expressed in tissues throughout the body in tissues such as heart, brain, kidney, lung, etc. (Liu *et al.*, 2006). These authors emphasize that the expression of αB crystallin is upregulated during conditions of stress that includes heat shock, low pH, and oxidative stress (Datta and Rao, 1999; Pasta *et al.*, 2003).

In this chapter we use fractal analysis to analyze the binding (and dissociation, if applicable) kinetics of (a) the binding of protease-sensitive prion protein (PrP-sen) in solution to PrP-res immobilized on a well surface (Maxson *et al.*, 2003), and (b) the binding and dissociation of αB crystallin in solution to αB crystallin subunits adsorbed on a surface adsorbed monolayer (SAM) surface (Liu *et al.*, 2006). Binding and dissociation rate coefficient values, as well as values of the fractal dimension (that indicate the degree of heterogeneity) on the biosensor surface are presented. This is one way of obtaining the binding and dissociation rate coefficient values, as well as the affinity values.

8.2 THEORY

Havlin (1989) has reviewed and analyzed the diffusion of reactants toward fractal surfaces. The details of the theory and the equations involved for the binding and the dissociation phases for analyte–receptor binding are available (Sadana, 2001). The details are not repeated here, except that just the equations are given to permit an easier reading. These equations have been applied to other biosensor systems (Sadana, 2001, 2005; Ramakrishnan and Sadana, 2001). For most applications, a single- or a dual-fractal analysis is often adequate to describe the binding and the dissociation kinetics. Peculiarities in the values of the binding and the dissociation rate coefficients, as well as in the values of the fractal dimensions with regard to the dilute analyte systems being analyzed will be carefully noted, if applicable.

8.2.1 Single-fractal analysis

Binding rate coefficient

Havlin (1989) indicates that the diffusion of a particle (analyte [Ag]) from a homogeneous solution to a solid surface (e.g., receptor [Ab]-coated surface) on which it reacts to form a product (analyte–receptor complex; Ab · Ag) is given by:

$$(\mathrm{Ab}\cdot\mathrm{Ag}) \approx \begin{cases} t^{(3-D_{\mathrm{f,bind}})/2} = t^{p} & t < t_{\mathrm{c}} \\ t^{1/2} & t > t_{\mathrm{c}} \end{cases} \tag{8.1a}$$

Here $D_{\mathrm{f,bind}}$ or D_{f} is the fractal dimension of the surface during the binding step. t_{c} is the cross-over value. Havlin (1989) indicates that the cross-over value may be determined by $r_{\mathrm{c}}^2 \sim t_{\mathrm{c}}$. Above the characteristic length, r_{c} the self-similarity of the surface is lost and the surface may be considered homogeneous. Above time, t_{c} the surface may be considered homogeneous, since the self-similarity property disappears, and 'regular' diffusion is now present. For a homogeneous surface where D_{f} is equal to 2, and when only diffusional limitations are present, $p = 1/2$ as it should be. Another way of looking at the $p = 1/2$ case (where $D_{\mathrm{f,bind}}$ is equal to two) is that the analyte in solution views the fractal object, in our case, the receptor-coated biosensor surface, from a 'large distance'. In essence, in the association process, the diffusion of the analyte from the solution to the receptor surface creates a depletion layer of width $(Đt)^{1/2}$ where $Đ$ is the diffusion constant. This gives rise to the fractal power law, (Analyte · Receptor) $\sim t^{(3-D_{\mathrm{f,bind}})/2}$. For the present analysis, t_c is arbitrarily chosen and we assume that the value of the t_c is not reached. One may consider the approach as an intermediate 'heuristic' approach that may be used in the future to develop an autonomous (and not time-dependent) model for diffusion-controlled kinetics.

Dissociation rate coefficient

The diffusion of the dissociated particle (receptor [Ab] or analyte [Ag]) from the solid surface (e.g., analyte [Ag]–receptor [Ab]) complex coated surface) into solution may be given, as a first approximation by:

$$(\mathrm{Ab \cdot Ag}) \approx -t^{(3-D_{\mathrm{f,diss}})/2} = -t^{\mathrm{p}}\,(t > t_{\mathrm{diss}}) \tag{8.1b}$$

Here $D_{\mathrm{f,diss}}$ is the fractal dimension of the surface for the dissociation step. This corresponds to the highest concentration of the analyte–receptor complex on the surface. Henceforth, its concentration only decreases. The dissociation kinetics may be analyzed in a manner 'similar' to the binding kinetics.

8.2.2 Dual-fractal analysis

Binding rate coefficient

Sometimes, the binding curve exhibits complexities and two parameters (k, D_{f}) are not sufficient to adequately describe the binding kinetics. This is further corroborated by low values of r^2 factor (goodness-of-fit). In that case, one resorts to a dual-fractal analysis (four parameters; k_1, k_2, D_{f1}, and D_{f2}) to adequately describe the binding kinetics. The single-fractal analysis presented above is thus extended to include two fractal dimensions. At present, the time ($t = t_1$) at which the 'first' fractal dimension 'changes' to the 'second' fractal dimension is arbitrary and empirical. For the most part, it is dictated by the data analyzed and experience gained by handling a single-fractal analysis. A smoother curve is obtained in the 'transition' region, if care is taken to select the correct number of points for the two regions. In this case, the product (antibody–antigen; or analyte–receptor complex, Ab · Ag or analyte–receptor) is given by:

$$(\mathrm{Ab \cdot Ag}) \approx \begin{cases} t^{(3-D_{\mathrm{f1,bind}})/2} = t^{\mathrm{p1}} & (t < t_1) \\ t^{(3-D_{\mathrm{f2,bind}})/2} = t^{\mathrm{p2}} & (t_1 < t < t_2 = t_{\mathrm{c}}) \\ t^{1/2} & (t > t_{\mathrm{c}}) \end{cases} \tag{8.1c}$$

In some cases, as mentioned above, a triple-fractal analysis with six parameters (k_1, k_2, k_3, D_{f1}, D_{f2}, and D_{f3}) may be required to adequately model the binding kinetics. This is when the binding curve exhibits convolutions and complexities in its shape due to perhaps the very dilute nature of the analyte (in some of the cases to be presented) or for some other reasons. Also, in some cases, a dual-fractal analysis may be required to describe the dissociation kinetics.

8.3 RESULTS

Maxson *et al.* (2003) have developed a solid-phase assay for the identification of modulators of prion protein interactions. These authors indicate that TSEs are a group of fatal neurodegenerative diseases (Kocisko *et al.*, 1994). Maxson *et al.* (2003) emphasize that although the precise infectious agent has not as yet been identified, a key step in TSEs is the conversion of the endogeneous prion protein (PrP-sen or PrP^{c}) which is fully digested by proteinase K (PK) to a partially PK-resistant isoform (PrP-res or PrP^{Sc}). McBride *et al.* (1988) and Brandner *et al.* (1996) indicate that the PrP-sen expression is necessary for infection. The primary component of deposits formed in the different brain areas is the PrP-res isoform. Maxson *et al.* (2003) indicate that phthalocyanine tetrasulfonate is a known modulator of the conversion from the PrP-sen to the PrP-res isoform. These authors emphasize that phthalocyanine tetrasulfonate interferes with the binding between the PrP-sen and the PrP-res forms, thereby inhibiting this conversion.

Figure 8.1a shows the binding of PrP-sen in solution to PRP-res immobilized on a well surface in the absence of a detergent (Maxson *et al.*, 2003). A single-fractal analysis is adequate to describe the binding kinetics. The values of the binding rate coefficient, k and the fractal dimension, D_f for a single-fractal analysis are given in Table 8.1. The values of the binding rate coefficient, k and the fractal dimension for the binding phase, D_f presented in Table 8.1 were obtained using Corel Quattro Pro 8.0 (1989). The binding rate coefficient values presented in Table 8.1 are within 95% confidence limits. For example, for the binding of PrP-sen in solution to the PrP-res form immobilized on a well surface, the binding rate coefficient, k value is equal to 9.237 ± 0.541. The 95% confidence limit indicates that the k value lies between 8.696 and 9.778. This indicates that the value is precise and significant.

Figure 8.1b shows the binding of PrP-sen in solution to PrP-res immobilized on a well surface in the presence of the detergent, Gdn^{+} (Maxson *et al.*, 2003). A single-fractal analysis is, once again, adequate to describe the binding kinetics. The values of the binding rate coefficient, k and the fractal dimension for the binding phase, D_f are given in Table 8.1. It is of interest to note that as one goes from the detergent-free case to the one where detergent is present (Gdn^{+}), the fractal dimension decreases by 19.34% from a value of D_f equal to 2.8320 to D_f equal to 2.2842, and the binding rate coefficient, k for a single-fractal analysis decreases by 34.3% from 9.237 to 6.0863. Note that changes in the fractal dimension or the degree of heterogeneity on the sensor chip surface and in the binding rate coefficient, k are in the same direction.

Figure 8.2a shows the binding and dissociation of 50 μg/ml αB crystallin in solution to αB crystallin units adsorbed on a SAM surface at 45 °C (Liu *et al.*, 2006). These authors

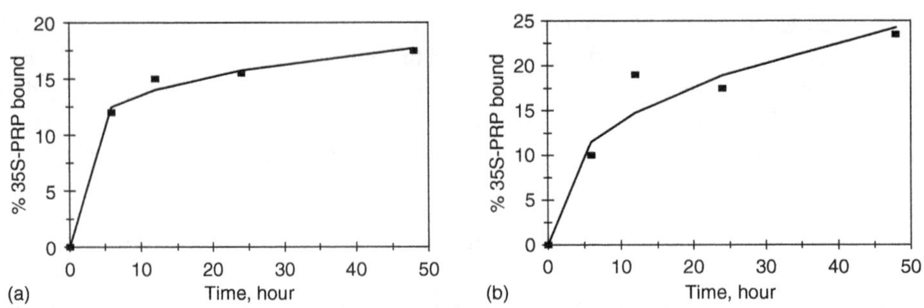

Figure 8.1 Binding of PrP-sen in solution to PrP-res immobilized on a well surface (Maxson *et al.*, 2003): (a) Gdn (detergent free); (b) Gdn^{+} (detergent present).

Table 8.1

Binding and dissociation rate coefficients and fractal dimensions in the binding and in the dissociation phases for (a) PrP-sen in solution to PrP-res immobilized on a well surface (Maxson *et al.*, 2003), and (b) αB crystallin subunits in solution to αB crystallin adsorbed on a SAM surface (Liu *et al.*, 2006)

Analyte in solution/ receptor on surface	k	k_d	D_f	D_{fd}	Reference
Prp-sen/PrP-res (detergent (Gdn) free)	9.237 ± 0.541	na	2.8320 ± 0.0734	na	Maxson *et al.* (2003)
Prp-sen/PrP-res (detergent (Gdn) present)	6.0683 ± 1.4405	na	2.2842 ± 0.2748	na	Maxson *et al.* (2003)
50 μg/ml αB crystallin at 45 °C/αB crystallin on a SAM surface	0.9444 ± 0.0498	0.4872 ± 0.0485	2.2446 ± 0.0454	2.1154 ± 0.08404	Liu *et al.* (2006)
50 μg/ml αB crystallin at 37 °C/αB crystallin on a SAM surface	0.1213 ± 0.0065	0.07651 ± 0.0203	1.7430 ± 0.0464	2.2548 ± 0.1354	Liu *et al.* (2006)
50 μg/ml αB crystallin at 24 °C/αB crystallin on a SAM surface	0.01049 ± 0.00038	0.002044 ± 0.00011	1.0568 ± 0.03178	0.9378 ± 0.0410	Liu *et al.* (2006)
αB crystallin subunit/αB crystallin on a $C_{11}NH_2$ surface at pH 6.0	0.6528 ± 0.0266	0.07451 ± 0.00450	1.9536 ± 0.04906	2.4214 ± 0.03302	Liu *et al.* (2006)
αB crystallin subunit/αB crystallin on a $C_{11}NH_2$ surface at pH 6.5	0.1618 ± 0.0125	0.05649 ± 0.00525	1.7558 ± 0.0822	2.6850 ± 0.0431	Liu *et al.* (2006)
αB crystallin subunit/αB crystallin on a $C_{11}NH_2$ surface at pH 6.8	0.06911 ± 0.0088	0.01257 ± 0.00135	1.2666 ± 0.1462	2.0726 ± 0.0552	Liu *et al.* (2006)
αB crystallin subunit/αB crystallin on a $C_{11}NH_2$ surface at pH 7.4	0.1059 ± 0.0053	0.1149 ± 0.0055	1.6568 ± 0.06690	2.7170 ± 0.02322	Liu *et al.* (2006)

indicate that αB crystallin is a major component of the vertebrate eye lens. This heat shock protein (αB crystallin) plays a critical role in maintaining lens transparency by preventing the unfolding and aggregation of proteins (Muchowaki *et al.*, 1997; Bloemendal *et al.*, 2004; Horwitz, 1992, 2003; Boelens and Dejong, 1995; Dejong *et al.*, 1993). Liu *et al.* (2006) emphasize that αB crystallin is ubiquitously expressed throughout the body in tissues, and is believed to be involved in the pathogenesis of diseases such as Alzheimer's, Parkinson's, and Alexander's (Dabir *et al.*, 2004; Renkawek *et al.*, 1994; Iwaki *et al.*, 1989; Iwaki, 1997).

In Figure 8.2a, a single-fractal analysis is adequate to describe the binding kinetics. The values of the binding rate coefficient, k and the fractal dimension, D_f for a single-fractal analysis are given in Table 8.1.

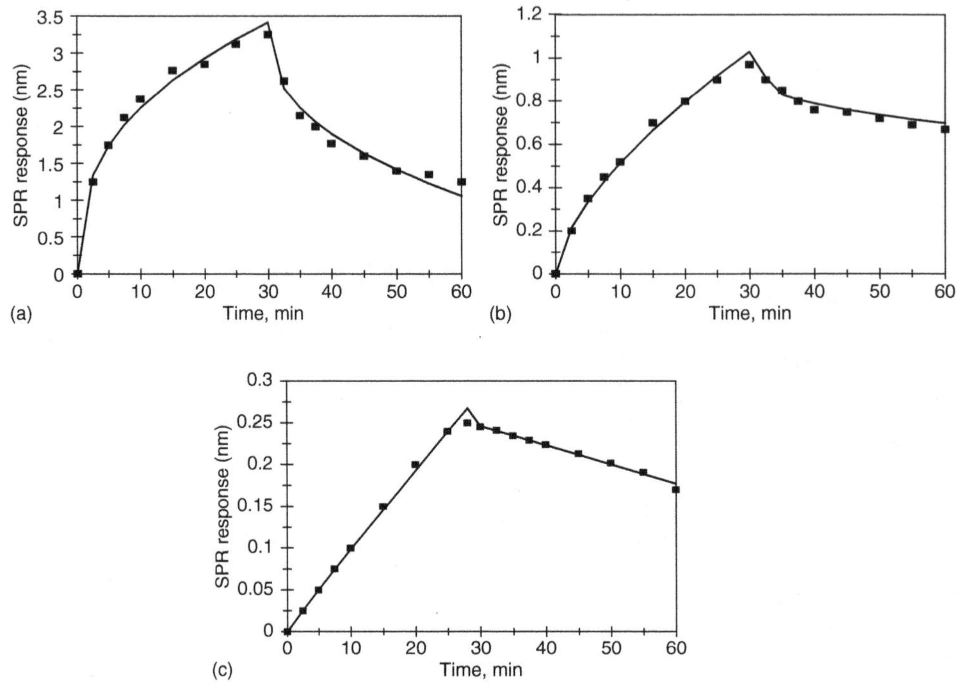

Figure 8.2 Binding and dissociation of 50 μg/ml αB crystallin in solution to αB crystallin units adsorbed on a SAM surface at different temperatures (in °C) (Liu *et al.*, 2006): (a) 45; (b) 37; (c) 24.

Figure 8.2b shows the binding and dissociation of 50 μg/ml αB crystallin in solution to αB crystallin units adsorbed on a SAM surface at 37°C (Liu *et al.*, 2006). Once again, a single-fractal analysis is adequate to describe the binding kinetics. The values of the binding rate coefficient, k and the fractal dimension, D_f for a single-fractal analysis are given in Table 8.1.

Figure 8.2c shows the binding and dissociation of 50 μg/ml αB crystallin in solution to αB crystallin units adsorbed on a SAM surface at 24°C (Liu *et al.*, 2006). Once again, a single-fractal analysis is adequate to describe the binding kinetics. The values of the binding rate coefficient, k and the fractal dimension, D_f for a single-fractal analysis are given in Table 8.1.

Figure 8.3a and Table 8.1 show the increase in the binding rate coefficient, k for a single-fractal analysis with an increase in the fractal dimension, D_f. For the data shown in Figure 8.3a, the binding rate coefficient, k is given by:

$$k = (0.06765 \pm 0.00368) D_f^{5.822 \pm 0.801} \quad (8.2a)$$

The fit is good. Only three data points are available. The availability of more data points would lead to a more reliable fit. The binding rate coefficient, k is very sensitive to the fractal dimension, D_f or the degree of heterogeneity that exists on the sensor chip surface as noted by the order of dependence between five and one-half and six (equal to 5.822) exhibited.

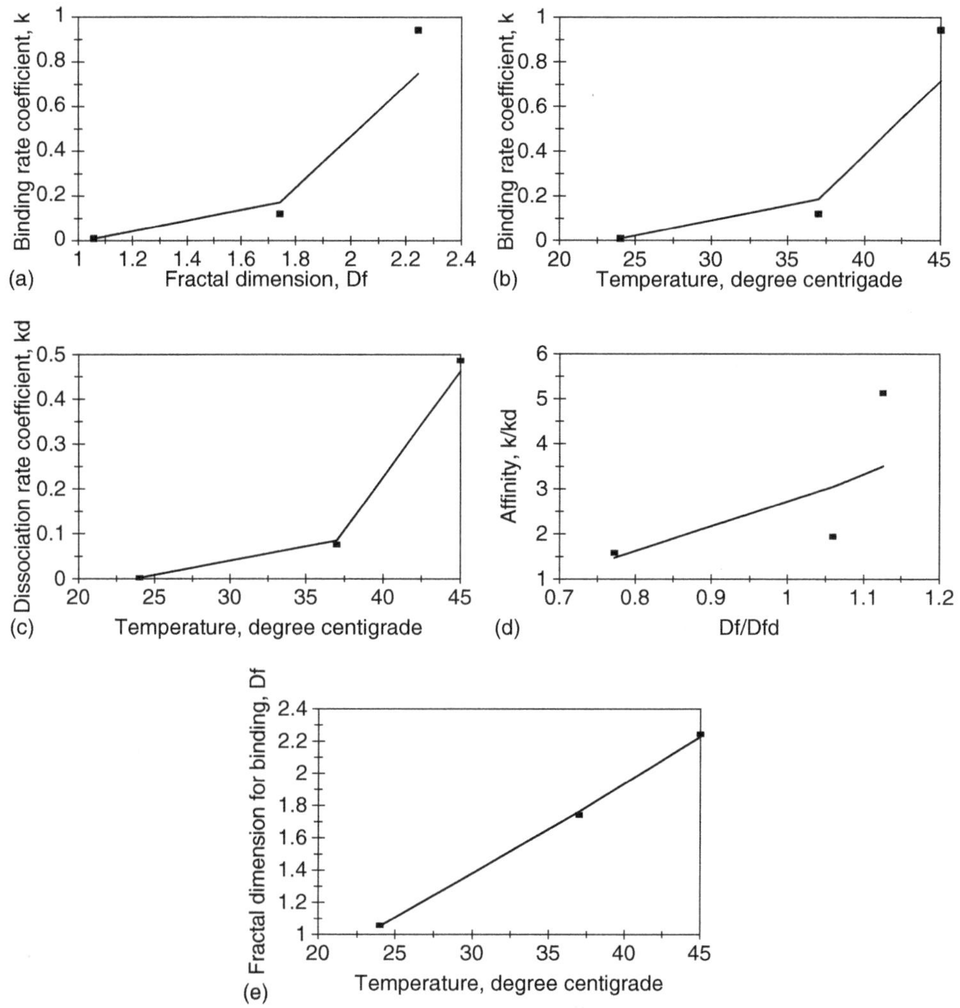

Figure 8.3 (a) Increase in the binding rate coefficient, k for a single-fractal analysis with an increase in the fractal dimension, D_f. (b) Increase in the binding rate coefficient, k for a single-fractal analysis with an increase in the temperature (in °C). (c) Increase in the dissociation rate coefficient, k_d for a single-fractal analysis with an increase in the temperature (in °C). (d) Increase in the affinity, $K(= k/k_d)$ with an increase in the ratio of the fractal dimensions, D_f/D_{fd}. (e) Increase in the fractal dimension for binding, D_f for a single-fractal analysis with an increase in the temperature (in °C).

Figure 8.3b and Table 8.1 show the increase in the binding rate coefficient, k for a single-fractal analysis with an increase in the temperature (in °C). For the data shown in Figure 8.3b, the binding rate coefficient, k is given by:

$$k = (2.7 \pm 1.8)\{T(\text{in } ^\circ\text{C})\}^{6.91\pm1.41} \tag{8.2b}$$

The fit is good. Only three data points are available. The availability of more data points would lead to a more reliable fit. The binding rate coefficient, k is very sensitive to the temperature (in °C) as noted by the order of dependence close to seventh (equal to 6.91) exhibited.

Figure 8.3c and Table 8.1 show the increase in the dissociation rate coefficient, k_d for a single-fractal analysis with an increase in the temperature (in °C). For the data shown in Figure 8.3c, the dissociation rate coefficient, k_d is given by:

$$k_d = (2.3 \pm 0.3)\{T(\text{in}\,^\circ\text{C})\}^{8.651 \pm 0.257} \tag{8.2c}$$

The fit is good. Only three data points are available. The availability of more data points would lead to a more reliable fit. The dissociation rate coefficient, k_d is extremely sensitive to the temperature (in °C) as noted by the order of dependence between eight and one half and nine (equal to 8.651) exhibited.

Figure 8.3d and Table 8.1 show the increase in the affinity, K $(=k/k_d)$ with an increase in the fractal dimension ratio, D_f/D_{fd}. For the data shown in Figure 8.3d, the affinity, K is given by:

$$K\left(=\frac{k}{k_d}\right) = (2.665 \pm 2.172)\left(\frac{D_f}{D_{fd}}\right)^{2.312 \pm 2.086} \tag{8.2d}$$

The fit is poor. Only three data points are available. There is scatter in the data, and this is reflected in the estimate(s) of the coefficient as well as in the order of dependence presented. The affinity, K exhibits an order of dependence between second and two and one-half (equal to 2.312) on the ratio of the fractal dimensions, D_f/D_{fd}.

Figure 8.3e and Table 8.1 show the increase in the fractal dimension, D_f for a single-fractal analysis with an increase in the temperature (in °C). For the data shown in Figure 8.3e, the fractal dimension, D_f is given by:

$$D_f = (0.0239 \pm 0.0003)\{T(\text{in}\,^\circ\text{C})\}^{1.191 \pm 0.052} \tag{8.2e}$$

The fit is very good. Only three data points are available. The availability of more data points would lead to a more reliable fit. The fractal dimension, D_f exhibits close to a first order (equal to 1.191) dependence on the temperature (in °C).

Liu *et al.* (2006) analyzed the binding of αB crystallin in solution to αB crystallin subunits adsorbed on a $C_{11}NH_2$ surface as a surface adsorbed layer. Figure 8.4a shows the binding and dissociation of αB crystallin in solution to αB crystallin adsorbed on a $C_{11}NH_2$ surface at pH 6.0. A single-fractal analysis is adequate to describe the binding and the dissociation kinetics. The values of (a) the binding rate coefficient, k and the fractal dimension, D_f for a single-fractal analysis, and (b) the dissociation rate coefficient, k_d and the fractal dimension for dissociation, D_{fd} for a single-fractal analysis are given in Table 8.1.

Figure 8.4b shows the binding and dissociation of αB crystallin in solution to αB crystallin adsorbed on a $C_{11}NH_2$ surface at pH 6.5. A single-fractal analysis is, once again, adequate to describe the binding and the dissociation kinetics. The values of (a) the binding

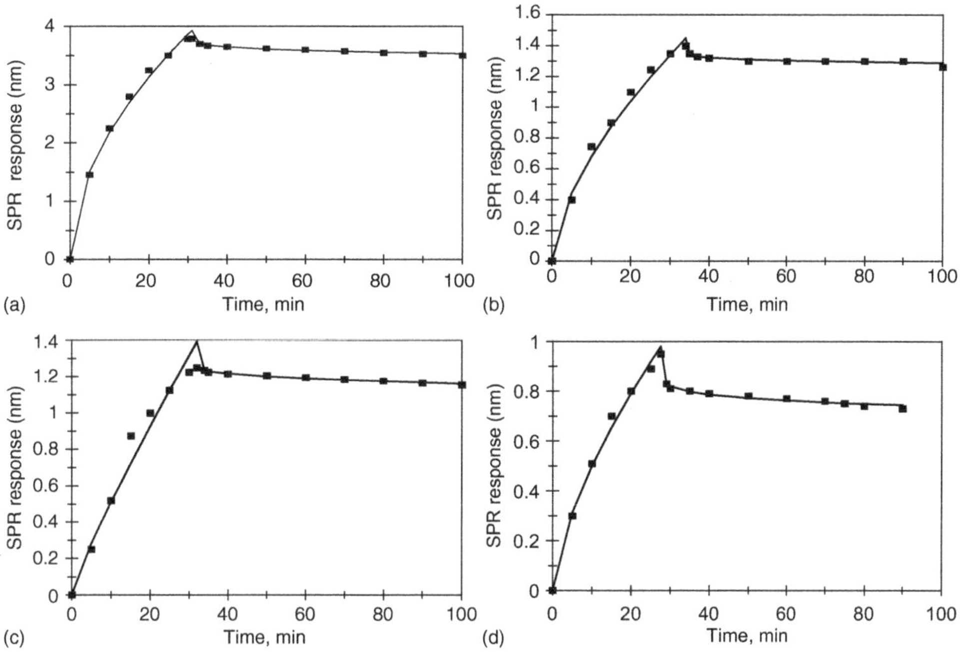

Figure 8.4 Binding and dissociation of αB crystallin in solution to αB crystallin adsorbed on a $C_{11}NH_2$ surface as a surface adsorbed layer at different pH (Liu *et al.*, 2006): (a) 6.0; (b) 6.5; (c) 6.8; (d) 7.4.

rate coefficient, k and the fractal dimension, D_f for a single-fractal analysis, and (b) the dissociation rate coefficient, k_d and the fractal dimension for dissociation, D_{fd} for a single-fractal analysis are given in Table 8.1.

Figure 8.4c shows the binding and dissociation of αB crystallin in solution to αB crystallin adsorbed on a $C_{11}NH_2$ surface at pH 6.8. A single-fractal analysis is, once again, adequate to describe the binding and the dissociation kinetics. The values of (a) the binding rate coefficient, k and the fractal dimension, D_f for a single-fractal analysis, and (b) the dissociation rate coefficient, k_d and the fractal dimension for dissociation, D_{fd} for a single-fractal analysis are given in Table 8.1.

Figure 8.4d shows the binding and dissociation of αB crystallin in solution to αB crystallin adsorbed on a $C_{11}NH_2$ surface at pH 7.4. A single-fractal analysis is, once again, adequate to describe the binding and the dissociation kinetics. The values of (a) the binding rate coefficient, k and the fractal dimension, D_f for a single-fractal analysis, and (b) the dissociation rate coefficient, k_d and the fractal dimension for dissociation, D_{fd} for a single-fractal analysis are given in Table 8.1.

Figure 8.5a and Table 8.1 show the decrease in the binding rate coefficient, k for a single-fractal analysis with an increase in the acidic pH (range 6–6.8). For the data shown in Figure 8.5a the binding rate coefficient, k is given by:

$$k = (5.4\text{E} + 13 \pm 0.2\text{E} + 13)[\text{pH}]^{-17.88 \pm 0.370} \quad (8.3a)$$

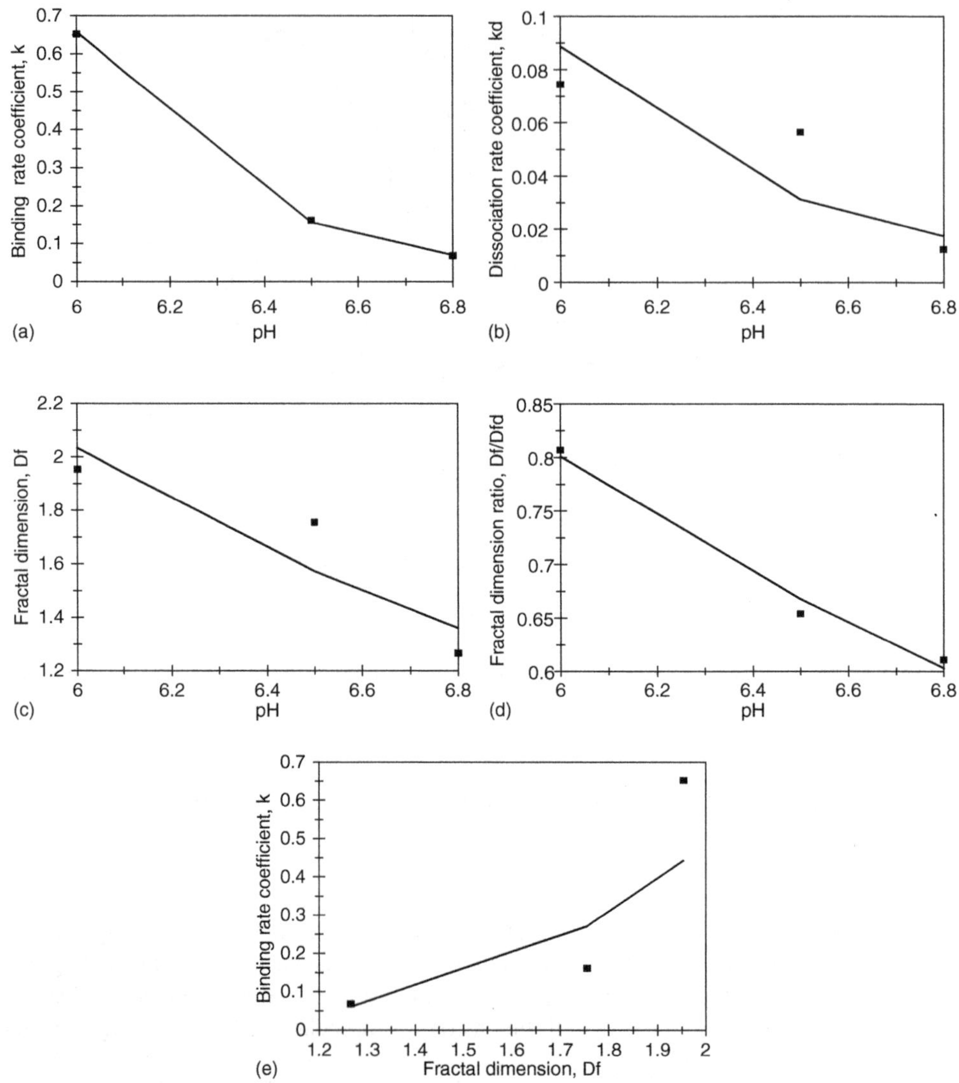

Figure 8.5 (a) Decrease in the binding rate coefficient, k for a single-fractal analysis with an increase in pH (acidic range 6–6.8). (b) Decrease in the dissociation rate coefficient, k_d for a single-fractal analysis with an increase in pH (acidic range 6–6.8). (c) Decrease in the fractal dimension, D_f for a single-fractal analysis with an increase in pH (acidic range 6–6.8). (d) Decrease in the fractal dimension ratio, D_f/D_{fd} for a single-fractal analysis with an increase in pH (acidic range 6–6.8). (e) Increase in the binding rate coefficient, k with an increase in the fractal dimension, D_f in the acidic pH (6–6.8) range.

The fit is good. Only three data points are available. The availability of more data points would lead to a more reliable fit. The binding rate coefficient, k is extremely sensitive to the pH as noted by the close to negative 18th (equal to -17.88) order of dependence exhibited on the pH in the acidic 6–6.8 range.

Figure 8.5b and Table 8.1 show the decrease in the dissociation rate coefficient, k_d for a single-fractal analysis with an increase in the acidic pH (range 6–6.8). For the data shown in Figure 8.5b the dissociation rate coefficient, k_d is given by:

$$k_d = (1.1\text{E}+9 \pm 1.2\text{E}+9)[\text{pH}]^{-12.971 \pm 7.744} \tag{8.3b}$$

The fit is good. Only three data points are available. The availability of more data points would lead to a more reliable fit. The dissociation rate coefficient, k_d is extremely sensitive to the pH as noted by the close to negative 13th (equal to -12.971) order of dependence exhibited on the pH in the acidic 6–6.8 range.

Figure 8.5c and Table 8.1 show the decrease in the fractal dimension, D_f in the binding phase for a single-fractal analysis with an increase in the acidic pH (range 6–6.8). For the data shown in Figure 8.5c the fractal dimension, D_f is given by:

$$D_f = (645.99 \pm 95.10)[\text{pH}]^{-3.22 \pm 1.53} \tag{8.3c}$$

The fit is good. Only three data points are available. The availability of more data points would lead to a more reliable fit. The fractal dimension, D_f is sensitive to the pH as noted by the close to negative third (equal to -3.22) order of dependence exhibited on the pH in the acidic 6–6.8 range.

Figure 8.5d and Table 8.1 show the decrease in the fractal dimension ratio, D_f/D_{fd} for a single-fractal analysis with an increase in the acidic pH (range 6–6.8). For the data shown in Figure 8.5d the fractal dimension ratio, D_f/D_{fd} for a single-fractal analysis with an increase in the acidic pH (range 6–6.8) is given by:

$$\frac{D_f}{D_{fd}} = (46.746 \pm 1.233)[\text{pH}]^{-2.270 \pm 0.290} \tag{8.3d}$$

The fit is good. Only three data points are available. The availability of more data points would lead to a more reliable fit. The fractal dimension, D_f/D_{fd} is sensitive to the pH as noted by the close to negative second (equal to -2.27) order of dependence exhibited on the pH in the acidic 6–6.8 range.

Figure 8.5e and Table 8.1 show the decrease in the binding rate coefficient, k for a single-fractal analysis with an increase in the fractal dimension, D_f in the acidic pH (range 6–6.8). For the data shown in Figure 8.5e the binding rate coefficient, k is given by:

$$k = (0.0206 \pm 0.0192)D_f^{4.576 \pm 2.066} \tag{8.3e}$$

The fit is reasonable. There is scatter in the data. This is reflected in the error in the estimated value of the binding rate coefficient. Only three data points are available. The availability of more data points would lead to a more reliable fit. The binding rate coefficient, k is very sensitive to the fractal dimension, D_f or the degree of heterogeneity that exists on the sensor chip surface as noted by the order of dependence between four and one-half and fifth (equal to 4.576) exhibited.

8.4 CONCLUSIONS

A fractal analysis is presented for prion-related interactions occurring on biosensor surfaces. The reactions analyzed include (a) the binding of PrP-sen in solution to PrP-res immobilized on a well surface (Maxson *et al.*, 2003), and (b) the binding and dissociation of αB crystallin in solution to αB crystallin subunits adsorbed on a SAM surface (Liu *et al.*, 2006). The fractal analysis provides a quantitative indication of the state of disorder (fractal dimension) and the binding (and dissociation, if applicable) rate coefficients on the sensor chip surface. As indicated in the chapter, though the precise mechanism of the infectious agent is not fully described, prions are involved in the onset of Alzheimer's disease (Maxson *et al.*, 2003). Furthermore, αB crystallin is ubiquitously expressed throughout the body in tissues (Liu *et al.*, 2006), and is believed to be important in the pathogenesis of Alzheimer's disease (Renkawek *et al.*, 1994; Iwaki *et al.*, 1997).

Both types of example are given wherein either a single- or a dual-fractal analysis is used. The dual-fractal analysis is used only when the single-fractal analysis does not provide an adequate fit. This was done by the regression analysis provided by Corel Quattro Pro 8.0 (Corel Corporation, 1997). In accord with the prefactor analysis of fractal aggregates (Sorenson and Roberts, 1997), quantitative (predictive) relations are developed for (a) the binding rate coefficient, k for a single-fractal analysis as a function of the fractal dimension for the binding of αB crystallin in solution to αB crystallin subunits adsorbed on a SAM surface in the temperature range 24–45 °C (Liu *et al.*, 2006), (b) the binding and dissociation rate coefficient(s), k and k_d for a single-fractal analysis as a function of temperature in the 24–45 °C range (Liu *et al.*, 2006), (c) the affinity, $K(=k/k_d)$ as a function of the ratio of fractal dimensions (D_f/D_{fd}) (Liu *et al.*, 2006), (d) the fractal dimension, D_f as a function of the temperature in the 24–45 °C range (Liu *et al.*, 2006), (e) the binding rate coefficient, k, the dissociation rate coefficient, k_d and the fractal dimension, D_f as a function of pH in the acidic range (pH 6.0–6.8) (Liu *et al.*, 2006), and (f) the binding rate coefficient, k as a function of the fractal dimension, D_f.

The fractal dimension is not a classical independent variable such as analyte (antigen, antibody, or other biological molecule) concentration in solution. Nevertheless, the expressions obtained for the binding (and the dissociation) rate coefficients as a function of the fractal dimension indicate a high sensitivity of these rate coefficients on their respective fractal dimensions on the SPR sensor chip surface.

It is suggested that the fractal surface (roughness) leads to turbulence, which enhances mixing, decreases diffusional limitations, and leads to an increase in the binding rate coefficient (Martin *et al.*, 1991). For this to occur the characteristic length of the turbulent boundary layer may have to extend a few monolayers above the sensor chip surface to affect bulk diffusion to and from the surface. However, given the extremely laminar flow regimes in most biosensors this may not actually take place. The sensor chip surface is characterized by grooves and ridges, and this surface morphology may lead to eddy diffusion. This eddy diffusion can then help to eliminate the mixing and extend the characteristic length of the boundary layer to affect the bulk diffusion to and from the surface.

The characterization of the surface by a fractal dimension provides extra flexibility and suggests an avenue whereby the nature of the surface (in this case the cell (tissue) surface) may be modulated in desired directions, and thereby simultaneously affecting or changing the binding and the dissociation rate coefficients in desired directions. Experimentalists as well as medical practitioners may find it worth their effort to pay a little more attention to the cellular (tissue) surface, and how it may be manipulated to control relevant parameters in desired directions. Alzheimer's, Parkinson's, and Alexander's and other similar ailments are slow forming, insidious, and intractable diseases. Any insights that may be obtained, no matter how small, in the onset and the progression of these types of diseases is invaluable. This is especially so if kinetic studies such as these assist in some small way to aid in the early detection and prognosis of these diseases.

REFERENCES

Bloemendal, H, WW De Jong, R Jaenicke, NH Lubsen, C Slingsby, and A Tardieu, Ageing and vision: Structure, stability, and function of lens crystallins. *Progress in Biophysics and Molecular Biology*, 2004, **86**, 407–485.

Boelens, WC and WW De Jong, α-Crystallins, versatile stress proteins. *Molecular Biology of Reproduction*, 1995, **21**, 75–80.

Brandner, S, S Isennman, A Raeber, M Fischer, A Sailer, Y Kobayashi, S Marino, and A Aguzzi, Normal host prion protein necessary for scrapie-induced neurotoxicity. *Nature*, 1996, **379**, 339–343.

Corel Corporation, Corel Quattro Pro 8.0, Ottawa, Canada, 1997.

Dabir, DV, JQ Trojanowski, C Richter-Landsberg, VMY Lee, and MS Forman, Expression of the small heat-shock protein αB-crystallin in tauopathies with glial pathology. *American Journal of Pathology*, 2004, **164**, 155–166.

Datta, SA and CM Rao, Differential temperature-dependent chaperone-like activity of αA- and αB-crystallin homoaggregates. *Journal of Biological Chemistry*, 1999, **274**, 34773–34778.

Dejong, WW, JAM Leunissen, and CEM Voorter, Evolution of the α-crystallin small heat-shock protein family. *Molecular Biology and Evolution*, 1993, **10**, 1–3–126.

Ghosh, JG and JL Clark, Insights into the domains required for dimerization and assembly of αB-crystallin. *Protein Science*, 2005, **14**, 684–695.

Havlin, S, Molecular diffusion and reaction. *The Fractal Approach to Heterogeneous Chemistry: Surfaces, Colloids, Polymers* (ed. D. Avnir), Wiley, New York, 1989, pp. 251–269.

Horwitz, J, α-Crystallin can function as a molecular chaperone. *Proceedings of the National Academy of Sciences, USA*, 1992, **89**, 10449–10453.

Horwitz, J, α-crystallin. *Experiments in Eye Research*, 2003, **76**, 145–153.

Iwaki, T, αB-crystallin in neurological diseases. *No to Shinkei*, 1997, **49**, 319–328.

Iwaki, T, A Kumeiwaki, RKH Liem, and JG Goldman, αB-crystallin is expressed in non-lenticular tissues and accumulates in Alexander's disease brain. *Cell*, 1989, **57**, 71–78.

Kim, KK, R Kim, and SH Kim, Crystal structure of a small heat shock protein. *Nature*, 1998, **394**, 595–599.

Kocisko, DA, JH Come, SA Priola, B Cheesbro, GJ Raymond, PT Lansbury Jr., and B Caughey, Cell-free formation of protease-resistant prion protein. *Nature*, 1994, **370**, 471–474.

Liu, L, JG Ghosh, JI Clark, and S Jiang, Studies of αB-crystallin subunit dynamics by surface plasmon resonance. *Analytical Biochemistry*, 2006, **350**, 186–195.

Martin, SJ, VE Granstaff, and GC Frye, Effect of surface roughness on the response of thickness-shear mode resonators in liquids. *Analytical Chemistry*, 1991, **65**, 2910–2922.

Maxson, L, C Wong, LM Herrmann, B Caughey, and GS Baron, A solid-phase assay for identification of modulators of prion protein interactions. *Analytical Biochemistry*, 2003, **323**, 54–64.

McBride, PA, ME Bruce, and H Fraser, Immunostaining of scrapie cerebral amyloid plaques with antisera raised to scrapie-associated fibrils (SAF). *Neuropathology and Applied Neurobiology*, 1988, **14**, 325–336.

Muchowaki, PJ, JA Bassuk, NH Lubsen, and JI Clark, Human αB-crystallin small heat-shock protein and molecular chaperone. *Journal of Biological Chemistry*, 1997, **272**, 2578–2582.

Pasta, SY, B Raman, T Ramakrishna, and CM Rao, Role of the conserved SRLFDQFFG region of α-crystallin, a small heat shock protein. *Journal of Biological Chemistry*, 2003, **278**, 51159–51166.

Pruisner, SB, Prions. *Proceedings of the National Academy of Sciences USA*, 1998, **95**, 13363–13383.

Ramakrishnan, A and A Sadana, A single-fractal analysis of cellular analyte-receptor binding kinetics using biosensors. *BioSystems*, 2001, **59**, 35–51.

Renkawek, K, CEM Voorter, G Bosman, FPA Vanworkum, and WW Dejong, Expression of αB-crystallin in Alzheimer's disease. *Acta Neuropathology*, 1994, **87**, 155–160.

Sadana, A, A fractal analysis for the evaluation of hybridization kinetics in biosensors. *Journal of Colloid and Interface Science*, 2001, **151**(1), 166–177.

Sadana, A, *Fractal Binding and Dissociation Kinetics for Different Biosensor Applications*, Elsevier, Amsterdam, 2005.

Sorenson, CM and GC Roberts The prefactor of fractal aggregates. *Journal of Colloid and Interface Science*, 1997, **186**, 447–452.

– 9 –

Fractal Analysis of Binding and Dissociation of DNA–Analyte Interactions on Biosensor Surfaces

9.1 INTRODUCTION

DNA sensors have been used in different areas, and these areas of application continue to increase. Liu *et al.* (2005) indicate that some of these areas where these sensors have been used effectively include disease diagnostics, detection of toxins (Lucarelli *et al.*, 2002), and the detection of pathogenic organisms (Baeumner *et al.*, 2004; Zhang *et al.*, 2003; Campbell *et al.*, 2002; Popovich *et al.*, 2002; Armistead and Thorp, 2002). Liu *et al.* (2005) have developed DNA sensors based on a DNA-modified electrode. These authors indicate that DNA oxidation sensors are sensors that monitor the degree of DNA oxidation, and are based on polymer-modified electrodes (Mugweru *et al.*, 2004).

Deng *et al.* (2006) indicate that DNA hybridization is the basic principle involved in DNA biosensors. In essence, the oligonucleotide probe (DNA probe) recognizes and binds to the nucleic acid target (target DNA). This leads to a double-stranded hybrid with its nucleic-acid complement. This process is highly specific and efficient. These authors emphasize that diffusional limitation is a problem in these micro-scale biological reactions. This coupled with the fact that the receptors are heterogeneously immobilized on the biosensor surface leads to the use of fractal analysis (used throughout this book) to help analyze the binding and dissociation of the analyte–receptor reactions occurring on biosensor surfaces. Mixing seems to be an obvious method to help alleviate diffusional limitations, and Deng *et al.* (2006) have developed a two-dimensional micro-bubble actuator array to help enhance the efficiency of DNA biosensors.

Tawa *et al.* (2005) have recently analyzed the kinetics of DNA–DNA hybridization using surface plasmon resonance (SPR) spectroscopy. These authors emphasize that the analysis of the duplex formation of two complementary strands provides insights into the physical–chemical reactions, besides being of medical and biological interest. SPR has also been used to analyze DNA hybridization (Su *et al.*, 2005), the conformational changes during the formation of DNA–protein complexes (Tsoi and Yang, 2004), and DNA–histone interactions (Barontes *et al.*, 2006).

In this chapter we use fractal analysis to analyze the binding and dissociation (if applicable) kinetics of (a) DNA (containing various mismatches)-protein complexes using SPR (Tsoi and Yang, 2004), (b) DNA–histone interactions using SPR (Barontes *et al.*, 2006), (c) a molecular beacon-based DNA micro-biosensor whose efficiency is enhanced using a two-dimensional micro-bubble actuator array (Deng *et al.*, 2006), (d) DNA assembly and hybridization using SPR and QCM, and (e) DNA–DNA hybridization using SPR spectroscopy (Tawa *et al.*, 2005).

The fractal analysis is presented below as an alternate method of analyzing the binding and dissociation kinetics of the above mentioned reactions. Other methods by which the kinetics of these types of DNA–DNA or DNA–protein interactions may be analyzed are also available in the literature. The fractal analysis, as indicated elsewhere in this book in the different chapters, takes into account both the diffusional limitations inherently present in these types of systems as well as the degree of heterogeneity of the receptors present on the biosensor surface.

9.2 THEORY

Havlin (1989) has reviewed and analyzed the diffusion of reactants towards fractal surfaces. The details of the theory and the equations involved for the binding and the dissociation phases for analyte–receptor binding are available (Sadana, 2001). The details are not repeated here except that just the equations are given to permit an easier reading. These equations have been applied to other biosensor systems (Sadana, 2001, 2005; Ramakrishnan and Sadana, 2001). For most applications, a single- or a dual-fractal analysis is often adequate to describe the binding and the dissociation kinetics. Peculiarities in the values of the binding and the dissociation rate coefficients, as well as in the values of the fractal dimensions with regard to the dilute analyte systems being analyzed will be carefully noted, if applicable.

9.2.1 Single-fractal analysis

Binding rate coefficient

Havlin (1989) indicates that the diffusion of a particle (analyte [Ag]) from a homogeneous solution to a solid surface (e.g., receptor [Ab]-coated surface) on which it reacts to form a product (analyte–receptor complex; (Ab·Ag)) is given by

$$(\mathrm{Ab}\cdot\mathrm{Ag}) \approx \begin{cases} t^{(3-D_{\mathrm{f,bind}})/2} = t^{\mathrm{p}} & t < t_{\mathrm{c}} \\ t^{1/2} & t > t_{\mathrm{c}} \end{cases} \tag{9.1}$$

Here $D_{\mathrm{f,bind}}$ or D_{f} is the fractal dimension of the surface during the binding step. t_{c} is the cross-over value. Havlin (1989) indicates that the cross-over value may be determined by $r_{\mathrm{c}}^2 \sim t_{\mathrm{c}}$. Above the characteristic length, r_{c}, the self-similarity of the surface is lost and the surface may be considered homogeneous. Above time, t_{c} the surface may be considered homogeneous, since the self-similarity property disappears, and 'regular' diffusion is now present. For a homogeneous surface where D_f is equal to 2, and when only diffusional limitations are present, $p = 1/2$ as it should be. Another way of looking at the $p = 1/2$ case (where $D_{f,bind}$ is equal to two) is that the analyte in solution views the fractal object, in our case, the receptor-coated biosensor surface, from a 'large distance.' In essence, in the association process, the diffusion of the analyte from the solution to the receptor surface creates a depletion layer of width $(Ðt)^{1/2}$ where $Ð$ is the diffusion constant. This gives rise to the fractal power law, $(\text{Analyte} - \text{Receptor}) \sim t^{(3-D_{f,bind})/2}$. For the present analysis, t_c is arbitrarily chosen and we assume that the value of the t_c is not reached. One may consider the approach as an intermediate 'heuristic' approach that may be used in the future to develop an autonomous (and not time-dependent) model for diffusion-controlled kinetics.

Dissociation rate coefficient

The diffusion of the dissociated particle (receptor [Ab] or analyte [Ag]) from the solid surface (e.g., analyte [Ag]-receptor [Ab]) complex coated surface) into solution may be given, as a first approximation by

$$(\text{Ab} \cdot \text{Ag}) \approx -t^{(3-D_{f,diss})/2} = -t^{p} \, (t > t_{diss}) \tag{9.2}$$

Here $D_{f,diss}$ is the fractal dimension of the surface for the dissociation step. This corresponds to the highest concentration of the analyte–receptor complex on the surface. Henceforth, its concentration only decreases. The dissociation kinetics may be analyzed in a manner 'similar' to the binding kinetics.

9.2.2 Dual-fractal analysis

Binding rate coefficient

Sometimes, the binding curve exhibits complexities and two parameters (k, D_f) are not sufficient to adequately describe the binding kinetics. This is further corroborated by low values of r^2 factor (goodness of fit). In that case, one resorts to a dual-fractal analysis (four parameters; k_1, k_2, D_{f1}, and D_{f2}) to adequately describe the binding kinetics. The single-fractal analysis presented above is thus extended to include two fractal dimensions. At present, the time ($t = t_1$) at which the 'first' fractal dimension 'changes' to the 'second' fractal dimension is arbitrary and empirical. For the most part, it is dictated by the data analyzed and experience gained by handling a single-fractal analysis. A smoother curve is obtained in the 'transition' region, if care is taken to select the correct number of points for the two regions. In this case, the product (antibody–antigen; or analyte–receptor complex, Ab·Ag or analyte–receptor) is given by

$$(\text{Ab} \cdot \text{Ag}) \approx \begin{cases} t^{(3-D_{f1,bind})/2} = t^{p1} & (t < t_1) \\ t^{(3-D_{f2,bind})/2} = t^{p2} & (t_1 < t < t_2 = t_c) \\ t^{1/2} & (t > t_c) \end{cases} \tag{9.3}$$

In some cases, as mentioned above, a triple-fractal analysis with six parameters (k_1, k_2, k_3, D_{f1}, D_{f2}, and D_{f3}) may be required to adequately model the binding kinetics. This is when the binding curve exhibits convolutions and complexities in its shape due to perhaps the very dilute nature of the analyte (in some of the cases to be presented) or for some other reasons. Also, in some cases, a dual-fractal analysis may be required to describe the dissociation kinetics.

9.3 RESULTS

At the outset it is appropriate to indicate that a fractal analysis will be applied to data obtained from the literature for the binding and dissociation (if applicable) kinetics of (a) DNA (containing various mismatches)-protein complexes using SPR (Tsoi and Yang, 2004), (b) DNA–histone interactions using SPR (Barontes *et al.*, 2006), (c) a molecular beacon based DNA micro-biosensor whose efficiency is enhanced using a two-dimensional micro-bubble actuator array (Deng *et al.*, 2006), (d) DNA assembly and hybridization using SPR and QCM, and (e) DNA–DNA hybridization using SPR spectroscopy (Tawa *et al.*, 2005). The above DNA–DNA or DNA–protein interactions may be considered as a sample of these types of interactions available in the literature.

Alternate expressions for fitting the data are available that include saturation, first-order reaction, and no diffusion limitations, but these expressions are apparently deficient in describing the heterogeneity that inherently exists on the surface. One might justifiably argue that the appropriate modeling may be achieved by using a Langmuirian or other approach. The Langmuirian approach may be used to model the data presented if one assumes the presence of discrete classes of sites (for example, double exponential analysis as compared with a single-fractal analysis). Lee and Lee (1995) indicate that the fractal approach has been applied to surface science, for example, adsorption and reaction processes. These authors emphasize that the fractal approach provides a convenient means to represent the different structures and morphology at the reaction surface. These authors also emphasize using the fractal approach to develop optimal structures and as a predictive approach. Another advantage of the fractal technique is that the analyte–receptor association (as well as the dissociation reaction) is a complex reaction, and the fractal analysis via the fractal dimension and the rate coefficient provide a useful lumped parameter(s) analysis of the diffusion-limited reaction occurring on a heterogeneous surface.

In a classical situation, to demonstrate fractality, one should make a log–log plot, and one should definitely have a large amount of data. It may be useful to compare the fit to some other forms, such as exponential, or one involving saturation, etc. At present, we do not present any independent proof or physical evidence of fractals in the examples presented. It is a convenient means (since it is a lumped parameter) to make the degree of heterogeneity that exists on the surface more quantitative. Thus, there is some arbitrariness in the fractal model to be presented. The fractal approach provides additional information about interactions that may not be obtained by conventional analysis of biosensor data.

There is no nonselective adsorption of the analyte. The present system (environmental pollutants in the aqueous or the gas phase) being analyzed may be typically very dilute. Nonselective adsorption would skew the results obtained very significantly. In these types of systems, it is imperative to minimize this nonselective adsorption. We also do recognize that, in some cases, this nonselective adsorption may not be a significant component of the adsorbed material and that this rate of association, which is of a temporal nature, would depend on surface availability. If we were to accommodate the nonselective adsorption into the model, there would be an increase in the heterogeneity on the surface, since, by its very nature, nonspecific adsorption is more homogeneous than specific adsorption. This would lead to higher fractal dimension values since the fractal dimension is a direct measure of the degree of heterogeneity that exists on the surface.

Tsoi and Yang (2004) indicate that the SPR biosensor has been used to analyze simple biomolecular interactions (Chaiken *et al.*, 1992; Myszka, 1997; Schuck, 1996) as well as DNA–protein interactions (Fisher *et al.*, 1994; Haruki *et al.*, 1979; Oda *et al.*, 1999). Tsoi and Yang (2004) have recently used the SPR biosensor to analyze a complex kinetic process such as DNA–polymerase interaction. This DNA–polymerase interaction involves a competitive and parallel binding process at two distinct sites. Furthermore, conformational changes within the DNA–protein complex are also involved. Tsoi and Yang (2004) used the SPR biosensor to analyze the binding and dissociation kinetics and conformational changes involved in the interaction between T7 DNA polymerase and various DNA duplexes. Kornberg and Baker (1991) have indicated that DNA polymerase has the ability to incorporate a correct base and to remove a mismatched base to ensure the high fidelity of the DNA replication process. Tsoi and Yang (2004) used the SPR biosensor to analyze the T7 polymerase and DNA duplex interactions containing different numbers of mismatches near the 3′-end of the primer.

Figure 9.1 shows the binding and dissociation of T7 DNA polymerase in solution to a single strand (ss) DNA immobilized on a sensor chip surface (Tsoi and Yang, 2004). A single-fractal analysis is adequate to describe the binding and the dissociation kinetics. The values of (a) the binding rate coefficient, k and the fractal dimension, D_f for a single-fractal analysis, and (b) the dissociation rate coefficient, k_d and the fractal dimension, D_{fd} for a single-fractal analysis are given in Table 9.1a. The values of the binding and the dissociation rate coefficients, and the fractal dimensions for the binding and the dissociation

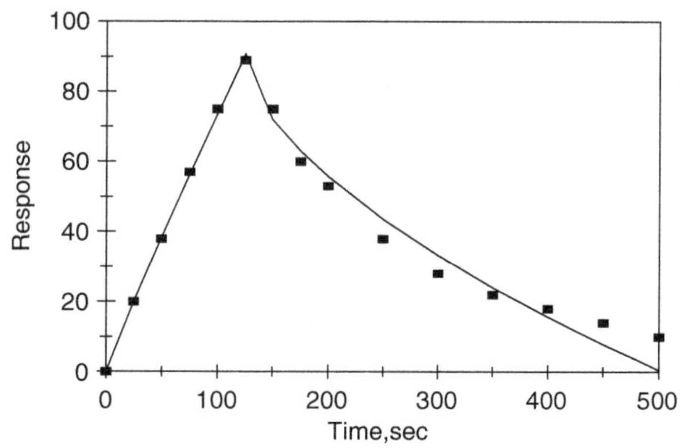

Figure 9.1 Binding and dissociation of T7 DNA polymerase in solution to DNA immobilized on a sensor chip surface (Tsoi and Yang, 2004).

Table 9.1a

Binding and dissociation rate coefficients for T7 DNA polymerase in solution to DNA immobilized on a sensor chip surface (Tsoi and Yang, 2004)

Analyte in solution/receptor on surface (nM T7 DNA polymerase/DNA)	K	k_1	k_2	k_d	k_{d1}	k_{d2}
(a) 40	0.9721 ± 0.0184	na	na	2.4089 ± 0.3063	na	na
(b) 100	6.1313 ± 0.1677	na	na	13.560 ± 0.358	na	na
80	6.5184 ± 0.0927	na	na	0.7989 ± 0.2127	0.3719 ± 0.109	4.137 ± 0.1014
60	4.8265 ± 0.1332	na	na	1.2406 ± 0.1215	na	na
40	2.5321 ± 0.093	na	na	1.7441 ± 0.3344	na	na
20	0.2711 ± 0.0118	na	na	0.08645 ± 0.0126	na	na

phase presented in Table 9.1 were obtained from a regression analysis using Corel Quattro Pro 8.0 (1997) to model the data using eqs. (9.1)–(9.3). The binding and the dissociation rate coefficients, and the fractal dimensions presented in Table 9.1 are within 95% confidence limits. For example, for the binding of 40 nM T7 DNA polymerase in solution to DNA immobilized on a sensor chip surface, the binding rate coefficient, k for a single-fractal analysis is equal to 0.9721 ± 0.0814. The 95% confidence limit indicates that the k value lies between 0.9537 and 0.9905. This indicates that the estimated k values are precise and significant.

Figure 9.2a shows the binding and dissociation of 100 nM T7 DNA polymerase in solution to a single-stranded DNA surface (Fc1). A single-fractal analysis is adequate to describe the binding and the dissociation kinetics. The values of (a) the binding rate coefficient, k and the fractal dimension, D_f for a single-fractal analysis, and (b) the dissociation rate coefficient, k_d and the fractal dimension in the dissociation phase, D_{fd} for a single-fractal analysis are given in Table 9.2. The value of the affinity K ($=k/k_d$) is 0.452.

Figure 9.2b shows the binding and dissociation of 80 nM T7 DNA polymerase in solution to a single-stranded DNA surface (Fc1). A single-fractal analysis is adequate to describe the binding kinetics. A dual-fractal analysis is required to describe the dissociation kinetics. The values of (a) the binding rate coefficient, k and the fractal dimension, D_f for a single-fractal analysis, (b) the dissociation rate coefficient, k_d and the fractal dimension for the dissociation phase, D_{fd} for a single-fractal analysis, and (c) the dissociation rate coefficients, k_{d1} and k_{d2} and the fractal dimensions, D_{fd1} and D_{fd2} for a dual-fractal analysis are given in Table 9.2.

Figure 9.2c shows the binding and dissociation of 60 nM T7 DNA polymerase in solution to a single-stranded DNA surface (Fc1). A single-fractal analysis is adequate to describe the binding and the dissociation kinetics. The values of (a) the binding rate

Table 9.1b

Fractal dimensions for the binding and the dissociation phase for T7 DNA polymerase in solution to DNA immobilized on a sensor chip surface (Tsoi and Yang, 2004)

Analyte in solution/ receptor on surface (nM T7 DNA polymerase/DNA)	D_f	D_{f1}	D_{f2}	D_{fd}	D_{fd1}	D_{fd2}
(a) 40	1.1206 ± 0.0295	na	na	1.7836 ± 0.0916	na	na
(b) 100	1.4432 ± 0.02932	na	na	2.0442 ± 0.0147	na	na
80	1.6086 ± 0.01533	na	na	1.1432 ± 0.1532	0.6778 ± 0.1902	1.7154 ± 0.0418
60	1.5978 ± 0.02956	na	na	1.3984 ± 0.0602	na	na
40	1.5470 ± 0.03918	na	na	1.7744 ± 0.1132	na	na
20	1.0412 ± 0.04628	na	na	0.8842 ± 0.08768	na	na

coefficient, k and the fractal dimension, D_f for a single-fractal analysis, and (b) the dissociation rate coefficient, k_d and the fractal dimension in the dissociation phase, D_{fd} for a single-fractal analysis are given in Table 9.2. The value of the affinity K ($=k/k_d$) is 3.89.

Figure 9.2d shows the binding and dissociation of 40 nM T7 DNA polymerase in solution to a single-stranded DNA surface (Fc1). A single-fractal analysis is adequate to describe the binding and the dissociation kinetics. The values of (a) the binding rate coefficient, k and the fractal dimension, D_f for a single-fractal analysis, and (b) the dissociation rate coefficient, k_d and the fractal dimension in the dissociation phase, D_{fd} for a single-fractal analysis are given in Table 9.2. The value of the affinity K ($=k/k_d$) is 1.452.

Figure 9.2e shows the binding and dissociation of 20 nM T7 DNA polymerase in solution to a single-stranded DNA surface (Fc1). A single-fractal analysis is adequate to describe the binding and the dissociation kinetics. The values of (a) the binding rate coefficient, k and the fractal dimension, D_f for a single-fractal analysis, and (b) the dissociation rate coefficient, k_d and the fractal dimension in the dissociation phase, D_{fd} for a single-fractal analysis are given in Table 9.2. The value of the affinity K ($=k/k_d$) is 3.135.

Figure 9.3a and Table 9.1 show for a single-fractal analysis the increase in the binding rate coefficient, k with an increase in the T7 DNA polymerase concentration (in nM) in solution. For the data presented in Figure 9.3a, the binding rate coefficient, k is given by

$$k = (0.00108 \pm 0.00086)[\text{T7 DNA polymerase}]^{1.9712 \pm 0.4962} \tag{9.4a}$$

The fit is quite good. Only four data points are available. The availability of more data points would lead to a more reliable fit. The binding rate coefficient, k exhibits close to a second (equal to 1.9712) order of dependence on the T7 DNA polymerase concentration

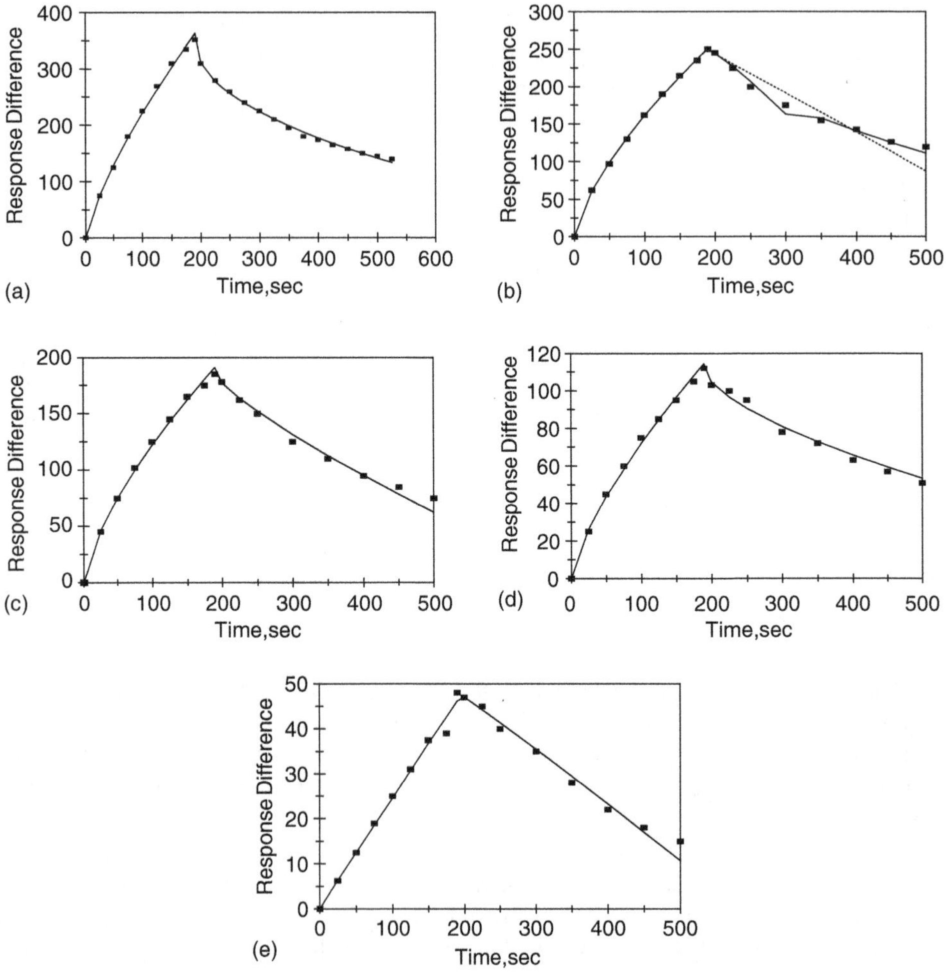

Figure 9.2 Binding and dissociation of different concentrations of T7 DNA polymerase (in nM) in solution to DNA immobilized on a sensor chip surface (Tsoi and Yang, 2004): (a) 100, (b) 80, (c) 60, (d) 40, (e) 20. (When only a solid line (____) is used, then a single-fractal analysis applies. When a dashed (---) and a solid line (____) is used, then the dashed line represents a single-fractal analysis, and the solid line represents a dual-fractal analysis. In this case the solid line provides the better fit).

in solution. The non-integer order of dependence exhibited by the binding rate coefficient, k on the T7 DNA polymerase concentration in solution lends support to the fractal nature of the system.

Figure 9.3b and Table 9.1 show for a single-fractal analysis the increase in the binding rate coefficient, k with an increase in the fractal dimension, D_f. For the data shown in Figure 9.3b, the binding rate coefficient, k is given by

$$k = (0.2409 + 0.2432)D_f^{6.633 \pm 2.051} \qquad (9.4b)$$

Table 9.2a

Binding and dissociation rate coefficients for DNA–histone interactions: (a) binding of histone to DNA immobilized on a sensor chip surface, and (b) binding and dissociation of DNA to histone immobilized on a sensor chip surface (Barontes *et al.*, 2006)

Analyte in solution/ receptor on surface	k	k_1	k_2	k_d
mg/ml histone/DNA				
(a) 0.01	0.008293 ± 0.00298	0.004927 ± 0.00065	0.02351 ± 0.000287	na
0.0025	0.001 ± 0.0	0	0	na
mg/ml DNA/histone				
(b) 0.008	0.007545 ± 0.000734	0.006029 ± 0.000298	0.02704 ± 0.00051	na
0.01	0.007431 ± 0.00175	0.004054 ± 0.000268	na	0.08468 ± 0.000938

Table 9.2b

Fractal dimensions for the binding and the dissociation phase for DNA–histone interactions (Barontes *et al.*, 2006)

Analyte in solution/ receptor on surface (mg/ml histone/DNA)	D_f	D_{f1}	D_{f2}	D_{fd}
(a) 0.01	1.4122 ± 0.2376	0.2946 ± 0.2330	2.3860 ± 0.1912	na
0.0025	2.0 ± 8.6E-16	0	0	na
(b) 0.008	1.8262 ± 0.1575	1.6270 ± 0.05748	2.5660 ± 0.07402	na
0.01	1.7378 ± 0.1519	1.1104 ± 0.07612	na	3.0 − 0.00576

The fit is reasonable. Only four data points are available. The availability of more data points would lead to a more reliable fit. There is scatter in the fit of the data, and this is reflected in the error in the binding rate coefficient, k. The binding rate coefficient k is extremely sensitive to the degree of heterogeneity present on the sensor surface, as noted by the higher than six and one-half order (equal to 6.633) of dependence exhibited.

Figure 9.3c and Table 9.1 show for a single-fractal analysis the increase in the dissociation rate coefficient, k_d with an increase in the T7 DNA polymerase concentration (in nM) in solution. For the data presented in Figure 9.3c, the dissociation rate coefficient, k_d is given by

$$k_d = (1.8 + 2.1\text{E} - 05)[\text{T7 DNA polymerase}]^{2.906 \pm 0.6653} \quad (9.4c)$$

The fit is good. Only four data points are available. The availability of more data points would lead to a more reliable fit. The dissociation rate coefficient, k_d exhibits close to a third (equal to 2.906) order of dependence on the T7 DNA polymerase concentration in solution. The non-integer order of dependence exhibited by the dissociation rate

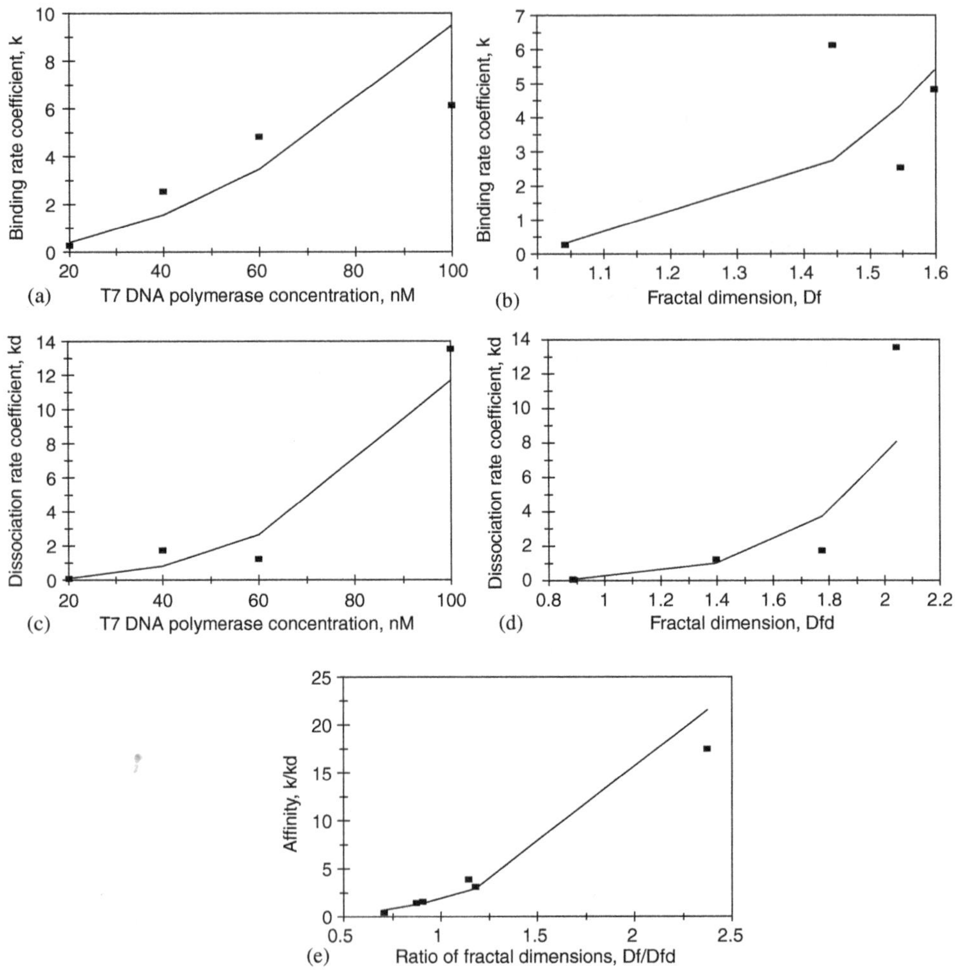

Figure 9.3 (a) Increase in the binding rate coefficient, k with an increase in the T7 DNA polymerase concentration (in nM) in solution. (b) Increase in the binding rate coefficient, k with an increase in the fractal dimension, D_f. (c) Increase in the dissociation rate coefficient, k_d with an increase in the T7 DNA polymerase concentration (in nM) in solution. (d) Increase in the dissociation rate coefficient, k_d with an increase in the fractal dimension, D_{fd}. (e) Increase in the affinity, K (= k/k_d) with an increase in the ratio of fractal dimensions, D_f/D_{fd}.

coefficient, k_d on the T7 DNA polymerase concentration in solution lends support to the fractal nature of the system.

Figure 9.3d and Table 9.1 show for a single-fractal analysis the increase in the dissociation rate coefficient, k_d with an increase in the fractal dimension, D_{fd}. For the data shown in Figure 9.3d, the dissociation rate coefficient, k_d is given by

$$k_d = (0.1629 \pm 0.1544) D_{fd}^{5.455 \pm 1.0471} \tag{9.4d}$$

The fit is reasonable. Only four data points are available. The availability of more data points would lead to a more reliable fit. There is scatter in the fit of the data, and this is reflected in the error in the binding rate coefficient, k. The dissociation rate coefficient, k_d is extremely sensitive to the degree of heterogeneity present on the sensor surface, as noted by the higher than fifth (equal to 5.4554) order of dependence exhibited.

Figure 9.3e and Table 9.1 show the increase in the affinity, K ($=k/k_d$) with an increase in the ratio of the fractal dimensions present in the binding and dissociation phases, D_f/D_{fd}. For the data presented in Figure 9.3e, the affinity, K is given by

$$K\left(=\frac{k}{k_d}\right) = (1.8737 \pm 0.6879)\left(\frac{D_f}{D_{fd}}\right)^{2.827\pm0.332} \quad (9.4e)$$

The fit is very good. Six data points are available. The affinity, K exhibits close to a third (equal to 2.827) order of dependence on the ratio of fractal dimensions, D_f/D_{fd}. The affinity is quite sensitive to the ratio of fractal dimensions as noted by the close to third order of dependence exhibited. This is one way of manipulating the affinity, K in desired directions by manipulating the degree of heterogeneity (in this case the ratio of fractal dimensions) present on the sensor chip surface.

Barontes *et al.* (2006) indicate that the analysis of histone–DNA interactions is essential to help understand the mechanisms governing DNA action. Though the SPR technique (Silin and Plant, 1997; Rich and Myszka, 2003; Homola, 1997; Kretschmann and Baker, 1968) has been used to analyze macromolecular complex formation, Barontes *et al.* (2006) indicate that it has not been used to analyze DNA–histone interactions. These authors (Barontes *et al.*, 2006) have recently proposed a DNA–histone immobilization method to analyze DNA–histone interactions using the SPR biosensor.

Figure 9.4a shows the binding of 0.01 mg/ml histone in solution to DNA immobilized on a sensor chip surface. A dual-fractal analysis is required to adequately describe the binding kinetics. The values of (a) the binding rate coefficient, k and the fractal dimension, D_f for a single-fractal analysis, and (b) the binding rate coefficients, k_1 and k_2, and the fractal dimensions, D_{f1} and D_{f2} for a dual-fractal analysis are given in Tables 9.2a and 9.2b. It is of interest to note that as the degree of heterogeneity on the sensor chip surface or the fractal dimension increases by a factor of 7.12 from a value of D_{f1} equal to 0.2946 to D_{f2} equal to 2.3860, the binding rate coefficient increases by a factor of 4.77 from a value of k_1 equal to 0.004927 to k_2 equal to 0.02351.

Figure 9.4b shows the binding of 0.0025 mg/ml histone in solution to DNA immobilized on a sensor chip surface. A single-fractal analysis is adequate to describe the binding kinetics. The values of the binding rate coefficient, k and the fractal dimension, D_f for a single-fractal analysis are given in Tables 9.2a and 9.2b. Note the change in the binding mechanism as one goes from the 0.01 mg/ml to the 0.0025 mg/ml histone in solution. At the higher histone concentration in solution a dual-fractal analysis is required to adequately describe the binding kinetics, whereas at the lower histone concentration in solution a single-fractal analysis is adequate to describe the binding kinetics.

Figure 9.5a shows the binding of 0.008mg/ml DNA in solution to histone immobilized on a sensor chip surface (Barontes *et al.*, 2006). A dual-fractal analysis is required to adequately describe the binding kinetics. The values of (a) the binding rate coefficient, k and

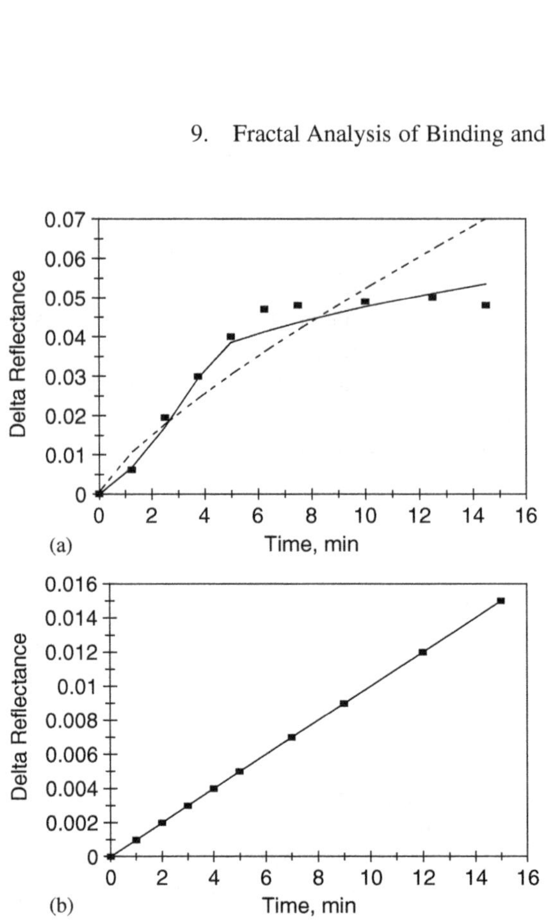

Figure 9.4 Binding of different concentrations (in mg/ml) of histone in solution to DNA immobilized on a sensor chip surface (Barontes et al., 2006): (a) 0.01, (b) 0.0025. (When only a solid line (____) is used, then a single-fractal analysis applies. When a dashed (---) and a solid line (____) is used, then the dashed line represents a single-fractal analysis, and the solid line represents a dual-fractal analysis. In this case the solid line provides the better fit).

the fractal dimension, D_f for a single-fractal analysis, and (b) the binding rate coefficients, k_1 and k_2 and the fractal dimensions, D_{f1} and D_{f2} for a dual-fractal analysis are given in Tables 9.2a and 9.2b.

Figure 9.5b shows the binding of 0.01 mg/ml DNA in solution to histone immobilized on a sensor chip surface (Barontes *et al.*, 2006). In this case, for the higher DNA concentration (0.01 mg/ml) compared to the lower 0.08 mg/ml in solution, a single-fractal analysis is adequate to describe the binding kinetics. The values of (a) the binding rate coefficient, k and the fractal dimension, D_f for a single-fractal analysis are given in Tables 9.2a and 9.2b.

Deng *et al.* (2006) recently indicate that hybridization of DNA probes is a diffusion-limited process. Pappaert *et al.* (2003) have indicated that the slow diffusive transport (i.e., a small number of target molecules reach the probe) leads to binding efficiencies less than 1% in micro-arrays. In micro-scale biological reactions, the problem is further exacerbated due to the limitations in the analyte concentration because of required reaction volume limits. Agitation of the sample solution is an obvious method to help minimize or eliminate diffusional limitations. These authors have used a two-dimensional micro-bubble actuator array to generate effective perturbations in the hybridization solution. These perturbations

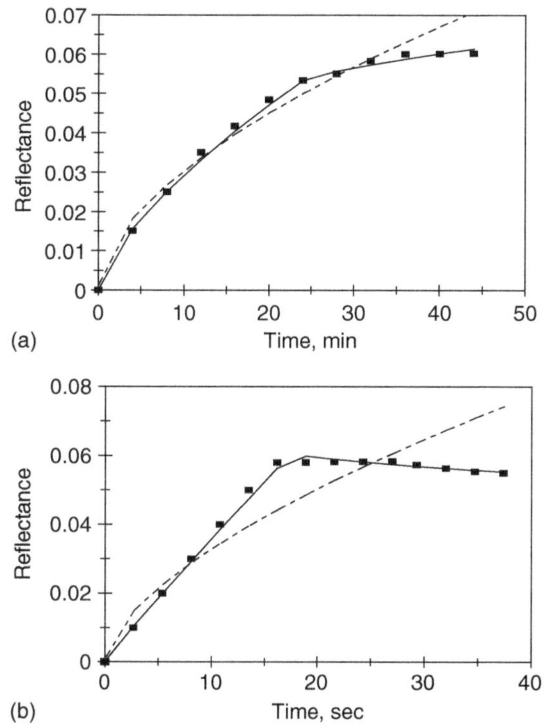

Figure 9.5 Binding of different concentrations (in mg/ml) of DNA in solution to histone immobilized on a sensor chip surface (Barontes#### *et al.*, 2006): (a) 0.008 (b) 0.01. (When only a solid line (___) is used, then a single-fractal analysis applies. When a dashed (---) and a solid line (___) is used, then the dashed line represents a single-fractal analysis, and the solid line represents a dual-fractal analysis. In this case the solid line provides the better fit).

enhance the diffusion rates of the DNA molecules to the probes, thereby increasing the DNA hybridization rate.

Figure 9.6 shows the molecular beacon hybridization without micro-bubble actuation (Deng *et al.*, 2006). The hybridization is between a molecular beacon (5′-/56/-FAM/CA-GTCGTATTAACTTACTCCCTCGACTG/3Dabcyl/-3′) where the underlined nucleotides are the stem sequences, and the beacon compound was 5′-TTAGGGAGTAAAGTTA-ATACGACTG-3′. The molecular beacon acts as a oligonucleotide probe. A dual-fractal analysis is required to adequately describe the binding kinetics. The values of (a) the binding rate coefficient, k and the fractal dimension, D_f for a single-fractal analysis, and (b) the binding rate coefficients, k_1 and k_2, and the fractal dimensions, D_{f1} and D_{f2} for a dual-fractal analysis are given in Tables 9.3a and 9.3b.

Figure 9.7a shows the binding and dissociation of the molecular beacon hybridization with micro-bubble agitation (Deng *et al.*, 2006). The micro-bubbles were generated from a 2×1 heater array. A single-fractal analysis is adequate to describe the binding and the dissociation kinetics. The values of (a) the binding rate coefficient, k and the fractal dimension, D_f for a single-fractal analysis, and (b) the dissociation rate coefficient, k_d and the fractal dimension, D_{fd} for the dissociation phase for a single-fractal analysis are given in Tables 9.3a and 9.3b.

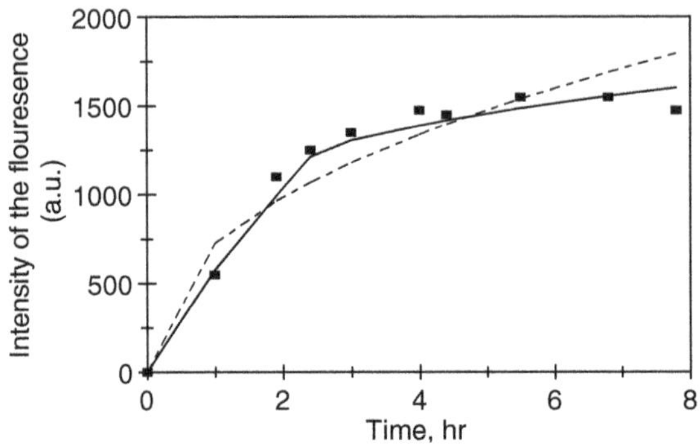

Figure 9.6 Binding (hybridization) of a molecular beacon (probe) and its complement without micro-bubble activation (Deng *et al.*, 2006). (When only a solid line (____) is used, then a single-fractal analysis applies. When a dashed (---) and a solid line (____) is used, then the dashed line represents a single-fractal analysis, and the solid line represents a dual-fractal analysis).

Figure 9.7b shows the binding and dissociation of the molecular beacon hybridization with micro-bubble agitation (Deng *et al.*, 2006). The micro-bubbles were generated from a 2×2 heater array. A single-fractal analysis is adequate to describe the binding and the dissociation kinetics. The values of (a) the binding rate coefficient, k and the fractal dimension, D_f for a single-fractal analysis, and (b) the dissociation rate coefficient, k_d and the fractal dimension, D_{fd} for the dissociation phase for a single-fractal analysis are given in Tables 9.3a and 9.3b.

It is of interest to compare the results when the micro-bubbles are generated by a 1×2 heater with a 2×2 heater. As one goes from the 1×2 heater array to a 2×2 heater array, the fractal dimension, D_f on the micro-array surface increases by a factor of 2.71 from a value of D_f equal to 0.4848 to 1.3214, and the binding rate coefficient, k increases by a factor of 1.215 from a value of k equal to 176.98 to 215.07. The changes in the binding rate coefficient, k and in the fractal dimension, D_f are in the same direction. Also, the micro-bubble actuator does lead to a decrease in the diffusional limitations, and subsequently to an increase in the binding rate coefficient.

Su *et al.* (2005) have recently compared SPR spectroscopy and QCM techniques for analyzing DNA assembly and hybridization. These authors indicate that both of these devices have been used for biological analysis and for clinical diagnosis (Cavic *et al.*, 1999; Englebienne *et al.*, 2003). Su *et al.* (2005) emphasize that these techniques, in a sense, complement each other. The SPR spectroscopy method is an optical technique that detects changes in the refractive index of thin films immobilized on a noble-metal surface. The QCM technique is an acoustic wave device. The technique measures thin films mechanically coupled to a metal electrode on a quartz disk. Su *et al.* (2005) indicate that the QCM oscillation frequency and quality are related to the mass loading. These authors followed the assembly of a biotinylated 30-mer oligonucleotide on a streptavidin-modified gold electrode used for hybridization.

Figure 9.8a shows the binding of 100 nM target DNA (3′-CGTGGACTGAGGACAC-CTCTTCAGACGGCA-5′) which is complementary to the probe DNA (density 1 μM) with a biotin label at the 5′-end (5′-biotin-GCACCTGACTCCTGTGGAGAAGTCTGCCGT-3′) immobilized on a sensor chip surface. A dual-fractal analysis is required to adequately

Table 9.3

Binding rate coefficients and fractal dimensions for enhancing the efficiencies of molecular beacon based DNA micro-sensors (a) without and (b) with micro-bubble actuation hybridization (Deng *et al.*, 2006)

Analyte in solution/receptor on surface	k	k_1	k_2	k_d	D_f	D_{f1}	D_{f2}	D_{fd}
(a) molecular beacon/DNA complement on micro-sensor chip surface	727.59 ± 132.23	576.95 ± 60.81	1031.20 ± 60.81	na	2.1204 ± 0.01794	1.3022 ± 0.2440	2.5702 ± 0.00862	na
(b) molecular beacon/DNA complement on sensor chip surface; micro-bubbles generated from a 2 × 1 heater array	176.98 ± 3.95	na	na	49.46 ± 8.58	0.4848 ± 0.05622	na	na	1.8790 ± 0.2770
(molecular beacon/DNA complement on sensor chip surface; micro-bubbles generated from a 2 × 2 heater array	215.07 ± 59.4	na	na	14.21 ± 9.79	1.3124 ± 0.6206	na	na	0 + 0.8682

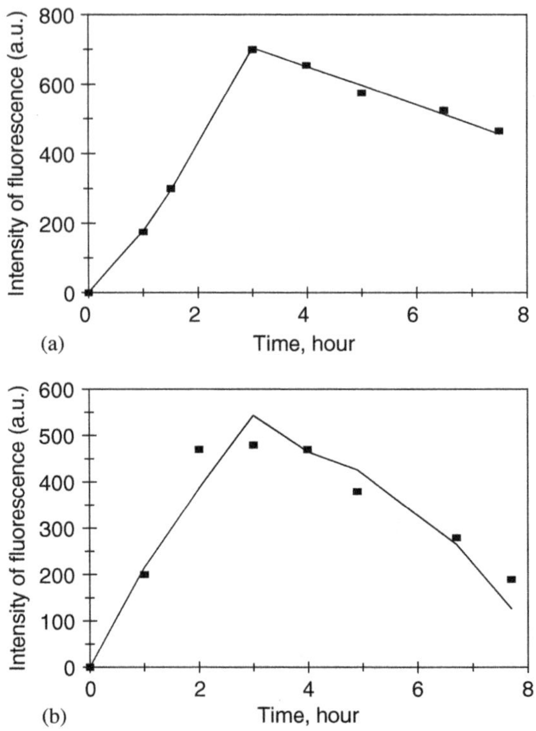

Figure 9.7 Binding (hybridization) of a molecular beacon (probe) and its complement with micro-bubble activation (Deng *et al.*, 2006): (a) micro bubbles generated from a 2 × 1 heater. (b) micro bubbles generated from a 2 × 2 heater.

describe the binding kinetics. The values of (a) the binding rate coefficient, k and the fractal dimension, D_f for a single-fractal analysis, and (b) the binding rate coefficients, k_1 and k_2, and the fractal dimensions, D_{f1} and D_{f2} for a dual-fractal analysis are given in Table 9.4. It is of interest to note that as the fractal dimension increases by a factor of 1.907 from a value of D_{f1} equal to 1.4502 to D_{f2} equal to 2.7650, the binding rate coefficient increases by a factor of 2.84 from a value of k_1 equal to 0.2552 to k_2 equal to 0.7248. The data presented in Figures 9.8a and b is normalized.

Figure 9.8b shows the binding of 1000 nM target DNA (3′-CGTGGACTGAGGACAC-CTCTTCAGACGGCA-5′) which is complementary to the probe DNA (density 1000 μM) with a biotin label at the 5′-end (5′-biotin-GCACCTGACTCCTGTGGAGAAGTCTGC-CGT-3′) immobilized on a sensor chip surface. Once again, a dual-fractal analysis is required to adequately describe the binding kinetics. The values of (a) the binding rate coefficient, k and the fractal dimension, D_f for a single-fractal analysis, and (b) the binding rate coefficients, k_1 and k_2, and the fractal dimensions, D_{f1} and D_{f2} for a dual-fractal analysis are given in Table 9.4. It is of interest to note that as the fractal dimension increases by a factor of 1.428 from a value of D_{f1} equal to 0.5664 to D_{f2} equal to 2.4238, the binding rate coefficient increases by a factor of 2.89 from a value of k_1 equal to 0.1163 to k_2 equal to 0.4524.

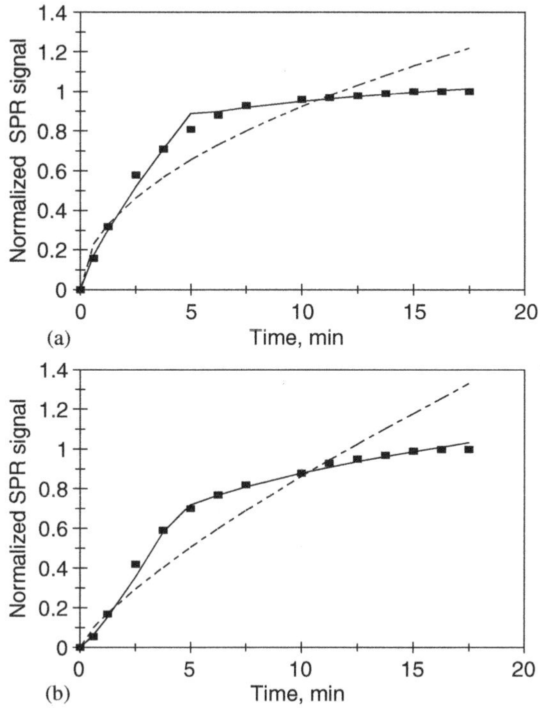

Figure 9.8 Binding of 1 μM target DNA to different concentrations (in μM) (probe density) of biotin–DNA immobilized on a sensor chip surface (Su *et al.*, 2005): (a) 100, (b) 1000. (When only a solid line (___) is used, then a single-fractal analysis applies. When a dashed (---) and a solid line (___) is used, then the dashed line represents a single-fractal analysis, and the solid line represents a dual-fractal analysis. In this case the solid line provides the better fit).

Note that as the probe density increases by an order of magnitude from 100 to 1000 nM biotin DNA, (a) the fractal dimensions D_{f1} and D_{f2} both exhibit decreases, and the binding rate coefficients, k_1 and k_2 also exhibit decreases. For example, the binding rate coefficient, k_1 exhibits a decrease by a factor of 2.19 from a value of 0.2552 to 0.1163. Similarly, the binding rate coefficient, k_2 exhibits a decrease by a factor of 1.60 from a value of k_2 equal to 0.7248 to 0.4524.

Figure 9.9a shows the binding of 100 nM target DNA (3′-CGTGGACTGAGGACAC-CTCTTCAGACGGCA-5′) which is complementary to the probe DNA (density 1 μM) with a biotin label at the 5′-end (5′-biotin-GCACCTGACTCCTGTGGAGAAGTCTGC-CGT-3′) immobilized on a sensor chip surface. A dual-fractal analysis is required to adequately describe the binding kinetics. The values of (a) the binding rate coefficient, k and the fractal dimension, D_f for a single-fractal analysis, and (b) the binding rate coefficients, k_1 and k_2, and the fractal dimensions, D_{f1} and D_{f2} for a dual-fractal analysis are given in Table 9.4. It is of interest to note that as the fractal dimension increases by a factor of 2.06 from a value of D_{f1} equal to 1.3838 to D_{f2} equal to 2.8525, the binding rate coefficient increases by a factor of 2.56 from a value of k_1 equal to 11.634 to k_2 equal to 29.741.

Table 9.4

Binding rate coefficients and fractal dimensions for (a) the binding of 1 μM DNA in solution to different probe densities (100 and 1000 nM biotin–DNA), and (b) binding of 1 μM target DNA to different concentrations of biotin–DNA assembly (in nM) (Su *et al.*, 2004)

Analyte in solution/ receptor on surface	k	k_1	k_2	D_f	D_{f1}	D_{f2}
(a) 1 μM target DNA/ 100 nM (probe density; biotin–DNA normalized data)	0.2955 ± 0.0609	0.2552 ± 0.0262	0.7248 ± 0.009	2.0096 ± 0.1014	1.4502 ± 0.1138	2.7650 ± 0.02492
1 μM target DNA/ 1000 nM (probe density; biotin–DNA normalized data)	0.1453 ± 0.0482	0.1163 ± 0.0212	0.4524 ± 0.0085	1.4490 ± 0.1551	0.5664 ± 0.1941	2.4238 ± 0.0297
(b) 1 μM target DNA/ 100 nM (probe density; biotin-DNA)	12.957 ± 4.919	11.634 ± 2.949	29.741 ± 0.491	2.1870 ± 0.1650	1.3838 ± 0.2896	2.8525 ± 0.0230
1 μM target DNA/ 200 nM (probe density; biotin–DNA)	29.966 ± 3.515	26.102 ± 1.709	42.850 ± 1.031	2.5602 ± 0.0708	2.494 ± 0.0928	2.8358 ± 0.0330
1 μM target DNA/ 1000 nM (probe density; biotin–DNA)	22.974 ± 3.624	21.761 ± 2.616	44.991 ± 1.745	2.1348 ± 0.07344	1.8120 ± 0.1125	2.6796 ± 0.0692

Figure 9.9b shows the binding of 200 nM target DNA (3′-CGTGGACTGAGGACAC-CTCTTCAGACGGCA-5′) which is complementary to the probe DNA (density 1 μM) with a biotin label at the 5′-end (5′-biotin-GCACCTGACTCCTGTGGAGAAGTCTGC-CGT-3′) immobilized on a sensor chip surface. In this case the data are not normalized. A dual-fractal analysis is, once again, required to adequately describe the binding kinetics. The values of (a) the binding rate coefficient, k and the fractal dimension, D_f for a single-fractal analysis, and (b) the binding rate coefficients, k_1 and k_2, and the fractal dimensions, D_{f1} and D_{f2} for a dual-fractal analysis are given in Table 9.4. It is of interest to note that as the fractal dimension increases by a factor of 1.137 from a value of D_{f1} equal to 2.494 to D_{f2} equal to 2.8358, the binding rate coefficient increases by a factor of 1.64 from a value of k_1 equal to 26.102 to k_2 equal to 42.850.

Figure 9.9c shows the binding of 1000 nM target DNA (3′-CGTGGACTGAGGACAC-CTCTTCAGACGGCA-5′) which is complementary to the probe DNA (density 1 μM) with a biotin label at the 5′-end (5′-biotin-GCACCTGACTCCTGTGGAGAAGTCTGCCGT-3′) immobilized on a sensor chip surface. A dual-fractal analysis is, once again, required to adequately describe the binding kinetics. The values of (a) the binding rate coefficient, k and the fractal dimension, D_f for a single-fractal analysis, and (b) the binding rate coefficients, k_1 and k_2, and the fractal dimensions, D_{f1} and D_{f2} for a dual-fractal analysis are given in Table 9.4. It is of interest to note that as the fractal dimension increases by a factor of 1.48 from a value of D_{f1} equal to 1.8120 to D_{f2} equal to 2.6796, the binding rate coefficient increases by a factor of 2.07 from a value of k_1 equal to 21.761 to k_2 equal to 44.991.

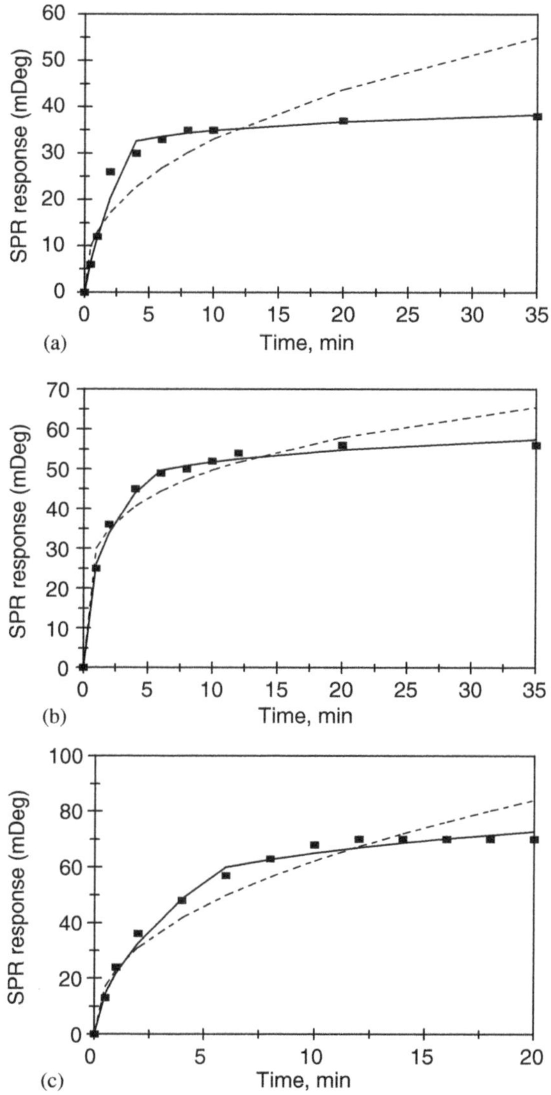

Figure 9.9 Binding of 1 μM target DNA to different concentrations (in μM) (probe density) of biotin–DNA immobilized on a sensor chip surface (Su *et al.*, 2005): (a) 100, (b) 200, (c) 1000. (When only a solid line (___) is used, then a single-fractal analysis applies. When a dashed (---) and a solid line (___) is used, then the dashed line represents a single-fractal analysis, and the solid line represents a dual-fractal analysis. In this case the solid line provides a better fit).

Table 9.4 and Figure 9.10a show for a dual-fractal analysis the increase in the fractal dimension, D_{f2} with an increase in the biotin–DNA concentration (in nM) in solution. For the data shown in Figure 9.10a, the fractal dimension, D_{f2} is given by

$$D_{f2} = (13.272 \pm 0.034)[\text{biotin} - \text{DNA concentration}]^{-0.0286 \pm 0.0062} \qquad (9.5a)$$

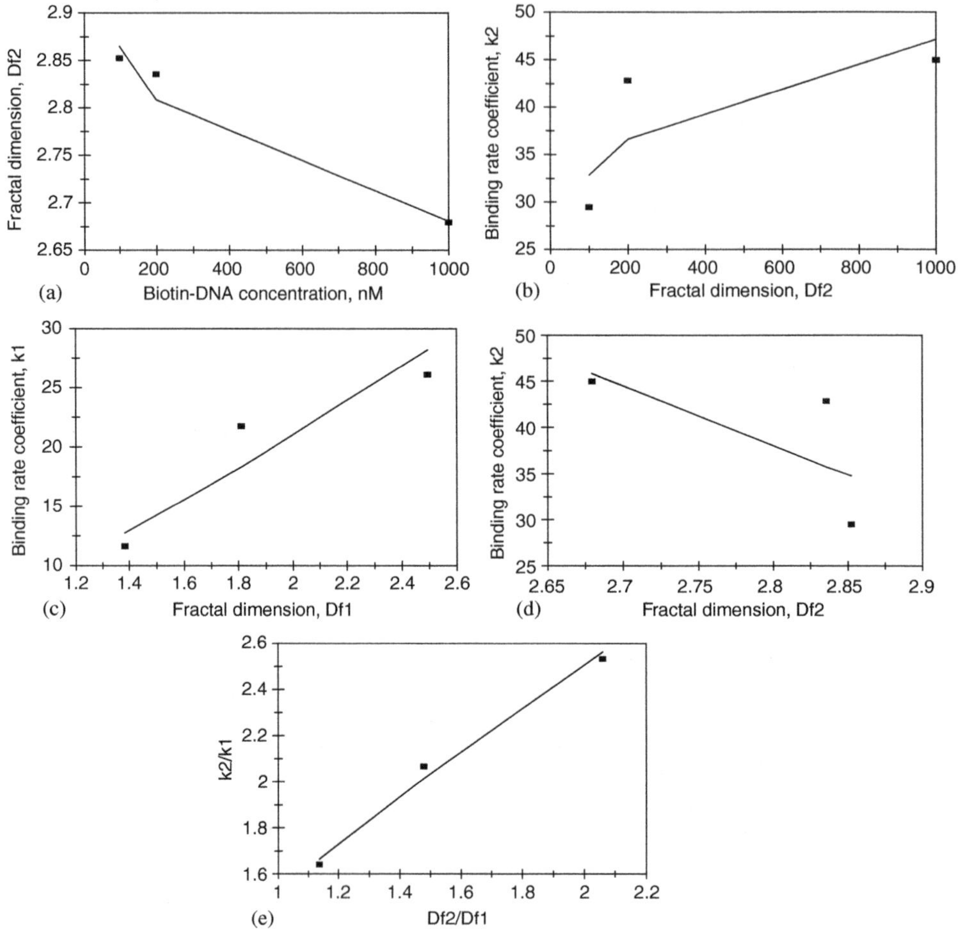

Figure 9.10 (a) Increase in the fractal dimension, D_{f2} with an increase in the biotin–DNA concentration (in nM) in solution. (b) Increase in the binding rate coefficient, k_2 with an increase in the fractal dimension, D_{f2}. (c) Increase in the binding rate coefficient, k_1 with an increase in the fractal dimension, D_{f1}. (d) Decrease in the binding rate coefficient, k_2 with an increase in the fractal dimension, D_{f2}. (e) Increase in the ratio of the binding rate coefficients, k_2/k_1 with an increase in the fractal dimension ratio, D_{f2}/D_{f1}.

The fit is good. Only three data points are available. The availability of more data points would lead to a more reliable fit. The fractal dimension, D_{f2} is only very mildly sensitive to the biotin–DNA concentration in solution as noted by the negative 0.0286 order of dependence exhibited. Note that the fractal dimension, D_{f2} is based on a log scale, and thus even small changes in the fractal dimension value lead to significant changes in the degree of heterogeneity on the sensor chip surface.

Table 9.4 and Figure 9.10b show for a dual-fractal analysis the increase in the binding rate coefficient, k_2 with an increase in the fractal dimension, D_{f2}. For the data shown in Figure 9.10b, the binding rate coefficient, k_2 is given by

$$k_2 = (15.975 \pm 3.469) D_{f2}^{0.1567 \pm 0.1176} \tag{9.5b}$$

The fit is reasonable. Only three data points are available. The availability of more data points would lead to a more reliable fit. The binding rate coefficient, k_2 is only very mildly sensitive to the fractal dimension, D_{f2} as noted by the very low order (equal to 0.1567) of dependence exhibited. In other words, in this case, the binding rate coefficient, k_2 is only mildly sensitive to the degree of heterogeneity present on the sensor chip surface.

Table 9.4 and Figure 9.10c show for a dual-fractal analysis the increase in the binding rate coefficient, k_1 with an increase in the fractal dimension, D_{f1}. For the data shown in Figure 9.10c, the binding rate coefficient, k_1 is given by

$$k_1 = (8.237 \pm 1.916) D_{f1}^{1.347 \pm 0.501} \tag{9.5c}$$

The fit is good. Only three data points are available. The availability of more data points would lead to a more reliable fit. The binding rate coefficient, k_1 exhibits an order of dependence between first and one and one-half (equal to 1.347) on the fractal dimension, D_{f1} or the degree of heterogeneity present on the sensor chip surface.

Table 9.4 and Figure 9.10d show for a dual-fractal analysis the decrease in the binding rate coefficient, k_2 with an increase in the fractal dimension, D_{f2}. For the data shown in Figure 9.10d, the binding rate coefficient, k_2 is given by

$$k_2 = (3468.74 \pm 973.35) D_{f2}^{-4.3908 + 5.065} \tag{9.5d}$$

The fit is reasonable. Only three data points are available. The availability of more data points would lead to a more reliable fit. There is scatter in the data, and this is reflected in the error in the binding rate coefficient, and also in the order of dependence exhibited. The binding rate coefficient, k_2 is very sensitive to the fractal dimension, D_{f2} as noted by the higher than negative fourth order (equal to –4.3908) of dependence exhibited, notwithstanding the large error in the order of dependence exhibited.

Table 9.4 and Figure 9.10e show for a dual-fractal analysis the increase in the ratio of the binding rate coefficients, k_2/k_1 with an increase in the ratio of fractal dimensions, D_{f2}/D_{f1}. For the data shown in Figure 9.10e, the ratio of the binding rate coefficients, k_2/k_1 is given by

$$\frac{k_2}{k_1} = (1.571 \pm 0.049) \left(\frac{D_{f2}}{D_{f1}} \right)^{0.7241 \pm 0.0749} \tag{9.5e}$$

The fit is very good. Only three data points are available. The availability of more data points would lead to a more reliable fit. The ratio of the binding rate coefficients, k_2/k_1 exhibits an order of dependence between one-half and first (equal to 0.7241) on the ratio of fractal dimensions, D_{f2}/D_{f1}.

Tawa *et al.* (2005) have recently analyzed the kinetics of DNA–DNA hybridization using surface plasmon fluorescence spectroscopy (SPFS). These authors indicate that the kinetics of association and dissociation of hybridization reactions in solution are influenced by the length of the matching base-pair sequence in a complementary target DNA. Tawa *et al.* (2005) have attempted to analyze the influence of the length of a matching base-pair sequence on surface hybridization reactions. These authors indicate that they have already reported on examples in the literature wherein they detected hybridization reactions on surface-attached probe oligo-DNA, and oligo-PNA and target DNA in solution by surface plasmon spectroscopy (SPS) (Knoll, 1998), and also by SPFS (Liebermann *et al.*, 2000; Liebermann and Knoll, 2000; Kamhampati *et al.*, 2001). These authors further indicate that high sensitivity is achievable for the mismatch discrimination (Liebermann and Knoll, 2000; Tawa and Knoll, 2004).

Figure 9.11a shows the binding and dissociation of P2-T2(14) (5′-biotin-TTT TTT TTT TTT TTT TGT ACA TCA CAA CTA-3′) in solution to T2(10) (3′-TAGTGTTGAT-Cy5-5′) immobilized on a sensor chip surface (Tawa *et al.*, 2005). The target DNAs are modified at the 5′ end with a fluorescent probe. The probe nucleotide is modified at the 5′-end by a biotin moiety. The molar concentration of the target DNA (analyte) is 2.5 nM. These authors obtained the kinetics of the hybridization parameters using an expanded Langmuir model. These authors presented and analyzed their results based on a model that included the association and dissociation processes involving a single type of catcher probe at the surface (in our language receptor on the surface). The analyte(s) were a binary mixed solution of oligonucleotide targets that were competing for the same binding sites (receptors on the surface), but with different affinities.

A single-fractal analysis is adequate to describe the binding and the dissociation kinetics. The values of (a) the binding rate coefficient, k and the fractal dimension, D_f for a single-fractal analysis, and (b) the dissociation rate coefficient, k_d and the fractal dimension for the dissociation phase, D_{fd} for a single-fractal analysis are given in Tables 9.5a and 9.5b.

Figure 9.11b shows the binding and dissociation of P2-T2(14) (5′-biotin-TTT TTT TTT TTT TTT TGT ACA TCA CAA CTA-3′) in solution to T2(10) (3′-TAGTGTTGAT-Cy5-5′) immobilized on a sensor chip surface (Tawa *et al.*, 2005). The target DNAs are modified at the 5′ end with a fluorescent probe. In this case, the molar concentration of the target DNA (analyte) is 10 nM.

A dual-fractal analysis is required to describe the binding kinetics. The dissociation kinetics is adequately described by a single-fractal analysis. The values of the binding rate coefficient, k and the fractal dimension, D_f for a single-fractal analysis, and (b) the binding rate coefficients, k_1 and k_2, and the fractal dimensions, D_{f1} and D_{f2} for a dual-fractal analysis, and (c) the dissociation rate coefficient, k_d and the fractal dimension for dissociation, D_{fd} for a single-fractal analysis are given in Table 9.5a. Note that as the fractal dimension increases by a factor of 1.39 from a value of D_{f1} equal to 2.0142 to D_{f2} equal to 2.8077 the binding rate coefficient increases by a factor of 35.3 from a value of k_1 equal to 46.15 to k_2 equal to 1629.76.

Figure 9.11c shows the binding and dissociation of P2-T2(14) (5′-biotin-TTT TTT TTT TTT TTT TGT ACA TCA CAA CTA-3′) in solution to T2(10) (3′-TAGTGTTGAT-Cy5-5′) immobilized on a sensor chip surface (Tawa *et al.*, 2005). The target DNAs are modified at the 5′ end with a fluorescent probe. In this case, the molar concentration of the

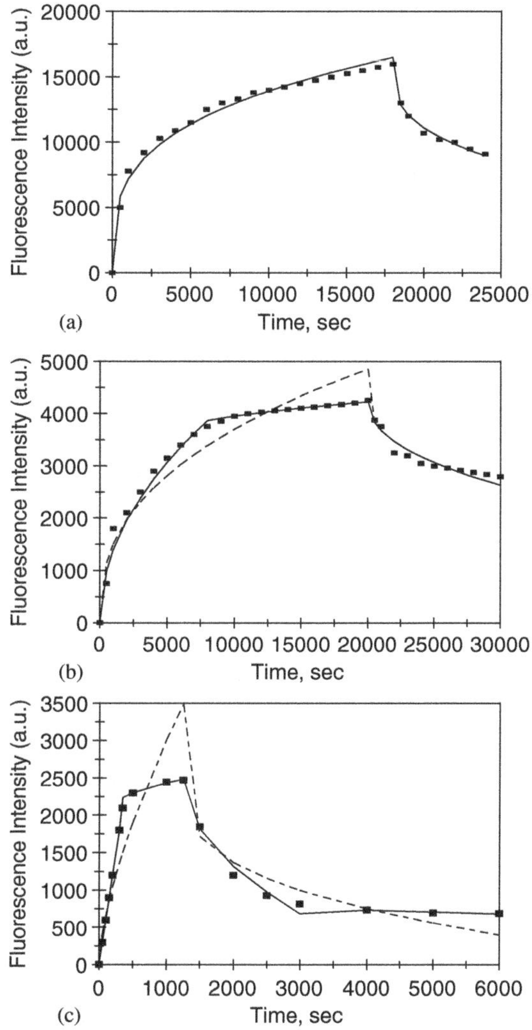

Figure 9.11 (a) Binding (hybridization) and dissociation of P2-T2(14) and T2(10) mixture (target) for different total concentrations (in nM). One single-stranded oligo-DNA (probe-DNA) immobilized on the substrate; the other one (target DNA) labeled with a fluorescent probe added to the flow cell (Tawa *et al.*, 2005). Molar ratio of P2-T2(14) and T2(10) was 0.08: (a) 2.5, (b) 10, (c) 250.

target DNA (analyte) is 250 nM. In this case, a dual-fractal analysis is required to describe the binding and the dissociation kinetics. The values of (a) the binding rate coefficient, k and the fractal dimension, D_f for a single-fractal analysis, (b) the binding rate coefficients, k_1 and k_2 and the fractal dimensions, D_{f1} and D_{f2} for a dual-fractal analysis, (c) the dissociation rate coefficient, k_d and the fractal dimension, D_{fd} for a single-fractal analysis, and (d) the dissociation rate coefficients, k_{d1} and k_{d2} and the fractal dimensions for the dissociation phase, D_{fd1} and D_{fd2} for a dual-fractal analysis are given in Table 9.5 a,b. Note that

Table 9.5a

Binding and dissociation rate coefficients for hybridization and dissociation of P2-T2(14) and T2(10) mixture for surface plasmon fluorescence spectroscopy (SPFS): one single-stranded oligo-DNA (probe-DNA) immobilized on substrate; the other one (target DNA) labeled with a fluorescent probe added to the flow cell. Molar ratio of P2-T2(14) and T2(10) was 0.08. Influence of different total concentrations (Tawa *et al.*, 2005)

Molar concentration (in nM) of analyte (target DNA)/probe DNA on receptor surface	k	k_1	k_2	k_d	k_{d1}	k_{d2}
2.5	971.16 ± 51.67	na	na	415.40 ± 19.29	na	na
10	100.87 ± 14.89	46.15 ± 6.72	1629.76 ± 4.73	25.56 ± 3.46	na	na
250	31.30 ± 9.85	6 ± 0	1377.50 ± 8.33	115.39 ± 19.85	37.94 ± 3.86	1141.80 ± 6.20

Table 9.5b

Fractal dimensions for the binding and dissociation phases for hybridization and dissociation of P2-T2(14) and T2(10) mixture for surface plasmon fluorescence spectroscopy (SPFS): one single-stranded oligo-DNA (probe-DNA) immobilized on substrate; the other one (target DNA) labeled with a fluorescent probe added to the flow cell. Molar ratio of P2-T2(14) and T2(10) was 0.08. Influence of different total concentrations (Tawa *et al.*, 2005)

Molar concentration (in nM) of analyte (target DNA)/probe DNA on receptor surface	D_f	D_{f1}	D_{f2}	D_{fd}	D_{fd1}	D_{fd2}
2.5	2.4218 ± 0.02450	na	na	2.3492 ± 0.0408	na	na
10	2.2176 ± 0.0614	2.0142 ± 0.08948	2.8077 ± 0.00964	2.100 ± 0.0833	na	na
250	1.6768 ± 0.1847	2.0 ± 5.4E-15	2.8347 ± 0.01786	2.3172 ± 0.1266	1.9680 ± 0.1315	2.8939 ± 0.0280

in this case increases in the fractal dimensions in the binding and in the dissociation phases lead to increases in the binding and in the dissociation rate coefficients, respectively.

Table 9.5a and 9.5b and Figure 9.12a show for a single- and a dual-fractal analysis the increase in the binding rate coefficient, (k, k_1, or k_2) with an increase in the fractal dimension (D_f, D_{f1}, or D_{f2}). Since very few points are available for the single- or the dual-fractal

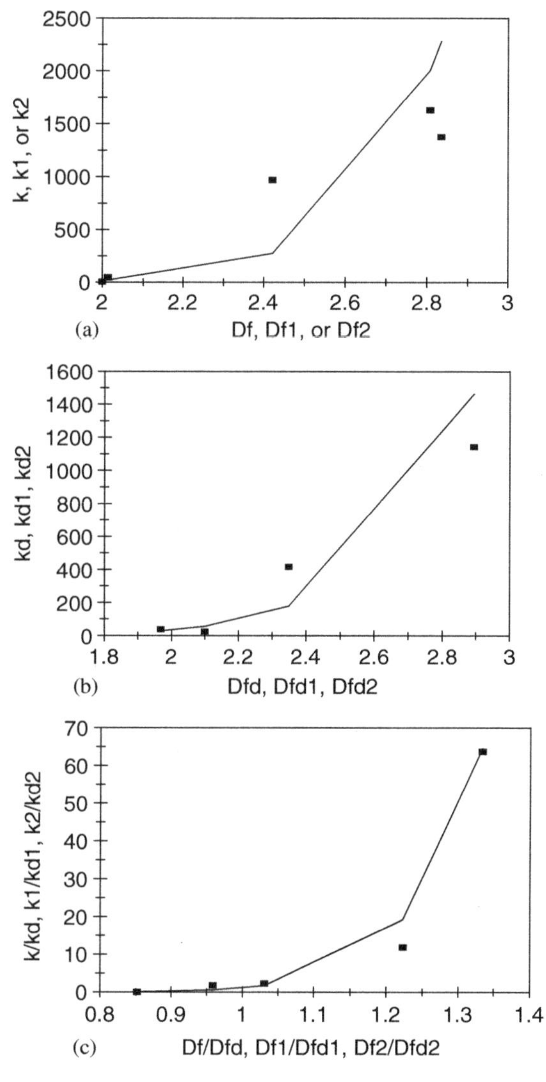

Figure 9.12 (a) Increase in the binding rate coefficient with an increase in the fractal dimension. (b) Increase in the dissociation rate coefficient with an increase in the fractal dimension for dissociation. (c) Increase in the affinity (ratio of the binding and the dissociation rate coefficients) with an increase in the ratio of the fractal dimensions in the binding and in the dissociation phases.

analysis taken separately, they are plotted together on the same plot. For the data presented in Figure 9.12a, the binding rate coefficient is given by

$$(k, k_1 \text{ or } k_2) = (0.00186 + 0.00399)[D_f, D_{f1}, \text{or } D_{f2}]^{13.455 \pm 3.359} \qquad (9.6a)$$

The fit is reasonable considering that the data are plotted for the single- as well as the dual-fractal analysis. There is scatter in the data, and this is reflected in the error in the estimated value of the rate coefficient. Note that only the positive value of the error is given since the binding rate coefficient cannot have a negative value. Only five data points are available. The availability of more data points for each of the two cases (single- and a dual-fractal analysis) presented together would lead to a more reliable fit. The binding rate coefficient is very sensitive to the fractal dimension or the degree of heterogeneity present on the biosensor chip surface as noted by the close to thirteen and one-half (equal to 13.455) order of dependence exhibited.

Table 9.5a and 9.5b and Figure 9.12b show for a single- and a dual-fractal analysis the increase in the dissociation rate coefficient, (k_d, k_{d1}, or k_{d2}) with an increase in the fractal dimension (D_{fd}, D_{fd1}, or D_{fd2}). Since very few points are available for the single- or the dual-fractal analysis taken separately, they are plotted together, once again, on the same plot. For the data presented in Figure 9.12b, the dissociation rate coefficient is given by

$$(k_d, k_{d1}, \text{or } k_{d2}) = (0.0332 + 0.0456)[D_{fd}, D_{fd1}, \text{or } D_{fd2}]^{10.062 \pm 2.948} \tag{9.6b}$$

The fit is good considering that the data are plotted for the single- as well as the dual-fractal analysis together. There is some scatter in the data, and this is reflected in the error in the estimated value of the rate coefficient. Note that only the positive value of the error is given since the dissociation rate coefficient cannot have a negative value. Only five data points are available. The availability of more data points for each of the two cases (single- and a dual-fractal analysis) presented together would lead to a more reliable fit. The dissociation rate coefficient is very sensitive to the fractal dimension or the degree of heterogeneity present on the biosensor chip surface as noted by the close to tenth (equal to 10.062) order of dependence exhibited.

Table 9.5a and 9.5b and Figure 9.12c show k for a single- and a dual-fractal analysis the increase in the affinity (k/k_d, k_1/k_{d1}, or k_2/k_{d2}) with an increase in the fractal dimension ratio (D_f/D_{fd}, D_{f1}/D_{fd1}, or D_{f2}/D_{fd2}). Since very few points are available for the single- or the dual-fractal analysis separately, once again, they are plotted together on the same plot. For the data presented in Figure 9.12c, the affinity is given by

$$\left(\frac{k}{k_d}, \frac{k_1}{k_{d1}}, \text{or } \frac{k_2}{k_{d2}}\right) = (1.1404 + 1.505)\left[\frac{D_f}{D_{fd}}, \frac{D_{f1}}{D_{fd1}}, \frac{D_{f2}}{D_{fd2}}\right]^{14.01 \pm 2.319} \tag{9.6c}$$

The fit is good considering that the data for the single- as well as the dual-fractal analysis is plotted together. There is some scatter in the data, and this is reflected in the error in the estimated value of the rate coefficient. Note that only the positive value of the error is given since the binding rate coefficient cannot have a negative value. Only five data points are available. The availability of more data points for each of the two cases (single- and a dual-fractal analysis) presented together would lead to a more reliable fit. The affinity is very sensitive to the fractal dimension or the degree of heterogeneity present on the biosensor chip surface as noted by the close to fourteen (equal to 14.01) order of dependence exhibited.

9.4 CONCLUSIONS

A fractal analysis is presented for the binding and dissociation of different DNA–analyte interactions occurring on biosensor surfaces. The binding and dissociation (if applicable) kinetics of (a) DNA (containing various mismatches)–protein complexes using SPR (Tsoi and Yang, 2004), (b) DNA–histone interactions using SPR (Barontes *et al.*, 2006), (c) a molecular beacon based DNA micro-biosensor whose efficiency is enhanced using a two-dimensional micro-bubble actuator array (Deng *et al.*, 2006), (d) DNA assembly and hybridization using SPR and QCM, and (e) DNA–DNA hybridization using surface plasmon fluoresecence spectroscopy (Tawa *et al.*, 2005) is presented using a fractal analysis. Note that, and as indicated elsewhere in this book the fractal analysis is presented as an alternate method of analyzing the binding and dissociation kinetics of the above mentioned reactions. The binding and dissociation kinetics may be adequately described by a single- or a dual-fractal analysis. A dual-fractal analysis is used only if a single-fractal analysis did not provide an adequate fit.

In accord with the prefactor analysis of aggregates (Sorenson and Roberts, 1997) quantitative and predictive equations are developed for (a) the binding rate coefficient, k as a function of the T7 DNA polymerase concentration in solution (Tsoi and Yang, 2004), (b) the binding rate coefficient, k as a function of the fractal dimension, D_f (Tsoi and Yang, 2004), (c) the dissociation rate coefficient, k_d as a function of the T7 DNA polymerase concentration in solution (Tsoi and Yang, 2004), (d) the dissociation rate coefficient, k_d as a function of the fractal dimension, D_{fd} (Tsoi and Yang, 2004), (e) the affinity, K $(=k/k_d)$ as a function of the ratio of the fractal dimensions present in the binding and in the dissociation phases (D_f/D_{fd}) (Tsoi and Yang, 2004), (f) the binding rate coefficients, k_1 and k_2 as function of the fractal dimensions, D_{f1} and D_{f2}, respectively for the binding of target DNA (3′-CGTGGACTG-AGGACACCTCTTCAGACGGCA-5′) which is complementary to the probe DNA with a biotin label at the 5′-end (5′-biotin-GCACCTGACTCCTGTGGAGAAGTCTGCCGT-3′) immobilized on a sensor chip (Su *et al.*, 2005), (g) the binding and dissociation rate coefficients as a function of the fractal dimension in the binding and in the dissociation phases, respectively, for DNA–DNA hybridization using SPFS (Tawa *et al.*, 2005), and (h) the affinity as a function of the ratio of fractal dimensions present in the binding and in the dissociation phases for DNA–DNA hybridization using SPFS (Tawa *et al.*, 2005).

The predictive equations presented above provide a means by which, for example, the binding rate coefficient may be manipulated by changing the analyte concentration or the degree of heterogeneity that exists on the biosensor surface. Note that, in general, the change in the degree of heterogeneity on the surface leads to a change in the binding rate coefficient in the same direction. Practicing biosensorists need to pay more attention to the nature of and the degree of heterogeneity present on the sensor chip surface. The quantitative relationships developed for the binding and the dissociation rate coefficients, and the affinity values as a function of the degree of heterogeneity present on the sensor chip surface will permit one to manipulate these parameters in desired directions. This type of flexibility with biosensors will permit one to increase biosensor performance parameters such as sensitivity, selectivity, stability, and other parameters of interest.

It is suggested that the fractal surface (roughness) leads to turbulence, which enhances mixing, decreases diffusional limitations, and leads to an increase in the binding rate coefficient

(Martin *et al.*, 1991). As indicated earlier, for this to occur the characteristic length of the turbulent boundary layer may have to extend a few monolayers above the sensor chip surface to affect bulk diffusion to and from the surface. However, given the extremely laminar flow regimes in most biosensors this may not actually take place. Often, the sensor chip surface is characterized by grooves and ridges, and this surface morphology may leads to eddy diffusion. This eddy diffusion can then help to enhance the mixing and extend the characteristic length of the boundary layer to affect the bulk duffusion to and from the surface.

DNA–DNA or DNA–analyte interactions are of primary importance for biological and medical applications. Any sets of studies that help provide additional physical insights into these type of interactions, such as the analysis of binding and dissociation kinetics using fractal analysis should prove beneficial to help better understand these types of reactions, especially if they are able to better predict the onset and manage especially insidious diseases. This has been a common theme throughout this book.

REFERENCES

Armistead, PM and HH Thorp, Electrochemical detection of gene expression in tumor samples: over expression of Rak nuclear tyrosine kinase. *Bioconjugate Chemistry*, 2002, **13**, 172–176.

Baeumner, AJ, J Pretz, and S Fang, A universal nucleic acid sequence biosensor with nanomolar detection limits. *Analytical Chemistry*, 2004, **76**, 888–894.

Barontes, A, PJ Navarro, MJ Benitez, and JS Jimenez, A DNA and histone immobilization method to study DNA–histone interactions by surface plasmon resonance. *Analytical Biochemistry*, 2006.

Campbell, CN, D Gal, N Gristler, C Banditrat, and A Heller, Enzyme amplified amperometric sandwich test for RNA and DNA. *Analytical Chemistry*, 2002, **74**, 158–162.

Cavic, BA, GL Hayward, and M Thompson, Accoustic waves and the study of biochemical macromolecules and cell at the sensor liquid interface. *Analyst*, 1999, **124**, 1405–1412.

Chaiken, I, S Rose, and R Karlsson, Analysis of macromolecular interactions using immobilized ligands. *Analytical Biochemistry*, 1992, **210**(2), 197–210.

Corel Quattro Pro 8.0, Corel Corporation, Ottawa, Canada, 1997.

Deng, P, Y-K Lee, and P Cheng, Two-dimensional micro-bubble actuator array to enhance the efficiency of molecular beacon based DNA micro-biosensors. *Biosensors and Bioelectronics*, 2006.

Englebienne, P, A Van Hoonacker, and M Verhas, Surface plasmon resonance: principles methods, and applications in biomedical sciences. *Spectroscopy International Journal*, 2003, **17**, 255–259.

Fisher, RJ, M Fivash, J Casas-Finet, JW Erickson, A Kondoh, SV Bladen, C Fisher, DK Watson, and T Papas, Real-time DNA binding measurements of the ETS1 recombinant oncoproteins reveal significant kinetic differences between the p42 and p51 isoforms. *Protein Science*, 1994, **3**(2), 257–266.

Haruki, M, E Noguchi, S Kanaya, and RJ Crouch, Kinetic and stoichiometric analysis for the binding of *Escherichia coli* ribonuclease HI to RNA-DNA hybrids using surface plasmon resonance. *Journal of Biological Chemistry*, 1979, **272**(35), 22015–22022.

Havlin, S, Molecular diffusion and reactions, in *The Fractal Approach to Heterogeneous Chemistry: Surfaces, Colloids, Polymers,* (ed. D. Avnir), Wiley, New York, 1989.

Homola, J, Present and future of surface plasmon resonance. *Tibtech*, 1997, **15**, 353–359.

Kambhampati, D, PE Nielsen, and W Knoll, Investigating the kinetics of DNA-DNA and PNA-DNA interactions using surface plasmon resonance-enhanced fluorescence spectroscopy. *Biosensors and Bioelectronics*, 2001, **16**, 1109–1118.

Knoll, W, Interfaces and thin films as seen by bound electromagnetic waves. *Annual Reviews in Physical Chemistry*, 1998, **49**, 569–638.

Kornberg, A and TA Baker, *DNA Replication*, Freeman, San Francisco, 1991.

Kretschmann, A and TA Baker, Radiative decay of nonradiative surface plasmons excited by light. *Z. Naturforsch*, 1968, **A23**, 2135–2136.

Lee, CK and SL Lee, Multi-fractal scaling analysis of reactions over fractal surfaces. *Surface Science*, 1995, **325**, 294–310.

Liebermann, T and W Knoll, Surface-plasmon field-enhanced fluorescence spectroscopy. *Colloids and Surfaces A*, 2000, **171**, 115–130.

Liebermann, T, W Knoll, P Sluka, and R Hermann, Complement hybridization from solution to surface-attached probe-nucleotides observed by surface-plasmon-field-enhanced fluorescence spectroscopy. *Colloids and Surfaces A*, 2000, **169**, 337–350.

Liu, J, C Roussel, G Lagger, P Tacchini, and HH Girault, Antioxidant sensors based on DNA-modified electrodes. *Analytical Chemistry,* 2005, **77**, 7687–7694.

Lucarelli, F, I Palchetti, G. Marrazza, and M Mascini, Electrochemical DNA biosensor as a screening tool for the detection of toxicants in water and wastewater sample. *Talanta*, 2002, **56**, 949–957.

Martin, JS, GC Frye, AJ Ricco, and AD Senturia, Effect of surface roughness on the response of thickness-shear mode resonators in liquids. *Analytical Chemistry*, 1991, **65**(20), 2910–2922.

Mugweru, A, BQ Wang, and J Rusling, Volumetric sensor for oxidized DNA using ultrathin films of osmium and ruthenium metallopolymers. *Analytical Chemistry*, 2004, **76**, 5557–5563.

Myszka, DG, Kinetic analysis of macromolecular interactions using surface plasmon resonance biosensors. *Current Opinion in Biotechnology*, 1997, **8**(1), 50–57.

Oda, M, K Furukawa, A Sarai, and H Nakamura, Kinetic analysis of DNA binding by the c-Myb DNA-binding domain using surface plasmon resonance. *FEBS Letters*, 1999, **454**(3), 288–292.

Pappaert, K *et al.*, Enhancement of DNA micro-array analysis using a shear-driven micro-channel flow system. *Journal of Chromatography*, 2003, **1014**, 1–9.

Popovich, ND, AE Eckhardt, JC Mikulecky, ME Napier, and RS Thomas, Electrochemical sensor for detection of unmodified nucleic acids. *Talanta*, 2002, **56**, 821–828.

Ramakrishnan, A and A Sadana, A fractal analysis for cellular analyte-receptor binding kinetics: biosensor application. *Automedica*, 2001, **20**, 313–340.

Rich, RL and DG Myszka, Advances in surface plasmon resonance biosensors. *Analytical and Bioanalytical Chemistry*, 2003, **377**, 528–539.

Sadana, A, A fractal analysis approach for the evaluation of hybridization kinetics in biosensors. *Journal of Colloid and Interface Science,* 2001, **234**, 9–18.

Sadana, A, *Fractal Binding and Dissociation Kinetics for Different Biosensor Applications*, Elsevier, Amsterdam, 2005.

Schuck, P, Kinetics of ligand binding to receptor immobilized in a polymer matrix, as detected with an evanescent wave biosensor. I. A computer simulation of the influence of mass transport. *Biophysical Journal*, 1996, **70**(3), 1230–1249.

Silin, V and A Plant, Biotechnological applications of surface plasmon resonance. *Tibtech*, 1997, **15**, 353–359.

Sorenson, CM and GC Roberts, The prefactor of fractal aggregates. *Journal of Colloid and Interface Science*, 1997, **186**, 447–453.

Su, X, Y-J Wu, and W Knoll, Comparison of surface plasmon resonance spectroscopy and quartz microbalance techniques for studying DNA assembly and hybridization. *Biosensors and Bioelectronics*, 2005, **21**, 719–726.

Su, X, R Robelek, Y Wu, G Wang, and W Knoll, Detection of point mutation and insertion mutations in DNA using a quartz crystal microbalance and Muts, a mismatch binding protein. *Analytical Chemistry*, 2004, **76**, 489–494.

Tawa, K and W Knoll, Mismatching base-pair dependence of the kinetics of DNA-DNA hybridization studied by surface plasmon resonance fluorescence spectroscopy. *Nucleic Acid Research*, 2004, **8**, 2372–2377.

Tawa, K, D Yao, and W Knoll, Matching base pair number dependence of the kinetics of DNA-DNA hybridization studied by surface plasmon fluorescence spectroscopy. *Biosensors and Bioelectronics*, 2005, **21**, 322–329.

Tsoi, PY and M Yang, Surface plasmon resonance study of the molecular recognition between polymerase and DNA containing various mismatches and conformational changes of DNA-protein complexes. *Biosensors and Bioelectronics*, 2004, **19**, 1209–1218.

Wu, Z-S, J-H Jiang, L Fu, G-L Shen, and RQ Yu, Optical detection of DNA hybridization based on fluorescence quenching of tagged oligonucleotide probes by gold nanoparticles. *Analytical Biochemistry*, 2006.

Zhang, YC, HH Kim, and A Heller, Enzyme-amplified amperometric detection of 3000 copies of DNA in a 10-μL droplet of 0.5 fM concentration. *Analytical Chemistry*, 2003, **75**, 3267–3269.

– 10 –

Fractal Analysis of Binding and Dissociation of Protein–Analyte Interactions on Biosensor Surfaces

10.1 INTRODUCTION

The analysis of protein–analyte interactions is an important and diversified area of application. Biosensors have often been employed to analyze these types of interactions. Some of the more recent biosensor applications of protein–analyte interactions include:

(a) Surface plasmon resonance (SPR) study of G protein receptor coupling in a lipid bilayer system (Komolov *et al.*, 2006). These authors indicate that G protein-coupled receptors (GPCRs) serve key functions in hormone, neurotransmitter, and sensory signaling.
(b) Poly(ethylene glycol)-based biosensor chip to analyze heparin–protein interactions (Munoz *et al.*, 2005). Capila and Linhardt (2002) indicate that heparin's role in biological processes is mediated by its interaction with proteins. They have used the SPR biosensor to monitor heparin–protein interactions.
(c) Biosensor recognition of thyroid-disrupting chemicals using transport proteins (Marchesini *et al.*, 2006). These authors indicate that organohalogen compounds may affect thyroid gland morphology and the hormonal status (Brouver *et al.*, 1995). Brouver *et al.* (1998) indicate that these compounds affect the thyroid hormone system at (a) the thyroid gland, (b) the thyroid transport proteins (TPs), and (c) the thyroid hormone metabolism levels.
(d) Zhang *et al.* (2005) have used a quartz crystal microbalance (QCM)-flow injection analysis (FIA) instrument with a poly(glycidyl methacrylate) (PGMA) coating to analyze the dynamic interactions of heparin and antithrombin III. These authors indicate that the QCM is an extremely sensitive surface mass sensor, and it has been used to analyze protein adsorption on a solid surface (Park *et al.*, 2000; Fung and Wang, 2001). Zhang *et al.* (2005) emphasize that combined with FIA, QCM permits online monitoring of the analyte binding. Liu *et al.* (2003a,b) have used the QCM-FIA technique to analyze protein–small molecule agent interactions.

In this chapter we use fractal analysis to analyze the binding and dissociation (if applicable) kinetics of: (a) recombinant transthyretin (rTTR) in solution to *L*-thyroxine (T4) immobilized on a sensor chip surface using different spacers (Marchesini *et al.*, 2006); (b) G protein transducin (G_t) in solution to different densities of rhodopsin (Rho) immobilized on a sensor chip surface (Komolov *et al.*, 2006); (c) different concentrations of Factor P in solution to heparin immobilized on a sensor chip surface (Munoz *et al.*, 2005); and (d) binding of transcription factors, rhSP1 and NF-κB in solution to complementary oligonucleotide immobilized on a sensor chip surface (Huber *et al.*, 2006). The fractal analysis, as indicated elsewhere in this book, is presented as an alternate analysis to obtain the binding and dissociation rate coefficient values, besides providing a quantitative measure of the degree of heterogeneity present on the biosensor chip surface.

10.2 THEORY

Havlin (1989) has reviewed and analyzed the diffusion of reactants toward fractal surfaces. The details of the theory and the equations involved for the binding and the dissociation phases for analyte–receptor binding are available (Sadana, 2001). The details are not repeated here, except that just the equations are given to permit an easier reading. These equations have been applied to other biosensor systems (Sadana, 2001, 2005; Ramakrishnan and Sadana, 2001). For most applications, a single- or a dual-fractal analysis is often adequate to describe the binding and the dissociation kinetics. Peculiarities in the values of the binding and the dissociation rate coefficients, as well as in the values of the fractal dimensions with regard to the dilute analyte systems being analyzed will be carefully noted, if applicable.

10.2.1 Single-fractal analysis

Binding rate coefficient

Havlin (1989) indicates that the diffusion of a particle (analyte [Ag]) from a homogeneous solution to a solid surface (e.g., receptor [Ab]-coated surface) on which it reacts to form a product (analyte–receptor complex; (Ab·Ag)) is given by:

$$(\mathrm{Ab}\cdot\mathrm{Ag}) \approx \begin{cases} t^{(3-D_{\mathrm{f,bind}})/2} = t^{p} & t < t_c \\ t^{1/2} & t > t_c \end{cases} \tag{10.1a}$$

Here $D_{\mathrm{f,bind}}$ or D_{f} (used later on in the chapter) is the fractal dimension of the surface during the binding step. t_c is the crossover value. Havlin (1989) indicates that the crossover value may be determined by $r_c^2 \sim t_c$. Above the characteristic length, r_c, the self-similarity of the surface is lost and the surface may be considered homogeneous. Above time, t_c the surface may be considered homogeneous, since the self-similarity property disappears, and 'regular' diffusion is now present. For a homogeneous surface where D_{f} is equal to 2, and when only diffusional limitations are present, $p = 1/2$ as it should be. Another way of looking at the $p = 1/2$ case (where $D_{\mathrm{f,bind}}$ is equal to two) is that the analyte in solution views the fractal object, in our case, the receptor-coated biosensor surface, from a 'large distance.' In essence, in the association process, the diffusion of the analyte from the solution to the receptor surface creates a depletion layer of width $(Ɖt)^{1/2}$ where $Ɖ$ is the diffusion constant. This gives rise to the fractal power law, $(\text{Analyte} - \text{Receptor}) \sim t^{(3-D_{\text{f,bind}})/2}$. For the present analysis, t_c is arbitrarily chosen and we assume that the value of the t_c is not reached. One may consider the approach as an intermediate 'heuristic' approach that may be used in the future to develop an autonomous (and not time-dependent) model for diffusion-controlled kinetics.

Dissociation rate coefficient

The diffusion of the dissociated particle (receptor [Ab] or analyte [Ag]) from the solid surface (e.g., analyte [Ag]–receptor [Ab]) complex coated surface) into solution may be given, as a first approximation by:

$$(\text{Ab} - \text{Ag}) \approx -t^{(3-D_{\text{f,diss}})/2} = -t^{p}\,(t > t_{\text{diss}}) \tag{10.1b}$$

Here $D_{\text{f,diss}}$ is the fractal dimension of the surface for the dissociation step. This corresponds to the highest concentration of the analyte–receptor complex on the surface. Henceforth, its concentration only decreases. The dissociation kinetics may be analyzed in a manner 'similar' to the binding kinetics.

10.2.2 Dual-fractal analysis

Binding rate coefficient

Sometimes, the binding curve exhibits complexities and two parameters (k, D_f) are not sufficient to adequately describe the binding kinetics. This is further corroborated by low values of r^2 factor (goodness-of-fit). In that case, one resorts to a dual-fractal analysis (four parameters; k_1, k_2, D_{f1}, and D_{f2}) to adequately describe the binding kinetics. The single-fractal analysis presented above is thus extended to include two fractal dimensions. At present, the time ($t = t_1$) at which the 'first' fractal dimension 'changes' to the 'second' fractal dimension is arbitrary and empirical. For the most part, it is dictated by the data analyzed and experience gained by handling a single-fractal analysis. A smoother curve is obtained in the 'transition' region, if care is taken to select the correct number of points for the two regions. In this case, the product (antibody–antigen; or analyte–receptor complex, Ab · Ag or analyte–receptor) is given by:

$$(\text{Ab} \cdot \text{Ag}) \approx \begin{cases} t^{(3-D_{\text{f1,bind}})/2} = t^{\text{p1}} & (t < t_1) \\ t^{(3-D_{\text{f2,bind}})/2} = t^{\text{p2}} & (t_1 < t < t_2 = t_c) \\ t^{1/2} & (t > t_c) \end{cases} \tag{10.1c}$$

In some cases, as mentioned above, a triple-fractal analysis with six parameters (k_1, k_2, k_3, D_{f1}, D_{f2}, and D_{f3}) may be required to adequately model the binding kinetics. This is when the binding curve exhibits convolutions and complexities in its shape due to perhaps the very dilute nature of the analyte (in some of the cases to be presented) or for some other reasons. Also, in some cases, a dual-fractal analysis may be required to describe the dissociation kinetics.

10.3 RESULTS

The fractal analysis will be applied to different protein–analyte interactions occurring on biosensor chip surfaces with the specific purpose of trying to help improve the different biosensor performance parameters such as selectivity, sensitivity, reproducibility, specificity, limit of detection, etc. Attempts will be made to relate particularly changes in the fractal dimension on the biosensor chip surface with the changes in the different biosensor performance parameters.

At the outset it should be indicated that alternate expressions for fitting the binding and dissociation data are available that include saturation, first-order reaction, and no diffusional limitations, but these expressions are deficient in describing the heterogeneity that inherently exists on the surface. It is this heterogeneity on the biosensor surface that one is attempting here to relate to the different biosensor performance parameters. More specifically the question we wish to answer is that how can one change the heterogeneity or the fractal dimension, D_f on the biosensor chip surface in order that one may be able to enhance the different biosensor performance parameters.

Other modeling attempts also need to be mentioned. One might justifiably argue that appropriate modeling may be achieved by using a Langmuirian or other approach. The Langmuirian approach may be used to model the data presented if one assumes the presence of discrete classes of sites, for example, double exponential analysis as compared with the single-fractal analysis. Lee and Lee (1995) indicate that the fractal approach has been applied to surface science, for example, adsorption and reaction processes. These authors emphasize that the fractal approach provides a convenient means to represent the different structures and morphology at the reaction surface. These authors also emphasize using the fractal approach to develop optimal structures and as a predictive approach. Another advantage of the fractal technique is that the analyte–receptor association is a complex reaction, and the fractal analysis via the fractal dimension and the rate coefficient provide a useful lumped parameter analysis of the diffusion-limited reaction occurring on a heterogeneous surface.

In a classical situation, to demonstrate fractality, one should make a log–log plot, and one should definitely have a large amount of data. It may be useful to compare the fit to some other forms, such as exponential, involving saturation, etc. At present, we do not present any independent proof or physical evidence of fractals in the examples presented. Nevertheless, we still use fractals and the degree of heterogeneity on the biosensor surface to gain insights into enhancing the different biosensor performance parameters. The fractal approach is a convenient means (since it is a lumped parameter) to make the degree of heterogeneity that exists on the surface more quantitative. Thus, there is some arbitrariness in the fractal approach to be presented. The fractal approach provides additional information about interactions that may not be obtained by a conventional analysis of biosensor data. In this chapter as mentioned above, we are attempting to relate the fractal dimension, D_f or the degree of heterogeneity on the biosensor surface with the different biosensor performance parameters. More specifically, we are interested in finding out how changes in the fractal dimension or the degree of heterogeneity on the biosensor chip surface affect the different biosensor parameters of interest.

Unless specifically mentioned there is no nonselective adsorption of the analyte. In other words, nonspecific binding is ignored. Nonselective adsorption would skew the results obtained very significantly. In these types of systems, it is imperative to minimize this nonselective adsorption. We also do recognize that, in some cases, this nonselective adsorption may not be a significant component of the adsorbed material and that the rate of association, which is of a temporal nature, would depend on surface availability.

Marchesini *et al.* (2006) have recently developed a novel SPR-based biosensor to screen chemicals which exhibit thyroid-disrupting activity. These authors detected two TPs, thyroxine binding globulin (TBG) and rTTR, in solution using a CM5 sensor chip coated with T4 in a Biacore 3000 biosensor system. These authors indicate that T4 is the main hormone of the thyroid system. Lans *et al.* (1993) and Hallgren and Darnerud (2002) indicate that polyhalogenated aromatic hydrocarbons (PHAHs) and several hydroxylated metabolites interact with high affinity with rTTR.

Figure 10.1a shows the binding of 18.2 nM rTTR in solution to T4 immobilized on a CM5 sensor chip using the spacer E (please see Figure 10.1e). A single-fractal analysis is adequate to describe the binding and dissociation kinetics. The values of the binding rate coefficient (k), the fractal dimension (D_f), the dissociation rate coefficient (k_d), and the fractal dimension for the dissociation phase (D_{fd}) are given in Table 10.1. The values of binding and dissociation rate coefficients presented in Table 10.1a were obtained from a regression analysis using Corel Quattro Pro 8.0 (1997) to model the data using eqs. (10.1a) and (10.1b), wherein $(\text{Analyte}-\text{Receptor}) = kt^{(3-D_f)/2}$ for a single-fractal analysis for the binding phase, and $(\text{Analyte}-\text{Receptor}) = kt^{(3-D_{f1})/2}$ and $kt^{(3-D_{f2})/2}$ for a dual-fractal analysis. The binding rate coefficient values presented in Table 10.1a are within 95% confidence values. For example, for the binding of 18.2 nM rTTR in solution to T4 immobilized on the CM5 sensor chip using the spacer E the binding rate coefficient, k is equal to 10.289 ± 0.352. The 95% confidence limit indicates that the k values lie between 9.937 and 10.641. This indicates that the values are precise and significant.

Figure 10.1b shows the binding of 18.2 nM rTTR in solution to T4 immobilized on a CM5 sensor chip using the spacer F (Please see Figure 10.1e). One again, a single-fractal analysis is adequate to describe the binding and the dissociation kinetics. The values of the binding rate coefficient (k), the fractal dimension (D_f), the dissociation rate coefficient (k_d), and the fractal dimension for the dissociation phase (D_{fd}) are given in Table 10.1.

Figure 10.1c shows the binding of 18.2 nM rTTR in solution to T4 immobilized on a CM5 sensor chip using the spacer D (Please see Figure 10.1e). Once again, a single-fractal analysis is adequate to describe the binding and the dissociation kinetics. The values of the binding rate coefficient (k), the fractal dimension (D_f), the dissociation rate coefficient (k_d), and the fractal dimension for the dissociation phase (D_{fd}) are given in Table 10.1.

Figure 10.1d shows the binding of 18.2 nM rTTR in solution to T4 immobilized on a CM5 sensor chip using the spacer C (Please see Figure 10.1e). Once again, a single-fractal analysis is adequate to describe the binding and the dissociation kinetics. The values of the binding rate coefficient (k), the fractal dimension (D_f), the dissociation rate coefficient (k_d), and the fractal dimension for the dissociation phase (D_{fd}) are given in Table 10.1.

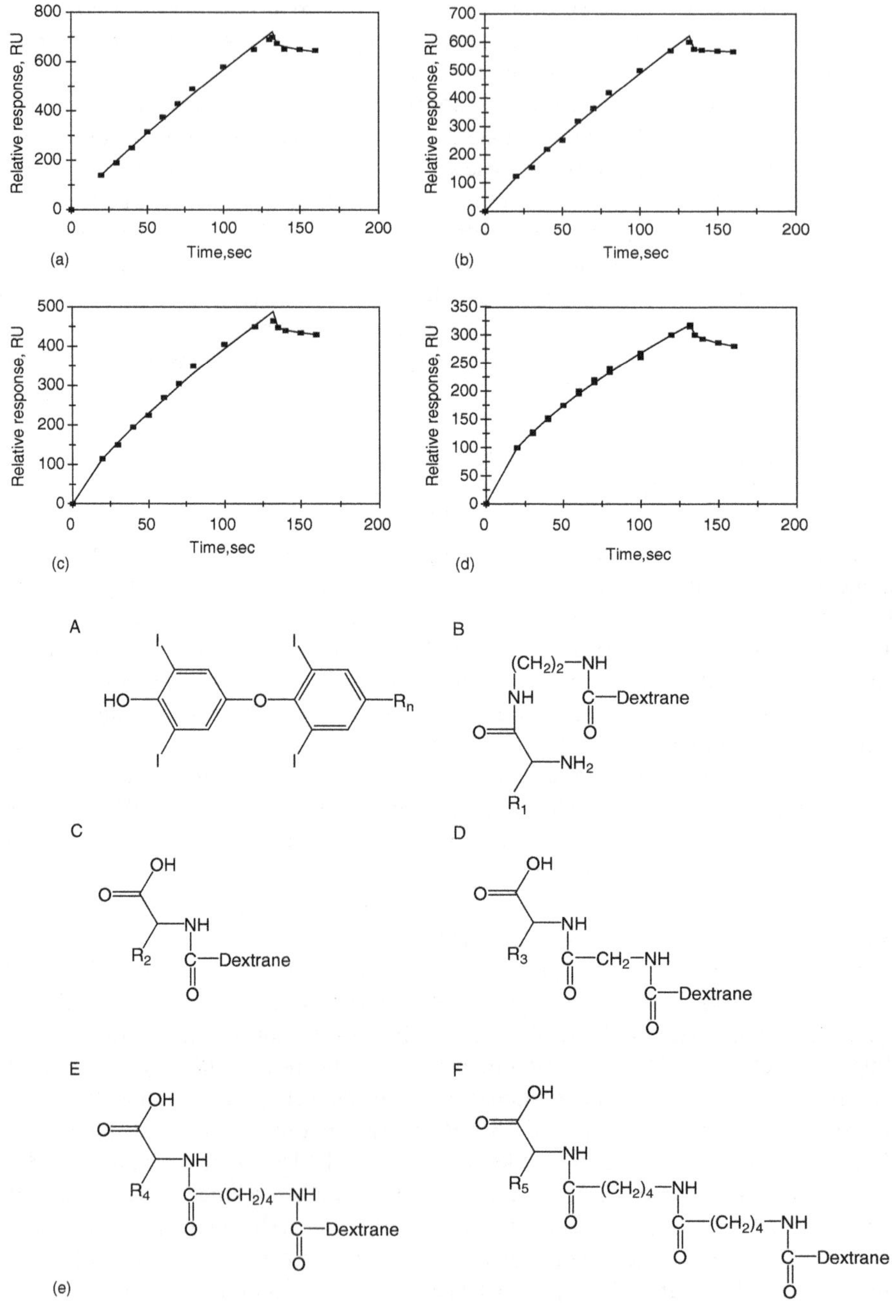

Figure 10.1 Binding and dissociation of 18.2 nM rTTR (recombinant transthyretin) in solution to T4 (*L*-thyroxine) immobilized on a sensor chip surface using different spacers (Marchesini *et al.*, 2006): (a) E; (b) F; (c) D; (d) C. (e) Structures of the different spacers used to immobilize T4 to the carboxymethylated dextran surface of the CM5 biosensor chip (Marchesini *et al.*, 2006).

Table 10.1

Binding and dissociation rate coefficients and fractal dimensions for the binding and the dissociation phases for 18.2 nM rTTR (recombinant transthyretin) in solution to T4 (*L*-thyroxine) immobilized on a CM5 sensor chip with different spacers (E, F, D, and C) (Marchesini *et al.*, 2006)

Spacer	k	k_d	D_f	D_{fd}
E	10.289 ± 0.352	19.760 ± 4.033	1.2586 ± 0.0342	2.3538 ± 0.2182
F	8.4600 ± 0.4197	21.399 ± 0.409	1.7614 ± 0.0529	2.7324 ± 0.0222
D	11.212 ± 0.378	12.334 ± 0.549	1.5466 ± 0.0362	2.3704 ± 0.0512
C	15.646 ± 0.328	9.977 ± 0.150	1.766 ± 0.0226	2.2516 ± 0.0176

Table 10.1 and Figure 10.2a show the increase in the binding rate coefficient, k with an increase in the fractal dimension, D_f. For the data shown in Figure 10.2a and in Table 10.1, the binding rate coefficient, k is given by:

$$k = (6.652 \pm 0.598) D_f^{1.505 \pm 0.302} \tag{10.2a}$$

The fit is good. Only four data points are available. The availability of more data points would lead to a more reliable fit. The binding rate coefficient, k is quite sensitive to the fractal dimension, D_f or the degree of heterogeneity that exists on the surface as noted by the close to one and one-half (equal to 1.505) order of dependence exhibited.

Table 10.1 and Figure 10.2b show the increase in the affinity (k/k_d) with an increase in the ratio of the fractal dimensions (D_f/D_{fd}). For the data shown in Figure 10.2b and in Table 10.1, the affinity (k/k_d) is given by:

$$\frac{k}{k_d} = (2.633 \pm 1.679) \left(\frac{D_f}{D_{fd}} \right)^{2.945 \pm 1.821} \tag{10.2b}$$

The fit is good. Only four data points are available. The availability of more data points would lead to a more reliable fit. The affinity (k/k_d) is sensitive to the ratio of the fractal dimensions, (D_f/D_{fd}) as noted by the close to third (equal to 2.945) order of dependence exhibited.

Komolov *et al.* (2006) have used SPR spectroscopy to analyze protein–protein interactions in real time. These authors developed a simple biosensor-based approach to monitor the interactions between G_t and Rho. Rhodopsin is a GPCR. These authors indicate that GPCRs serve key functions in hormone, neurotransmitter, and sensory signaling. They do indicate that evanescent wave-based biosensor systems have been used previously to analyze G_t–Rho interactions (Bieri *et al.*, 1999; Clark *et al.*, 2001; Karlsson and Lofas, 2002; Minic *et al.*, 2005). However, the results obtained were heterogeneous. Komolov *et al.* (2006) indicate that they have developed a robust biosensor system to analyze G_t–Rho interactions.

Figure 10.3a shows the binding of 0.7 μM G_t in solution to 8.05×10^{10} Rho/mm^2 (density) immobilized on a sensor chip surface. These authors wanted to analyze the influence of different density of Rho immobilized on the sensor surface on the binding of G_t. A dual-

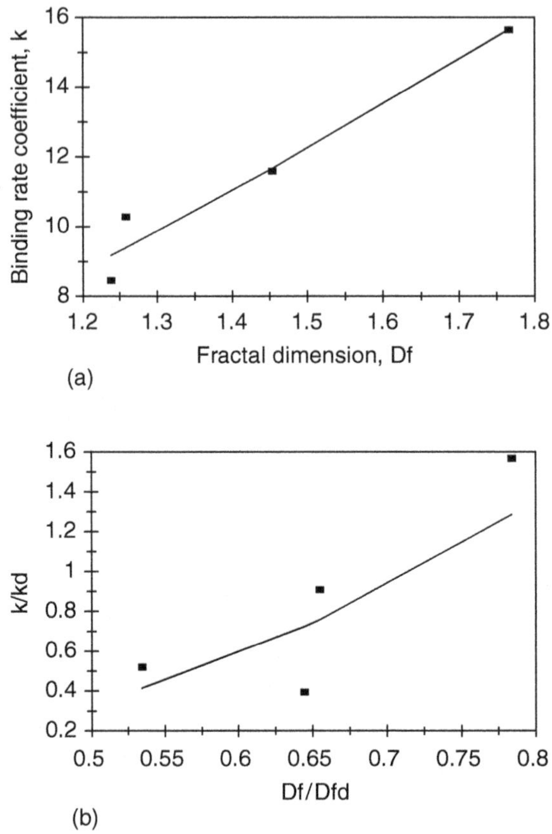

Figure 10.2 (a) Increase in the binding rate coefficient, k with an increase in the fractal dimension, D_f. (b) Increase in the affinity ($=k/k_d$) with an increase in the fractal dimension ratio, D_f/D_{fd}.

fractal analysis is required to adequately describe the binding kinetics. The values of (a) the binding rate coefficient, k and the fractal dimension, D_f for a single-fractal analysis and (b) the binding rate coefficients, k_1 and k_2 and the fractal dimensions, D_{f1} and D_{f2} for a dual-fractal analysis are given in Table 10.2.

It is of interest to note that for a dual-fractal analysis, as the fractal dimension increases by a factor of 1.85 from a value of D_{f1} equal to 1.2082 to D_{f2} equal to 2.2388, the binding rate coefficient value increases by a factor of 14.0 from a value of k_1 equal to 7.445 to k_2 equal to 104.298. Note that changes in the degree of heterogeneity on the biosensor surface (or the fractal dimension value) and in the binding rate coefficient are in the same direction.

Figure 10.3b shows the binding of 0.7 μM G_t in solution to 2.98×10^{10} Rho/mm^2 (density) immobilized on a sensor chip surface. Once again, a dual-fractal analysis is required to adequately describe the binding kinetics. The values of (a) the binding rate coefficient, k and the fractal dimension, D_f for a single-fractal analysis and (b) the binding rate coefficients, k_1 and k_2 and the fractal dimensions, D_{f1} and D_{f2} for a dual-fractal analysis are given in Table 10.2.

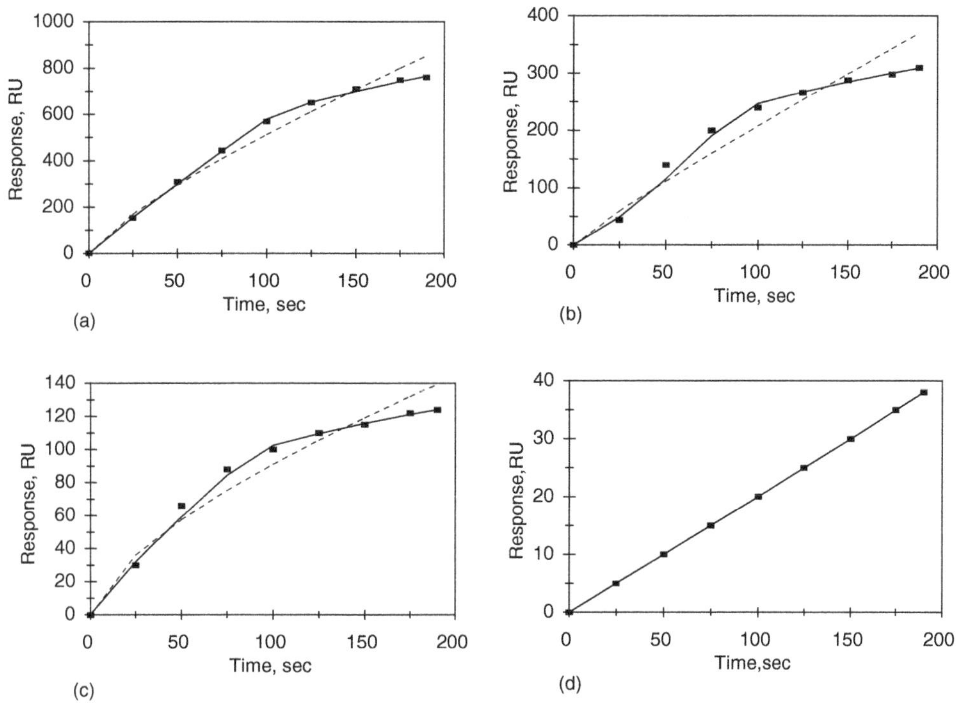

Figure 10.3 Binding of G_t (G protein transducin) in solution to different densities of Rho (rhodopsin) immobilized on a sensor chip: (a) 8.05×10^{10}Rho/mm^2; (b) 2.98×10^{10}Rho/mm^2; (c) 1.4×10^{10}Rho/mm^2; (d) 0.54×10^{10}Rho/mm^2.

Table 10.2

Binding rate coefficients and fractal dimensions for the binding of 0.7 μM G_t (G protein transducin) in solution to different densities of Rho (rhodposin, a prototypical G protein-coupled receptor (GPCR)) immobilized on a sensor chip surface (Komolov *et al.*, 2006)

Rho density (Rho/mm^2)	k	k_1	k_2	D_f	D_{f1}	D_{f2}
8.05×10^{10}	13.445 ± 1.313	7.445 ± 0.199	104.298 ± 1.396	1.4178 ± 0.101	1.2082 ± 0.0508	2.2388 ± 0.0832
2.98×10^{10}	3.287 ± 0.777	0.8957 ± 0.1795	49.399 ± 0.525	1.200 ± 0.231	0.5152 ± 0.0351	2.301 ± 0.0662
1.4×10^{10}	4.297 ± 0.628	1.8504 ± 0.2215	26.106 ± 0.184	1.6734 ± 0.148	1.2294 ± 0.2172	2.4054 ± 0.0442
0.54×10^{10}	0.2 ± 0	na	na	2.0 ± 1.3E−14	na	na

Once again for a dual-fractal analysis, as the fractal dimension increases by a factor of 4.466 from a value of D_{f1} equal to 0.5152 to D_{f2} equal to 2.301, the binding rate coefficient value increases by a factor of 55.15 from a value of k_1 equal to 0.8957 to k_2 equal to 49.399. Note that changes in the degree of heterogeneity on the biosensor surface (or the fractal dimension value) and in the binding rate coefficient are, once again, in the same direction.

Figure 10.3c shows the binding of 0.7 μM G_t in solution to 1.4×10^{10} Rho/mm^2 (density) immobilized on a sensor chip surface. Once again, a dual-fractal analysis is required to adequately describe the binding kinetics. The values of (a) the binding rate coefficient, k and the fractal dimension, D_f for a single-fractal analysis and (b) the binding rate coefficients, k_1 and k_2 and the fractal dimensions, D_{f1} and D_{f2} for a dual-fractal analysis are given in Table 10.2.

Figure 10.3d shows the binding of 0.7 μM G_t in solution to 0.54×10^{10} Rho/mm^2 (density) immobilized on a sensor chip surface. Once again, a dual-fractal analysis is required to adequately describe the binding kinetics. The values of (a) the binding rate coefficient, k and the fractal dimension, D_f for a single-fractal analysis and (b) the binding rate coefficients, k_1 and k_2, and the fractal dimensions, D_{f1} and D_{f2} for a dual-fractal analysis are given in Table 10.2.

Figure 10.4a and Table 10.2 show the increase in the binding rate coefficient, k_1 with an increase in the Rho density (in Rho/mm^2) on the sensor chip surface. For the data shown in Figure 10.4a, the binding rate coefficient, k_1 is given by:

$$k_1 = (0.841 + 1.595)[\text{density of rhodopsin, in Rho/mm}^2]^{0.849 \pm 0.830} \tag{10.3a}$$

There is scatter in the data. This is reflected in the error in the binding rate coefficient. Only the positive value is presented, since the binding rate coefficient cannot have a negative value. Only three data points are available. The availability of more data points would lead to a more reliable fit. The binding rate coefficient, k_1 exhibits less than a first (equal to 0.849) order of dependence on the density of Rho (in Rho/mm^2) on the sensor chip surface.

Figure 10.4b and Table 10.2 show the increase in the binding rate coefficient, k_2 with an increase in the Rho density (in Rho/mm^2) on the sensor chip surface. For the data shown in Figure 10.4a, the binding rate coefficient, k_2 is given by:

$$k_2 = (20.636 + 0.99)[\text{density of rhodopsin, in Rho/mm}^2]^{0.765 \pm 0.0366} \tag{10.3b}$$

The fit is good. Only three data points are available. The availability of more data points would lead to a more reliable fit. The binding rate coefficient, k_2 exhibits less than a first (equal to 0.756) order of dependence on the density of Rho (in Rho/mm^2) on the sensor chip surface. Note that the binding fate coefficient, k_1 exhibits a slightly higher order of dependence than k_2 on the Rho density on the sensor chip surface.

Figure 10.4c and Table 10.2 show the decrease in the fractal dimension, D_{f2} with an increase in the Rho density (in Rho/mm^2) on the sensor chip surface. For the data shown in Figure 10.4c, the fractal dimension, D_{f2} is given by:

$$D_{f2} = (2.424 \pm 0.028)[\text{density of rhodopsin, in Rho/mm}^2]^{-0.0389 \pm 0.00914} \tag{10.3c}$$

The fit is good. Only three data points are available. The availability of more data points would lead to a more reliable fit. The fractal dimension, D_{f2} exhibits only a very slight negative order

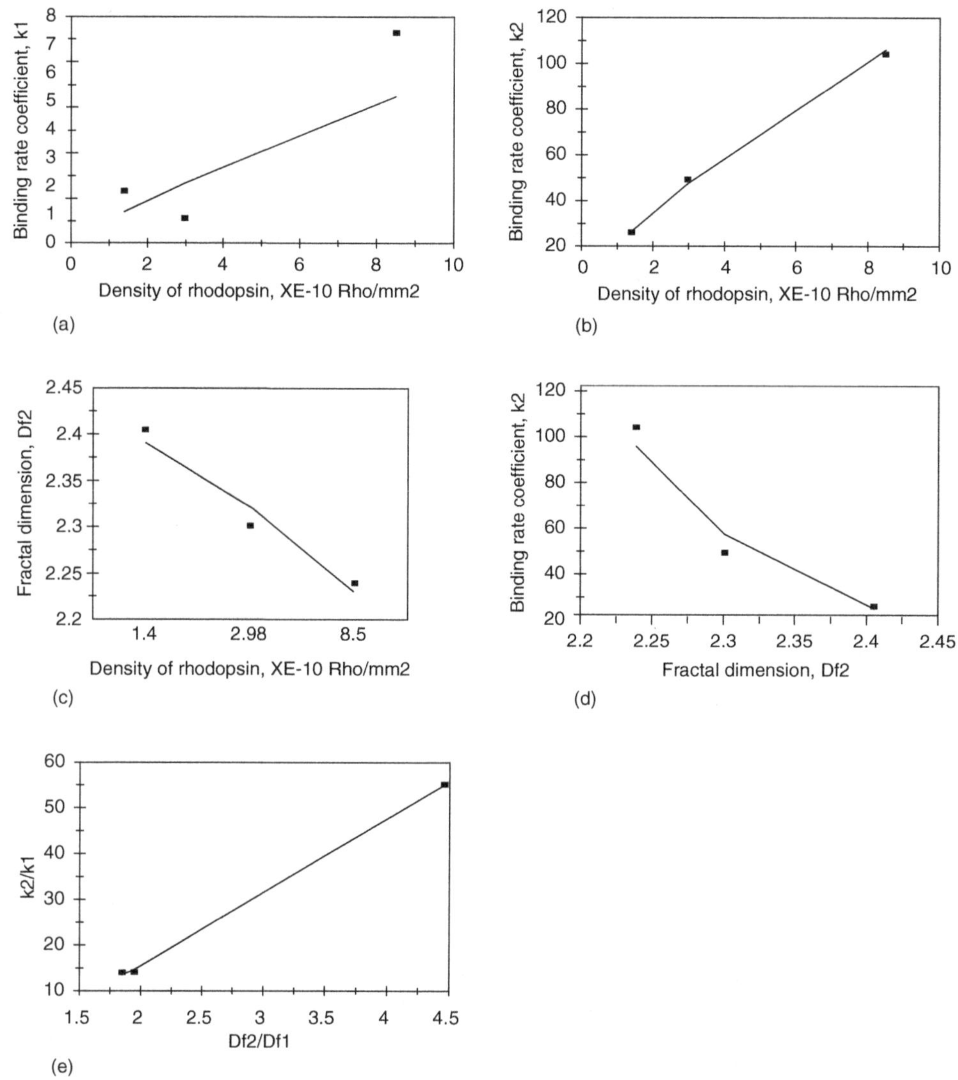

Figure 10.4 (a) Increase in the binding rate coefficient, k_1 with an increase in the rhodopsin density. (b) Increase in the binding rate coefficient, k_2 with an increase in the rhodopsin density. (c) Increase in the fractal dimension, D_{f2} with an increase in the rhodopsin density. (d) Decrease in the binding rate coefficient, k_2 with an increase in the fractal dimension, D_{f2}. (e) Increase in the binding rate coefficient ratio, k_2/k_1 with an increase in the fractal dimension ratio, D_{f2}/D_{f1}.

(equal to -0.0389) of dependence on the Rho density (in Rho/mm^2) on the sensor chip surface. Note that the fractal dimension is based on a log scale, and even a small change in the fractal dimension value on the sensor chip surface leads to a significant change in the degree of heterogeneity on the sensor chip surface.

Figure 10.4d show the decrease in the binding rate coefficient, k_2 with an increase in the fractal dimension, D_{f2}. For the data shown in Figure 10.4d the binding rate coefficient, k_2 is given by:

$$k_2 = (3.7\text{E}+8 \pm 0.7\text{E}+8) D_{f2}^{-18.82 \pm 3.48} \tag{10.3d}$$

The fit is good. Only three data points are available. The availability of more data points would lead to a more reliable fit. The binding rate coefficient, k_2 is extremely sensitive to the fractal dimension, D_{f2} or the degree of heterogeneity that exists on the surface as noted by the close to negative 19th (equal to -18.82) order of dependence exhibited.

Figure 10.4e and Table 10.2 show the increase in the ratio of the binding rate coefficients, k_2/k_1 with an increase in the ratio of the fractal dimensions, D_{f2}/D_{f1}. For the data shown in Figure 10.4e, the ratio of the binding rate coefficients, k_2/k_1 is given by:

$$\frac{k_2}{k_1} = (5.050 \pm 0.285) \left(\frac{D_{f2}}{D_{f1}} \right)^{1.597 \pm 0.0787} \tag{10.3e}$$

The fit is good. Only three data points are available. The availability of more data points would lead to a more reliable fit. The ratio of the binding rate coefficients, k_2/k_1 is sensitive to the ratio of the fractal dimensions, D_{f2}/D_{f1} as noted by the close to one and one-half (equal to 1.597) order of dependence exhibited.

Komolov *et al.* (2006) have also recently used a SPR biosensor to analyze the binding kinetics between G_t in solution to Rho, a prototypical GPCR immobilized on a sensor chip surface. These authors indicate that GPCRs exhibit the property to catalyze GDP/GTP-exchange in downstream interacting heterotrimeric G proteins (Helmrich and Hofman, 1996; Ji *et al.*, 1998; Pierce *et al.*, 2002). Komolov *et al.* (2006) emphasize that GPCRs play a significant role in basic biological and pharmaceutical research.

Figure 10.5a shows the binding of 0.0175 μM G_t in solution to 2.88×10^{10} molecules/mm^2 Rho (density) immobilized on a sensor chip surface. A dual-fractal analysis is adequate to describe the binding kinetics. The values of (a) the binding rate coefficient, k and the fractal dimension, D_f for a single-fractal analysis and (b) the binding rate coefficients, k_1 and k_2 and the fractal dimensions, D_{f1} and D_{f2} for a dual-fractal analysis are given in Table 10.3. It is of significant interest to note that an increase in the fractal dimension from a value of D_{f1} equal to zero (surface acts like a Cantor-like dust) to D_{f2} equal to 1.5498 leads to an increase in the binding rate coefficient by a factor of 22.9 from a value of k_1 equal to 0.00616 to k_2 equal to 0.1409. Note that as indicated elsewhere in different chapters in the book, an increase in the degree of heterogeneity on the sensor chip surface (increase in the fractal dimension) leads to an increase in the binding rate coefficient.

Figure 10.5b shows the binding of 0.035 μM G_t in solution to 2.88×10^{10} molecules /mm^2 Rho (density) immobilized on a sensor chip surface. A single-fractal analysis is adequate to describe the binding kinetics. The values of the binding rate coefficient, k and the fractal dimension, D_f for a single-fractal analysis are given in Table 10.3. No explanation is offered, at present, why a single-fractal analysis is adequate to describe the binding kinetics when 0.035 μM G_t is present in solution wherein a dual-fractal analysis is required to describe the binding kinetics when 0.0175 μM G_t is present in solution.

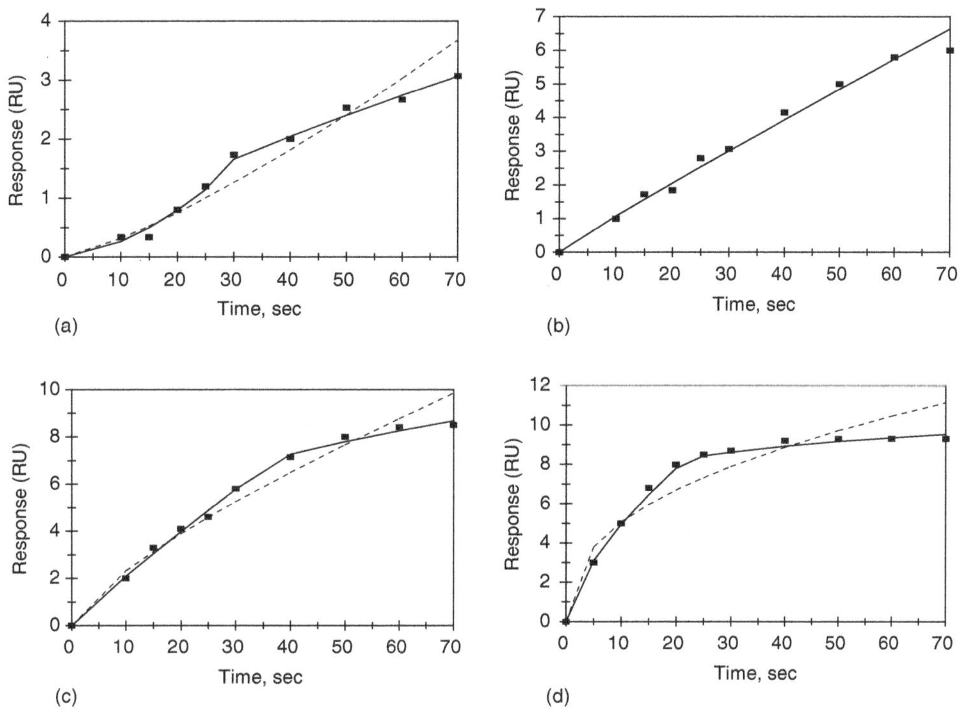

Figure 10.5 Binding of different concentrations (in μM) of G_t (G protein receptor) to 2.8×10^{10} molecules/mm^2 Rho immobilized on a sensor chip surface (Komolov *et al.*, 2006): (a) 0.0175; (b) 0.035; (c) 0.07; (d) 0.14.

Figure 10.5c shows the binding of 0.07 μM G_t in solution to 2.88×10^{10}molecules/mm^2 Rho (density) immobilized on a sensor chip surface. Once again, a dual-fractal analysis is adequate to describe the binding kinetics. The values of (a) the binding rate coefficient, k and the fractal dimension, D_f for a single-fractal analysis and (b) the binding rate coefficients, k_1 and k_2 and the fractal dimensions, D_{f1} and D_{f2} for a dual-fractal analysis are given in Table 10.3. It is of significant interest to note that an increase in the fractal dimension by a factor of 2.05 from a value of D_{f1} equal to 1.1566 to D_{f2} equal to 2.3694 leads to an increase in the binding rate coefficient by a factor of 9.04 from a value of k_1 equal to 0.2515 to k_2 equal to 2.2739.

Figure 10.5d shows the binding of 0.14 μM G_t in solution to 2.88×10^{10}molecules/mm^2 Rho (density) immobilized on a sensor chip surface. Once again, a dual-fractal analysis is adequate to describe the binding kinetics. The values of (a) the binding rate coefficient, k and the fractal dimension, D_f for a single-fractal analysis, and (b) the binding rate coefficients, k_1 and k_2 and the fractal dimensions, D_{f1} and D_{f2} for a dual-fractal analysis are given in Table 10.3. It is of significant interest to note that an increase in the fractal dimension by a factor of 1.66 from a value of D_{f1} equal to 1.6630 to D_{f2} equal to 2.7636 leads to an increase in the binding rate coefficient by a factor of 5.47 from a value of k_1 equal to 1.0543 to k_2 equal to 5.771.

Table 10.3

Binding rate coefficients and fractal dimensions in the binding phases for different concentrations (in μM) of G_t (G protein transducin) in solution to 2.88 × 10^{10} molecules/mm^2 Rho immobilized on a sensor chip surface (Komolov *et al.*, 2006)

G_t concentration in solution, μM	k	k_1	k_2	D_f	D_{f1}	D_{f2}
0.0175	0.0169 ± 0.00462	0.006163 ± 0.002034	0.1409 ± 0.0064	0.4664 ± 0.2606	0+ 0.6576	1.5498 ± 0.2098
0.035	0.1265 ± 0.0111	na	na	1.1350 ± 0.0915	na	na
0.07	0.4194 ± 0.0447	0.2515 ± 0.0168	2.2739 ± 0.0662	1.5142 ± 0.1093	1.1566 ± 0.1495	2.3694 ± 0.1374
0.14	1.9789 ± 0.3276	1.0543 ± 0.0588	5.7710 ± 0.1353	2.1864 ± 0.1220	1.6630 ± 0.08544	2.7636 ± 0.0406

Table 10.3 and Figure 10.6a show for a dual-fractal analysis the increase in the binding rate coefficient, k_1 with an increase in the G_t concentration (in μM) in solution. For the data shown in Figure 10.6a, the binding rate coefficient, k_1 is given by:

$$k_1 = (162.80 \pm 41.14)[G_t, \text{in}\,\mu\text{M}]^{2.502\pm0.150} \tag{10.4a}$$

The fit is good. Only three data points are available. The availability of more data points would lead to a more reliable fit. The binding rate coefficient, k_1 exhibits a two and one-half (equal to 2.502) order of dependence on the G_t concentration in solution. The non-integral order of dependence exhibited, once again, lends support to the fractal nature of the system.

Table 10.3 and Figure 10.6b show for a dual-fractal analysis the increase in the binding rate coefficient, k_2 with an increase in the G_t concentration (in μM) in solution. For the data shown in Figure 10.6b, the binding rate coefficient, k_2 is given by:

$$k_2 = (234.23 \pm 65.17)[G_t, \text{in}\,\mu\text{M}]^{1.817\pm0.164} \tag{10.4b}$$

The fit is good. Only three data points are available. The availability of more data points would lead to a more reliable fit. The binding rate coefficient, k_1 exhibits an order of dependence between one and one-half and two (equal to 1.817) on the G_t concentration in solution. The non-integral order of dependence exhibited, once again, lends support to the fractal nature of the system.

Figure 10.6c shows the decrease in the ratio of the binding rate coefficients, k_2/k_1 with an increase in the G_t concentration (in μM) in solution. For the data given in Figure 10.6c, the ratio of the binding rate coefficients, k_2/k_1 is given by:

$$\frac{k_2}{k_1} = (1.437 \pm 0.03)[G_t, \text{in}\,\mu\text{M}]^{-0.685\pm0.0137} \tag{10.4c}$$

The fit is good. Only three data points are available. The availability of more data points would lead to a more reliable fit. The ratio of the binding rate coefficients, k_2/k_1 exhibits a

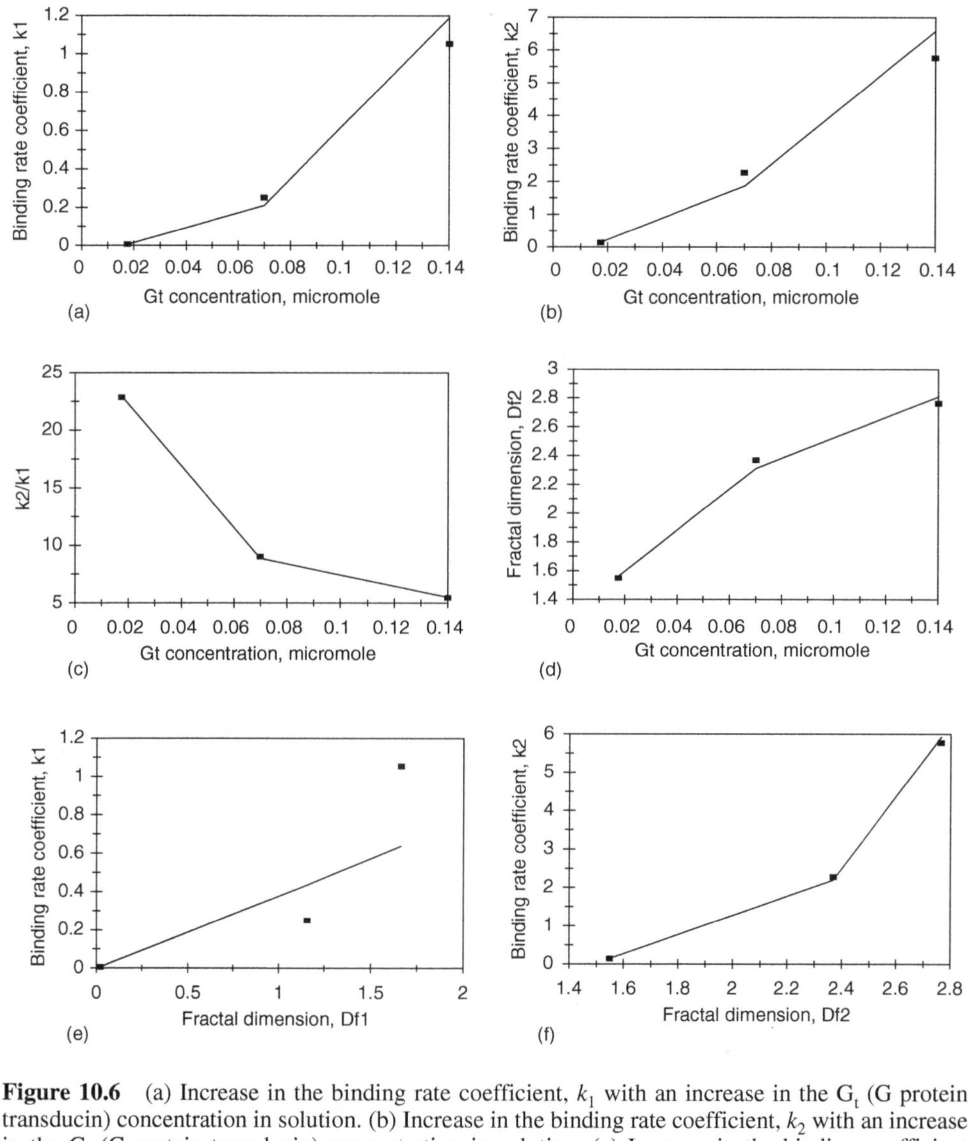

Figure 10.6 (a) Increase in the binding rate coefficient, k_1 with an increase in the G_t (G protein transducin) concentration in solution. (b) Increase in the binding rate coefficient, k_2 with an increase in the G_t (G protein transducin) concentration in solution. (c) Increase in the binding coefficient ratio, k_2/k_1 with an increase in the G_t concentration in solution. (d) Increase in the fractal dimension, D_{f2} with an increase in the G_t concentration in solution. (e) Increase in the binding rate coefficient, k_1 with an increase in the fractal dimension, D_{f1}. (f) Increase in the binding rate coefficient, k_2 with an increase in the fractal dimension, D_{f2}.

negative order between one-half and first (equal to -0.685) on the G_t concentration (in μM) in solution. Once again, the non-integral order of dependence exhibited by the ratio of the binding rate coefficients, k_2/k_1 on the G_t concentration lends support to the fractal nature of the system.

Figure 10.6d shows the increase in the fractal dimension, D_{f2} with an increase in the G_t concentration (in μM) in solution. For the data given in Figure 10.6d, the fractal dimension, D_{f2} is given by:

$$D_{f2} = (4.894 \pm 0.155)[G_t, \text{in}\,\mu\text{M}]^{0.282 \pm 0.021} \tag{10.4d}$$

The fit is good. Only three data points are available. The availability of more data points would lead to a more reliable fit. The fractal dimension, D_{f2} exhibits only a mild order of dependence (equal to 0.282) on the G_t concentration (in μM) in solution. Note that the fractal dimension, D_{f2} is based on a log scale, and even small changes in the fractal dimension may lead to significant changes in the degree of heterogeneity on the sensor chip surface.

Figure 10.6e and Table 10.3 show the increase in the binding rate coefficient, k_1 with an increase in the fractal dimension, D_{f1}. For the data shown in Figure 10.6e, the binding rate coefficient, k_1 is given by:

$$k_1 = (0.427 + 0.578)D_{f1}^{0.618 \pm 0.145} \tag{10.4e}$$

There is scatter in the data. The fit is poor. This is reflected in the error in the estimate of the binding rate coefficient, k_1 presented. Only the positive error is presented, since the binding rate coefficient, k_1 cannot have a negative value. The binding rate coefficient, k_1 exhibits slightly more than one-half (equal to 0.618) order of dependence on the fractal dimension, D_{f1} or the degree of heterogeneity on the sensor chip surface.

Figure 10.6f and Table 10.3 show the increase in the binding rate coefficient, k_2 with an increase in the fractal dimension, D_{f2}. For the data shown in Figure 10.6f, the binding rate coefficient, k_2 is given by:

$$k_2 = (0.00844 \pm 0.00038)D_{f2}^{6.45 \pm 0.105} \tag{10.4f}$$

The fit is good. The binding rate coefficient, k_2 exhibits an extremely high (equal to 6.45) order of dependence on the fractal dimension, D_{f2} or the degree of heterogeneity on the sensor chip surface.

Munoz *et al.* (2005) have recently used a poly(ethylene glycol)-based biosensor chip to analyze heparin–protein interactions. These authors have previously used the SPR biosensor to analyze heparin–protein interactions (Barth *et al.*, 2003; Zhang *et al.*, 2002; Rathore *et al.*, 2001; Dong *et al.*, 2001). Munoz *et al.* (2005) used the Poly(ethylene glycol) (PEG)-based chips since these types of chips show reduced protein adsorption when compared with dextran-based sensor chips (Masson *et al.*, 2004).

Figure 10.7a shows the binding of 60 nM complement protein Factor P in solution to heparin immobilized on a sensor chip surface (Munoz *et al.*, 2005). Factor P is called properdin. These authors indicate that Factor P functions as an enhancing regulator in the complement alternative pathway, and apparently binds to heparin (Holt *et al.*, 1990; Wilson *et al.*, 1984). A single-fractal analysis is adequate to describe the binding and the dissociation kinetics. The values of the binding rate coefficient, k and the fractal dimension, D_f are 4.126 ± 0.374 and 1.2840 ± 0.105, respectively. These values are given in Table 10.4. The dissociation rate coefficient, k_d and the fractal dimension in the dissociation phase, D_{fd} are

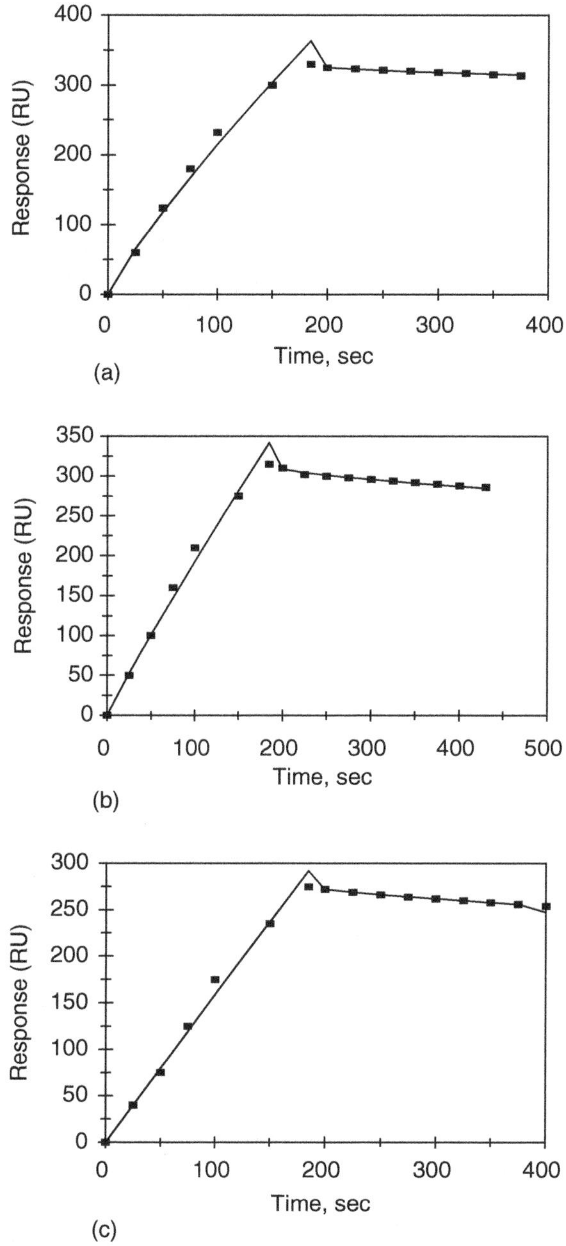

Figure 10.7 Binding of different concentrations of Factor P (in nM) to heparin immobilized on a sensor chip surface (Munoz *et al.*, 2005): (a) 60 nM; (b) 50 nM; (c) 40 nM.

Table 10.4a

Binding and dissociation rate coefficients and fractal dimensions in the binding and in the dissociation phases for different concentrations of Factor P in solution to heparin immobilized on a sensor chip surface (Munoz *et al.*, 2005)

Factor P concentration in solution, nM/heparin	k	k_d	D_f	D_{fd}
60	4.126 ± 0.374	1.245 ± 0.105	1.284 ± 0.105	2.050 ± 0.072
50	2.625 ± 0.214	1.239 ± 0.136	1.134 ± 0.095	1.844 ± 0.080
40	1.664 ± 0.215	0.422 ± 0.007	1.020 ± 0.084	1.551 ± 0.015

Table 10.4b

Affinity, K ($=k/k_d$) values for different concentrations (in nM) of Factor P in solution to heparin immobilized on a sensor chip surface (Munoz *et al.*, 2005)

Factor P concentration (in nM) in solution/heparin	D_f/D_{fd}	K
60	0.626	3.314
50	0.658	3.914
40	0.615	2.114

1.2450 ± 0.105, and 2.0504 ± 0.07172, respectively. In this case, the affinity, K ($=k/k_d$) is equal to 3.314. The values of the binding and dissociation rate coefficients are given in Table 10.4.

Figure 10.7b shows the binding of 50 nM complement protein Factor P in solution to heparin immobilized on a sensor chip surface (Munoz *et al.*, 2005). A single-fractal analysis is, once again, adequate to describe the binding and the dissociation kinetics. The values of the binding rate coefficient, k and the fractal dimension, D_f are 2.6246 ± 0.2145 and 1.1344 ± 0.0952, respectively. The dissociation rate coefficient, k_d and the fractal dimension in the dissociation phase, D_{fd} are 1.2385 ± 0.137 and 1.8438 ± 0.080, respectively. In this case, the affinity, K ($=k/k_d$) is equal to 2.12. These values are given in Table 10.4 above.

Figure 10.7c shows the binding of 40 nM complement protein Factor P in solution to heparin immobilized on a sensor chip surface (Munoz *et al.*, 2005). A single-fractal analysis is, once again, adequate to describe the binding and the dissociation kinetics. The values of the binding rate coefficient, k and the fractal dimension, D_f are 1.6637 ± 0.1197 and 1.0202 ± 0.0842, respectively. The dissociation rate coefficient, k_d and the fractal dimension in the dissociation phase, D_{fd} are 0.4221 ± 0.0074 and 1.5510 ± 0.0155, respectively. In this case, the affinity, K ($=k/k_d$) is equal to 3.941. These values are given in Table10.4 above.

Figure 10.8a shows for a single-fractal analysis the increase in the binding rate coefficient, k with an increase in the complement Factor P concentration in solution. For the data shown in Figure 10.8a, the binding rate coefficient, k is given by:

$$k = (0.000345 \pm 0.000016)[\text{Factor P, in nM}]^{2.232 \pm 0.125} \quad (10.5a)$$

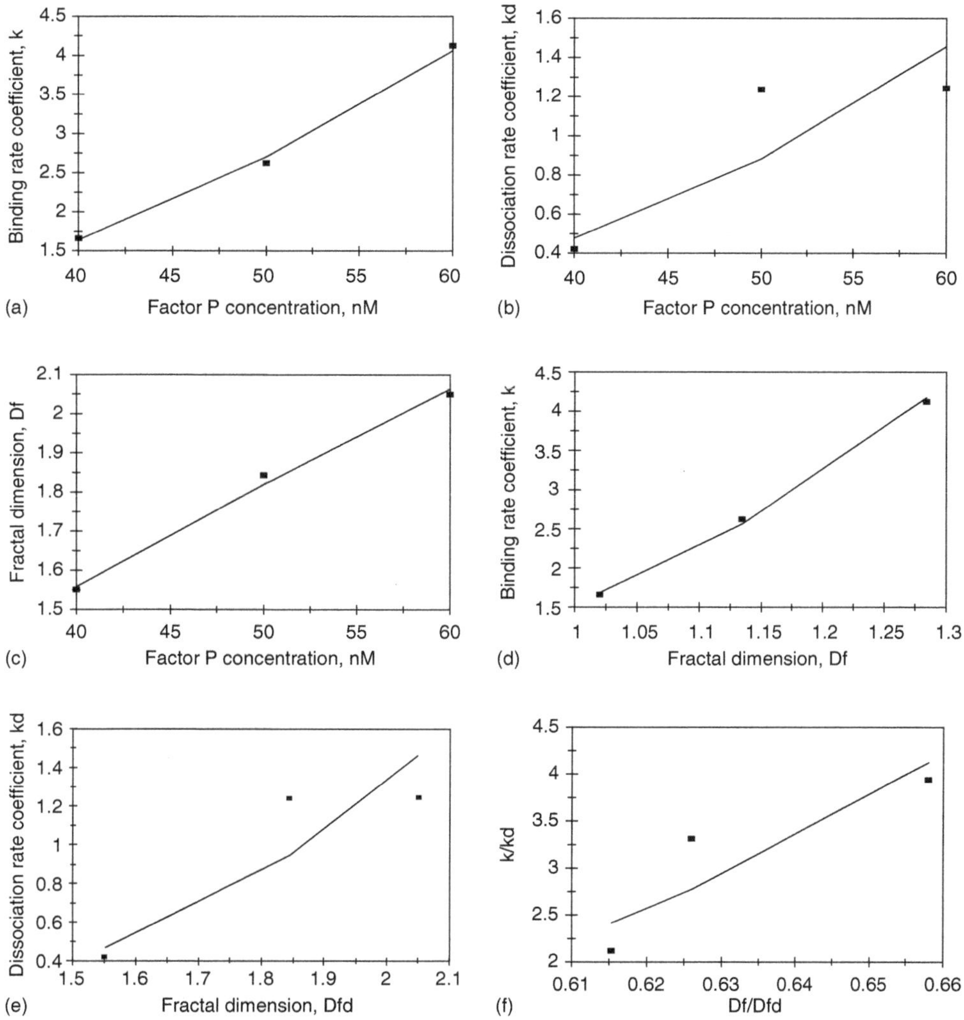

Figure 10.8 (a) Increase in the binding rate coefficient, k with an increase in the Factor P concentration (in nM) in solution. (b) Increase in the dissociation rate coefficient, k_d with an increase in the Factor P concentration (in nM) in solution. (c) Increase in the fractal dimension, D_f with an increase in the Factor P concentration (in nM) in solution. (d) Increase in the binding rate coefficient, k with an increase in the fractal dimension, D_f. (e) Increase in the dissociation rate coefficient, k_d with an increase in the fractal dimension, D_f. (f) Increase in the affinity, K $(=k/k_d)$ with an increase in the ratio of fractal dimensions, D_f/D_{fd}.

The fit is good. Only three data points are available. The availability of more data points would lead to a more reliable fit. The binding rate coefficient, k exhibits an order of dependence between two and two and one-half (equal to 2.232) on the Factor P concentration in solution. The non-integer order of dependence exhibited supports the fractal nature of the system.

Figure 10.8b shows for a single-fractal analysis the increase in the dissociation rate coefficient, k_d with an increase in the complement Factor P concentration in solution. For the data shown in Figure 10.8b, the dissociation rate coefficient, k_d is given by:

$$k_d = (1.9\text{E}-05 \pm 1.0\text{E}-05)[\text{Factor P, in nM}]^{2.747\pm1.366} \tag{10.5b}$$

The fit is reasonable. Only three data points are available. The availability of more data points would lead to a more reliable fit. The dissociation rate coefficient, k_d exhibits an order of dependence between two and one-half and three (equal to 2.747) on the Factor P concentration in solution. The non-integer order of dependence exhibited once again supports the fractal nature of the system.

Figure 10.8c shows for a single-fractal analysis the increase in the fractal dimension, D_f with an increase in the complement Factor P concentration (in nM) in solution. For the data shown in Figure 10.8c, the fractal dimension, D_f is given by:

$$D_f = (0.1216 \pm 0.0019)[\text{Factor P, in nM}]^{0.692\pm0.055} \tag{10.5c}$$

The fit is good. Only three data points are available. The availability of more data points would lead to a more reliable fit. The fractal dimension, D_f is only mildly sensitive to the Factor P concentration in solution as noted by the order of dependence between one-half and first (equal to 0.692) exhibited. Note that, and as indicated previously in this and other chapters in the book, the fractal dimension is based on a log scale, and even small changes in the fractal dimension, D_f indicate a significant change in the degree of heterogeneity on the sensor chip surface.

Figure 10.8d shows the increase in the binding rate coefficient, k with an increase in the fractal dimension, D_f. For the data shown in Figure 10.8d, the binding rate coefficient, k is given by:

$$k = (1.558 \pm 0.048) D_f^{3.941\pm0.185} \tag{10.5d}$$

The fit is very good. Only three data points are available. The availability of more data points would lead to a more reliable fit. The binding rate coefficient, k is very sensitive to the fractal dimension, D_f or the degree of heterogeneity that exists on the sensor chip surface as noted by the close to fourth (equal to 3.941) order of dependence exhibited.

Figure 10.8e shows the increase in the dissociation rate coefficient, k_d with an increase in the fractal dimension, D_{fd}. For the data shown in Figure 10.8e, the dissociation rate coefficient, k_d is given by:

$$k_d = (0.0771 \pm 0.030) D_{fd}^{4.103\pm1.649} \tag{10.5e}$$

The fit is very good. Only three data points are available. The availability of more data points would lead to a more reliable fit. The dissociation rate coefficient, k_d is very sensitive to the fractal dimension, D_{fd} or the degree of heterogeneity that exists on the sensor chip surface in the dissociation phase as noted by the slightly higher than fourth (equal to 4.103) order of dependence exhibited.

Figure 10.8f shows the increase in the affinity, K ($=k/k_d$) with an increase in the ratio of fractal dimensions present in the binding and in the dissociation phases, respectively. For the data given in Figure 10.8f, the affinity, K ($=k/k_d$) is given by:

$$K\left(=\frac{k}{k_d}\right)=(115.42\pm29.3)\left(\frac{D_f}{D_{fd}}\right)^{7.96\pm4.59} \tag{10.5f}$$

The fit is reasonable. There is some scatter in the data. Only three data points are available. The availability of more data points would lead to a more reliable fit. The affinity, K is extremely sensitive to the ratio of the fractal dimensions, D_{f1}/D_{fd} as noted by the close to eighth (equal to 7.96) order of dependence exhibited.

Huber *et al.* (2006) have recently used microcantilevers to analyze the binding of transcription factors using complementary oligonucleotides. Figure 10.9a shows the binding of

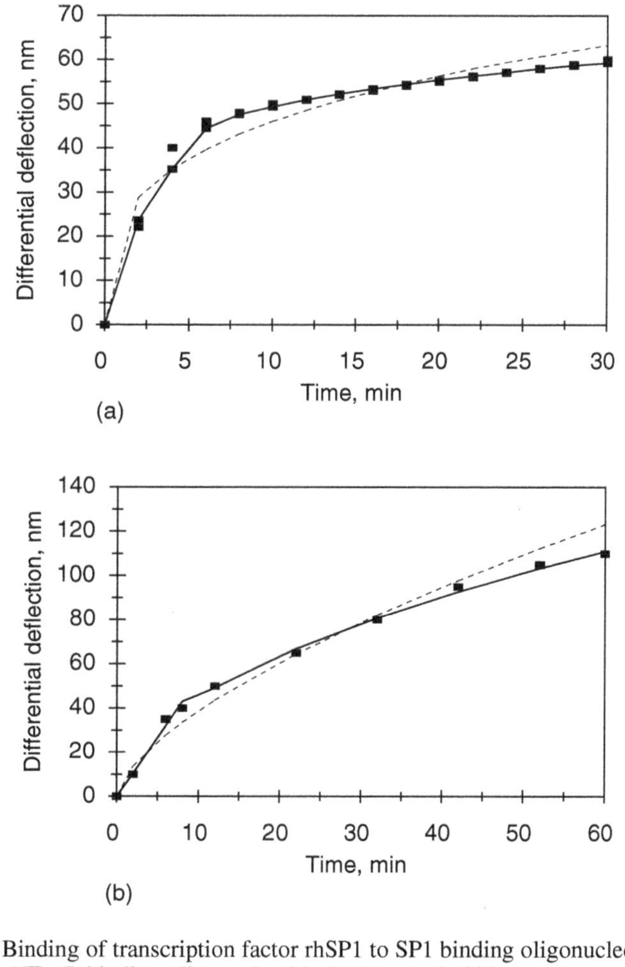

Figure 10.9 (a) Binding of transcription factor rhSP1 to SP1 binding oligonucleotide. (b) Binding of rhNF-κB to the NF-κB binding oligonucleotide (Huber *et al.*, 2006).

Table 10.5

Fractal dimensions and binding rate coefficients for the binding of (a) rhSP1 in SP1 binding buffer to the double-stranded SP1 binding oligonucleotide and rhNF-κB in NF-κB binding buffer to the double-stranded NF-κB binding oligonucleotide, (b) comparison of binding of 80 nM rhSP1 with rhNF-κB, and (c) different concentrations of rhNF-κB in NF-κB binding buffer to the double-stranded NF-κB binding oligonucleotide immobilized on a microarray surface (Huber *et al.*, 2006)

Analyte in solution/receptor on surface	k	k_1	k_2	D_f	D_{f1}	D_{f2}
(a) rhSP1/SP1 oligonucleotide	23.616 ± 2.615	15.829 ± 2.078	33.576 ± 0.272	2.4196 ± 0.0718	1.8464 ± 0.2370	2.6654 ± 0.4228
rhNF-κB/NF-κB oligonucleotide	8.766 ± 1.702	4.961 ± 0.628	13.933 ± 0.322	1.7092 ± 0.109	0.9222 ± 0.231	1.9848 ± 0.0399
(b) 100 nM rhNF-κB/NF-κB oligonucleotide (cantilever functionalized with SP1 binding oligonucleotide used as reference)	20.978 ± 0.226	na	na	0.5924 ± 0.0376	na	na
80 nM rhSP1/SP1 oligonucleotide (cantilever functionalized with rhNF-κB binding oligonucleotide used as reference)	33.081 ± 2.720	na	na	2.4752 ± 0.0527	na	na
(c) 400 nM nM rhNF-κB/NF-κB oligonucleotide	22.545 ± 0.640	na	na	2.153 ± 0.0204	na	na
200 nM nM rhNF-κB/NF-κB oligonucleotide	15.946 ± 0.465	na	na	2.2802 ± 0.0208	na	na
100 nM nM rhNF-κB/NF-κB oligonucleotide	20.00 ± 1.121	na	na	2.5966 ± 0.0434	na	na

transcription factor rhSP1 in solution to the complementary SP1 binding oligonucleotide (5′-GAC ATT CGA <u>TCG GGG CGG GGC</u> GAG CAA AAA GCT <u>CGC CCC GCC CCG</u> ATC GAA TGT-3′) immobilized on a microcantilever array surface. A dual-fractal analysis is required to adequately describe the binding kinetics. The values of (a) the binding rate coefficient, k and the fractal dimension, D_f for a single-fractal analysis and (b) the binding rate coefficients, k_1 and k_2 and the fractal dimensions, D_{f1} and D_{f2} for a dual-fractal analysis are given in Table 10.5. Note that an increase in the fractal dimension by a factor of 1.443 from a value of D_{f1} equal to 1.8464 to D_{f2} equal to 2.6654 leads to an increase in the binding rate coefficient by a factor of 2.12 from a value of k_1 equal to 15.828 to k_2 equal to 33.576. As noted elsewhere in the book, here too, an increase in the fractal dimension or the degree of heterogeneity on the sensor chip surface leads to an increase in the binding rate coefficient.

Figure 10.9b shows the binding of transcription factor rhNF-κB in solution to the complementary NF-κB binding oligonucleotide (5′-GAC ACT TGA <u>GGG GAC TTT CCC</u> AGG CAA AAA GCC TG<u>G GAACAGT CCC</u> CTC AAC TGT C-3′) immobilized on a microcantilever array surface. A dual-fractal analysis is required to adequately describe the binding kinetics. The values of (a) the binding rate coefficient, k and the fractal dimension, D_f for a single-fractal analysis and (b) the binding rate coefficients, k_1 and k_2, and the fractal dimensions, D_{f1} and D_{f2} for a dual-fractal analysis are given in Table 10.5. Note that an increase in the fractal dimension by a factor of 2.15 from a value of D_{f1} equal to 0.922 to D_{f2} equal to 1.9848 leads to an increase in the binding rate coefficient by a factor of 2.81 from a value of k_1 equal to 4.961 to k_2 equal to 13.933. Changes in the fractal dimension or the degree of heterogeneity on the sensor chip surface and in the binding rate coefficient are in the same direction.

Figure 10.10a shows the binding of 100 nM of the transcription factor rhNF-κB in solution to the SP1 binding oligonucleotide (5′-GAC ATT CGA TCG GGG CGG GGC GAG CAA AAA GCT CGC CCC GCC CCG ATC GAA TGT-3′) immobilized on a microcantilever array surface. In this case a single-fractal analysis is adequate to describe the binding kinetics. The values of the binding rate coefficient, k and the fractal dimension, D_f for a single-fractal analysis are given in Table 10.5.

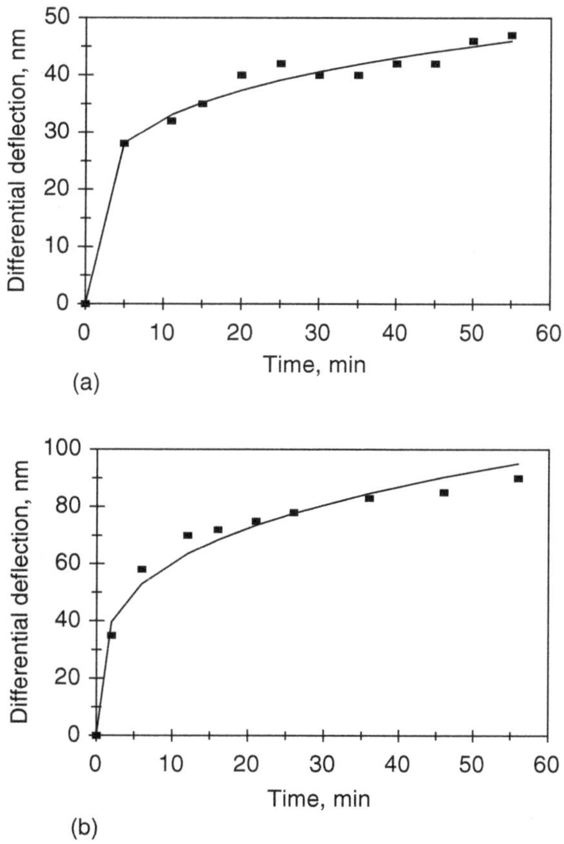

Figure 10.10 (a) Binding of 100 nM rhNF-κB (SP1-binding nucleotide used as reference.) (b) Binding of 80 nM rhSP1 (rhNF-κB binding oligonucleotide used as reference.) (Huber *et al.*, 2006)

Figure 10.10b shows the binding of 80 nM of the transcription factor rhSP1 to the rhNF-κB binding oligonucleotide (5′-GAC ACT TGA GGG GAC TTT CCC AAA GCC TGG GAA AGT CCC CTC AAC TGT C-3′) immobilized on a microcantilever array surface. Once again, a single-fractal analysis is adequate to describe the binding kinetics. The values of the binding rate coefficient, k and the fractal dimension, D_f are given in Table 10.5.

It is of interest to note compare the fractal dimension and the binding rate coefficient values for Figures 10.10a and b. The binding of the transcription factor rhNF-κB to the SP1 oligonucleotide exhibits a higher fractal dimension value (D_f = 2.5924) than that exhibited by the binding of the transcription factor rhSP1 to its complementary oligonucleotide rhNF (D_f = 2.4752). However, as expected, the binding rate coefficient, k exhibited by rhNF to its complementary rhNF binding oligonucleotide (k = 33.081) is higher than that exhibited by rhNF to the SP1 binding oligonucleotide (k = 20.978) by 57.7%. This is in spite of the fact that the fractal dimension (D_f = 2.4752) for the binding to the complementary oligonucleotide rhNF is lower by 4.73% (D_f = 2.5924) for the binding to the SP1 oligonucleotide.

Huber *et al.* (2006) have also analyzed the influence of the transcription factor rhNF-κB concentration in the (100–400 nM) range in solution on its binding kinetics to its target sequence in the complementary NF-κB binding oligonucleotide (5′-GAC ACT TGA GGG GAC TTT CCC AGG CAA AAA GCC TGG GAACAGT CCC CTC AAC TGT C-3′) immobilized on a microcantilever array surface.

Figure 10.11a shows the binding of 400 nM rhNF-κB in solution to its target sequence in the complementary NF-κB binding oligonucleotide (5′-GAC ACT TGA GGG GAC TTT CCC AGG CAA AAA GCC TGG GAACAGT CCC CTC AAC TGT C-3′) immobilized on a microcantilever array surface. A single-fractal analysis is adequate to describe its binding kinetics. The values of the binding rate coefficient, k and its fractal dimension, D_f are given in Table 10.5.

Figure 10.11b shows the binding of 200 nM rhNF-κB in solution to its target sequence in the complementary NF-κB binding oligonucleotide (5′-GAC ACT TGA GGG GAC TTT CCC AGG CAA AAA GCC TGG GAACAGT CCC CTC AAC TGT C-3′) immobilized on a microcantilever array surface. Once again, a single-fractal analysis is adequate to describe its binding kinetics. The values of the binding rate coefficient, k and its fractal dimension, D_f are given in Table 10.5.

Figure 10.11c shows the binding of 100 nM rhNF-κB in solution to its target sequence in the complementary NF-κB binding oligonucleotide (5′-GAC ACT TGA GGG GAC TTT CCC AGG CAA AAA GCC TGG GAACAGT CCC CTC AAC TGT C-3′) immobilized on a microcantilever array surface. Once again, a single-fractal analysis is adequate to describe its binding kinetics. The values of the binding rate coefficient, k and its fractal dimension, D_f are given in Table 10.5.

Figure 10.12 and Table 10.5 show the decrease in the fractal dimension, D_f with an increase in the rhNF-κB concentration (in the 100–400 nM range) in solution. For the data shown in Figure 10.12, the fractal dimension, D_f is given by:

$$D_f = (4.780 \pm 0.144)[\text{rhNF - }\kappa B,\text{in nM}]^{-0.135 \pm 0.030} \tag{10.6}$$

The fit is good. Only three data points are available. The availability of more data points would lead to a more reliable fit. The fractal dimension, D_f is only mildly dependent on

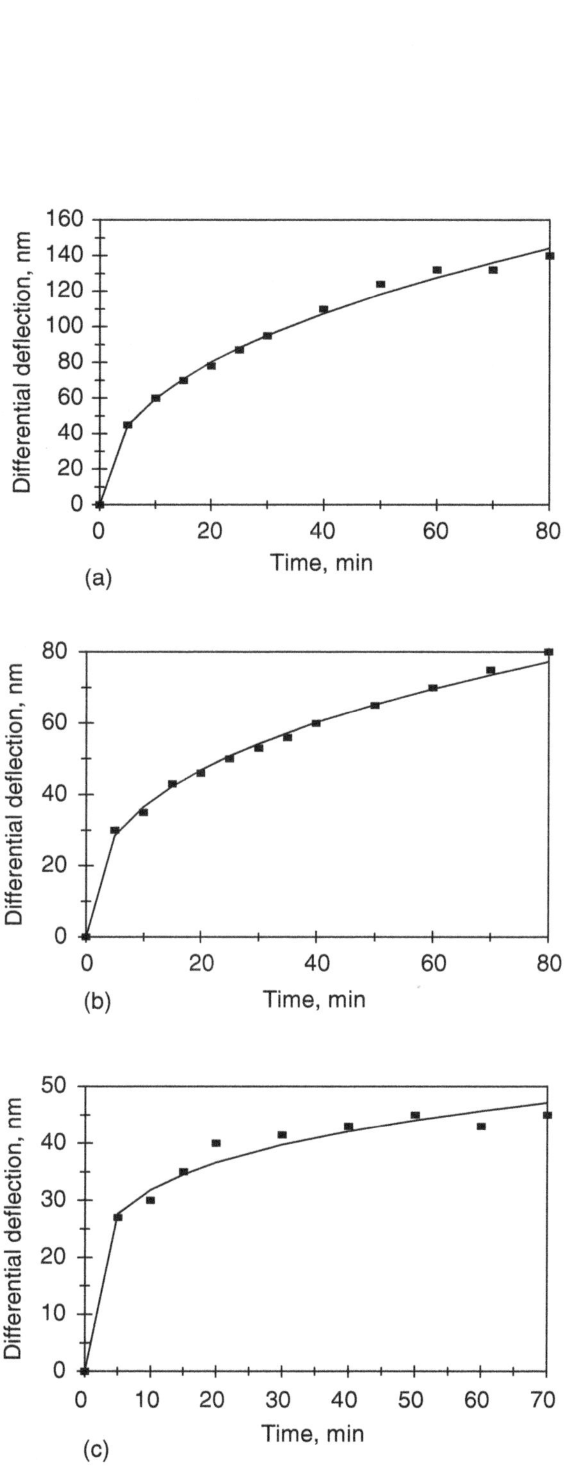

Figure 10.11 (a) Binding of 400 nM rhNF-κB in solution to its target sequence. (b) Binding of 200 nM rhNF-κB in solution to its target sequence. (c) Binding of 100 nM rhNF-κB in solution to its target sequence.

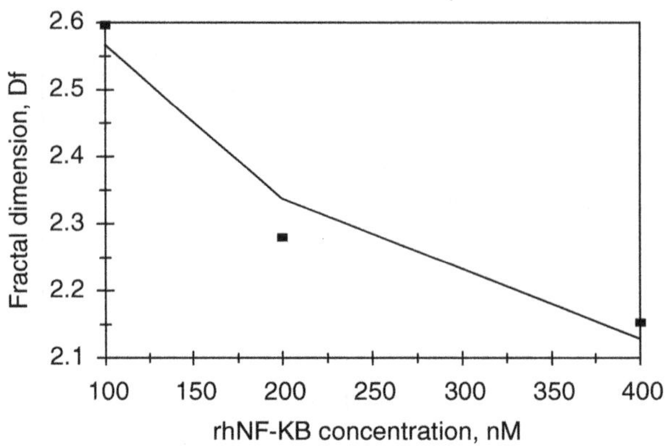

Figure 10.12 Decrease in the fractal dimension, D_f with an increase in the rhNF-κB concentration (in nM) in solution.

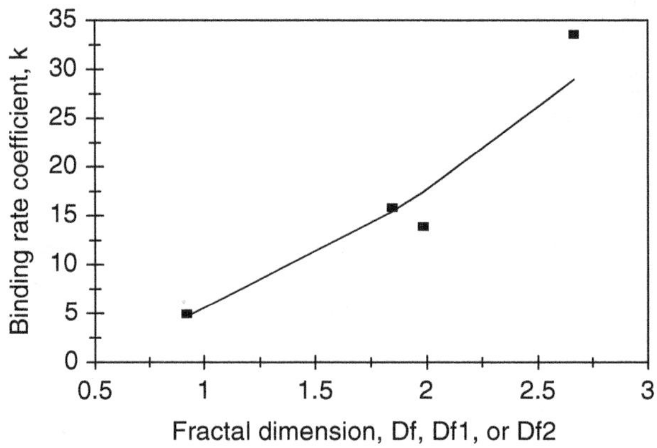

Figure 10.13 Increase in the binding rate coefficient with an increase in the fractal dimension.

the rhNF-κB concentration in solution as noted by the −0.135 order of dependence exhibited. Note that, and as indicated elsewhere in the book, the fractal dimension is based on a log scale, and even small changes in the fractal dimension lead to significant changes in the degree of heterogeneity on the sensor chip surface.

Figure 10.13 and Table 10.5 show the increase in the binding rate coefficient, k with an increase in the fractal dimension, D_f. Since very little data is available, the binding rate coefficients for both the single- and the dual-fractal analysis are plotted together. For the data shown in Figure 10.13, the binding rate coefficient, k is given by:

$$k = (5.398 \pm 1.168)(\text{fractal dimension}, D_f, D_{f1}, \text{or } D_{f2})^{1.714 \pm 0.252} \tag{10.7}$$

The fit is good considering that the binding rate coefficients obtained for a single- and a dual-fractal analysis are plotted together. The binding rate coefficient is sensitive to the fractal dimension or the degree of heterogeneity that exists on the surface as noted by the 1.714 order of dependence exhibited.

10.4 CONCLUSIONS

A fractal analysis is presented for the binding and dissociation (wherever applicable) of protein–analyte interactions occurring on biosensor surfaces. Both, a single- and a dual-fractal analysis are used to model the binding and dissociation kinetics. As indicated elsewhere in the different chapters in the book, the dual-fractal analysis is used only when the single-fractal analysis did not provide an adequate fit, and the regression coefficient was, in general, less than 0.97 (Corel Quattro Pro 8.0, 1997).

Predictive and quantitative relationships are provided for: (a) the binding rate coefficient as a function of the analyte concentration in solution; (b) the ratio of the binding rate coefficients, k_2/k_1 as a function of the analyte concentration in solution; (c) the fractal dimension, D_{f2} as a function of the analyte concentration in solution; (d) the binding rate coefficient, k_1 as a function of the fractal dimension, D_{f1} or the degree of heterogeneity that exists on the sensor chip surface; (e) the binding rate coefficient, k_2 as a function of the fractal dimension, D_{f2} or the degree of heterogeneity that exists on the sensor chip surface; (f) the dissociation rate coefficient, k_d as a function of the analyte concentration in solution; (g) the dissociation rate coefficient, k_d as a function of the fractal dimension in the dissociation phase, D_{fd}; and (h) the affinity ($K = k/k_d$) as a function of the ratio of the fractal dimensions, D_f/D_{fd}. The non-integer order of dependence exhibited by the above relationships on the analyte concentration lends support to the fractal nature of the system.

The fractal analysis presented provides an alternate analysis for determining the binding and the dissociation kinetics of analyte–receptor reactions occurring on biosensor surfaces. The predictive relationships developed for the different analyte–receptor interactions occurring on biosensor surfaces provide a means by which to manipulate the binding and the dissociation rate coefficients and affinity values in desired directions. Biosensorists should pay more attention to the nature of the surface, since it significantly influences biosensor performance parameters such as rate coefficients and affinity values. The biosensor surface represents another source by which one may very significantly affect the kinetics of protein–analyte and other types of reactions of interest occurring on biosensor surfaces. It would be of significant interest to be able to make quantitative the degree of heterogeneity present on the biosensor surface. At present, it is presented as a lumped parameter, and the cause for the heterogeneity on the surface may be due to different reasons, such as surface morphology, heterogeneity in the analyte, heterogeneity in the analyte–receptor complex as the reaction progresses, or for some other reason(s). These types of heterogeneities may be identified and linked to the different biosensor performance parameters of interest.

REFERENCES

Barth, H, C Schaefer, MI Adah, F Zhang, RJ Linhardt, H Toyoda, A Kinoshita-Toyoda, T Toida, TH Van Kuppevelt, E Depla, F Von Weiszsacker, HE Blum, and T Baumert, Cellular binding of hepatitis C virus envelope glycoprotein E2 requires cell surface heparan sulfate. *Biophysical Chemistry*, 2003, **278**, 41003–41012.

Bieri, C, O Ernst, S Heyse, KP Hofmann, and H Vogel, *Nature Biotechnology*, 1999, **17**, 1105–1108.

Brouver, A, UG Ahlborg, M Van den Berg, LS Birnbaum, ER Boersma, B Bosveld, MS Denison, LE Gray, L Hagmar, and E Holene, *European Journal of Pharmacology*, 1995, **293**, 1–40.

Brouver, A, DC Morse, MC Lans, AG Schuur, AJ Mark, E Klasson-Wehler, A Bergman, and TJ Visser, *Toxicology and Industrial Health*, 1998, **14**, 59–84.

Capilla, I, and RJ Linhardt, Heparin–protein interactions, *Angew. Chem. Int. Ed. Engl.*, 2002, **41**, 390–412.

Clark, WA, X Jian, L Chen, and JK Northup, *Biochemical Journal*, 2001, **358**, 389–397.

Corel Quattro Pro 8.0, Corel Corporation, Ottawa, Canada, 1997.

Dong, J, CA Peters-Libeu, KH Weisgarber, BW Segelke, B Rupp, I Capila, MJ Hernaiz, LA LeBrun, and RJ Linhardt, Interaction of the *N*-terminal domain of apoliprotein E4 with heparin. *Biochemistry*, 2001, **40**, 2826–2834.

Fung, YS and YY Wang, Self-assembled monolayer as the coating in a quartz piezoelectric crystal immunosensor to detect *Salmonella* in aqueous solution. *Analytical Chemistry*, 2001, **73**, 5302–5309.

Hallgren, S and PO Darnerud, *Toxicology*, 2002, **177**, 227–243.

Havlin, S, Molecular diffusion and reaction, in *The Fractal Approach to Heterogeneous Chemistry: Surfaces, Colloids, Polymers* (ed. D. Avnir), Wiley, New York, 1989, pp. 251–269.

Helmrich, EJM and KP Hofman, *Biochimica et Biophysica Acta*, 1996, **1286**, 285–322.

Holt, GD, MK Pamgburn, and V Ginsburg, Properdin binds to sulfatide [Gal(3-SO4)β 1-1 Cer] and has a sequence homology with other proteins that bind sulfated glycoconjugates. *Journal of Biological Chemistry*, 1990, **265**, 2852–2855.

Huber, F, M Hegner, C Gerber, HJ Guntherodt, and HP Lang, Label free analysis of transcription factors using microcantilever arrays. *Biosensors and Bioelectronics*, 2006, **21**, 1599–1605.

Ji, TH, M Grossman, and I Ji, *Journal of Biological Chemistry*, 1998, **273**, 17299–17302.

Karlsson, OP and S Lofas, *Analytical Biochemistry*, 2002, **300**, 1332–138.

Komolov, KE, II Senin, PP Philippov, and KW Koch, Surface plasmon resonance study of G-protein receptor coupling in a lipid bilayer-free system. *Analytical Chemistry*, 2006, **78**, 1228–1234.

Lans, MC, E Klasson-Wehler, M Willemsen, E Meussen, S Safe, and A Brouwer, *Chemico-Biological Interactions*, 1993, **88**, 7–21.

Lee, CK and SL Lee, Multi-fractal scaling analysis of reactions over fractal surfaces. *Surface Science*, 1995, **325**, 294–310.

Liu, Y, X Yu, R Zhao, DH Shangguan, ZY Bo, and GQ Liu, Real-time kinetic analysis of the interaction between immunoglobulin G and histidine using quartz crystal microbalance in solution. *Biosensors and Bioelectronics*, 2003a, **18**, 1419–1427.

Liu, Y, X Yu, R Zhao, DH Shangguan, ZY Bo, and GQ Liu, Quartz crystal microbalance biosensor for real-time monitoring of molecular recognition between protein and small molecular medicinal agents. *Biosensors and Bioelectronics*, 2003b, **19**, 9–19.

Marchesini, GR, E Meulenberg, W Haasnoot, M Miziguchi, and H Irth, Biosensor recognition of thyroid-disrupting chemicals using transport proteins. *Analytical Chemistry*, 2006, **78**, 1107–1114.

Masson, JF, TM Battaglia, YC Kim, A Prakash, S Beaudoin, and KS Booksh, Preparation of analyte-sensitve supports for biochemical sensors. *Talanta*, 2004, **64**, 716–725.

Minic, J, J Grosclaude, J Aioun, MA Persuy, T Gorojankina, R Salesse, E Pajot-Augy, Y Hou, S Helali, N Jaffrezic-Renault, F Bessueille, A Errachid, G Gomila, O Ruiz, and J Samitier, *Biochimica et Biophysica Acta*, 2005, **1724**, 324–332.

Munoz, EM, H Yu, J Hallock, RE Edens, and RJ Linhardt, Poly(ethylene glycol)-based biosensor chip to study heparin–protein interactions. *Analytical Biochemistry*, 2005, **343**, 176–178.

Park, IS, WY Kim, and N Kim, Operational characteristics of an antibody-immobilized QCM system detecting *Salmonella* spp. *Biosensors and Bioelectronics*, 2000, **15**, 167–172.

Pierce, KL, RJ Premont, and RJ Lefokowitz, *Journal of Molecular and Cellular Biology*, 2002, **3**, 639–650.

Ramakrishnan, A and A Sadana, A single-fractal analysis of cellular analyte-receptor binding kinetics using biosensors. *BioSystems*, 2001, **59**, 35–51.

Rathore, D, TF McCutchan, DN Garboczi, T Toida, MJ Hernaiz, LA LeBrun, SC Lang, and RJ Linhardt, Direct measurement of the interactions of glycosaminoglycans and a heparin decasachharide with malaria circumsporozoite protein. *Biochemistry*, 2001, **40**, 11518–11524.

Sadana, A,A fractal analysis for the evaluation of hybridization kinetics in biosensors. *Journal of Colloid and Interface Science*, 2001, **151**(1), 166–177.

Sadana, A, *Fractal Binding and Dissociation Kinetics for Different Biosensor Applications*, Elsevier, Amsterdam, 2005.

Wilson, JG, DT Fearon, RL Stevens, N Seno, and KF Austen, Inhibition of the function of activated properdin by squid chondroitin sulfate E glycosaminoglycan and murine bone marrow-derived mast cell choondroitin sulfate E proteoglycan. *Journal of Immunology*, 1984, **132**, 3058–3063.

Zhang, F, M Fath, and RJ Linhardt, A highly stable covalent conjugated heparin biochip for heparin–protein interactions studies. *Analytical Biochemistry*, 2002, **304**, 271–273.

Zhang, H, R Zhao, Z Chen, DH Shangguan, and G Liu, QCM-FIA with PGMA coating for dynamic interaction study of heparin and antithrombin III. *Biosensors and Bioelectronics*, 2005, **21**, 121–127.

– 11 –

Fractal Analysis of Different Compounds Binding and Dissociation Kinetics on Biosensor Surfaces

11.1 INTRODUCTION

Biosensors are being used effectively in an increasing number of applications. In order to emphasize the different applications in which biosensors may be effectively used, we will analyze the kinetics of binding and dissociation (if applicable) of different analytes in solution to receptors on biosensor surfaces with no commonality among them unlike the common theme presented in Chapters 4–10. The cases presented in this chapter should be treated as just examples, with no specific disposition toward their selection and subsequent kinetic analysis presented in this chapter. In effect, the seven sets of examples analyzed were selected at random. The intent is to provide an 'idea' or 'perspective' of the different types of areas where biosensors have been effectively used to detect appropriate analytes of interest. Hopefully, the examples presented in this chapter may together help stimulate further interest in the application of biosensors to other areas of interest. Surely, the inclusion of more carefully selected examples in this chapter would further widen the applications of biosensors to different areas.

Some of the more recent biosensor applications that have appeared in the biosensor literature in the first part of the year 2007 include: (a) silicate hybrid sol–gel membrane for glucose and ATP detection (Liu and Sun, 2007); (b) detection of food-borne pathogenic bacteria using a novel 16S rDNA-based oligonucleotide signature chip (Eom *et al.*, 2007); (c) carbon nanotube sensor for the detection of aliphatic hydrocarbons (Padigi *et al.*, 2007); (d) surface plasmon resonance (SPR) biosensor with rolling circle amplification (RCA) and nanogold-modified tags for protein detection (Huang *et al.*, 2007); (e) an amperometric phenol biosensor based on laponite clay–chitosan nanocomposite matrix (Fan *et al.*, 2007); and (f) an immunosensor based on optical waveguide lightmode spectroscopy (OWLS) technique for the detection of alfatoxin B1 and ochratoxin A (Adanyi *et al.*, 2007). None of these examples are analyzed using fractal analysis in this chapter.

In this chapter we use fractal analysis to analyze the binding and the dissociation (if applicable) kinetics of: (a) C-reactive protein (CRP) in solution to anti-CRP immobilized on a dual-polarization interferometer (DPI) sensor chip surface (Lin *et al.*, 2006); (b) 100 nM free hK1 (kallikrein1) in solution to M0097-G11 Fab immobilized on a sensor chip surface (Wassaf *et al.*, 2006); (c) anti-CA (carbohydrate antigen) 15-3 antibody in solution to CA 15-3 antigen immobilized on an electrochemical protein chip (Wilson and Nie, 2006); (d) heparanase in solution to biotinylated heparin sulfate glycosaminogen (platelet extract) covalently linked to the surface of a 96-well immunoassay plate (Behzad and Brenchley, 2003); (e) morphine in solution to an imprinted self-assembled molecular thin film (i-SAM) prepared from 1 g/l diethanoilamine (DEA) at pH 7 to a gold substrate sensor surface (Tappura *et al.*, 2006); (f) 1.5×10^{-5} M ricin-Cy3 in solution to lactose immobilized on a sensor chip surface (Dyukova *et al.*, 2005); and (g) human chemokine receptor, CCR5 eluted from a gp120 column to monoclonal 1D4 immobilized on a CM4 sensor chip surface (Navratilova *et al.*, 2006).

Values of the binding and dissociation (if applicable) rate coefficients along with affinity values (if applicable) are provided. Note that the fractal analysis, as indicated earlier, includes the effects of diffusion and the heterogeneity present on the biosensor surface. The fractal analysis may be considered as an alternate analysis to the kinetic analysis presented in the above-mentioned references.

11.2 THEORY

11.2.1 Single-fractal analysis

Binding rate coefficient

Havlin (1989) indicates that the diffusion of a particle (analyte [Ag]) from a homogeneous solution to a solid surface (e.g., receptor [Ab]-coated surface) on which it reacts to form a product (analyte–receptor complex; Ab · Ag) is given by:

$$(\mathrm{Ab}\cdot\mathrm{Ag}) \approx \begin{cases} t^{(3-D_{\mathrm{f,bind}})/2} = t^{\mathrm{p}} & t < t_{\mathrm{c}} \\ t^{1/2} & t > t_{\mathrm{c}} \end{cases} \tag{11.1}$$

Here $D_{\mathrm{f,bind}}$ or D_{f} is the fractal dimension of the surface during the binding step. t_{c} is the crossover value. Havlin (1989) indicates that the crossover value may be determined by $r_{\mathrm{c}}^2 \sim t_{\mathrm{c}}$. Above the characteristic length, r_{c}, the self-similarity of the surface is lost and the surface may be considered homogeneous. Above time, t_{c} the surface may be considered homogeneous, since the self-similarity property disappears, and 'regular' diffusion is now present. For a homogeneous surface where D_{f} is equal to 2, and when only diffusional limitations are present, $p = 1/2$ as it should be. Another way of looking at the $p = 1/2$ case (where $D_{\mathrm{f,bind}}$ is equal to 2) is that the analyte in solution views the fractal object, in our case, the receptor-coated biosensor surface, from a 'large distance.' In essence, in the association process, the diffusion of the analyte from the solution to the receptor surface creates a depletion layer of width $(Đt)^{1/2}$ where $Đ$ is the diffusion constant. This gives rise to the fractal power law, $(\mathrm{Analyte} - \mathrm{Receptor}) \sim t^{(3-D_{\mathrm{f,bind}})/2}$. For the present analysis, t_{c} is arbitrarily chosen and we assume that the value of the t_{c} is not reached. One may consider the approach as an intermediate 'heuristic' approach that may be used in the future to develop an autonomous (and not time-dependent) model for diffusion-controlled kinetics.

Dissociation rate coefficient

The diffusion of the dissociated particle (receptor [Ab] or analyte [Ag]) from the solid surface (e.g., analyte [Ag]–receptor [Ab]) complex coated surface) into solution may be given, as a first approximation by:

$$(\mathrm{Ab}\cdot\mathrm{Ag}) \approx -t^{(3-D_{\mathrm{f,diss}})/2} = -t^{\mathrm{p}}(t > t_{\mathrm{diss}}) \tag{11.2}$$

Here $D_{\mathrm{f,diss}}$ is the fractal dimension of the surface for the dissociation step. This corresponds to the highest concentration of the analyte–receptor complex on the surface. Henceforth, its concentration only decreases. The dissociation kinetics may be analyzed in a manner 'similar' to the binding kinetics.

11.2.2 Dual-fractal analysis

Binding rate coefficient

Sometimes, the binding curve exhibits complexities and two parameters (k, D_{f}) are not sufficient to adequately describe the binding kinetics. This is further corroborated by low values of r^2 factor (goodness-of-fit). In that case, one resorts to a dual-fractal analysis (four parameters; k_1, k_2, D_{f1}, and D_{f2}) to adequately describe the binding kinetics. The single-fractal analysis presented above is thus extended to include two fractal dimensions. At present, the time ($t = t_1$) at which the 'first' fractal dimension 'changes' to the 'second' fractal dimension is arbitrary and empirical. For the most part, it is dictated by the data analyzed and experience gained by handling a single-fractal analysis. A smoother curve is obtained in the 'transition' region, if care is taken to select the correct number of points for the two regions. In this case, the product (antibody–antigen; or analyte–receptor complex, Ab · Ag or analyte–receptor) is given by:

$$(\mathrm{Ab}\cdot\mathrm{Ag}) \approx \begin{cases} t^{(3-D_{\mathrm{f1,bind}})/2} = t^{\mathrm{p1}} & (t < t_1) \\ t^{(3-D_{\mathrm{f2,bind}})/2} = t^{\mathrm{p2}} & (t_1 < t < t_2 = t_{\mathrm{c}}) \\ t^{1/2} & (t > t_{\mathrm{c}}) \end{cases} \tag{11.3}$$

In some cases, as mentioned above, a triple-fractal analysis with six parameters (k_1, k_2, k_3, D_{f1}, D_{f2}, and D_{f3}) may be required to adequately model the binding kinetics. This is when the binding curve exhibits convolutions and complexities in its shape due to perhaps the very dilute nature of the analyte (in some of the cases to be presented) or for some other reasons. Also, in some cases, a dual-fractal analysis may be required to describe the dissociation kinetics.

11.3 RESULTS

A fractal analysis is applied to the binding and dissociation (if applicable) kinetics of different analyte–receptor reactions occurring on different biosensor surfaces. No particular attempt has been made to focus on any particular analyte–receptor interaction or on any particular sensing surface. In essence, the examples presented below are selected in a random fashion.

Lin *et al.* (2006) have recently used a DPI biosensor to analyze homo-polyvalent antibody–antigen interaction kinetic studies. These authors indicate that different types of biosensors have been developed to analyze the binding interactions on sensing surfaces (Ramsay, 1998). Lin *et al.* (2006) emphasize that even with the SPR (a popular) technique one is not able to determine the absolute magnitude or the nature of the molecular events occurring on the sensing surface (Gestwicki *et al.*, 2001; Cross *et al.*, 2000).

Lin *et al.* (2006) indicate that the DPI technique is a new method that permits the analysis of bioaffinity interactions (Cross *et al.*, 2000; Biehle *et al.*, 2004). Lin *et al.* (2006) emphasize that the density and thickness of adsorbed protein layers at the sensor surface (solid)–liquid interface may be determined by addressing the waveguide structure with alternate polarizations. These authors further indicate that the DPI method has recently been used to analyze different types of analyte–receptor interactions (Cross *et al.*, 2000, 2004; Biehle *et al.*, 2004; Swann *et al.*, 2004; Armstrong *et al.*, 2004). Lin *et al.* (2006) have used the DPI method to analyze the binding kinetics of CRP. Tracy *et al.* (2002) have indicated that CRP is an important risk factor for atherosclerosis and coronary heart disease. Blake *et al.* (2003) indicate that CRP is a useful prognostic indicator in acute coronary syndromes, and in predicting future cardiovascular events in even healthy men and women.

Figure 11.1a shows the binding of 40 μ g/ml CRP (a homo-polyvalent antigen) to a monoclonal anti-CRP antibody at the silicon/water interface of DPI (Lin *et al.*, 2006). Homo-polyvalent implies that there are two identical antigenic determinants (epitopes, the regions of an antigen that bind to an antibody) on one polymeric antigen. A dual-fractal analysis is required to adequately describe the binding kinetics. The values of (a) the binding rate coefficient, k and the fractal dimension, D_f for a single-fractal analysis, and (b) the binding rate coefficients, k_1 and k_2 and the fractal dimensions, D_{f1} and D_{f2} for a dual-fractal analysis are given in Table 11.1.

It is of interest to note that as the fractal dimension value increases by a factor of 1.476 from a value of D_{f1} equal to 1.8804 to D_{f2} equal to 2.7752, the binding rate coefficient value increases by a factor of 5.974 from a value of k_1 equal to 0.07769 to k_2 equal to 0.4641. As noted elsewhere in different chapters in this book, in general, an increase in the degree of heterogeneity on the biosensor surface leads to an increase in the binding rate coefficient.

Figure 11.1b shows the binding of 35 μg/ml CRP to a monoclonal anti-CRP antibody at the silicon/water interface of DPI (Lin *et al.*, 2006). Once again, a dual-fractal analysis is required to adequately describe the binding kinetics. The values of (a) the binding rate coefficient, k and the fractal dimension, D_f for a single-fractal analysis, and (b) the binding rate coefficients, k_1 and k_2 and the fractal dimensions, D_{f1} and D_{f2} for a dual-fractal analysis are given in Table 11.1.

It is of interest to note that as the fractal dimension value increases by a factor of 1.398 from a value of D_{f1} equal to 1.9796 to D_{f2} equal to 2.7686, the binding rate coefficient value increases by a factor of 4.92 from a value of k_1 equal to 0.07796 to k_2 equal to

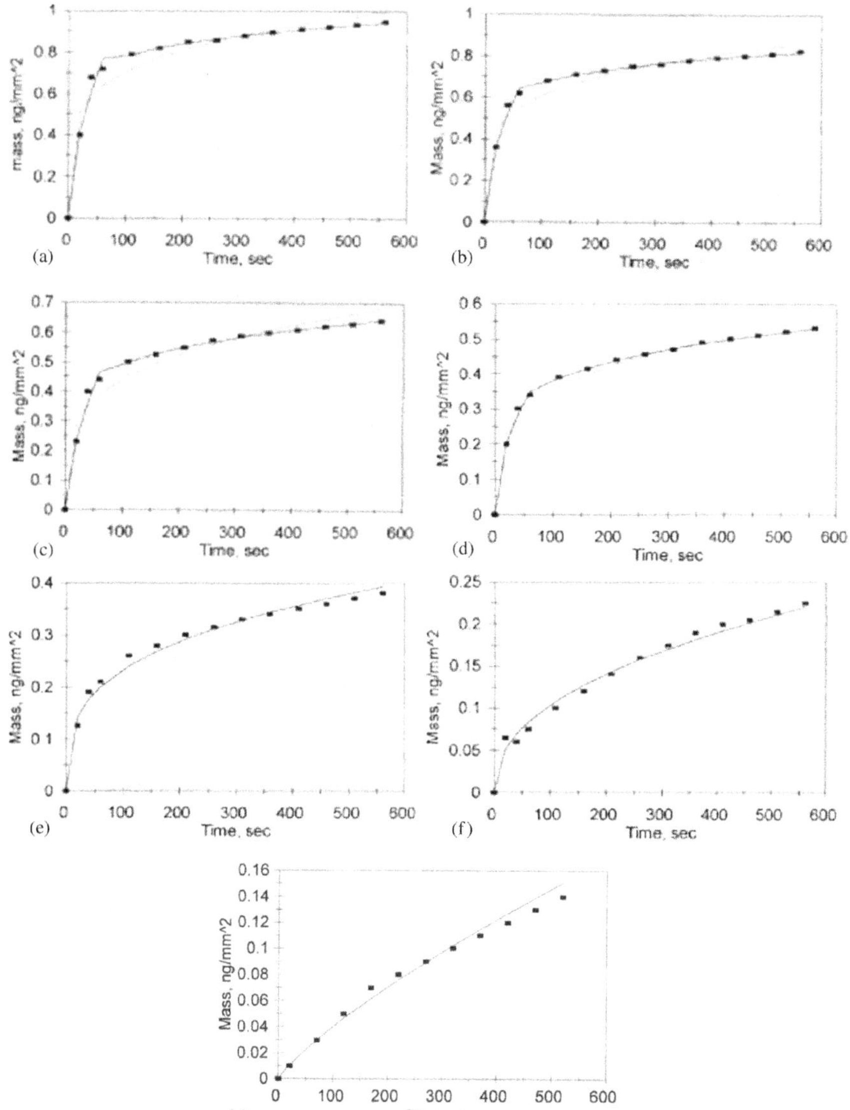

Figure 11.1 Binding of different concentrations (in μg/ml) of CRP (C-reactive protein) in solution to anti-CRP immobilized on a DPI (dual-polarization interferometer) sensor chip (Lin *et al.*, 2006): (a) 40; (b) 35; (c) 30; (d) 25; (e) 20; (f) 15; (g) 10. When only a solid line (___) is used then a single-fractal analysis applies. When a dashed (---) and a solid (___) line is used, then the dashed line represents a single-fractal analysis, and the solid line represents a dual-fractal analysis. In this case the solid line provides the better fit.

0.3938. Once again, an increase in the degree of heterogeneity on the biosensor surface leads to an increase in the binding rate coefficient.

Figures 11.1c and d show the binding of 30 and 25 μg/ml CRP, respectively, in solution to a monoclonal anti-CRP antibody at the silicon/water interface of DPI (Lin *et al.*, 2006).

Table 11.1

Binding rate coefficients and fractal dimensions for different concentrations of CRP (in μg/ml) in solution to anti-CRP antibody immobilized on a DPI (dual-polarization interferometer) sensor chip (Lin *et al.*, 2006)

Analyte in solution (μg/ml)	k	k_1	k_2	D_f	D_{f1}	D_{f2}
40	0.2847 ± 0.0308	0.07769 ± 0.01070	0.4641 ± 0.0020	2.6038 ± 0.0562	1.8804 ± 0.3284	2.7752 ± 0.0005456
35	0.2460 ± 0.0225	0.07996 ± 0.00664	0.3938 ± 0.0016	0.6040 ± 0.04678	1.9796 ± 0.2032	2.7686 ± 0.000494
30	0.1409 ± 0.0154	0.03797 ± 0.00468	0.2446 ± 0.0014	2.5024 ± 0.05676	1.7744 ± 0.2962	2.6938 ± 0.000494
25	0.1094 ± 0.0075	0.04632 ± 0.00272	0.1582 ± 0.0007	2.4890 ± 0.03650	2.0120 ± 0.1452	2.6182 ± 0.005744
20	0.05840 ± 0.00357	na	na	2.3974 ± 0.03236	na	na
15	0.01389 ± 0.00142	na	na	2.1266 ± 0.05296	na	na
10	0.001016 ± 0.000103	na	na	1.4012 ± 0.06232	na	na

In each of these cases too, a dual-fractal analysis is required to adequately describe the binding kinetics. Here too, an increase in the fractal dimension or the degree of heterogeneity on the biosensor surface leads to an increase in the binding rate coefficient.

Figure 11.1e shows the binding of 20 μg/ml CRP in solution to a monoclonal anti-CRP antibody at the silicon/water interface of DPI (Lin *et al.*, 2006). A single-fractal analysis is adequate to describe the binding kinetics. The values of the binding rate coefficient, k and the fractal dimension, D_f for a single-fractal analysis are given in Table 11.1. It is of interest to note that at this lower concentration (20 μg/ml) a single-fractal analysis is adequate to describe the binding kinetics, whereas at the higher CRP concentrations in solution a dual-fractal analysis is required to describe the binding kinetics. This indicates that there is a change in the binding mechanism when comparing the binding of 20 μg/ml CRP in solution with the binding of 25–40 μg/ml CRP in solution. Presumably, at the higher end of the CRP concentration in solution, there is a saturation of the binding sites of the anti-CRP on the air/water interface, and subsequently this leads to a change in the binding mechanism.

Figure 11.1f shows the binding of 15 μg/ml CRP in solution to a monoclonal anti-CRP antibody at the silicon/water interface of DPI (Lin *et al.*, 2006). Once again, a single-fractal analysis is adequate to describe the binding kinetics. The values of the binding rate coefficient, k and the fractal dimension, D_f for a single-fractal analysis are given in Table 11.1.

Figure 11.1g shows the binding of 15 μg/ml CRP in solution to a monoclonal anti-CRP antibody at the silicon/water interface of DPI (Lin *et al.*, 2006). Once again, a single-fractal analysis is adequate to describe the binding kinetics. The values of the binding rate coefficient, k and the fractal dimension, D_f for a single-fractal analysis are given in Table 11.1.

Figure 11.2a and Table 11.1 show for a dual-fractal analysis the increase in the binding rate coefficient, k_1 with an increase in the CRP concentration (in μg/ml) in solution.

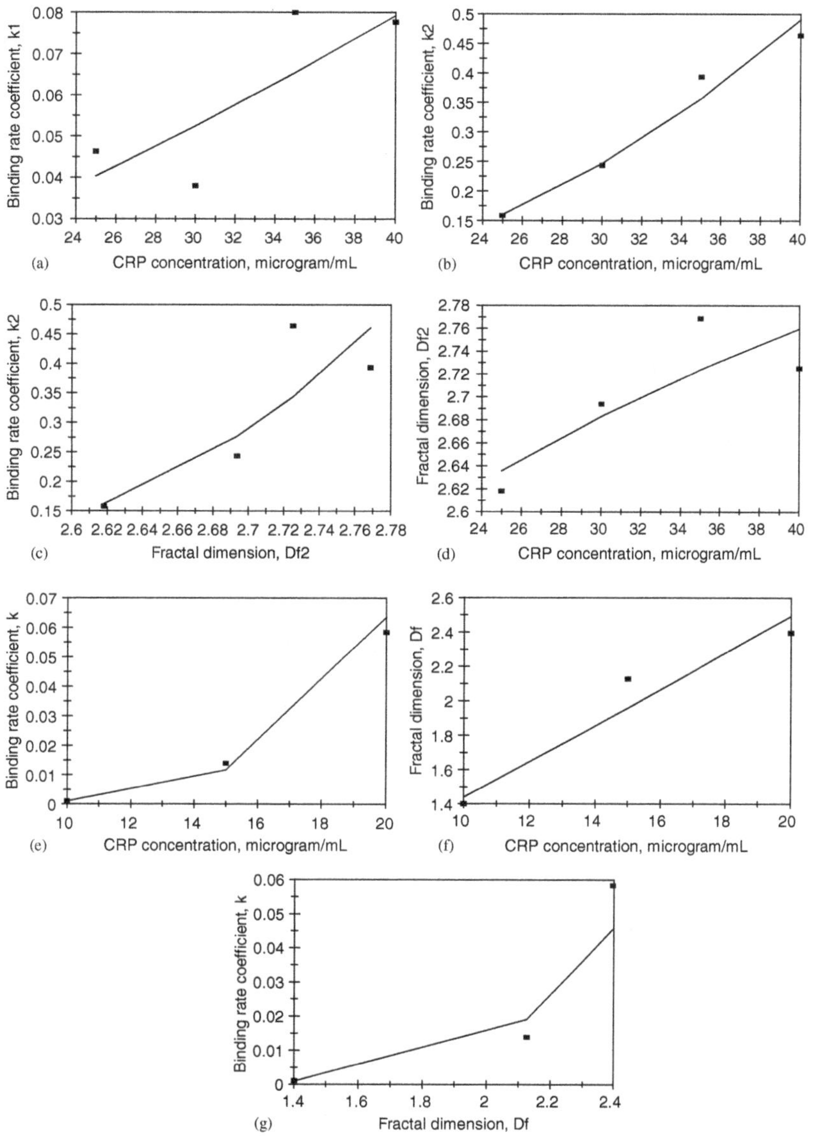

Figure 11.2 (a) Increase in the binding rate coefficient, k_1 for a dual-fractal analysis with an increase in the CRP concentration (in μg/ml) in solution. (b) Increase in the binding rate coefficient, k_2 for a dual-fractal analysis with an increase in the CRP concentration (in μg/ml) in solution. (c) Increase in the binding rate coefficient, k_2 for a dual-fractal analysis with an increase in the fractal dimension, D_{f2}. (d) Increase in the fractal dimension, D_{f2} for a dual-fractal analysis with an increase in the CRP concentration (in μg/ml) in solution. (e) Increase in the binding rate coefficient, k for a single-fractal analysis with an increase in the CRP concentration (in μg/ml) in solution. (f) Increase in the fractal dimension, D_f for a single-fractal analysis with an increase in the CRP concentration (in μg/ml) in solution. (g) Increase in the binding rate coefficient, k for a single-fractal analysis with an increase in the fractal dimension, D_f.

For the 25–40 μg/ml CRP concentration in solution, the binding rate coefficient, k_1 is given by:

$$k_1 = (0.000393 \pm 0.000130)[\text{CRP, in } \mu\text{g/ml}]^{1.44 \pm 0.82} \quad (11.4a)$$

The fit is reasonable. There is scatter in the data. Only four data points are available. The availability of more data points would lead to a more reliable fit. The binding rate coefficient, k_1 exhibits an order of dependence slightly less than one and one-half (equal to 1.44) on the CRP concentration in solution. The fractional order of dependence exhibited by the binding rate coefficient, k_1 on the CRP concentration in solution lends support to the fractal nature of the system.

Figure 11.2b and Table 11.1 show for a dual-fractal analysis the increase in the binding rate coefficient, k_2 with an increase in the CRP concentration (in μg/ml) in solution. For the 25–40 μg/ml CRP concentration in solution, the binding rate coefficient, k_2 is given by:

$$k_2 = (7.4\text{E}-05 \pm 0.6\text{E}-05)[\text{CRP, in } \mu\text{g/ml}]^{2.386 \pm 0.231} \quad (11.4b)$$

The fit is good. Only four data points are available. The availability of more data points would lead to a more reliable fit. The binding rate coefficient, k_2 exhibits an order of dependence between two and two and one-half (equal to 2.386) on the CRP concentration in solution. The fractional order of dependence exhibited by the binding rate coefficient, k_2 on the CRP concentration in solution, once again, lends support to the fractal nature of the system.

Figure 11.2c and Table 11.1 show for a dual-fractal analysis the increase in the binding rate coefficient, k_2 with an increase in the fractal dimension, D_{f2}. For the 25–40 μg/ml CRP concentration in solution, the binding rate coefficient, k_2 is given by:

$$k_2 = (2.6\text{E}-09 \pm 0.8\text{E}-09)D_{f2}^{18.65 \pm 6.31} \quad (11.4c)$$

The fit is reasonable. Only four data points are available. The availability of more data points would lead to a more reliable fit. The binding rate coefficient, k_2 is extremely sensitive to the fractal dimension, D_{f2} or the degree of heterogeneity that exists on the surface as noted by the higher than 18.5 (equal to 18.65) order of dependence exhibited. This extremely high order of dependence exhibited is unusual. No explanation is offered at present to help explain this extremely high order of dependence exhibited.

Figure 11.2d and Table 11.1 show for a dual-fractal analysis the increase in the fractal dimension, D_{f2} with an increase in the CRP concentration (in μg/ml) in solution. For the 25–40 μg/ml CRP concentration in solution, the fractal dimension, D_{f2} is given by:

$$D_{f2} = (1.922 \pm 0.030)[\text{CRP, in } \mu\text{g/ml}]^{0.098 \pm 0.044} \quad (11.4d)$$

The fit is reasonable. Only four data points are available. The availability of more data points would lead to a more reliable fit. The fractal dimension, D_{f2} is only mildly sensitive to the CRP concentration in solution as noted by the 0.098 order of dependence exhibited. Note that the fractal dimension is based on a log scale, and even very small changes in the fractal dimension lead to significant changes in the degree of heterogeneity on the biosensor surface.

Figure 11.2e and Table 11.1 show for a single-fractal analysis the increase in the binding rate coefficient, k with an increase in the CRP concentration (in μg/ml) in solution. For the 25–40 μg/ml CRP concentration in solution, the binding rate coefficient, k is given by:

$$k = (1.4\text{E}-09 \pm 0.3\text{E}-09)[\text{CRP, in}\ \mu\text{g/ml}]^{5.885 \pm 0.405} \tag{11.4e}$$

The fit is good. Only three data points are available. The availability of more data points would lead to a more reliable fit. The binding rate coefficient, k is very sensitive to the to the CRP concentration (in the 0–20 μg/ml range) in solution as noted by the order between five and one-half and six (equal to 5.885) exhibited. Once again, the non-integral order of dependence exhibited by the binding rate coefficient, k on the CRP concentration in solution lends support to the fractal nature of the system.

Figure 11.2f and Table 11.1 show for a single-fractal analysis the increase in the fractal dimension, D_f with an increase in the CRP concentration (in μg/ml) in solution. For the 0–20 μg/ml CRP concentration in solution, the fractal dimension, D_f is given by:

$$D_f = (0.2329 \pm 0.0203)[\text{CRP, in}\ \mu\text{g/ml}]^{0.791 \pm 0.170} \tag{11.4f}$$

The fit is good. Only three data points are available. The availability of more data points would lead to a more reliable fit. The fractal dimension, D_f is somewhat sensitive to the CRP concentration (in the 0–20 μg/ml range) in solution as noted by the order of dependence between one-half and first (equal to 0.791) exhibited. Note that the fractal dimension is based on a log scale, and as indicated above, even very small changes in the fractal dimension lead to significant changes in the degree of heterogeneity on the sensor surface.

Figure 11.2g and Table 11.1 show for a single-fractal analysis the increase in the binding rate coefficient, k with an increase in the fractal dimension, D_f. For the 0–20 μg/ml CRP concentration in solution, the binding rate coefficient, k is given by:

$$k = (8.3\text{E}-05 \pm 4.3\text{E}-09)D_f^{7.212 \pm 1.038} \tag{11.4g}$$

The fit is reasonable. Only three data points are available. There is scatter in the data. This is reflected in the estimated value of the binding rate coefficient. The availability of more data points would lead to a more reliable fit. The binding rate coefficient, k is extremely sensitive to the fractal dimension, D_f or the degree of heterogeneity that exists on the surface as noted by the order of dependence between seven and seven and one-half (equal to 7.212) exhibited. This extremely high order of dependence exhibited is unusual. No explanation is offered at present to help explain this extremely high order of dependence exhibited.

Wassaf *et al.* (2006) have very recently developed a procedure using SPR microarrays to rapidly identify high-affinity human antibodies from phage display library selection outputs. These authors indicate that phage display serves as a useful display tool in affinity chromatography (Kelley *et al.*, 2004; Sato *et al.*, 2002), small-molecule therapeutics (Rodi *et al.*, 2001; Hyde-DeRuyschev *et al.*, 2000), proteomics (Liu *et al.*, 2002; Hust and Dubel, 2004), enzyme engineering (Fernandez-Gacco *et al.*, 2003), and in the development of fully human therapeutic antibodies (Stockwin and Holmes, 2003; Brekke and Loset, 2003). Wassaf *et al.* (2006) emphasize that their procedure combines automated Fab purification using protein A PhyTip columns with high throughput determination of the kinetic constants of Fabs using the Flexchip SPR technology. These PhyTip columns consist of pipette tips that contain 5 μl of a purification resin. This resin such as protein A-Sepharose is encased in hydrophobic screens at the ends of the tips (Chapman, 2005; Bhikabhai *et al.*, 2005). Wassaf *et al.* (2006) have used this approach to identify Fabs from a phage solution campaign against human tissue kallikrein 1 (hK1, KLK1, gene product). Clements *et al.* (2004) indicate that hK1 is a serine protease, and is responsible for the production of Lys-bradykinin. Lys-bradykinin is a potent mediator of inflammation.

Figure 11.3a shows the binding and dissociation of free hK1 (kallikrein 1) in solution to M0097-G11 Fab immobilized on a sensor chip (Wassaf *et al.*, 2006). A dual-fractal analysis is required to adequately describe the binding kinetics. A single-fractal analysis is adequate to describe the dissociation kinetics. The values of (a) the binding rate coefficient, k and the fractal dimension, D_f for a single-fractal analysis, and (b) the binding rate coefficients, k_1 and k_2 and the fractal dimensions, D_{f1} and D_{f2} for a dual-fractal analysis are given in Tables 11.2a and 11.2b. It is of interest to note that as the fractal dimension for binding increases by 63.2% from a value of D_{f1} equal to 1.6216 to D_{f2} equal to 2.6464, the binding

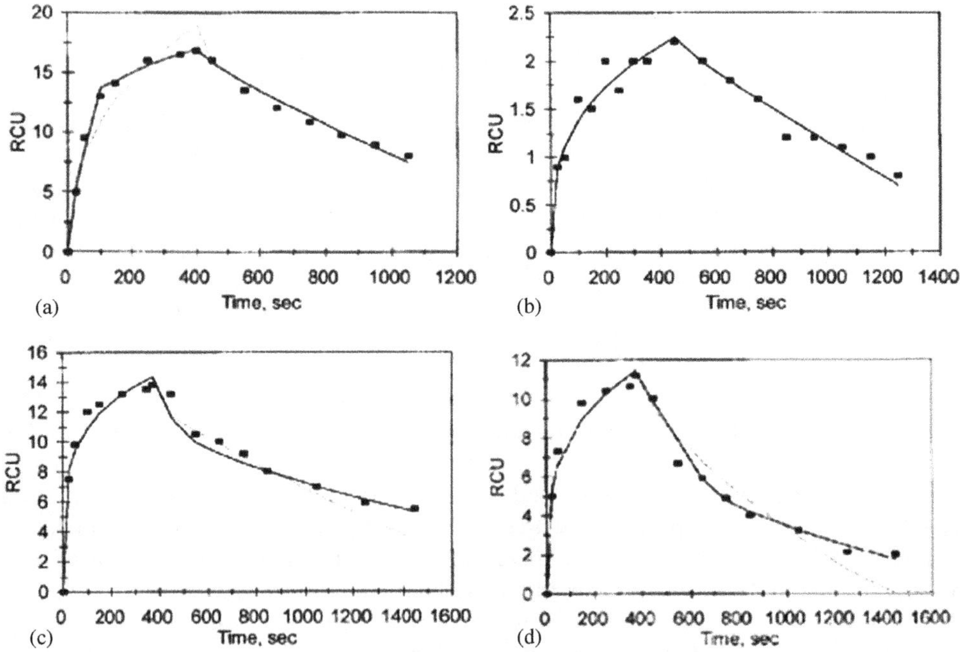

Figure 11.3 (a) Binding of 100 nM free hK1 (kallikrein1) in solution to M0097-G11 Fab immobilized on a sensor chip (Wassaf *et al.*, 2006). (b) Binding of 100 nM hK1-aprotinin complex to M0097-G11 Fab immobilized on a sensor chip (Wassaf *et al.*, 2006). (c) Binding of 100 nM free hK1 (kallikrein1) in solution to M0135 F03 Fab immobilized on a sensor chip (Wassaf *et al.*, 2006). (d) Binding of 100 nM hK1-aprotinin in solution to M0135 F03 Fab immobilized on a sensor chip (Wassaf *et al.*, 2006). When only a solid line (____) is used then a single-fractal analysis applies. When a dashed (---) and a solid (____) line is used, then the dashed line represents a single-fractal analysis, and the solid line represents a dual-fractal analysis. In this case the solid line provides the better fit.

Table 11.2a

Binding and dissociation rate coefficients for free hK1 (kallikrein 1) and hK1-aprotinin in solution to (a) M0097-G11 Fab and (b) M0135-F03 Fab immobilized on a sensor chip surface (Wassaf *et al.*, 2006)

Analyte in solution/ receptor on surface	k	k_1	k_2	k_d	k_{d1}	k_{d2}
(a) 100 nM free hK1/ M0097-G11 Fab	1.7354 ± 0.3072	0.5744 ± 0.0823	5.8814 ± 0.1224	0.03446 ± 0.00674	na	na
100 nM hK1-aprotinin complex/ M0097-G11 Fab	0.3199 ± 0.0323	na	na	0.002877 ± 0.000356	na	na
(b) 100 nM free hK1/ M0135-F03 Fab	4.1913 ± 0.3034	na	na	0.02556 ± 0.009	0.002971 ± 0.009	0.1926 ± 0.0035
100 nM hK1-aprotinin complex/ M0135-F03 Fab	2.3000 ± 0.2107	na	na	0.08504 ± 0.0246	0.08504 ± 0.02461	0.7345 ± 0.0243

Table 11.2b

Fractal dimensions for the binding and the dissociation phase for free hK1 (kallikrein 1) and hK1-aprotinin in solution to (a) M0097-G11 Fab and (b) M0135-F03 Fab immobilized on a sensor chip surface (Wassaf *et al.*, 2006)

Analyte in solution/ receptor on surface	D_f	D_{f1}	D_{f2}	D_{fd}	D_{fd1}	D_{fd2}
(a) 100 nM free hK1/ M0097-G11 Fab	2.2022 ± 0.1289	1.6216 ± 0.2734	2.6464 ± 0.0544	1.2702 ± 0.1464	na	na
100 nM hK1-aprotinin complex/ M0097-G11 Fab	2.3622 ± 0.07114	na	na	1.1290 ± 0.1254	na	na
(b) 100 nM free hK1/ M0135-F03 Fab	2.5940 ± 0.05590	na	na	1.2990 ± 0.2428	0.4490 + 0.6024	1.9288 ± 0.0372
100 nM hK1-aprotinin complex/ M0135-F03 Fab	2.4566 ± 0.0756	na	na	1.5928 ± 0.2170	0.9234 ± 0.4274	2.2668 ± 0.07554

rate coefficient increases by a factor of 10.24 from a value of k_1 equal to 0.5744 to k_2 equal to 5.8814. Once again, as noticed before in this chapter and in general, throughout the book, changes in the degree of heterogeneity on the sensor chip surface and in the binding rate coefficient are in the same direction.

Figure 11.3b shows the binding and dissociation of aprotinin complex in solution to M0097-G11 Fab (Wassaf *et al.* 2006). Hoffman *et al.* (1989) indicate that aprotinin is a Kunitz domain. It is a known site inhibitor of hK1 and other serine proteases. Wassaf *et al.* (2006) indicate that since aprotinin is an active site inhibitor of hK1, Fabs that do not bind to the hK1–aprotinin complex with hK1 may be active site binders and potential inhibitors of hK1 activity. A single-fractal analysis is adequate to describe the binding and the dissociation kinetics. The values of (a) the binding rate coefficient, k and the fractal dimension, D_f for a single-fractal analysis, and (b) the binding rate coefficients, k_1 and k_2 and the fractal dimensions, D_{f1} and D_{f2} for a dual-fractal analysis are given in Tables 11.2a and 11.2b.

Figure 11.3c shows the binding and dissociation of 100 nM free hK1 in solution to an M0135-F03 Fab immobilized on a sensor chip surface (Wassaf *et al.*, 2006). A single-fractal analysis is adequate to describe the binding kinetics. A dual-fractal analysis is required to describe the dissociation kinetics. The values of (a) the binding rate coefficient, k and the fractal dimension, D_f for a single-fractal analysis, (b) the dissociation rate coefficient, k_d and the fractal dimension in the dissociation phase, D_{fd} for a single-fractal analysis, and (c) the dissociation rate coefficients, k_{d1} and k_{d2} and the fractal dimensions in the dissociation phases, D_{fd1} and D_{fd2} are given in Tables 11.2a and 11.2b. It is of interest to note that as the fractal dimension in the dissociation phase increases by a factor of 4.296 from a value of D_{fd1} equal to 0.4490 to D_{fd2} equal to 1.9288, the dissociation rate coefficient increases by a factor of 64.82 from a value of k_{d1} equal to 0.002971 to k_{d2} equal to 0.1926. Note that changes in the degree of heterogeneity in the dissociation phase and in the dissociation rate coefficient are in the same direction.

Note the changes in the binding mechanism as one goes from the binding of 100 nM free hK1 in solution to M0097-G11 Fab immobilized on a sensor chip surface to the binding of 100 nM free hK1 in solution to M0135-F03 Fab immobilized on a sensor chip surface. In the first case a dual-fractal analysis is adequate to describe the binding kinetics, whereas in the second case a single-fractal analysis is adequate to describe the binding kinetics. Note that in both cases, a single-fractal analysis is adequate to describe the dissociation kinetics.

Figure 11.3d shows the binding of 100 nM hK1–aprotinin complex in solution to M0135-F03 Fab immobilized on a sensor chip surface. A single-fractal analysis is adequate to describe the binding and the dissociation kinetics. The values of (a) the binding rate coefficient, k and the fractal dimension, D_f for a single-fractal analysis, and (b) the dissociation rate coefficient, k_d and the fractal dimension in the dissociation phase, D_{fd} for a single-fractal analysis are given in Tables 11.2a and 11.2b.

Figure 11.4a and Tables 11.2a and 11.2b show the increase in the binding rate coefficient, k for a single-fractal analysis with an increase in the fractal dimension, D_f. For the data shown in Figure 11.4a and in Table 11.2a, the binding rate coefficient, k is given by:

$$k = (6E-11 \pm 6E-11)D_f^{26.45 \pm 10.95} \quad (11.5a)$$

The fit is reasonable. Only three data points are available. The availability of more data points would lead to a more reliable fit. The binding rate coefficient, k is extremely sensitive to the fractal dimension, D_f or the degree of heterogeneity that exists on the sensor chip surface as noted by the extremely high order of dependence close to 26.5 (equal to 26.45) exhibited.

Figure 11.4b and Tables 11.2a and 11.2b show the increase in the dissociation rate coefficient, k_{d1} and k_{d2} with an increase in the fractal dimension, D_{fd1} or D_{fd2}, respectively. Very few data points are available for each of these two cases, and thus they are presented

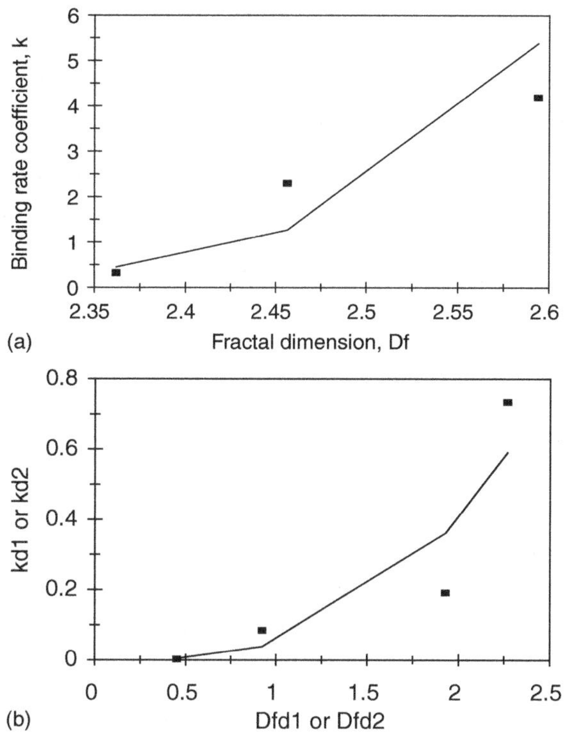

Figure 11.4 (a) Increase in the binding rate coefficient, k with an increase in the fractal dimension, D_f. (b) Increase in the dissociation rate coefficient, k_{d1} or k_{d2} with an increase in the fractal dimension, D_{fd1} or D_{fd2}.

together. For the data shown in Figure 11.4b, the dissociation rate coefficient, k_{d1} or k_{d2} is given by:

$$k_{d1} \text{ or } k_{d2} = (0.0492 + 0.0578)(D_{fd1} \text{ or } D_{fd2})^{3.038 \pm 0.6033} \quad (11.5b)$$

There is scatter in the data, and this is reflected in the error in the dissociation rate coefficient. Only the positive error is presented since the dissociation cannot have a negative value. Note that, and as indicated above, both k_{d1} and k_{d2} are plotted on the same graph due to the lack of points present in each case. This could contribute to the scatter in the data and in the estimated error in the data and in the estimated value of the dissociation rate coefficient. The dissociation rate coefficient, k_{d1} and k_{d2} is quite sensitive to the degree of heterogeneity present in the dissociation phase which is denoted by the D_{fd1} or D_{fd2}, respectively, due to the slightly higher than third (equal to 3.038) order of dependence exhibited.

Wilson and Nie (2006) have recently developed an electrochemical protein chip for the multiplex measurement of seven tumor markers. Maruvuda *et al.* (2005) indicate that the measurement of tumor markers is useful for the early detection of cancer and for the differentiation between benign and malignant conditions. Wilson and Nie (2006) emphasize that the measurement of a single marker is not sufficient to diagnose cancer. These authors, however, emphasize that more than one marker may be associated with the incidence of cancer. The measurement of panels of tumor markers can improve the diagnosis of cancer (Carpelan-Holmstom *et al.*, 2002; Louhimo *et al.*, 2002; Hayakawa *et al.*, 1999; Tsao *et al.*, 2006). Wilson and Nie (2006) emphasize that electrochemical sensors (EIS) offer advantages for performing multi-analyte protein assays. These authors have developed a multiplex EIS that can detect seven tumor markers: AFP (α-fetoprotein), ferritin, CEA (carcinoembryonic antigen), hCG-β (β-human choriogonadotropin), CA 15-3 (carbohydrate antigen), CA 125, and CA 19-9 by a simple to use, robust, sensitive, precise, and accurate method.

Wilson and Nie (2006) have developed a simultaneous electrochemical immunoassay (SEMI). An array of immunosensing electrodes was formulated on a glass substrate. These authors observed that each electrode contained a different immobilized antigen. This antigen was capable of measuring a specific tumor marker using an electrochemical enzyme-based competitive immunoassay.

Figure 11.5a shows the binding of 15 ng/ml anti-CA 15-3 antibody in solution to the CA 15-3 antigen immobilized on the SEMI sensing surface. A single-fractal analysis is adequate to describe the binding kinetics. The values of the binding rate coefficient, k and the fractal dimension, D_f for a single-fractal analysis are given in Table 11.3.

Figure 11.5b shows the binding of 70 ng/ml anti-ferritin antibody in solution to the ferritin antigen immobilized on the SEMI sensing surface. A dual-fractal analysis is required to adequately describe the binding kinetics. The values of (a) the binding rate coefficient, k and the fractal dimension, D_f for a single-fractal analysis, and (b) the binding rate coefficients, k_1 and k_2 and the fractal dimensions, D_{f1} and D_{f2} for a dual-fractal analysis are given in Table 11.3.

Figure 11.5c shows the binding of 15 ng/ml anti-CEA antibody in solution to the CE antigen immobilized on the SEMI sensing surface. A single-fractal analysis is adequate to describe the binding kinetics. The values of the binding rate coefficient, k and the fractal dimension, D_f for a single-fractal analysis are given in Table 11.3.

Figure 11.5d shows the binding of 15 ng/ml anti-CA 125 antibody in solution to the CA 125 antigen immobilized on the SEMI sensing surface. A single-fractal analysis is adequate to describe the binding kinetics. The values of the binding rate coefficient, k and the fractal dimension, D_f for a single-fractal analysis are given in Table 11.3.

Figure 11.5e shows the binding of 300 ng/ml anti-AFP antibody in solution to the AFP antigen immobilized on the SEMI sensing surface. A dual-fractal analysis is required to adequately describe the binding kinetics. The values of (a) the binding rate coefficient, k and the fractal dimension, D_f for a single-fractal analysis, and (b) the binding rate coefficients, k_1 and k_2 and the fractal dimensions, D_{f1} and D_{f2} for a dual-fractal analysis are given in Table 11.3.

It is of interest to note that as the fractal dimension increases by 34.17% from a value of D_{f1} equal to 2.236 to D_{f2} equal to 3.0 (the maximum value possible), the binding rate coefficient increases by a factor of 9.79 from a value of k_1 equal to 6.1742 to k_2 equal to 60.451. Note that changes in the degree of heterogeneity on the sensor surface (or the fractal dimension value) and in the binding rate coefficient are in the same direction.

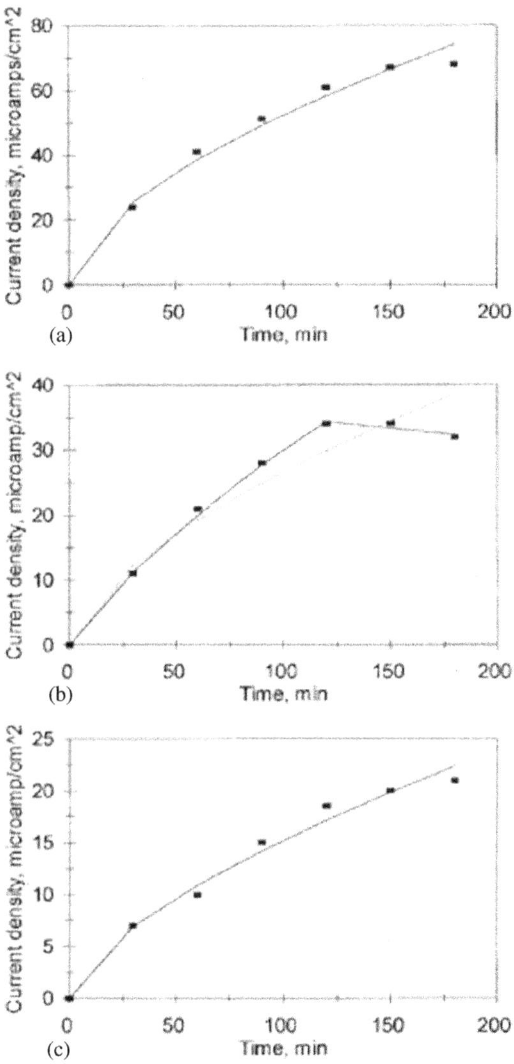

Figure 11.5 (a) Binding of 15 ng/ml anti-CA15-3 antibody in solution to CA15-3 antigen immobilized on an electrochemical protein chip (Wilson and Nie, 2006). (b) Binding of 70 ng/ml anti-Ferritin antibody in solution to Ferritin antigen immobilized on an electrochemical protein chip. (c) Binding of 15 ng/ml anti-CEA antibody in solution to CEA antigen immobilized on an electrochemical protein chip. (d) Binding of 15 ng/ml anti-CA 125 antibody in solution to CA 125 antigen immobilized on an electrochemical protein chip. (e) Binding of 300 ng/ml anti-AFP antibody in solution to AFP antigen immobilized on an electrochemical protein chip. (f) Binding of 25 ng/ml anti-CA19-9 antibody in solution to CA 19-9 antigen immobilized on an electrochemical protein chip. (g) Binding of 12.5 ng/ml anti-hCG-β antibody in solution to hCG-β antigen immobilized on an electrochemical protein. When only a solid line (—) is used then a single-fractal analysis applies. When a (---) and a solid (—) line is used, then the dashed line represents a single-fractal analysis, and the solid line represents a dual-fractal analysis. In this case, the solid line represents the best fit.

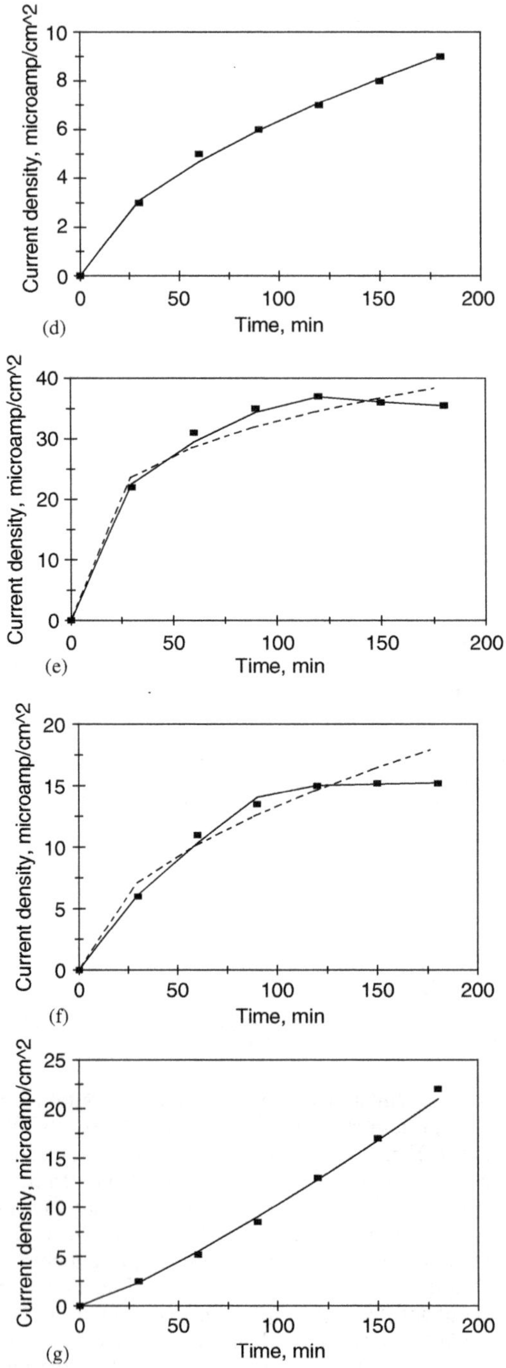

Figure 11.5

Table 11.3

Binding rate coefficients and fractal dimensions for the seven tumor markers on an electrochemical protein chip (Wilson and Nie, 2006)

Analyte in solution/ receptor on surface	k	k_1	k_2	D_f	D_{f1}	D_{f2}
15.3 Anti-CA 15-3 Antibody/CA 15-3	3.376 ± 0.239	na	na	1.8100 ± 0.0923	na	na
70 ng/ml Anti-Ferritin antibody/Ferritin	1.4469 ± 0.232	0.7012 ± 0.0346	68.429 ± 1.886	1.7346 ± 0.200	1.3650 ± 0.9242	0 + 0.1894
15 ng/ml Anti-CEA antibody/CEA	0.7402 ± 0.0546	na	na	1.6868 ± 0.0959	na	na
15 ng/ml Anti-CA 125 antibody/CA 125	0.4098 ± 0.0156	na	na	1.8088 ± 0.503	na	na
300 ng/ml Anti-AFP antibody/AFP	9.4683 ± 0.880	6.1742 ± 0.316	60.451 ± 0.228	2.4548 ± 0.1199	2.236 ± 0.0960	3.0 − 0.0262
25 ng/ml anti-CA19-9 Antibody/CA19-9	1.1542 ± 0.158	0.4743 ± 0.0376	12.791 ± 0.062	1.9564 ± 0.173	1.4944 ± 0.194	2.9327 ± 0.0338
12.5 ng/ml anti-hCG-β/ hCG-β	0.0372 ± 0.00214	na	na	0.5588 ± 0.0750	na	na

Figure 11.5f shows the binding of 25 ng/ml anti-CA 19-9 antibody in solution to the CA 19-9 antigen immobilized on the SEMI sensing surface. A dual-fractal analysis is required to adequately describe the binding kinetics. The values of (a) the binding rate coefficient, k and the fractal dimension, D_f for a single-fractal analysis, and (b) the binding rate coefficients, k_1 and k_2 and the fractal dimensions, D_{f1} and D_{f2} for a dual-fractal analysis are given in Table 11.3.

It is of interest to note that as the fractal dimension increases by 96.2 % from a value of D_{f1} equal to 1.4944 to D_{f2} equal to 2.9327, the binding rate coefficient increases by a factor of 26.97 from a value of k_1 equal to 0.4743 to k_2 equal to 12.791. Note that, once again, changes in the degree of heterogeneity on the sensor surface (or the fractal dimension value) and in the binding rate coefficient are in the same direction.

Figure 11.5g shows the binding of 12.5 ng/ml anti-hCG-β antibody in solution to the hCG-β antigen immobilized on the SEMI sensing surface. A single-fractal analysis is adequate to describe the binding kinetics. The values of the binding rate coefficient, k and the fractal dimension, D_f for a single-fractal analysis are given in Table 11.3.

Figure 11.6a and Table 11.3 show the increase in the binding rate coefficient, k for a single-fractal analysis with an increase in the fractal dimension, D_f. For the data shown in Figure 11.6a the binding rate coefficient, k is given by:

$$k = (0.196 + 0.373) D_f^{2.878 \pm 1.065} \tag{11.6a}$$

The fit is poor. There is scatter in the data. This is reflected in the error in the estimated value of the binding rate coefficient. Only the positive error is presented since the binding rate coefficient cannot have a negative value. Only four data points are available.

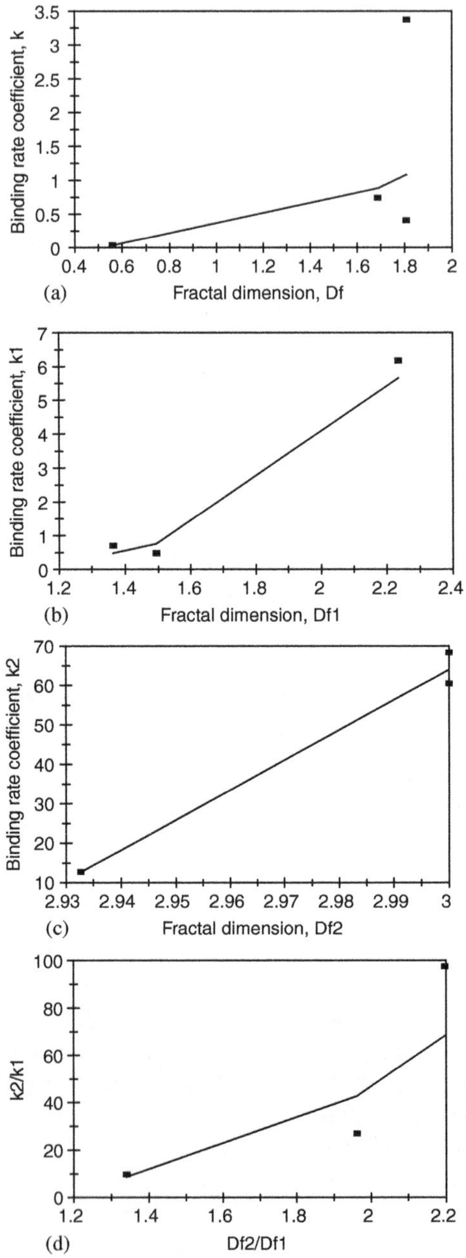

Figure 11.6 (a) Increase in the binding rate coefficient, k with an increase in the fractal dimension, D_f. (b) Increase in the binding rate coefficient, k_1 with an increase in the fractal dimension, D_{f1}. (c) Increase in the binding rate coefficient, k_2 with an increase in the fractal dimension, D_{f2}. (d) Increase in the binding rate coefficient ratio, k_2/k_1 with an increase in the fractal dimension ratio, D_{f2}/D_{f1}.

The availability of more data points would lead to a more reliable fit. The binding rate coefficient, k is quite sensitive to the fractal dimension, D_f or the degree of heterogeneity on the biosensor surface as noted by the order of dependence between two and one-half and three (equal to 2.878) exhibited.

Figure 11.6b and Table 11.3 show the increase in the binding rate coefficient, k_1 for a dual-fractal analysis with an increase in the fractal dimension, D_{f1}. For the data shown in Figure 11.6b, the binding rate coefficient, k_1 is given by:

$$k_1 = (0.1012 \pm 0.0842) D_{f1}^{5.00 \pm 1.63} \tag{11.6b}$$

The fit is good. Only three data points are available. The availability of more data points would lead to a more reliable fit. The binding rate coefficient, k_1 is very sensitive to the fractal dimension, D_{f1} or the degree of heterogeneity that exists on the biosensor surface as noted by the fifth order of dependence exhibited.

Figure 11.6c and Table 11.3 show the increase in the binding rate coefficient, k_2 for a dual-fractal analysis with an increase in the fractal dimension, D_{f2}. For the data shown in Figure 11.6c, the binding rate coefficient, k_2 is given by:

$$k_2 = (7\text{E}-33 \pm 0.6\text{E}-33) D_{f2}^{71.81 \pm 4.73} \tag{11.6c}$$

The fit is good. Only three data points are available. The availability of more data points would lead to a more reliable fit. The binding rate coefficient, k_2 is extremely sensitive to the fractal dimension, D_{f2} or the degree of heterogeneity that exists on the biosensor surface as noted by the higher than 70th (equal to 71.81) order of dependence exhibited.

Figure 11.6d and Table 11.3 show the increase in the ratio of the binding rate coefficients, k_2/k_1 for a dual-fractal analysis with an increase in the ratio of fractal dimensions, D_{f2}/D_{f1}. For the data shown in Figure 11.6d, ratio of the binding rate coefficients, k_2/k_1 is given by:

$$\frac{k_2}{k_1} = (2.605 \pm 2.102) \left(\frac{D_{f2}}{D_{f1}} \right)^{4.152 \pm 1.617} \tag{11.6d}$$

The fit is reasonable. Only three data points are available. The availability of more data points would lead to a more reliable fit. The, ratio of the binding rate coefficients, k_2/k_1 is very sensitive to the ratio of the fractal dimensions, D_{f2}/D_{f1} present on the biosensor surface as noted by the higher than fourth (equal to 4.152) order of dependence exhibited.

Behzad and Brenchley (2003) have developed a multi-well format assay for heparanase. These authors used a biotinylated heparin sulfate glycosaminoglycan (HSGAG) substrate that was covalently linked to the surface of a 96-well immunoassay plate. Behzad and Brenchley (2003) indicate that heparanase is a specific modulator of HSGAG, and thus it plays an important role in normal physiology including embryogeneisis (Ikeda, 2001), angiogenesis (Iozzo and San Antonio, 2001; Vlodavsky and Friedmann, 2001), and the control of cytokine/growth factor expression in the extracellular region (Matsuda *et al.*, 2001; El Assal *et al.*, 2001). Behzad and Brenchley (2003) further indicate that heparanase also appears to be involved in tumor invasion, angiogenesis, and metastasis (El Assal *et al.*, 2001; Vlodarvsky *et al.*, 2000). Behzad and Brenchley (2003) indicate that cleavage of

HSGAG results in the loss of biotin. This biotin is then detected by the reduction in binding of streptavidin-horseradish peroxidase.

Figure 11.7a shows the binding of platelet extract in sodium acetate buffer, pH 5.5, at 37 °C with a dilution factor of 1 (Behzad and Brenchley, 2003). These authors indicate that the decrease in the optical density (OD) represents the removal of the substrate (HSGAG) from the plate due to the enzyme (heparanase) activity. A dual-fractal analysis is required to adequately describe the binding kinetics. The values of (a) the binding rate coefficient, k and the fractal dimension, D_f for a single-fractal analysis, and (b) the binding rate coefficients, k_1 and k_2 and the fractal dimensions, D_{f1} and D_{f2} for a dual-fractal analysis are given in Table 11.4. Note that an increase in the fractal dimension or the degree of heterogeneity on the multi-well surface by 13.47% from a value of D_{f1} equal to 2.6438 to D_{f2} equal to 3.0 (highest possible value) leads to an increase in the binding rate coefficient by a factor of 3.49 from a value of k_1 equal to 0.8460 to k_2 equal to 2.9546. Once again, an increase in the fractal dimension value or the degree of heterogeneity on the multi-well surface leads to an increase in the binding rate coefficient.

Figure 11.7b shows the binding of platelet extract in sodium acetate buffer, pH 5.5, at 37 °C with a dilution factor of 2 (Behzad and Brenchley, 2003). A dual-fractal analysis is,

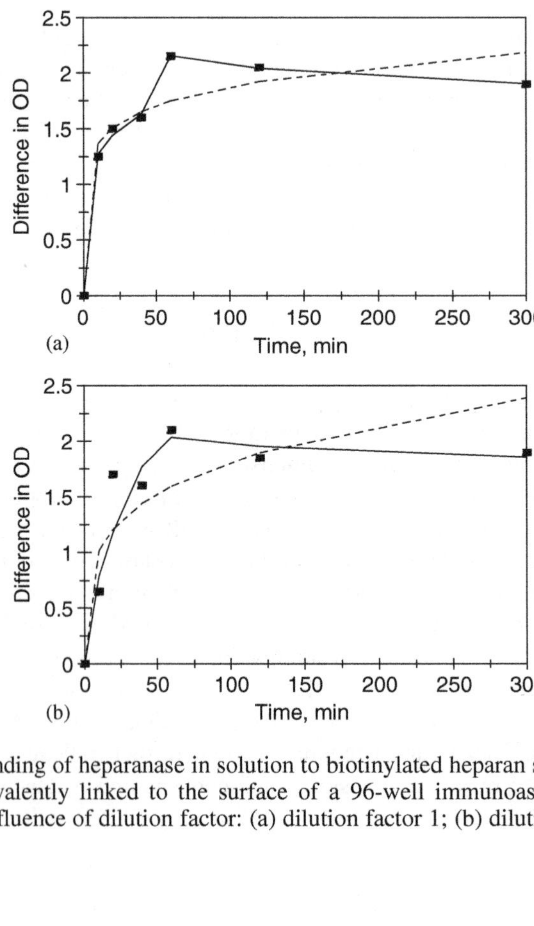

Figure 11.7 (a) Binding of heparanase in solution to biotinylated heparan sulfate glycosaminogen (platelet extract) covalently linked to the surface of a 96-well immunoassay plate (Behzad and Brenchley, 2003). Influence of dilution factor: (a) dilution factor 1; (b) dilution factor 2.

Table 11.4

Effect of dilution factor on the binding rate coefficients and fractal dimensions for heparanase in solution to biotinylated heparin sulfate glycosaminogen (HSGAG) covalently linked to a 96-well immunoassay plate. Dilution factor of platelet extract (Behzad and Brenchley, 2003)

Dilution factor	k	k_1	k_2	D_f	D_{f1}	D_{f2}
1	0.9977 ± 0.1464	0.8460 ± 0.0417	2.9546 ± 0.0132	2.7254 ± 0.100	2.6438 ± 0.0991	3 − 0.0079
2	0.5713 ± 0.0742	0.2094 ± 0.0742	2.5740 ± 0.1813	2.4970 ± 0.2460	1.8434 ± 0.4442	3 − 0.1193

once again, required to adequately describe the binding kinetics. The values of (a) the binding rate coefficient, k and the fractal dimension, D_f for a single-fractal analysis, and (b) the binding rate coefficients, k_1 and k_2 and the fractal dimensions, D_{f1} and D_{f2} for a dual-fractal analysis are given in Table 11.4. Note that an increase in the fractal dimension or the degree of heterogeneity on the multi-well surface by 62.71% from a value of D_{f1} equal to 1.8434 to D_{f2} equal to 3.0 (highest possible value) leads to an increase in the binding rate coefficient by a factor of 12.29 from a value of k_1 equal to 0.2094 to k_2 equal to 2.5740. Once again, an increase in the fractal dimension value or the degree of heterogeneity on the multi-well surface leads to an increase in the binding rate coefficient.

Tappura *et al.* (2006) have recently used lipoate-based imprinted self-assembled molecular thin films to detect morphine. These authors emphasize that molecularly imprinted polymers (MIPs) are being used as more stable, sensitive sensor materials (Wulff, 1995; Haupt and Mosbach, 2000). Gabl *et al.* (2004) indicate that i-SAMs may be considered as a suitable method for using thinner receptor layers. This facilitates the development of more sensitive sensors using the SPR or SAW (surface acoustic wave) sensors. Tappura *et al.* (2006) used bifunctional self-assembling ligands as the basic building blocks for producing mixed monolayers. These authors further add that lipoate was used since it is a very suitable linking group to gold and exhibits no long-range interactions with biomolecules due to the neutral charge in its dithioline ring. Furthermore, morphine is the least substituted drug with the characteristic opiate structure.

Figure 11.8a shows the binding of 0.1 mM morphine in solution to an i-SAM prepared from 1 g/l DEA at pH 7 on a SPR sensor chip surface (Tappura *et al.*, 2006). A single-fractal analysis is adequate to describe the binding kinetics. The values of the binding rate coefficient, k and the fractal dimension, D_f for a single-fractal analysis are given in Table 11.5.

Figure 11.8b shows the binding of 1 mM morphine in solution to an i-SAM prepared from 1 g/l DEA at pH 7 on a SPR sensor chip surface (Tappura *et al.*, 2006). A single-fractal analysis is adequate to describe the binding kinetics. The values of the binding rate coefficient, k and the fractal dimension, D_f for a single-fractal analysis are given in Table 11.5.

Figure 11.8c shows the binding of 10 mM morphine in solution to an i-SAM prepared from 1 g/l DEA at pH 7 on a SPR sensor chip surface (Tappura *et al.*, 2006). A single-fractal analysis is adequate to describe the binding kinetics. The values of the binding

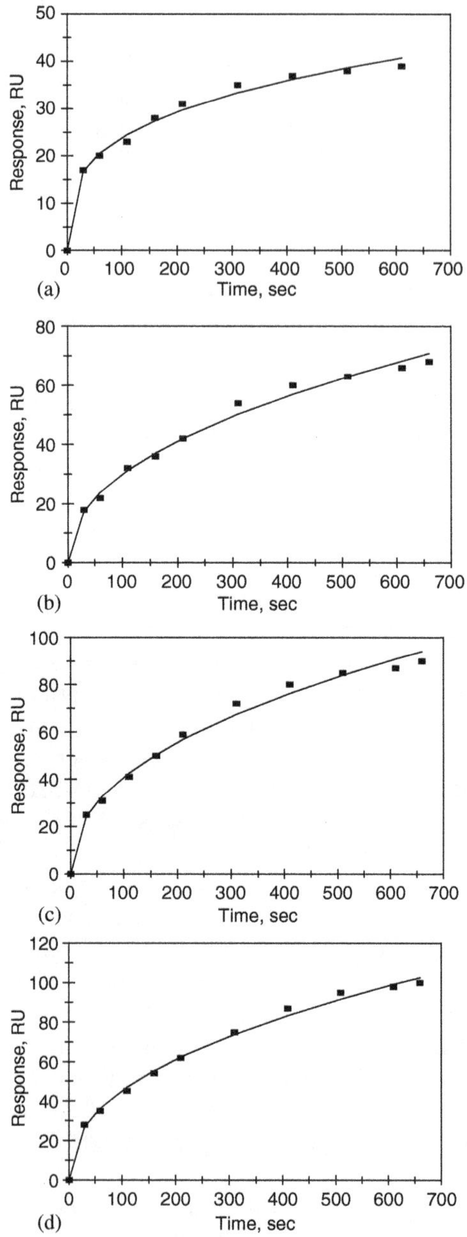

Figure 11.8 Binding of different concentrations of morphine (in mM) in solution to an i-SAM (imprinted self-assembled molecular thin film) prepared from 1 g/l DEA (diethanolamine) at pH 7 to a gold substrate sensor surface (Tappura *et al.*, 2006): (a) 0.1; (b) 1; (c) 10; (d) 100.

Table 11.5

Binding rate coefficients and fractal dimensions for different concentrations (0.1–100 nM) of morphine in solution to i-SAM prepared from 1 g/l DEA at pH 7 (Tappura *et al.*, 2006)

Morphine concentration in solution (mM)	k	D_f
0.1	6.0843 ±0.2607	2.4068 ± 0.0293
1	3.6885 ± 0.1871	2.0898 ± 0.03182
10	5.4037 ± 0.2622	2.1198 ± 0.0305
100	6.0504 ± 0.2163	2.1278 ± 0.2258

rate coefficient, k and the fractal dimension, D_f for a single-fractal analysis are given in Table 11.5.

Figure 11.8d shows the binding of 100 mM morphine in solution to an i-SAM prepared from 1 g/l DEA at pH 7 on a SPR sensor chip surface (Tappura *et al.*, 2006). A single-fractal analysis is adequate to describe the binding kinetics. The values of the binding rate coefficient, k and the fractal dimension, D_f for a single-fractal analysis are given in Table 11.5.

Figure 11.9a and Table 11.5 show the increase in the binding rate coefficient, k for a single-fractal analysis with an increase in the fractal dimension, D_f. For the data shown in Figure 11.9a, the binding rate coefficient, k is given by:

$$k = (1.1323 \pm 0.3106) D_f^{1.9544 \pm 0.1766} \tag{11.7a}$$

The fit is poor. Only four data points are available. The availability of more data points would lead to a more reliable fit. In spite of the poor fit of the data, the binding rate coefficient, k for a single-fractal analysis is sensitive to the fractal dimension, D_f or the degree of heterogeneity on the sensor chip surface as noted by the close to second (equal to 1.9544) order of dependence exhibited.

Figure 11.9b and Table 11.5 show the decrease in the fractal dimension, D_f for a single-fractal analysis with an increase in the 0.1–10 mM morphine concentration in solution. For the data shown in Figure 11.9b the fractal dimension, D_f is given by:

$$D_f = (2.221 \pm 0.132)[\text{morphine, in mM}]^{-0.0153 \pm 0.0112} \tag{11.7b}$$

The fit is reasonable. Only four data points are available. The availability of more data points would lead to a more reliable fit. The fractal dimension, D_f exhibits a very slight dependence on the morphine concentration in the 0.1–10 mM range in solution as noted by the -0.0153 order of dependence exhibited.

Dyukova *et al.* (2005) have analyzed the influence of mixing on the binding of 1.5×10^{-5} M ricin-Cy3 in solution to lactose immobilized on a sensor chip surface. Mixing should minimize or help eliminate diffusional limitations in the system, if they are present, and thereby lead to an increase in the binding rate coefficient. Figure 11.10a shows the binding of 1.5×10^{-5} M ricin-Cy3 in solution to lactose immobilized on a sensor chip surface in the absence of any mixing. A dual-fractal analysis is required to adequately describe the binding kinetics. The values of (a) the binding rate coefficient, k and the fractal

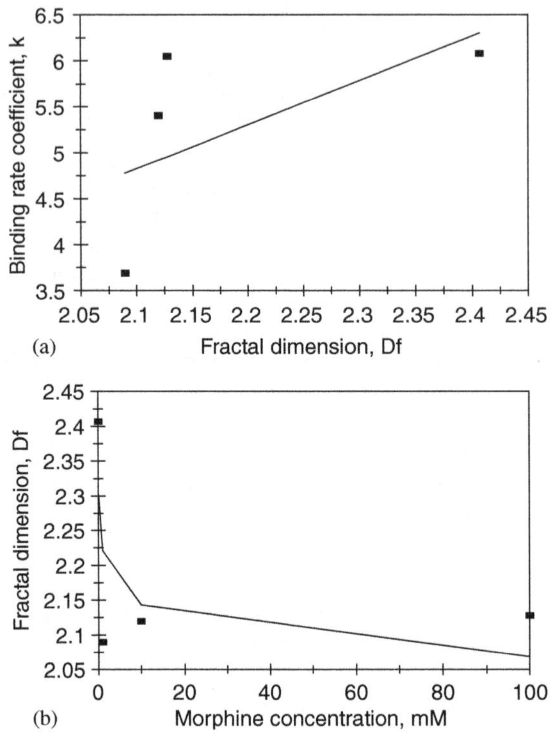

Figure 11.9 (a) Increase in the binding rate coefficient, k for a single-fractal analysis with an increase in the fractal dimension, D_f. (b) Decrease in the fractal dimension, D_f for a single-fractal analysis with an increase in the morphine concentration (in mM) in solution.

dimension, D_f for a single-fractal analysis, and (b) the binding rate coefficients, k_1 and k_2 and the fractal dimensions, D_{f1} and D_{f2} for a dual-fractal analysis are given in Table 11.6. Note that for a dual-fractal analysis, an increase in the fractal dimension by 44.86% from a value of D_{f1} equal to 1.68 to D_{f2} equal to 2.4336 leads to an increase in the binding rate coefficient by a factor of 8.89 from a value of k_1 equal to 4.654 to k_2 equal to 41.385. An increase in the degree of heterogeneity on the sensor chip surface leads to an increase in the binding rate coefficient.

Figure 11.10b shows the binding of 1.5×10^{-5}M ricin-Cy3 in solution to lactose immobilized on a sensor chip surface in the presence of mixing. Once again, a dual-fractal analysis is required to adequately describe the binding kinetics. The values of (a) the binding rate coefficient, k and the fractal dimension, D_f for a single-fractal analysis, and (b) the binding rate coefficients, k_1 and k_2 and the fractal dimensions, D_{f1} and D_{f2} for a dual-fractal analysis are given in Table 11.6. Note that for a dual-fractal analysis, an increase in the fractal dimension by 31% from a value of D_{f1} equal to 2.1 to D_{f2} equal to 2.7510 leads to an increase in the binding rate coefficient by a factor of 4.34 from a value of k_1 equal to 33.345 to k_2 equal to 144.643. Once again, an increase in the degree of heterogeneity on the sensor chip surface leads to an increase in the binding rate coefficient.

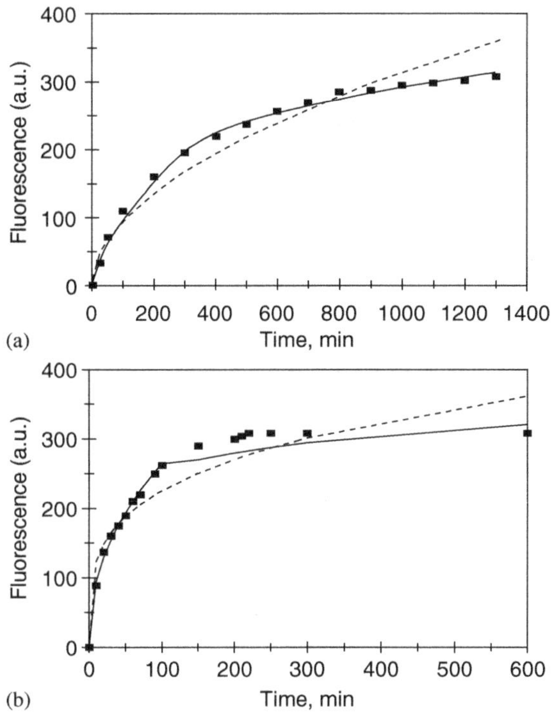

Figure 11.10 Binding of 1.5×10^{-5} M ricin-Cy3 in solution to lactose immobilized on a sensor chip surface (Dyukova *et al.*, 2005): (a) without mixing; (b) with mixing.

Table 11.6

Binding rate coefficients and fractal dimensions for ricin-Cy3 in solution to lactose immobilized on a sensor chip surface (Dyukova *et al.*, 2005). Influence of mixing

Analyte in solution/ receptor on surface	k	k_1	k_2	D_f	D_{f1}	D_{f2}
1.5×10^{-9} M ricin-Cy3 (without mixing)/lactose on a sensor chip surface	8.629 ± 1.336	4.654 ± 0.665	41.385 ± 0.947	1.9590 ± 0.0634	1.68 ± 0.111	2.4336 ± 0.3870
1.5×10^{-9} M ricin-Cy3 (with mixing)/lactose on a sensor chip surface	67.291 ± 11.542	33.345 ± 0.499	144.643 ± 13.672	2.4746 ± 0.0638	2.100 ± 0.0524	2.7510 ± 0.0613

Note that when the case with mixing is compared with the case without mixing, all of the corresponding fractal dimension and the binding rate coefficient values are higher for the mixing case. The higher values of the binding rate coefficient is to be expected since, and as indicated earlier, mixing minimizes the diffusional limitations, which subsequently leads to an increase in the binding rate coefficient value. For example, the binding rate coefficient, k_1 value is higher by a factor of 7.16 when mixing is present compared to when it is absent. Please see Table 11.6. Similarly, the binding rate coefficient, k_2 value is higher by a factor of 3.49 when mixing is present compared to when it is absent. Similarly, the values of the fractal dimensions, D_{f1} and D_{f2} are higher when mixing is present compared to when it is absent. For example, Table 11.6 indicates that the D_{f1} value is higher by 25% (2.1 and 1.68), and the D_{f2} value is higher by 13% (2.7510 and 2.4336).

Figure 11.11a shows the binding and dissociation of fraction #5 of the human chemokine receptor, CCR5 eluted from the gp120 column to monoclonal antibody 1D4 immobilized on a sensor chip surface (Navratilova *et al.*, 2006). Fractions # 4, #5, and #6 contained the highest concentration of the receptor (Navratilova *et al.*, 2006). A single-fractal analysis is adequate to describe the binding and dissociation kinetics. The values of the binding rate coefficient, k and the fractal dimension, D_f for a single-fractal analysis are given in Table 11.7.

Figure 11.11b shows the binding and dissociation of fractions #4 or 6 of the human chemokine receptor, CCR5 eluted from the gp120 column to monoclonal antibody 1D4 immobilized on a sensor chip surface (Navratilova *et al.*, 2006). Once again, a single-fractal analysis is adequate to describe the binding and dissociation kinetics. The values of the binding rate coefficient, k and the fractal dimension, D_f for a single-fractal analysis are given in Table 11.7.

Figure 11.11c shows the binding and dissociation of fraction #7 of the human chemokine receptor, CCR5 eluted from the gp120 column to monoclonal antibody 1D4 immobilized on a sensor chip surface (Navratilova *et al.*, 2006). A single-fractal analysis is adequate to describe the binding and dissociation kinetics. The values of the binding rate coefficient, k and the fractal dimension, D_f for a single-fractal analysis are given in Table 11.7.

Figure 11.11d shows the binding and dissociation of fractions #3 or 8 of the human chemokine receptor, CCR5 eluted from the gp120 column to monoclonal antibody 1D4 immobilized on a sensor chip surface (Navratilova *et al.*, 2006). A single-fractal analysis is adequate to describe the binding and dissociation kinetics. The values of the binding rate coefficient, k and the fractal dimension, D_f for a single-fractal analysis are given in Table 11.7.

Figure 11.11e shows the binding and dissociation of fraction #9 of the human chemokine receptor, CCR5 eluted from the gp120 column to monoclonal antibody 1D4 immobilized on a sensor chip surface (Navratilova *et al.*, 2006). A single-fractal analysis is adequate to describe the binding and dissociation kinetics. The values of the binding rate coefficient, k and the fractal dimension, D_f for a single-fractal analysis are given in Table 11.7.

Figure 11.11f shows the binding and dissociation of fraction #10 of the human chemokine receptor, CCR5 eluted from the gp120 column to monoclonal antibody 1D4 immobilized on a sensor chip surface (Navratilova *et al.*, 2006). A single-fractal analysis is adequate to describe the binding and dissociation kinetics. The values of the binding rate coefficient, k and the fractal dimension, D_f for a single-fractal analysis are given in Table 11.7.

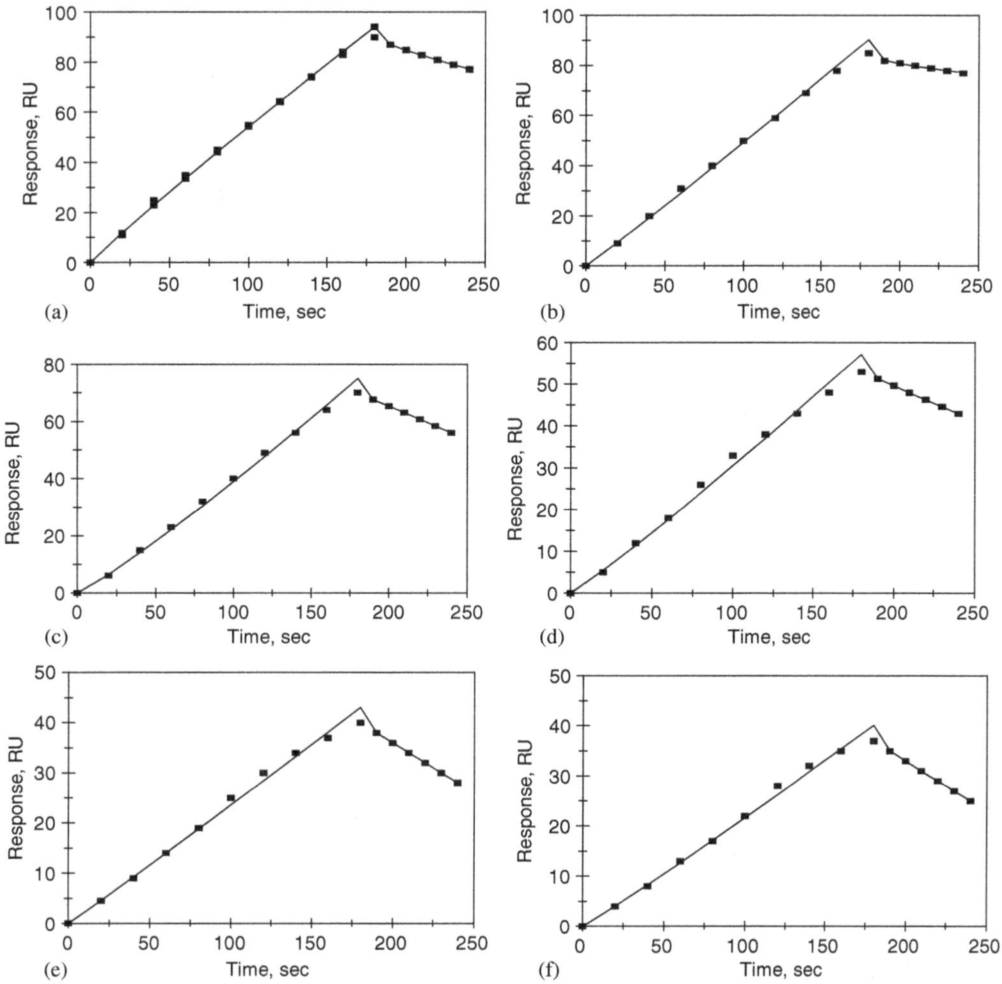

Figure 11.11 Binding and dissociation of different fractions of the human chemokine receptor, CCR5 eluted from the gp120 column to monoclonal 1D4 immobilized on a CM4 sensor chip surface (Navratilova *et al.*, 2006): (a) #5; (b) #4 or 6; (c) #7; (d) #3 or 8; (e) #9; (f) #10.

Figure 11.12a and Table 11.7 show the increase in the binding rate coefficient, k for a single-fractal analysis with an increase in the fractal dimension, D_f. For the data shown in Figure 11.12a, the binding rate coefficient, k is given by:

$$k = (0.3793 \pm 0.193) D_f^{3.271 \pm 1.506} \tag{11.8a}$$

The fit is good. Only six data points are available. The availability of more data points would lead to a more reliable fit. The binding rate coefficient, k is sensitive to the fractal

Table 11.7

Binding and dissociation rate coefficients and fractal dimensions in the binding and in the dissociation phase of fractions of cells expressing the chemokine receptor activity and eluted from the gp120 column in solution to a 1D4 antibody surface (Navratilova *et al.*, 2006). No biosensor detection here; just binding of the different eluted fractions to a 1D4 antibody surface

Fraction number	k	k_d	Affinity, $K = k/k_d$	D_f	D_{fd}	D_f/D_{fd}
5	0.7228 ± 0.0390	0.4415 ± 0.0124	1.637	1.1244 ± 0.0516	1.3606 ± 0.0376	0.826
4 or 6	0.4362 ± 0.0169	0.8109 ± 0.0390	0.538	0.9444 ± 0.0405	1.9048 ± 0.0635	0.496
7	0.2351 ± 0.0135	0.8109 ± 0.0390	1.01	0.7798 ± 0.0548	2.0 ± 0.00076	0.3899
3 or 8	0.2209 ± 0.0154	0.1673 ± 0.0001	1.32	0.8590 ± 0.0660	2.0 ± 0.00106	0.4295
9	0.2127 ± 0.0098	0.2 ± 0	1.0635	0.9550 ± 0.0438	1.0 ± 3.4E-15	0.9550
10	0.1741 ± 0.0081	0.2 ± 0	0.8705	0.9050 ± 0.0449	1.0 ± 3.4E-15	0.9050

dimension, D_f or the degree of heterogeneity on the sensor chip surface as noted by the greater than third (equal to 3.271) order of dependence exhibited.

Figure 11.12b and Table 11.7 show the increase in the dissociation rate coefficient, k_d for a single-fractal analysis with an increase in the fractal dimension, D_{fd}. For the data shown in Figure 11.12b, the dissociation rate coefficient, k_d is given by:

$$k_d = (0.201 \pm 0.028) D_{fd}^{2.236 \pm 0.216} \tag{11.8b}$$

The fit is very good. Only six data points are available. The availability of more data points would lead to a more reliable fit. The binding rate coefficient, k is sensitive to the fractal dimension, D_f or the degree of heterogeneity on the sensor chip surface as noted by the greater than third (equal to 3.271) order of dependence exhibited.

Figure 11.12c and Table 11.7 show the increase in the affinity, K ($=k/k_d$) with an increase in the ratio of the fractal dimensions, D_f/D_{fd}. For the data shown in Figure 11.12c, the affinity, K is given by:

$$K = (1.354 \pm 0.467) \left(\frac{D_f}{D_{fd}} \right)^{1.156 \pm 0.556} \tag{11.8c}$$

The fit is reasonable. There is scatter in the data. Only six data points are available. The availability of more data points would lead to a more reliable fit. The affinity, K exhibits slightly higher than first (equal to 1.156) order of dependence on the ratio of fractal dimensions, D_f/D_{fd} present on the biosensor chip surface.

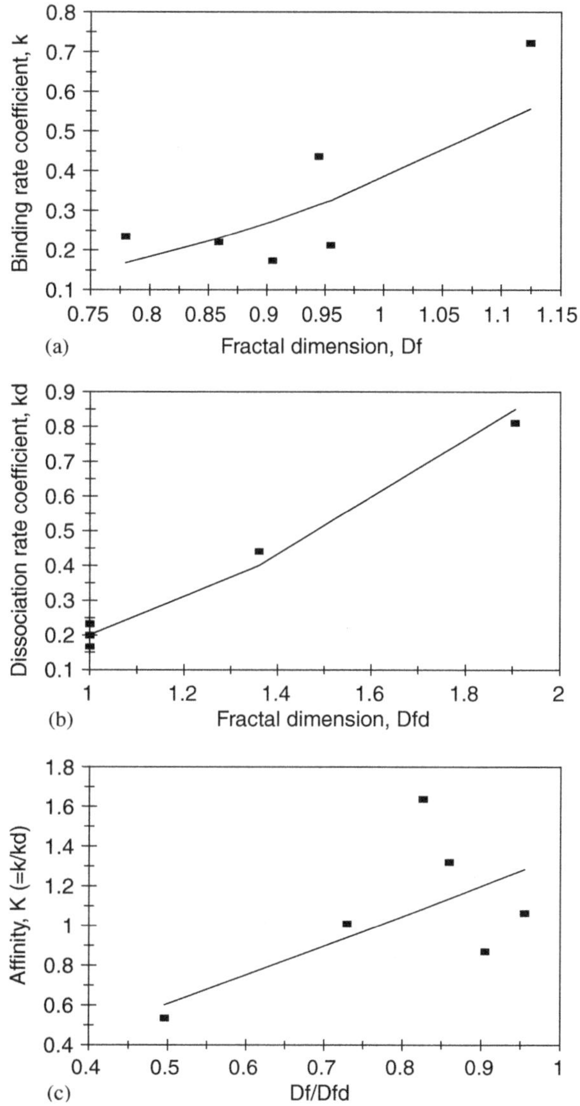

Figure 11.12 (a) Increase in the binding rate coefficient, k for a single-fractal analysis with an increase in the fractal dimension, D_f. (b) Increase in the dissociation rate coefficient k_d for a single-fractal analysis with an increase in the fractal dimension, D_{fd}. (c) Increase in the affinity, K ($=k/k_d$) with an increase in the ratio of the fractal dimensions, D_f/D_{fd}.

Figure 11.13a shows the binding of 100 nM YU2120 (a gp120 mutant) in solution to the human chemokine receptor in the presence of SCD4 (normalized to account for the mutant's molecular weight) immobilized on a CM4 sensor chip surface (Navratilova *et al.*, 2006). Myszka *et al.* (2000) have indicated that CD4 promotes the gp120/chemokine interaction. Furthermore, Navratilova *et al.* (2006) indicate that CD4 affects the association or

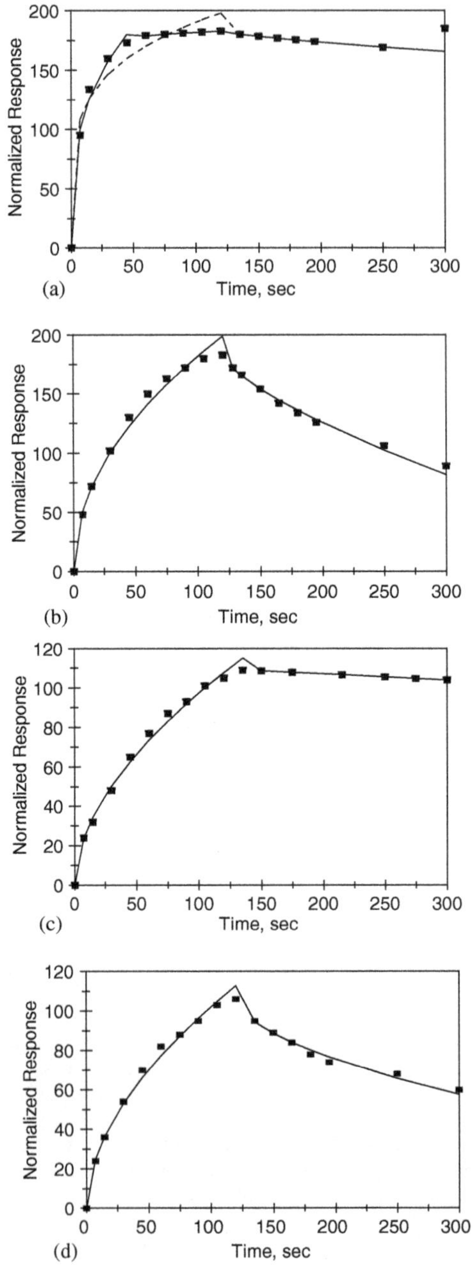

Figure 11.13 Binding of different gp120 mutants (100 nM) in solution to the human chemokine receptor in the presence of SCD4 (normalized to account for the mutants' different molecular weights) immobilized on a CM4 sensor chip surface (Navratilova *et al.*, 2006): (a) YU2120; (b) YU2DV1V2; (c) YU2120GCN4 (d) YU2DV1V2GCN4.

binding rates of the gp120 mutant/CCR5 interactions; however, it does not seem to affect the dissociation rates. The binding kinetics shown in Figure 11.13c is adequately described by a dual-fractal analysis. A single-fractal analysis is required to describe the dissociation kinetics. The values of (a) the binding rate coefficient, k and the fractal dimension, D_f for a single-fractal analysis, (b) the binding rate coefficients, k_1 and k_2 and the fractal dimensions, D_{f1} and D_{f2} for a dual-fractal analysis, and (c) the dissociation rate coefficient, k_d for a single-fractal analysis are given in Tables 11.8a and 11.8b. It is of interest to note that as the fractal dimension or the degree of heterogeneity increases by a factor of 1.255 from a value of D_{f1} equal to 2.3382 to D_{f2} equal to 2.9365, the binding rate coefficient increases by a factor of 3.075 from a value of k_1 equal to 51.057 to k_2 equal to 157.04.

Figure 11.13b shows the binding of 100 nM YU2DV1V2 (a gp120 mutant) in solution to the human chemokine receptor in the presence of SCD4 (normalized to account for

Table 11.8a

Binding rate coefficients and affinity values for 100 nM gp120 mutants in solution to CCR5 in the presence of SCD4 on a sensor chip surface (normalized to account for the mutants' different molecular weights (Navratilova *et al.*, 2006))

gp120 mutant	k	k_1	k_2	k_d	Affinity, $K\ (=k/k_d)$	$K_1\ (=k_1/k_d)$	$K_2\ (=k_2/k_d)$
YU2120	69.594 ± 5.979	51.057 ± 3.471	157.04 ± 0.16	0.3849 ± 0.0185	180.8	132.65	408.42
YU2DV1V2	18.748 ± 1.105	na	na	2.655 ± 0.173	7.061	na	na
YU2120GCN4	7.629 ± 0.353	na	na	0.02924 ± 0.0004	260.9	na	na
YU2DV1V2GCN4	8.296 ± 0.365	na	na	2.438 ± 0.174	3.402	na	na

Table 11.8b

Fractal dimensions for the binding of 100 nM gp120 mutants in solution to CCR5 in the presence of SCD4 on a sensor chip surface (normalized to account for the mutants' different molecular weights (Navratilova *et al.*, 2006))

gp120 mutant	D_f	D_{f1}	D_{f2}	D_{fd}	D_f/D_{fd}	D_{f1}/D_{fd}	D_{f2}/D_{fd}
YU2120	2.5632 ± 0.0618	2.3382 ± 0.09632	2.9365 ± 0.00363	1.5332 ± 0.0452	1.67	1.525	1.915
YU2DV1V2	2.1032 ± 0.0396	na	na	1.5976 ± 0.456	1.260	na	na
YU2120GCN4	1.894 ± 0.0316	na	na	0.9844 ± 0.0133	1.924	na	na
YU2DV1V2GCN4	1.9096 ± 0.0324	na	na	1.8482 ± 0.0066	1.033	na	na

the mutant's molecular weight) immobilized on a CM4 sensor chip surface (Navratilova *et al.*, 2006). A single-fractal analysis is adequate to describe the binding kinetics. The values of the binding rate coefficient, k and the fractal dimension, D_f for a single-fractal analysis are given in Tables 11.8a and 11.8b. There is a change in the binding mechanism as one goes from the mutant YU2120 to the mutant YU2DV1V2 since a dual-fractal mechanism is required to describe the binding kinetics for the first case, whereas a single-fractal analysis is adequate to describe the binding kinetics for the second case.

Figure 11.13c shows the binding of 100 nM YU2120GCN4 (a gp120 mutant) in solution to the human chemokine receptor in the presence of SCD4 (normalized to account for the mutant's molecular weight) immobilized on a CM4 sensor chip surface (Navratilova *et al.*, 2006). Once again, a single-fractal analysis is adequate to describe the binding kinetics. The values of the binding rate coefficient, k and the fractal dimension, D_f for a single-fractal analysis are given in Tables 11.8a and 11.8b.

Figure 11.13d shows the binding of 100 nM YU2DV1V2GCN4 (a gp120 mutant) in solution to the human chemokine receptor in the presence of SCD4 (normalized to account for the mutant's molecular weight) immobilized on a CM4 sensor chip surface (Navratilova *et al.*, 2006). Once again, a single-fractal analysis is adequate to describe the binding kinetics. The values of the binding rate coefficient, k and the fractal dimension, D_f for a single-fractal analysis are given in Tables 11.8a and 11.8b.

Figure 11.14a shows the increase in the binding rate coefficient, k with an increase in the fractal dimension, D_f for a single-fractal analysis. For the data shown in Figure 11.14a, the binding rate coefficient, k is given by:

$$k = (0.09857 \pm 0.0303) D_f^{7.035 \pm 1.0905} \tag{11.8d}$$

The fit is good. Only four data points are available. The availability of more data points would lead to a better fit. The binding rate coefficient, k is very sensitive to the fractal dimension, D_f or the degree of heterogeneity present on the sensor surface as noted by the higher than seventh (equal to 7.035) order of dependence exhibited.

Figure 11.14b shows the increase in the dissociation rate coefficient, k_d with an increase in the fractal dimension, D_{fd} in the dissociation phase for a single-fractal analysis. For the data shown in Figure 11.14b, the dissociation rate coefficient, k_d is given by:

$$k_d = (0.0331 + 0.0434)\ D_{fd}^{7.367 \pm 1.781} \tag{11.8e}$$

The fit is reasonable. Only four data points are available. The availability of more data points would lead to a better fit. The dissociation rate coefficient, k_d is very sensitive to the fractal dimension, D_{fd} or the degree of heterogeneity present on the sensor surface as noted by the order of dependence between seven and seven and one-half (equal to 7.367) order of dependence exhibited.

Figure 11.14c shows the increase in the affinity, $K(=k/k_d)$ for a single-fractal analysis with an increase in the ratio of fractal dimensions, D_f/D_{fd}. For the data shown in Figure 11.14e, the affinity is given by:

$$K\left(=\frac{k}{k_d}\right) = (2.0437 \pm 1.614)\left(\frac{D_f}{D_{fd}}\right)^{7.745 \pm 1.204} \tag{11.8f}$$

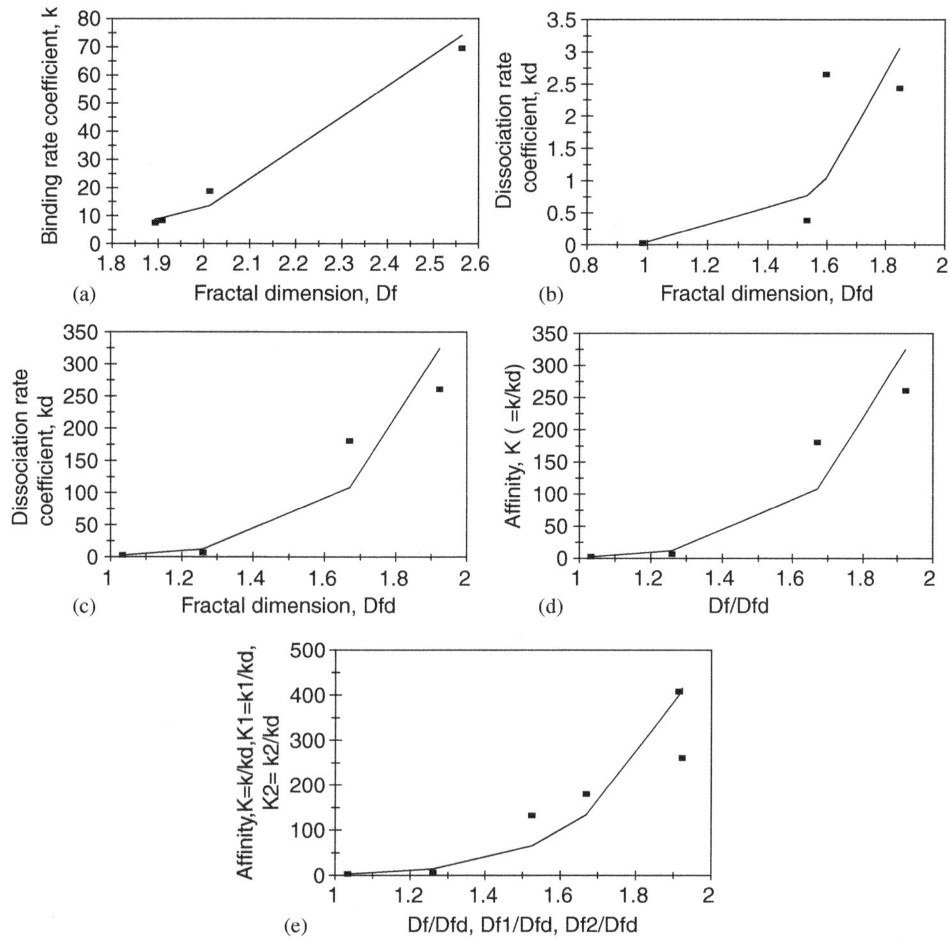

Figure 11.14 (a) Increase in the binding rate coefficient, k for a single-fractal analysis with an increase in the fractal dimension, D_f. (b) Increase in the dissociation rate coefficient, k_d for a single-fractal analysis with an increase in the fractal dimension in the dissociation phase, D_{fd}. (c) Increase in the dissociation rate coefficient, k_d with an increase in the fractal dimension, D_{fd}. (d) Increase in the affinity, K with an increase in the ratio of the fractal dimensions, D_f/D_{fd} (single-fractal analysis only). (e) Increase in the affinity with an increase in the ratio of the fractal dimensions present in the binding and in the dissociation phases (single- and dual-fractal analysis).

The fit is good. Only four data points are available. The availability of more data points would lead to a more reliable fit. The affinity, K is very sensitive to the ratio of the fractal dimensions, D_f/D_{fd} as noted by the higher than seven and one-half(equal to 7.745) order of dependence exhibited.

Affinity values are of interest to practicing biosensorists. Define affinity, K ($=k/k_d$) as the ratio of the binding rate coefficient to the dissociation rate coefficient. Tables 11.8a and 11.8b and Figure 11.14d show the increase in the affinity, K ($=k/k_d$) with an increase in the ratio of the fractal dimensions present in the binding and in the dissociation phases, D_f/D_{fd} when a single-fractal analysis applies. For the data shown in Figure 11.14d, the affinity, K is given by:

$$K\left(=\frac{k}{k_d}\right) = (2.044 \pm 1.615)\left(\frac{D_f}{D_{fd}}\right)^{7.745 \pm 1.204} \tag{11.8g}$$

The fit is good. Only four data points are available. The availability of more data points would lead to a more reliable fit. The affinity, K is extremely sensitive to the ratio of the fractal dimensions, D_f/D_{fd} as noted by the order of dependence between seven and one-half and eight (equal to 7.745) exhibited.

Tables 11.8a and 11.8b and Figure 11.14e show the increase in the affinity with an increase in the ratio of the fractal dimensions present in the binding and in the dissociation phases. In this case, and as noted above due to the lack of points available for a single-fractal analysis, the points for the single- and the dual-fractal analysis are plotted on the same figure. For the data shown in Figure 11.14e, the affinity, K or K_1 or K_2 is given by:

$$\text{Affinity}, K, K_1, K_2 = (4.030 \pm 1.764)(D_{fd}, D_{fd}, D_{fd})^{7.964 \pm 1.047} \tag{11.8h}$$

The fit is good considering that the data sets for a single- and a dual-fractal analysis are plotted together on the same figure. The affinity is, once again, very sensitive to the ratio of the fractal dimension present in the binding and in the dissociation phases, as noted by the very close to eighth (equal to 7.964) order of dependence exhibited.

11.4 CONCLUSIONS

A fractal analysis is used to analyze the binding and dissociation (if applicable) kinetics of different compounds on biosensor surfaces. The systems analyzed include: (a) binding of CRP in solution to anti-CRP immobilized on a DPI sensor chip surface (Lin *et al.*, 2006); (b) binding of 100 nM free hK1 (kallikrein1) in solution to M0097-G1 Fab and M0135 F03 Fab immobilized on a sensor chip surface (Wassaf *et al.*, 2006), and the binding of 100 nM hK1-aprotinin in solution to M0135 F03 Fab immobilized on a sensor chip surface (Wassaf *et al.*, 2006); (c) binding of 15 ng/ml anti-CA15-3 antibody in solution to CA15-3 antigen immobilized on an electrochemical protein chip and the binding of 70 ng/ml anti-Ferritin antibody in solution to Ferritin antigen immobilized on an electrochemical protein chip (Wilson and Nie, 2006); (d) binding of heparanase in solution to biotinylated heparin sulfate glycosaminogen (platelet extract) covalently linked to the surface of a 96-well immunoassay plate (Behzad and Brenchley, 2003); (e) binding of different concentrations of morphine (in mM) in solution to i-SAM at pH 7 to a gold substrate (Tappura *et al.*, 2006); (f) binding of 1.5×10^{-5} M ricin-Cy3 in solution to lactose immobilized on a sensor chip surface (Dyukova *et al.*, 2005); (g) binding and dissociation of different fractions of the human chemokine receptor, CCR5 eluted from the gp120 column to monoclonal 1D4 immobilized on a CM4 sensor chip surface (Navratilova *et al.*, 2006); and (h) binding of different gp120 mutants (100 nM) in solution to the human chemokine receptor in the presence of SCD4 (normalized to account for the mutant's different molecular weights) immobilized on a CM4 sensor chip surface (Navratilova *et al.*, 2006).

The binding kinetics is described by either a single- or dual-fractal analysis. A dual-fractal analysis is used only when a single-fractal analysis does not provide an adequate fit. This was done using Corel Quattro Pro 8.0 (1997). The fractal dimension provides a quantitative measure of the degree of heterogeneity present on the biosensor chip surface. Note that, and as indicated earlier in the different chapters in the book, the fractal dimension for the binding and the dissociation phase, D_f and D_{fd}, respectively, is not a typical independent variable, such as analyte concentration, that may be directly manipulated. It is estimated from eqs. (11.1)–(11.3), and one may consider it as a derived variable.

An increase in the fractal dimension value or the degree of heterogeneity on the surface leads, in general, to an increase in the binding and in the dissociation rate coefficient(s). For example, for the binding of CRP in solution to anti-CRP immobilized on a sensor chip surface, and for a dual-fractal analysis, the binding rate coefficient, k_2 exhibits an order of dependence between second and two and one-half (equal to 2.386) on the fractal dimension, D_{f2} or the degree of heterogeneity at the silicon/water interface of a DPI biosensor. This indicates that the binding rate coefficient, k_2 is sensitive to the fractal dimension or the degree of heterogeneity present on the sensor chip surface.

Predictive relations are also developed, for example, for: (a) the binding rate coefficients, k_1 and k_2 as a function of the CRP concentration (in μg/ml) in solution (Lin *et al.*, 2006); (b) the fractal dimension, D_{f2} as a function of the CRP concentration (in μg/ml) in solution; (c) the binding rate coefficient, k for a single-fractal analysis as a function of the CRP concentration (in μg/ml) in solution; (d) the binding rate coefficient, k as a function of the fractal dimension, D_f for the binding of hK1 (kallikrein 1) in solution to different Fab fragments immobilized on a sensor chip surface (Wassaf *et al.*, 2006); (e) the dissociation rate coefficients, k_{d1} or k_{d2} as a function of the fractal dimension in the dissociation phases, D_{fd1} or D_{fd2} (Wassaf *et al.*, 2006); (f) the binding rate coefficient, k for a single-fractal analysis as a function of the fractal dimension, D_f; (g) the binding rate coefficients, k_1 and k_2 as functions of the fractal dimensions, D_{f1} and D_{f2}, respectively; (h) the ratio of the binding rate coefficients, k_2/k_1 as a function of ratio of fractal dimensions, D_{f2}/D_{f1}; (i) the fractal dimension, D_f for a single-fractal analysis as a function of the morphine concentration (in mM) in solution (Tappura *et al.*,2006); (j) the dissociation rate coefficient, k_d as a function of the fractal dimension in the dissociation phase, D_{fd} for the binding and dissociation of different fractions of the human chemokine receptor, CCR5 eluted from the gp120 column to monoclonal antibody 1D4 immobilized on a sensor chip surface (Navratilova *et al.*, 2006); and (*k*) the affinity, K as a function of the ratio of the fractal dimensions present in the binding and the dissociation phases, respectively (Navratilova *et al.*, 2006).

Quite a few different examples are presented in this chapter, emphasizing that the degree of heterogeneity that exists on the biosensor surface does significantly affect, in general, the rate coefficient and affinity values, and subsequently the kinetics in general. More such studies are required to determine whether the binding and the dissociation rate coefficient(s), and subsequently the affinity values are sensitive to their respective fractal dimensions on the biosensor chip surface.

A better understanding of all of the possible parameters that influence the kinetics of binding and dissociation of different analyte–receptor systems on biosensor surfaces is critical. This will be of considerable assistance, for example, to help select the correct drug of choice from a list of possible candidates. More often than not, the influence of diffusion and heterogeneity on the biosensor surface is neglected. As indicated in this chapter and elsewhere in the book, the degree of heterogeneity significantly influences, in general, the binding as well as the dissociation kinetics occurring on biosensor surfaces. It would behoove the practicing biosensorists to start paying more attention to this aspect of kinetics on biosensor surfaces. One may perhaps argue that the influence of diffusional limitations may be minimized or perhaps even be eliminated if the biosensor is run properly.

In fact, ideally one should really analyze the influence of the degree of heterogeneity and the diffusional aspects separately on the kinetics of analyte–receptor reactions occurring on biosensor surfaces. This is not possible in present time by the manner in which the fractal analysis is presented. If one is able to separate the influence of diffusional limitations and heterogeneities on the biosensor surface, and analyze the influence of each separately on the analyte–receptor reactions occurring on biosensor surfaces, then one may be able to better manage these analyte–receptor interactions to advantage. This should very significantly impact the different biosensor performance parameters such as sensitivity, selectivity, stability, and permit one to optimize these parameters in desired directions for the different analyte–receptor reactions occurring on biosensor surfaces.

REFERENCES

Adanyi, N, IA Levkovets, S Rodriguez-Gil, A Ronald, M Varadi, and I Szendro, Development of immunosensor based on OWLS technique for determining Alfatoxin B1 and Ochratoxin A. *Biosensors and Bioelectronics*, 2007, **22**(6), 797–802.

Armstrong, J, HJ Salacinski, Q Mu, AM Seifalian, L Peel, N Freeman, CM Holt, and JR Lu, *Journal of Physics: Condensed Matter*, 2004, **16**, S2483–S2491.

Behzad, F and PEC Brenchley, A multiwell format assay for heparanase. *Analytical Biochemistry*, 2003, **320**, 207–213.

Bhikabhai, R, A Sjoberg, L Hedkvist, M Galin, P Liljedahl, T Frizard, N Petersson, M Nilsson, JA Sigrell-Simon, and C Markeland-Johannson, Production of milligram quantities of affinity-tagged proteins using automated multistep chromatographic purification. *Journal of Chromatography A*, 2005, **1080**, 83–92.

Biehle, SJ, J Carrozzezella, R Shukla, J Popplewell, M Swann, N Freeman, and JF Clark, *Biochimica et Biophysica Acta*, 2004, **1689**(3), 244–251.

Blake, GJ, N Rifai, JE Buring, and PM Ridker, *Circulation*, 2003, **108**(24), 2993–2999.

Brekke, OH and GA Loset, New techniques in therapeutic antibody development. *Current Opinion in Pharmacology*, 2003, **31**, 433–436.

Carpelan-Holmstrom, M, J Louhimo, UH Steman, H Alfthan, and C Hagland, *Anticancer Research*, 2002, **22**, 2311–2316.

Chapman, T, Protein purification: pure but not simple. *Nature*, 2005, **434**, 795–798.

Clements, JA, NM Willemsen, SA Myers, and Y Dong, The tissue kallikrien family of serine proteases: functional roles in human disease and potential as chemical biomarkers. *Critical Reviews in Clinical Laboratory Science*, 2004, **41**, 265–312.

Corel Quattro Pro 8.0 (1997). Corel Corporation, Canada.

Cross, GH, AA Reeves, S Brand, MJ Swann, LL Peel, NJ Freeman, and JR Lu, *Journal of Physics D: Applied Physics*, 2004, **37**(1), 74–80.

Cross, GH, Y Ren, and MJ Swann, *The Analyst*, 2000, **125**(12), 2173–2175.

Dyukova VI, EI Dementieva, DA Zubtsov, OE Galanina, NV Bovin, and A Yu Rubina, Hydrogel glycan microarrays. *Analytical Biochemistry*, 2005, **347**, 94–105.

El Assal, O, A Yamanoi, T Ono, H Kohno, and N Nagasue, The clinicopathological significance of heparanase and basic fibroblast growth factor expressions in hepatocellular carcinoma. *Clinical Cancer Research*, 2001, **7**, 1299–1305.

Eom, HS, BH Hwang, D-H Kim, I-B Lee, Y H Kim, and HJ Cha, Multiple detecton of food-borne pathogenic bacteria using a novel 16S rDNA-based oligonuleotide signature chip. *Biosensors and Bioelectronics*, 2007, **22**(6), 845–853.

Fan, Q, D Shan, H Xue, Y He, and S Cosnier, Amperometric phenol biosensor based on laponite clay-chitosan nanocomposite matrix. *Biosensors and Bioelectronics*, 2007, **22**(6), 816–821.

Fernandez-Gacco, A, M Uguen, and J Fastrez, Phage display as a tool for the directed evolution of enzymes. *Trends in Biotechnology*, 2003, **21**, 408–414.

Gabl, R, HD Feucht, H Zeininger, G Eckstein, M Schreiter, R Primig, D Pitzer, and W Wessing, *Biosensors and Bioelectronics*, 2004, **19**, 615–620.

Gestwicki, JE, HV Hsieh, and JB Pitner, *Analytical Chemistry*, 2001, **73**, 5732–5737.

Haupt, K, and K Mosbach, *Chemical Reviews*, 2000, **100**, 2495–1504.

Havlin, S, Molecular diffusion and reaction, in *The Fractal Approach to Heterogeneous Chemistry: Surfaces, Colloids, Polymers* (ed. D. Avnir), Wiley, New York, 1989, pp. 251–269.

Hayakawa, T, S Naruse, M Kitagawa, H Ishiguro, T Kondo, K Kurimoto, M Fukushima, T Takayama, Y Horiguchi, N Kuno, A Nodu, and T Furukawa, *International Journal of Pancreatology*, 1999, **25**, 23–29.

Hoffman, H, M Siebeck, O Thetter, M Jochum, and H Fritz, Aprotinin concentrations effective for the inhibition of tissue kallikrein and plasma kallikrien in vitro and in vivo. *Advances in Experimental Medicine and Biology*, 1989, **B247**, 35–42.

Huang, YY, HY Hsu, and CJC Huang, A protein detection technique by using surface plasmon resonance (SPR) with rolling circle amplification (RCA) and nanogold-modified tags. *Biosensors and Bioelectronics*, 2007, **22**(60), 980–985.

Hust, M, and S Dubel, Mating antibody phage display with proteomics. *Trends in Biotechnology*, 2004, **22**, 8–14.

Hyde-DeRuyschev, R, LA Paige, DJ Christensen, N Hyde-DeRuyscher, A Lim, ZL Fredericks, J Krantz, P Gallant, J Zhang, SM Rocklage, DM Fowlkes, PA Wendler, and PT Hamilton, Detection of small-molecule enzyme inhibitors with peptides isolated from phage-displayed combinatorial libraries. *Chemical Biology*, 2000, **7**, 17–25.

Ikeda, M, Expression of *N*-deacylase/*N*-sulfotransferase (NDST) during early rat embryogenesis. *Kokubyo Gakkai Zasshi*, 2001, **68**, 184–192.

Iozzo, RV, and JD San Antonio, Heparan sulfate proteoglycans: heavy hitters in the angiogenesis arena. *The Journal of Clinical Investigation*, 2001, **108**, 341–347.

Kelley, BD, J Broth, M Tannatt, QL Wub, R Ladner, J Yuc, D Potter, and A Ley, Isolation of a peptide ligand for affinity purification of factor VIII using phage display. *Journal of Chromatography A*, 2004, **1038**, 121–130.

Lin, S, CK Lee, YH Lin, SY Lee, BC Shen, JC Tsai, and SM Hsu, Homopolyvalent antibody-antigen interaction kinetic studies with use of a dual-polarization interferometric biosensor. *Biosensors and Bioelectronics*, 2006.

Liu, B, C Huang, C Sihlbom, A Burlingame, D Marks, Towards proteome-wide production of monoclonal antibody by phage display. *Journal of Molecular Biology*, 2002, **315**, 1063–1073.

Liu, S and Y Sun, Co-immobilization of glucose oxidase and hexokinase on silicate hybrid sol-gel membrane for glucose and ATP detections. *Biosensors and Bioelectronics*, 2007, **22**(6), 905–911

Louhimo, J, P Finne, H Alfthan, UH Stenman, and C Hagland, *Anticancer Research*, 2002, **22**, 1759–1764.

Maruvuda, P, W Wang, PD Wagner, and S Srivastava, *Biotechniques*, 2005, Suppl 1, 9–15.

Matsuda, K, H Maruyama, G Guo, J Kleef, J Itakura, Y Matsumoto, AD Ladner, and M Korc, Glypican-1 is over-expressed in human breast cancer and modulates the mitogenic effects of multiple heparin-binding growth factors in breast cancer cells. *Cancer Cells*, 2001, **61**, 1299–1305.

Myszka, DG, RW Sweet, P Hensley, M Brighham-Burke, PD Kwong, WA Hebderrickson, R Wyatt, J Sodorski, ML Doyle, Energetics of the HIV gp1200CD4 binding reaction. *Proceedings of the National Academy of Sciences*, 2000, **97**, 5639–5644.

Navratilova, I, M Pancera, RT Wyatt, and DG Myszka, A biosensor-based approach toward purification and crystallization of G protein-coupled receptors. *Analytical Biochemistry*, 2006 in press.

Padigi, S, RKK Reddy, and S Prasad, Carbon nanotube based aliphatic hydrocarbon sensor. *Biosensors and Bioelectronics*, 2007, **22**(6), 829–837.

Ramsay, G (ed.), *Commercial Biosensor*, Wiley Interscience, New York, 1998.

Rodi, DJ, GE Agoston, R Manon, R Lapcevich, SJ Green, and L Makowski, Identification of small molecule binding sites within proteins using phage display technology. *Combinatorial Chemistry and High Throughput Screening*, 2001.

Sato, AK, DJ Sexton, LA Morganelli, EH Cohen, QL Wu, GP Conley, Z Streltsova, SW Lee, M Devlin, DB Oliveira, J Enright, RB Kent, CRWescott, TC Ransohoff, AC Ley, and RC Ladner, Development of mammalian serum albumin affinity purification media by peptide phage display. *Biotechnology Progress*, 2002, **18**, 182–192.

Stockwin, LH and S Holmes, The role of therapeutic antibodies in drug discovery. *Biochemical Society Transactions*, 2003, **31**, 433–436.

Swann, MJ, LL Peel, S Carrington, and NJ Freeman, *Analytical Biochemistry*, 2004, **329**(2), 190–198.

Tappura, K, I Vikholm-Lunidn, and WM Albers, Lipoate-based imprinted self-assembled molecular thin films for biosensor applications. *Biosensors and Bioelectronics*, 2006.

Tracy, WL, HT Chang, SW Hsu, RJ Chen, and S Lin, *Electrophoresis*, 2002, **23**(6), 836–846.

Tsao, KC, TL Wu, PY Chang, JH Hong, and JT Wu, *Journal of Clinical Laboratory Analysis* 2006, **20**, 42–46.

Vlodarvsky, I, M Elkin, O Pappo, H Aingorn, R Atzmon, R Ishai-Michael;I, A Avivi, I Pecker, and Y Friedmann, Mammalian heparanase as mediator of tumor metastasis and angiogenesis. *Israel Medical Association Journal*, 2000, **2**(Supplement), 37–45.

Vlodavsky I and Y Friedmann, Molecular properties and involvement of heparanase in cancer metastasis and angiogenesis. *Journal of Clinical Investigation*, 2001, **108**, 341–347.

Wassaf, D, G Kuang, K Kopacz, QL Wu, Q Nguyen, M Toews, J Cosic, J Jacques, S Wiltshire, J Lambert, CC Pazmany, S Hogen, RC Ladner, AE Nixon, and DJ Sexton, High-throughput affinity ranking of antibodies using surface plasmon resonance bioassays. *Analytical Biochemistry*, 2006, **351**, 241–253.

Wilson, MS and W Nie, Multiplex measurement of seven tumor markers using an electrochemical protein chip. *Analytical Chemistry*, 2006, **78**, 6476–6483.

Wulff, G, *Angewandte Chemistry International Edition in English*, 1995, **34**, 1812–1832.

– 12 –

Fractal Analysis of Binding and Dissociation Kinetics of Environmental Contaminants and Explosives on Biosensor Surfaces

12.1 INTRODUCTION

Biosensors are finding increasing application in the detection of harmful pollutants in both the atmospheric and aquatic environments. Some of the common analytes detected recently by biosensors include atrazine (Cummins *et al.*, 2006; Ciumasu *et al.*, 2005), acetonitrile (Hakansson and Mattiasson, 2004), acetylcholine (Hai *et al.*, 2006), TNT (Ciumasu *et al.*, 2005), copper (Yamasaki *et al.*, 2004), carbamate pesticides (Suwansard *et al.*, 2005), arsenic (Ivandini *et al.*, 2006), organophosphorous compounds (Orbulescu *et al.*, 2006), organophosphate pesticides (Liu and Lin, 2006), trace phenolics (Shiddiky *et al.*, 2006), nerve agents (Liu and Lin, 2006), polychlorinated biphenyl analytes (Glass *et al.*, 2006), dioxins (You *et al.*, 2006), catechol (Zucolotto *et al.*, 2006), toxic agent monitoring (Sharma *et al.*, 2005), dioxins (You *et al.*, 2006), and ethanol (Weng *et al.*, 2004).

The detection and monitoring of environmental contaminants is also quite popular as session themes for conferences. Some of the environmental-related agent detection recent presentations include the detection of toxins in source water (Greenbaum and Rodriguez, 2006), cyanide (Hawkridge *et al.*, 2006), environmental application (Lao *et al.*, 2006), organophosphorous compounds (LeBlanc and Wang, 2006), heavy metals (Mulchandani *et al.*, 2006a; White and Holcombe, 2006), estrogens (Mulchandani *et al.*, 2006b), toxic chemicals (Oh *et al.*, 2006; Zhang and Suslick, 2006), inhalation hazards (Riley and Lucas, 2006), and dissolved organics (Zhang *et al.*, 2006).

Various biosensor detection schemes have been used to detect the environmental pollutants presented above. Very few, if any, references in the open literature provide the kinetics of binding and dissociation of environmental pollutant-related analytes in solution to receptors immobilized on biosensor surfaces. These kinetics are important to enhance biosensor-related performance parameters. In this chapter we use fractal analysis to analyze the binding and dissociation (if applicable) kinetics of (a) anti-atrazine antibody (IgG) in solution to atrazine (rabbit IgG) coated on a 96-well plate (Cummins *et al.*, 2006), acetylcholine (ACh) in solution to acetylcholinesterase immobilized on an ion-selective field-effect transistor (ISFET) biosensor (Hai *et al.*, 2006), and different concentrations (in nM) of catechol in solution to a 10-bilayer PAMAM (poly(amidoamine)) dendrimer/CCD (C1 catechol 1,2-deoxygenase) film on a nanostructured surface (Zuculotto *et al.*, 2006). Binding and dissociation rate coefficient values, as well as affinity (wherever applicable) values are provided. As indicated throughout the different chapters in the book, the fractal analysis method is one possible way of providing the binding and the dissociation rate coefficient values of analyte–receptor reactions occurring on structured surfaces.

12.2 THEORY

12.2.1 Single-fractal analysis

Binding rate coefficient

Havlin (1989) indicates that the diffusion of a particle (analyte [Ag]) from a homogeneous solution to a solid surface (e.g., receptor [Ab]-coated surface) on which it reacts to form a product (analyte–receptor complex; (Ab–Ag)) is given by

$$(\mathrm{Ab}\cdot\mathrm{Ag}) \approx \begin{cases} t^{(3-D_{\mathrm{f,bind}})/2} = t^{\mathrm{p}} & t < t_{\mathrm{c}} \\ t^{1/2} & t > t_{\mathrm{c}} \end{cases} \tag{12.1}$$

Here $D_{\mathrm{f,bind}}$ or D_{f} (used later on in the chapter) is the fractal dimension of the surface during the binding step. t_{c} is the crossover value. Havlin (1989) indicates that the crossover value may be determined by $r_{\mathrm{c}}^2 \sim t_{\mathrm{c}}$. Above the characteristic length, r_{c}, the self-similarity of the surface is lost and the surface may be considered homogeneous. Above time, t_{c} the surface may be considered homogeneous, since the self-similarity property disappears, and 'regular' diffusion is now present. For a homogeneous surface where D_{f} is equal to 2, and when only diffusional limitations are present, $p = 1/2$ as it should be. Another way of looking at the $p = 1/2$ case (where $D_{\mathrm{f,bind}}$ is equal to two) is that the analyte in solution views the fractal object, in our case, the receptor-coated biosensor surface, from a 'large distance.' In essence, in the association process, the diffusion of the analyte from the solution to the receptor surface creates a depletion layer of width $(Ðt)^{1/2}$ where $Ð$ is the diffusion constant. This gives rise to the fractal power law, $(\mathrm{Analyte} - \mathrm{Receptor}) \sim t^{(3-D_{\mathrm{f,bind}})/2}$. For the present analysis, t_{c} is arbitrarily chosen and we assume that the value of the t_{c} is not reached. One may consider the approach as an intermediate 'heuristic' approach that may be used in the future to develop an autonomous (and not time-dependent) model for diffusion-controlled kinetics.

Dissociation rate coefficient

The diffusion of the dissociated particle (receptor [Ab] or analyte [Ag]) from the solid surface (e.g., analyte [Ag] –receptor [Ab] complex-coated surface) into solution may be given, as a first approximation by

$$(\mathrm{Ab}\cdot\mathrm{Ag}) \approx -t^{(3-D_{\mathrm{f,diss}})/2} = -t^{\mathrm{p}}\,(t > t_{\mathrm{diss}}) \tag{12.2}$$

Here $D_{f,diss}$ is the fractal dimension of the surface for the dissociation step. t_{diss} corresponds to the highest concentration of the analyte–receptor complex on the surface. Henceforth, its concentration only decreases. The dissociation kinetics may be analyzed in a manner 'similar' to the binding kinetics.

12.2.2 Dual-fractal analysis

Binding rate coefficient

Sometimes, the binding curve exhibits complexities and two parameters (k, D_f) are not sufficient to adequately describe the binding kinetics. This is further corroborated by low values of r^2 factor (goodness of fit). In that case, one resorts to a dual-fractal analysis (four parameters; k_1, k_2, D_{f1}, and D_{f2}) to adequately describe the binding kinetics. The single-fractal analysis presented above is thus extended to include two fractal dimensions. At present, the time ($t = t_1$) at which the 'first' fractal dimension 'changes' to the 'second' fractal dimension is arbitrary and empirical. For the most part, it is dictated by the data analyzed and experience gained by handling a single-fractal analysis. A smoother curve is obtained in the 'transition' region, if care is taken to select the correct number of points for the two regions. In this case, the product (antibody–antigen; or analyte–receptor complex, Ab·Ag or analyte–receptor) is given by

$$(\mathrm{Ab} \cdot \mathrm{Ag}) \approx \begin{cases} t^{(3-D_{f1,\mathrm{bind}})/2} = t^{p1} & (t < t_1) \\ t^{(3-D_{f2,\mathrm{bind}})/2} = t^{p2} & (t_1 < t < t_2 = t_c) \\ t^{1/2} & (t > t_c) \end{cases} \tag{12.3}$$

In some cases, as mentioned above, a triple-fractal analysis with six parameters (k_1, k_2, k_3, D_{f1}, D_{f2}, and D_{f3}) may be required to adequately model the binding kinetics. This is when the binding curve exhibits convolutions and complexities in its shape due to perhaps to the very dilute nature of the analyte (in some of the cases to be presented) or for some other reasons. Also, in some cases, a dual-fractal analysis may be required to describe the dissociation kinetics.

12.3 RESULTS

The fractal analysis will be applied to (a) the binding of IgG (anti-atrazine antibody) in solution to rabbit IgG (atrazine) coated on a 96-well plate (Cummins *et al.*, 2006), (b) the binding of acetylcholine (ACh) in solution to acetylcholinesterase immobilized on an ISFET biosensor (Hai *et al.*, 2006), and (c) the binding of different concentrations (in nM) of catechol in solution to a 10-bilayer PAMAN (poly(amidoamine)) dendrimer/CCD (C1 catechol 1,2-deoxygenase) film on a nanostructured surface.

Alternate expressions for fitting the data are available that include saturation, first-order reaction, and no diffusion limitations, but these expressions are apparently deficient in describing the heterogeneity that inherently exists on the surface. One might justifiably argue that the appropriate modeling may be achieved by using a Langmuirian or other approach. The Langmuirian approach may be used to model the data presented if one assumes the presence of discrete classes of sites (e.g., double exponential analysis as compared with a single-fractal analysis). Lee and Lee (1995) indicate that the fractal approach has been applied to surface science, for example, adsorption and reaction processes. These authors emphasize that the fractal approach provides a convenient means to represent the different structures and morphology at the reaction surface. These authors also emphasize using the fractal approach to develop optimal structures and as a predictive approach. Another advantage of the fractal technique is that the analyte–receptor association (as well as the dissociation reaction) is a complex reaction, and the fractal analysis via the fractal dimension and the rate coefficient provide a useful lumped parameter(s) analysis of the diffusion-limited reaction occurring on a heterogeneous surface.

In a classical situation, to demonstrate fractality, one should make a log–log plot, and one should definitely have a large amount of data. It may be useful to compare the fit to some other forms, such as exponential, or one involving saturation, etc. At present, we do not present any independent proof or physical evidence of fractals in the examples presented. It is a convenient means (since it is a lumped parameter) to make the degree of heterogeneity that exists on the surface more quantitative. Thus, there is some arbitrariness in the fractal model to be presented. The fractal approach provides additional information about interactions that may not be obtained by conventional analysis of biosensor data.

There is no nonselective adsorption of the analyte. The present system (environmental pollutants in the aqueous or the gas phase) being analyzed may be typically very dilute. Nonselective adsorption would skew the results obtained very significantly. In these types of systems, it is imperative to minimize this nonselective adsorption. We also do recognize that, in some cases, this nonselective adsorption may not be a significant component of the adsorbed material and that this rate of association, which is of a temporal nature, would depend on surface availability. If we were to accommodate the nonselective adsorption into the model, there would be an increase in the heterogeneity on the surface, since, by its very nature, nonspecific adsorption is more homogeneous than specific adsorption. This would lead to higher fractal dimension values since the fractal dimension is a direct measure of the degree of heterogeneity that exists on the surface.

Cummins *et al.* (2006) have recently used an optical indium tin oxide (ITO) (quartz) waveguide to develop a fluoroimmunoassay for atrazine. They used fluorescent europium (III) chelate-dyed nanoparticle labels (Seradyn) to formulate a competitive atrazine immunoassay. These authors evaluated the influence of three biconjugated particle labels (107, 304, and 396 nm) on the binding kinetics of rabbit IgG/anti-IgG system in a 96-well plate array. As expected, a decrease in particle size resulted in faster binding which was due primarily to diffusion kinetics.

Figure 12.1a shows the binding of IgG (anti-atrazine antibody) in solution to rabbit IgG (atrazine) coated on a 96-well plate array (Cummins *et al.*, 2006). The biconjugated label was 107 nm. A dual-fractal analysis is required to adequately describe the binding kinetics. The values of (a) the binding rate coefficient, k and the fractal dimension, D_f for a single-fractal analysis, and (b) the binding rate coefficients, k_1 and k_2 and the fractal dimensions, D_{f1} and D_{f2} for a dual-fractal analysis are given in Table 12.1a.

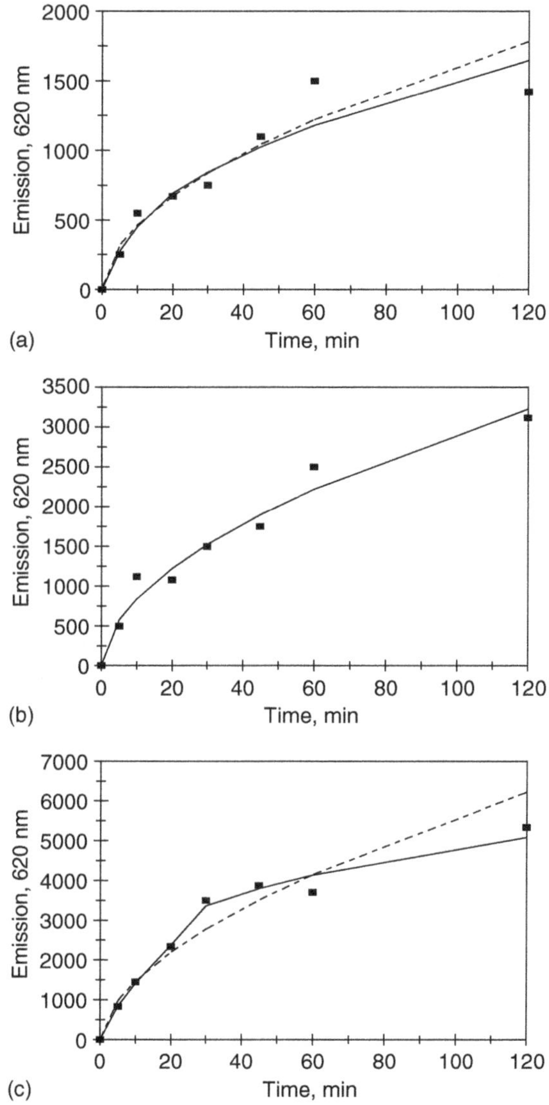

Figure 12.1 Binding of IgG (anti-atrazine antibody) in solution to rabbit IgG (atrazine) coated on a 96-well plate array. Influence of particle size (in nm) (Cummins *et al.*, 2006): (a) 107, (b) 304, (c) 396. When only a solid (___) line is used then a single-fractal analysis applies. When both a solid (___) line and a (---) line are used then the dashed line is for a single-fractal analysis, and the solid line is for a dual-fractal analysis. The solid line is then the best-fit line.

The values of the binding rate coefficient and the fractal dimension presented in Table 12.1 were obtained from a regression analysis using Corel Quattro Pro 8.0 (1997) to model the data using eq. 12.1 wherein the $(\text{analyte} - \text{receptor}) = k^{(3-D_f)/2}$, for a single-fractal analysis, and $(\text{analyte} - \text{receptor}) = k_1 t^{(3-D_{f1})/2}$, and $= k_2 t^{(3-D_{f2})/2}$, for a dual-fractal analysis,

Table 12.1

Fractal dimensions and binding rate coefficients for (a) the binding of goat anti-rabbit IgG (anti-atrazine antibody) in solution to rabbit IgG (atrazine) coated on a 96-well plate array. Influence of particle size (in nm) (Cummins *et al.*, 2006), and (b) influence of particle size on number of particles in solution bound to the sensor surface (Cummins *et al.*, 2006)

Analyte in solution/ receptor on surface	k	k_1	k_2	D_f	D_{f1}	D_{f2}
(a) Goat anti-rabbit IgG (anti-atrazine antibody)/ rabbit IgG (atrazine); 107 nm particles	125.50 ± 27.12	87.37 ± 24.44	163.87 ± 32.81	1.8950 ± 0.1474	1.5778 ± 0.4922	2.0352 ± 0.2672
Goat anti-rabbit IgG (anti-atrazine antibody)/ rabbit IgG (atrazine); 304 nm particles	242.15 ± 44.28	na	na	1.9178 ± 0.1264	na	na
Goat anti-rabbit IgG (anti-atrazine antibody)/ rabbit IgG (atrazine); 396 nm particles	378.07 ± 65.99	255.32 ± 8.33	1222.62 ± 115.11	1.8314 ± 0.1212	1.5142 ± 0.0656	2.4048 ± 0.1778
(b) Goat anti-rabbit IgG (anti-atrazine antibody)/ rabbit IgG (atrazine); 107 nm particles	73141.02 ± 15093.2	65273.92 ± 12299.18	96819.72 ± 20209.08	1.8896 ± 0.1412	1.8150 ± 0.2528	2.0354 ± 0.2776
Goat anti-rabbit IgG (anti-atrazine antibody)/ rabbit IgG (atrazine); 304 nm particles	887.20 ± 182.91	na	na	1.8836 ± 0.1372	na	na
Goat anti-rabbit IgG (anti-atrazine antibody)/ rabbit IgG (atrazine); 396 nm particles	33.456 ± 0.019	na	na	1.0016 ± 0.0004	na	na

respectively. The analyte–receptor(t) versus time data is regressed to obtain the values of k and p (single-fractal analysis), and k_1 and k_2 and p_1 and p_2 (dual-fractal analysis). Note that $p_1 = (3 - D_{f1})/2$ and $p_2 = (3 - D_{f1})/2$ from which relations the fractal dimension values were obtained. Only those values of the binding rate coefficient and the fractal dimension will be analyzed for which the sum of least squares (r^2) obtained regression is equal to or greater than 0.97. In fact, this is one of the criteria used to go from a single-fractal analysis to a dual-fractal analysis.

The binding rate coefficients presented in Table 12.1a are within 95 confidence limits. For example, for the binding of goat anti-rabbit IgG (anti-atrazine antibody) in solution (107 nm particles) to rabbit IgG (atrazine) immobilized on a 96-well plate the binding rate coefficient, k_1 is equal to 87.37 ± 24.44. The 95% confidence limit indicates that the k_1 lies between 62.93 and 111.81. It is of interest to note that an increase in the fractal dimension by a factor of 1.29 from a value of D_{f1} equal to 1.5778 to D_{f2} equal to 2.0352 leads to an increase in the binding rate coefficient by a factor of 1.875 from a value of k_1 equal to 87.37 to k_2 equal to 163.87. Note that changes in the degree of heterogeneity on the surface (fractal dimension) and in the binding rate coefficient are in the same direction.

Figure 12.1b shows the binding of IgG (anti-atrazine antibody) in solution to rabbit IgG (atrazine) coated on a 96-well plate array. In this case, the biconjugated label was 304 nm. A single-fractal analysis is adequate to describe the binding kinetics. The values of the binding rate coefficient, k and the fractal dimension, D_f for a single-fractal analysis is given in Table 12.1a. No explanation is offered at present why a dual-fractal analysis is required to describe the binding kinetics when a 107 nm biconjugated label was used, and a single-fractal analysis is adequate to describe the binding kinetics when a 304 nm biconjugated label was used. This is especially so since, once again a dual-fractal analysis is required to adequately describe the binding kinetics when a 396 nm conjugated label is used. Please see Figure 12.1c that follows.

Figure 12.1c shows the binding of IgG (anti-atrazine antibody) in solution to rabbit IgG (atrazine) coated on a 96-well plate array (Cummins *et al.*, 2006). The biconjugated label was 396 nm. A dual-fractal analysis is, once again and as mentioned above, required to adequately describe the binding kinetics. The values of (a) the binding rate coefficient, k and the fractal dimension, D_f for a single-fractal analysis, and (b) the binding rate coefficients, k_1 and k_2 and the fractal dimensions, D_{f1} and D_{f2} for a dual-fractal analysis are given in Table 12.1a. Once again, an increase in the fractal dimension by a factor of 1.59 from a value of D_{f1} equal to 1.5142 to D_{f2} equal to 2.4048 leads to an increase in the binding rate coefficient by a factor of 4.79 from a value of k_1 equal to 255.32 to k_2 equal to 1222.62. Note that changes in the degree of heterogeneity on the surface (fractal dimension) and in the binding rate coefficient are in the same direction.

Cummins *et al.* (2006) also investigated the influence of biconjugate particle size (107, 304, and 396 nm) on the number of particles bound (analyte in solution) for the binding of IgG (anti-atrazine antbody) in solution to rabbit IgG (atrazine) coated on a 96-well plate. Figure 12.2a shows the number of particles bound when the 107 nm particle size was used. A dual-fractal analysis is required to adequately describe the binding kinetics and the number of particles bound. Note that for the smallest size of particles used (107 nm), the number of particles bound is the highest. This indicates the presence of increasing diffusional limitations when higher particle sizes (304 and 396 nm) are used.

The values of (a) the binding rate coefficient, k and the fractal dimension, D_f for a single-fractal analysis, and (b) the binding rate coefficients, k_1 and k_2 and the fractal dimensions, D_{f1} and D_{f2} for a dual-fractal analysis are given in Table 12.1b. Once again, an increase in the fractal dimension by a factor of 1.12 from a value of D_{f1} equal to 1.8150 to D_{f2} equal to 2.0354 leads to an increase in the binding rate coefficient by a factor of 1.483 from a value of k_1 equal to 65273.92 to k_2 equal to 96819.72. Note that, once again, changes in the degree of heterogeneity on the surface (fractal dimension) and in the binding rate coefficient are in the same direction.

Figure 12.2b shows the number of particles bound when the 304 nm particle size was used. A single-fractal analysis is required to describe the binding kinetics and the number of particles bound. The values of (a) the binding rate coefficient, k and the fractal dimension, D_f for a single-fractal analysis are given in Table 12.1b. As indicated above, the number of

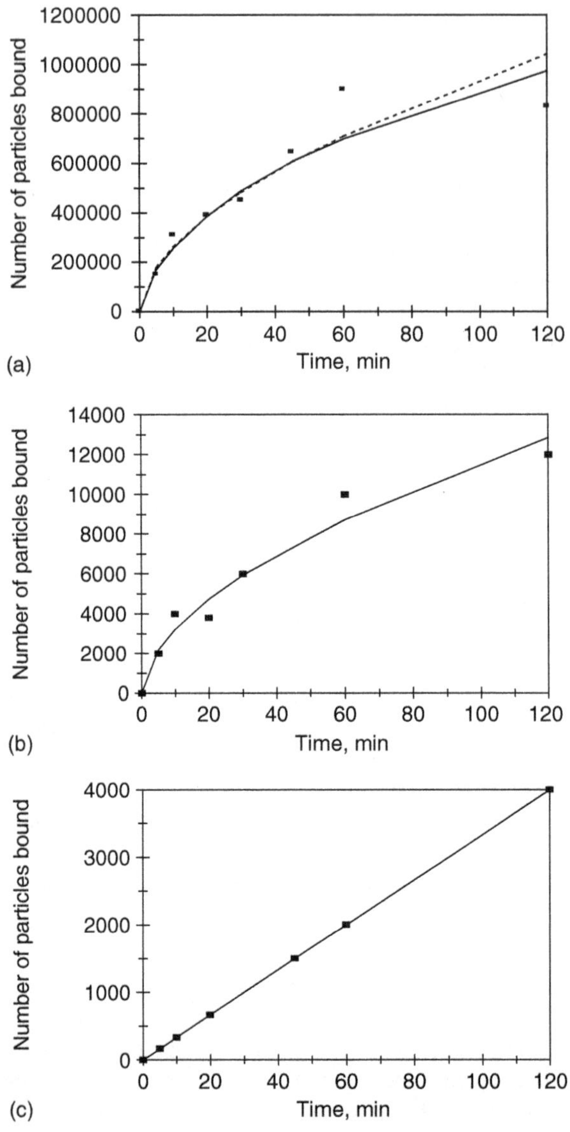

Figure 12.2 Influence of particle size (in nm) on the number of particles bound for the binding of IgG (anti-atrazine antibody) in solution to rabbit IgG (atrazine) coated on a 96-well plate array (Cummins *et al.*, 2006): (a) 107, (b) 304, (c) 396. When only a solid (___) line is used then a single-fractal analysis applies. When both a solid (___) line and a (---) line are used then the dashed line is for a single-fractal analysis, and the solid line is for a dual-fractal analysis. The solid line is then the best-fit line.

particles bound in this case is less than that observed when the 104 nm particle size biconjugated label was used.

Figure 12.2c shows the number of particles bound when the 396 nm particle size was used. Once again, a single-fractal analysis is required to describe the binding kinetics and the number of particles bound. The values of (a) the binding rate coefficient, k and the fractal dimension, D_f for a single-fractal analysis are given in Table 12.1b. As indicated above, the number of particles bound in this case is less than that observed when the 104 nm particle size biconjugated label was used. It is of interest to note that when the smaller particle size is used (107 nm) a dual-fractal analysis required to describe the binding kinetics, whereas, when the larger particle sizes are used (304 and 396 nm), a single-fractal analysis is adequate to describe the binding kinetics.

Hai *et al.* (2006) have very recently developed an acetylcholinesterase-ISFET for the detection of acetylcholine and acetylcholinesterase inhibitors. These authors indicate that ISFET technology provides a means whereby the normal metal-oxide-silicon field-effect transistor (MOSFET) gate electrode may be replaced by an ion-selective surface. Bergveld (1970, 1972) has indicated that this ion-sensitive surface permits the detection of ion concentrations in solution. Furthermore, Khartinov *et al.* (2000) indicate that the ISFET technology permits high quality performance especially with regard to the detection of enzyme-substrate reactions. Hai *et al.* (2006) indicate that acetylcholine (ACh) is a ubiquitous neurotransmitter. It is found in the peripheral and central nervous system. These authors emphasize that the response time of the ISFET to bath or ionophoretic application of the ACh from a micropipette is of the order of a second. They further indicate that *in vivo* acetylcholinestaerase (AChE) resides in the presynaptic cleft between presynaptic cholinergic neurons and their post-synaptic counterparts. This enzyme catalyzes the hydrolysis of acetylcholine.

Hai *et al.* (2006) have very recently analyzed the influence of ionophoretic charge (nC) on the binding and the dissociation of acetylcholine (Ach) in solution to acetylcholinesterase immobilized on an ISFET biosensor. Figure 12.3 shows the ACh iontophoresis onto an AChE-ISFET biosensor. Figure 12.3a shows the influence of 846.6 nC (ionophoretic charge) on the 'binding' and 'dissociation' kinetics of ACh in solution to the acetylcholinestearse immobilized on an ISFET biosensor. A single-fractal analysis is adequate to describe the binding and dissociation kinetics. The values of the binding rate coefficient, k and the fractal dimension, D_f for a single-fractal analysis, and the dissociation rate coefficient, k_d and the fractal dimension for the dissociation phase, D_{fd} for a single-fractal analysis are given in Table 12.2. In this case the affinity, K $(=k/k_d)$ is equal to 13.35.

Figure 12.3b shows the influence of 495.3 nC (ionophoretic charge) on the binding and dissociation kinetics of ACh in solution to the acetylcholinesterase immobilized on an ISFET biosensor (Hai *et al.*, 2006). In this case, a single-fractal analysis is adequate to describe the binding kinetics. However, a dual-fractal analysis is required to describe the dissociation kinetics. The values of (a) the binding rate coefficient, k and the fractal dimension, D_f for a single-fractal analysis, (b) the dissociation rate coefficient, k_d and the fractal dimension, D_{fd} for a single-fractal analysis, and (c) the dissociation rate coefficients, k_{d1} and k_{d2} and the fractal dimensions, D_{fd1} and D_{fd2} for a dual-fractal analysis are given in Table 12.2. In this case, the affinity values are, K_1 $(=k/k_{d1}) = 75{,}000$ (very high, ignore), and K_2 $(=k/k_{d2}) = 1.0582$. The K_1 value is very high since the value of $D_{fd1} = 0.0$, and the corresponding value of k_{d1} is very low.

Figure 12.3c shows the influence of 350.6 nC (ionophoretic charge) on the 'binding' and 'dissociation' kinetics of ACh in solution to the acetylcholinestearse immobilized on an

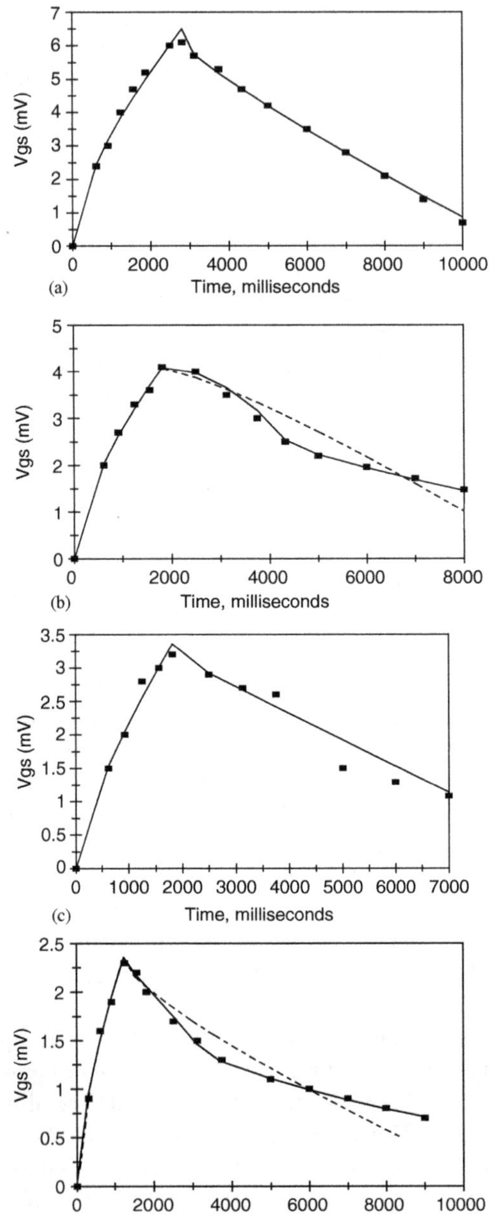

Figure 12.3 Binding of acetylcholine (ACh) in solution to acetylcholinesterase immobilized on an ISFET biosensor. Influence of the ionophoretic charge (nC) (Hai *et al.*, 2006): (a) 846.6, (b) 495.3, (c) 350.6, (d) 258.2.

Table 12.2

Binding and dissociation rate coefficients and fractal dimensions for the binding and the dissociation phase for acetylcholine (AC) in solution to acetylcholinesterase (AChE) immobilized to the gate surface of an ion-sensitive field-effect transistor (ISFET) (Hai *et al.*, 2006). Influence of ionophoretic charge

Ionophoretic charge	D_f	D_{fd}	D_{fd1}	D_{fd2}	k	k_d	k_{d1}	k_{d2}
846.6	1.6896 ± 0.0695	1.2992 ± 0.0467	na	na	0.03671 ± 0.0797	0.002749 ± 0.000192	na	na
495.3	1.5176 ± 0.06952	0.5604 ± 0.3358	0 + 0.4712	1.9612 ± 0.0343	0.0300 ± 0.009	7.3E-05 ± 3.6E-05	4.0E-07 ± 1.3E-07	0.02835 ± 0.00039
350.6	1.5176 ± 0.1473	1.0314 ± 0.1653	na	na	0.02193 ± 0.00084	0.000456 ± 0.000091	na	na
258.2	1.6610 ± 0.1204	1.4340 ± 0.1501	0.8848 ± 0.2508	2.2086 ± 0.0295	0.01993 ± 0.00128	0.001761 ± 0.000488	0.000286 ± 0.000069	0.04589 ± 0.00039

ISFET biosensor (Hai *et al.*, 2006). A single-fractal analysis is adequate to describe the binding and dissociation kinetics. The values of the binding rate coefficient, k and the fractal dimension, D_f for a single-fractal analysis, and the dissociation rate coefficient, k_d and the fractal dimension for the dissociation phase, D_{fd} for a single-fractal analysis are given in Table 12.2. In this case the affinity, K ($=k/k_d$) is equal to 48.092.

Figure 12.3d shows the influence of 258.2 nC (ionophoretic charge) on the binding and dissociation kinetics of ACh in solution to the acetylcholinesterase immobilized on an ISFET biosensor (Hai *et al.*, 2006). In this case, a single-fractal analysis is adequate to describe the binding kinetics. However, a dual-fractal analysis is required to describe the dissociation kinetics. The values of (a) the binding rate coefficient, k and the fractal dimension, D_f for a single-fractal analysis, (b) the dissociation rate coefficient, k_d and the fractal dimension, D_{fd} for a single-fractal analysis, and (c) the dissociation rate coefficients, k_{d1} and k_{d2} and the fractal dimensions, D_{fd1} and D_{fd2} for a dual-fractal analysis are given in Table 12.2. In this case, the affinity values are, K_1 ($=k/k_{d1}$) $=69.68$, and K_2 ($=k/k_{d2}$) $= 0.434$.

Figure 12.4a and Table 12.2 show the increase in the binding rate coefficient, k for a single-fractal analysis with an increase in the ionophoretic charge (nC). For the data shown in Figure 12.4a and in Table 12.2, the binding rate coefficient, k is given by

$$k = (0.000382 \pm 0.000225)[\text{ionophoretic charge, nC}]^{0.6632 \pm 0.5320} \tag{12.4a}$$

The fit is reasonable. Only three data points are available. The availability of more data points would lead to a more reliable fit. There is scatter in the data, and this is reflected in the error in the estimated value of the rate coefficients. The binding rate coefficient, k exhibits an order of dependence between one-half and first (equal to 0.6632) on the ionophoretic charge, nC. The non-integer dependence exhibited by the binding rate coefficient, k lends support to the fractal nature of the system.

Figure 12.4b and Table 12.2 show the increase in the binding rate coefficient, k for a single-fractal analysis with an increase in the fractal dimension, D_f. For the data shown in Figure 12.4b and in Table 12.2, the binding rate coefficient, k is given by

$$k = (0.000416 \pm 0.000208) D_f^{8.202 \pm 5.338} \tag{12.4b}$$

The fit is reasonable. Only three data points are available. The availability of more data points would lead to a more reliable fit. There is scatter in the data, and this is reflected in the error in the estimated value of the rate coefficients. The binding rate coefficient, k is extremely sensitive to the degree of heterogeneity on the sensor surface or the fractal dimension, D_f that exists on the sensor surface as noted by the higher than eighth (equal to 8.202) order of dependence exhibited.

Figure 12.4c and Table 12.2 show the increase in the dissociation rate coefficient, k_d for a single-fractal analysis with an increase in the fractal dimension, D_{fd}. For the data shown in Figure 12.4c and in Table 12.2, the dissociation rate coefficient, k_d is given by

$$k_d = (0.000462 \pm 0.000447) D_{fd}^{4.754 \pm 2.827} \tag{12.4c}$$

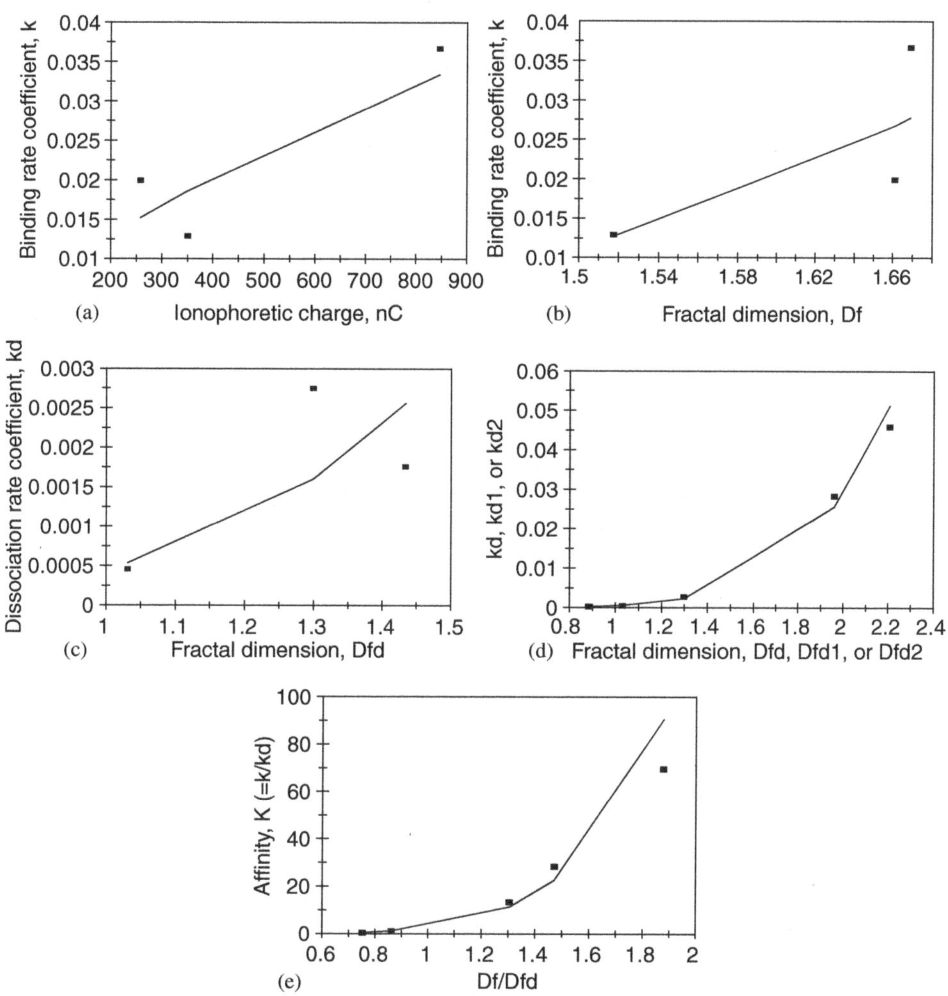

Figure 12.4 (a) Increase in the binding rate coefficient, k for a single-fractal analysis with an increase in the ionophoretic charge, nC. (b) Increase in the binding rate coefficient, k for a single-fractal analysis with an increase in the fractal dimension, D_f. (c) Increase in the dissociation rate coefficient, k_d for a single-fractal analysis with an increase in the fractal dimension, D_{fd}. (d) Increase in the dissociation rate coefficient, k_d, k_{d1}, or k_{d2} with an increase in the fractal dimension, D_{fd}, D_{fd1}, or D_{fd2}. (e) Increase in the affinity, K $(=k/k_d)$ with an increase in the fractal dimension ratio for the binding and the dissociation phases (D_f/D_{fd}).

The fit is reasonable. Only three data points are available. The availability of more data points would lead to a more reliable fit. There is scatter in the data, and this is reflected in the error in the estimated value of the rate coefficients. The dissociation rate coefficient, k_d is very sensitive to the degree of heterogeneity on the sensor surface or the fractal dimension, D_{fd} that exists on the sensor surface during the dissociation phase as noted by the order of dependence between four and one-half and five (equal to 4.754) exhibited.

It is perhaps instructive to plot the dissociation rate coefficient as a function of the degree of heterogeneity that exists on the sensor surface, irrespective whether a single- or a dual-fractal analysis is required to fit the data. This is also done primarily since only three data points were available for a single-fractal analysis. For the data shown in Table 12.2 and in Figure 12.4d, the dissociation rate coefficient, k_d is given by

$$k_d, k_{d1}, k_{d2} = (0.00051 \pm 0.00128)[D_{fd}, D_{fd1}, D_{fd2}]^{5.816 \pm 0.283} \tag{12.4d}$$

The fit is very good. Note that points obtained for a single- and a dual-fractal analysis are plotted together. Only five data points are available. The dissociation rate coefficient is very sensitive to the fractal dimension or the degree of heterogeneity that exists on the ISFET sensor surface as noted by the close to sixth (equal to 5.816) order of dependence exhibited.

Affinity values are of interest to practicing biosensorists. Figure 12.4e and Table 12.2 show the increase in the affinity, K with an increase in the ratio of the fractal dimensions present in the binding and in the dissociation phases, respectively. For the data shown in Figure 12.4e and in Table 12.2, the affinity, K is given by

$$K = (2.489 \pm 0.65)[\text{ratio of fractal dimensions present in the binding and in the dissociation phase}]^{5.708 \pm 0.305} \tag{12.4e}$$

The fit is very good. Only five data points are available. The availability of more data points would lead to a more reliable fit. The affinity, K is very sensitive to the ratio of fractal dimensions present in the binding and in the dissociation phases as noted by the order of dependence between five and one-half and six (equal to 5.708) exhibited. The above equation is useful since it permits a means by which the affinity, K on the ISFET sensor may be manipulated. This may require a little ingenuity since by changing the degree of heterogeneity on the ISFET sensor surface one may change both of the fractal dimensions present in the binding and in the dissociation phases, respectively.

Zucolotto *et al.* (2006) have recently developed a highly sensitive biosensor for the detection of catachol by immobilizing C1-catechol 1,2 dehydrogenase (CCD) in nanostructured films. These authors used CCD layers alternated with poly(amidoamine) generation 4 (PAMAM G4) dendrimer using the electrostatic layer-by-layer (Lbl) technique. They used either an optical or an electrical approach to detect catechol in dilute solutions using their PAMAM/CCD 10-bilayer films. These authors emphasize that they were able to detect catechol in solutions at concentrations as low as 10^{-10}M. Lvov *et al.* (1993, 1995) had initially developed the Lbl technique. Broderick and O'Hallaran (1991) emphasize the importance of detecting aromatic compounds and pesticides in wastewater streams that withstand chemical oxidation and biological degradation.

Figure 12.5a shows the binding of 10^{-7}M catechol in solution to a 10-bilayer PAMAM/CCD nanostructured film deposited on a quartz surface (Zucolotto *et al.*, 2006). These authors indicate that the increase in absorbance observed is due to the formation of *cis, cis*-muconic acid with the Lbl film. A single-fractal analysis is adequate to describe the binding kinetics. The values of the binding rate coefficient, k and the fractal dimension, D_f for a single-fractal analysis are given in Table 12.3.

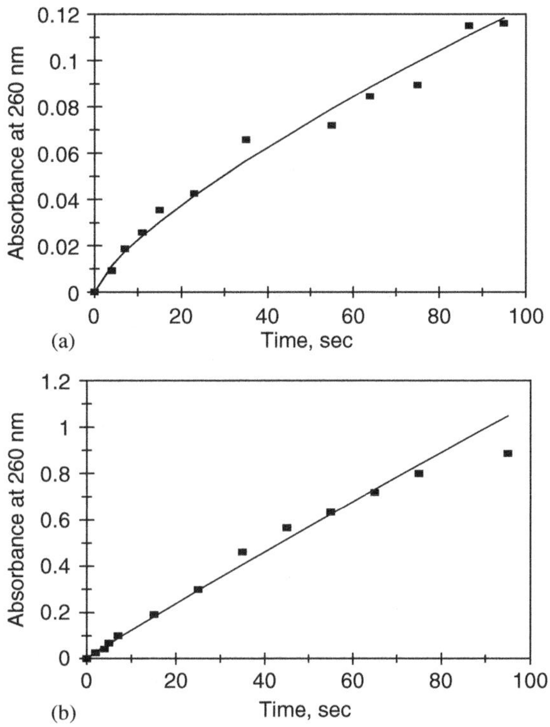

Figure 12.5 Binding of different concentrations (in M) of catechol in solution to a 10-bilayer PAMAM (poly(amidoamine)) dendrimer/CCD (C1 catechol 1,2-deoxygenase) film on a nanostructured surface (Zucolotto *et al.*, 2006): (a) 10^{-2}, (b) 10^{-7}.

Table 12.3

Binding rate coefficients and fractal dimensions for the binding of two different concentrations of catechol in solution (in M) to a 10-bilayer PAMAM/CCD nanostructured film deposited on a quartz surface (Zucolotto *et al.*, 2006)

Catechol concentration (M)	k	D_f
10^{-7}	1.5290 ± 0.06788	0.004157 ± 0.000518
10^{-2}	1.1048 ± 0.04908	0.01402 ± 0.00156

Figure 12.5b shows the binding of 10^{-2} M catechol in solution to a 10-bilayer PAMAM/CCD nanostructured film deposited on a quartz surface (Zucolotto *et al.*, 2006). Once again, a single-fractal analysis is adequate to describe the binding kinetics. The values of the binding rate coefficient, k and the fractal dimension, D_f for a single-fractal analysis are given in Table 12.3. It is of interest to note that as the catechol concentration in solution increases by five orders of magnitude from 10^{-7} to 10^{-2} M, the fractal dimension decreases by a 27.7% from a value of D_f equal to 1.5290 to 1.1048, and the binding rate coefficient, k increases by a factor of 3.37 from a value of k equal to 0.004157 to

0.01402. Note that in this case, the changes in the binding rate coefficient, k and in the fractal dimension, D_f or the degree of heterogeneity on the sensor surface are in opposite directions.

12.4 CONCLUSIONS

A fractal analysis is used to analyze the binding and dissociation (if applicable) of analytes of environmental pollution interest on biosensor surfaces. The analytes analyzed include (a) the binding of anti-atrazine antibody (IgG) in solution to atrazine (rabbit IgG) coated on a 96-well plate array (Cummins *et al.*, 2006), (b) the binding of ACh in solution to acetylcholinesterase immobilized on an ISFET biosensor, and (c) the binding of different concentrations (in nM) of catechol in solution to a 10-bilayer PAMAM poly(amidoamine) dendrimer/CCD (C1 catechol 1,2-deoxygenase) film on a nanostructured surface (Zucolotto *et al.*, 2006).

The binding kinetics is described by either a single- or a dual-fractal analysis. The dual-fractal analysis is used only when a single-fractal analysis did not provide an adequate fit. This was done using Corel Quattro Pro 8.0 (1997). The fractal dimension provides a quantitative measure of the degree of heterogeneity present on the ISFET biosensor surface. As indicated previously in the different chapters in the book, the fractal dimension is not a typical independent variable, such as analyte concentration that may be directly manipulated. It is estimated from eqs. (12.1)–(12.3), and one may consider it as a derived variable.

An increase in the fractal dimension value or the degree of heterogeneity on the biosensor surface leads to an increase in the binding and the dissociation rate coefficient. This is noted for the binding of acetylcholine (ACh) in solution to acetylcholinesterase immobilized on an ISFET biosensor (Hai *et al.*, 2006). The binding rate coefficient, k for a single-fractal analysis is extremely sensitive to the degree of heterogeneity or the fractal dimension on the ISFET biosensor surface as noted by the greater than eighth (equal to 8.202) order of dependence exhibited. Similarly, the dissociation rate coefficient, k_d is very sensitive to the degree of heterogeneity or the fractal dimension in the dissociation phase, D_{fd} on the ISFET biosensor surface as noted by the order of dependence between four and one-half and five (equal to 4.754) exhibited.

A predictive equation for affinity, K $(= k/k_d)$ as a function of the ratio of fractal dimensions present in the binding and in the dissociation phases is also presented. This equation exhibits an order of dependence between five and one-half and six (equal to 5.708) on the ratio of fractal dimensions present in the binding and in the dissociation phases. In accord with the prefactor analysis for fractal aggregates (Sorenson and Roberts, 1997), predictive relations are also developed for (a) the binding rate coefficient, k for a single-fractal analysis as a function of the ionophoretic charge (nC) for the binding of acetylcholine (ACh) in solution to the acetylcholinesterase immobilized on an ISFET biosensor surface (Hai *et al.*, 2006).

The binding and dissociation (if applicable) kinetics of environmental pollutants present either in the gas phase or in solution need to be analyzed. This is especially so since quite a few of these pollutants are resistant to chemical action and to biological degradation, and have been linked to the onset of intractable and insidious diseases, even if present in rather dilute solutions or concentrations. The identification of these harmful environmental pollutants is the first step in their elimination from either water or air environments. The accurate determination of these harmful environmental pollutants requires the continuous development of more sensitive, reliable, and accurate biosensors. The analysis of their kinetics of binding and dissociation is a step in this direction by providing a means by which biosensor performance parameters may be improved. Furthermore, environmental contamination detection will continue to remain an important part of biosensor applications in the near future.

REFERENCES

Bergveld, P, Development of an ion-sensitive solid-state device for neurophysiological measurements. *IEEE Transactions in Biomedical Engineering*, 1970, **17**(1), 70–71.

Bergveld, P, Development, operation, and application of ion-sensitive field-effect transistor as a tool for electrophysiology. *IEEE Transactions in Biomedical Engineering*, 1972, **19**(5), 342–351.

Broderick, JB and TV O'Hallaran, Overproduction, purification, and characterization of chlorocatechol dioxygenase, a non-heme iron dioxygenase with broad substrate tolerance. *Biochemistry*, 1991, **30**, 7349–7358.

Ciumasu, IM, PM Kramer, CM Weber, G Kolb, D Tiemann, S Windisch, I Frese, and AA Kettrup, A new, versatile field immunosensor for environmental pollutants. Development and proof of principle with TNT, diuron, and atrazine. *Biosensors and Bioelectronics*, 2005, **21**, 354–364.

Corel Quattro Pro 8.0, Corel Corporation, Ottawa, Canada, 1989.

Cummins, CM, ME Koivunen, A Stephanian, SJ Gee, BD Hammock, and IM Kennedy, Application of europium (III) chelate-dyed nanoparticle labels in a competitive atrazine fluoroimmunoassay on an ITO waveguide. *Biosensor and Bioelectronics*, 2006, **21**, 1077–1085.

Glass, TR, N Ohumura, K Morita, K Sasaki, H Saiki, Y Takagi, C Kataoka, and A Ando, Improving an immunoassay response to related polychlorinated biphenyl analytes by mixing antibodies. *Analytical Chemistry*, 2006, **78**, 7240–7247.

Greenbaum, E and M Rodriguez, *Aquasentinel: a Real Time Reagentless Biosensor System for Standoff Detection and Classification of Toxins in Source Water*. Paper #24, SERMACS 2006, The Southeastern Regional Meeting of the American Chemical Society, Augusta, GA, November 1–4, 2006.

Hai, A, D Ben-Haim, N Korbakkov, A Cohen, J Shappir, R Oren, ME Spira, and S Yitzchaik, Acetylcholineesterase-ISFET based system for the detection of acetylcholine and acetylcholinesterase inhibitors. *Biosensors and Bioelectronics*, 2006.

Hakansson, K and B Mattiasson, A biosensor for the analysis of acetonitrile. *Biosensors and Bioelectronics*, 2004, **19**, 721–726.

Havlin, S, Molecular diffusion and reaction. in: *The Fractal Approach to Heterogeneous Chemistry: Surfaces, Colloids, Polymers* (ed. D. Avnir), Wiley, New York, 1989, pp. 251–269.

Hawkridge, FM, L Su, and MC Rhoten, *An Oxidase Modified Electrode for Cyanide Sensing*. Paper #100. SERMACS 2006, The Southeastern Regional Meeting of the American Chemical Society, Augusta, GA, November 1–4, 2006.

Ivandini, TA, R Sato, Y Makide, A Fujishima, and Y Einaga, Electrochemical detection of arsenic (III) using iridium-implanted boron-doped diamond electrodes. *Analytical Chemistry*, 2006, **78**, 6291–6298.

Khartinov, AB, M Zayats, A Lichtenstein, E Katz, and I Willner, Enzyme monolayer-functionalized field-effect transistors for biosensor application. *Sensors and Actuators B: Chemical*, 2000, **70**, 222–231.

Lao, UL, P Hehn, G Jenikova, A Mulchandani, and W Chen, *Potential Environmental Application of Self-Assembled Protein Hydrogel Based on Elastin-Like Protein Sequences.* Paper 447l, Annual American Institute of Chemical Engineers Meeting, San Francisco, CA, November 12–17, 2006.

LeBlanc, RM and C Wang, *Fluorescent Bioassays Based on Organophosphorous Hydrolase (OPH).* Paper #103, SERMACS 2006, The Southeastern Regional Meeting of the American Chemical Society, Augusta, GE, November 1–4, 2006.

Lee, CK and SL Lee, Multi-fractal scaling analysis of reactions over fractal surfaces. *Surface Science*, 1995, **325**, 294–310.

Liu, G and Y Lin, Biosensor based on self-assembling acetylcholinesterase on carbon nanotubes for flow injection/amperometric detection of organophosphate pesticides and nerve agents. *Analytical Chemistry*, 2006, **78**, 835–843.

Lvov, Y, K Ariga, I Ichinose, and T Kunitake, Assembly of multicomponent protein films by means of electrostatic layer-by-layer adsorption. *Journal of the American Chemical Society*, 1995, **117**, 6117–6123.

Lvov, Y, G Decher, and G Sukhorukov, Assembly of thin films by means of successive deposition of alternate layers of DNA and poly(allyl amine). *Macromolecules*, 1993, **26**, 5396–5399.

Mulchandani, A, Z Meng, and W Chen, *Molecularly Imprinted Polymers for Estrogenic Pollutants Removal.* Paper 560f, Annual American Institute of Chemical Engineers Meeting, San Francisco, CA, November 12–17, 2006a.

Mulchandani, A, NV Myung, J Wu, and W Chen, *Chelator-Doped Conducting Polymer Thin Films for Heavy Metal Ion Detection.* Paper 545d, Annual American Institute of Chemical Engineers Meeting, San Francisco, CA, November 12–17, 2006b.

Oh, I, CN Monty, MA Shannon, and RI Masel, *Electrochemical Sensor for Toxic Chemicals Based on Nanodeposit and Enzyme.* Paper 165a, Annual American Institute of Chemical Engineers Meeting, San Francisco, CA, November 12–17, 2006.

Orbulescu, J, CA Constantine, VK Rastogi, SS Shah, JJ DeFrank, and RM LeBlanc, Detection of organophosphorous compounds by covalently immobilized organophosphorous hydrolase. *Analytical Chemistry*, 2006, **77**, 8115–8121.

Riley, MR and P Lucas, *Cell-Based Non-Invasive Sensing of Inhalation Health Hazards.* Paper 545a, Annual American Institute of Chemical Engineers Meeting, San Francisco, CA, November 12–17, 2006.

Sharma, V, A Narayanan, T Rengachari, GC Temes, F Chaplen, and UK Moon, A low-cost, portable generic biotoxicity assay for environmental monitoring applications. *Biosensors and Bioelectronics*, 2005, **20**, 2218–2227.

Shiddiky, MJA, H Park, and YB Shim, Direct analysis of trace phenols with a microchip: in-channel preconcentration, separation, and electrochemical detection. *Analytical Chemistry*, 2006, **78**, 6809–6817.

Sorenson, CM and GC Roberts, The prefactor of fractal aggregates. *Journal of Colloid and Interface Science*, 1997, **186**, 447–452.

Suwansaard, S, P Kanatharana, P Asawatreratanakul, C Limsakul, B Wongkittisuksa, and P Thavarungkul, Semi disposable reactor biosensors for detecting carbamate pesticides in water. *Biosensor and Bioelectronics*, 2005, **21**, 445–454.

Weng, YC, JF Rick, and TC Chou, A sputtered thin film of nanostructured Ni/Pt/Ti on Al_2O_3 substrate for ethanol sensing. *Biosensors and Bioelectronics*, 2004, **20**, 41–51.

White, BR and JA Holcombe, *Peptide Sensors Utilizing Fluorescence Resonance Energy Transfer for the Selective Detection of Heavy Metals.* Paper #30, SERMACS 2006, The Southeastern Regional Meeting of the American Chemical Society, Augusta, GA, November 1–4, 2006.

Yamasaki, A, MASDA Cunha, JABP Oliveira, AC Duarte, and MTSR Gomes, Assessment of copper toxicity using an acoustic wave sensor. *Biosensors and Bioelectronics*, 2004, **19**, 1203–1208.

You, F, YF Zhou, XE Zhang, Z Huang, LJ Bi, ZP Zhang, JK Wen, YY Chen, GB Jiang, and MH Zheng, Cell-free bioassay for measurement of dioxins based on fluorescence enhancement of fluorescein isothiocyanate-labeled DNA probe. *Analytical Chemistry*, 2006, **78**, 7138–7144.

Zhang, C and KS Suslick, *A Calorimeter Sensor Array for the Detection of Toxic Chemicals (Tics)*. Paper 545e, Annual American Institute of Chemical Engineers Meeting, San Francisco, CA, November 12–17, 2006.

Zhang, J, N Liu, J Hui, H Xiao, and J Dong, *Zeolite Thin Film-Based Fiber Intrinsic Fabry-Perot Interferometric Sensor for Detection of Dissolved Organics and Ions in Water*. Paper 545c, Annual American Institute of Chemical Engineers Meeting, San Francisco, CA, November 12–17, 2006.

Zucolotto, V, APA Pinto, T Tumolo, ML Moraes, MS Baptista, A Riul, Jr., APAU Araujo, and ON Oliveira, Jr., Catechol biosensing using a nanostructured layer-by-layer film containing Cl-catechol 1,2-dioxygenase. *Biosensors and Bioelectronics*, 2006, **21**, 1320–1326.

– 13 –

Market Size and Economics for Biosensors

13.1 INTRODUCTION

Biosensors are gradually being used in an increasing number of applications. Heffner (2006) very recently emphasizes that though medical applications still remain the major area of application, biosensors are gradually penetrating a wide variety of non-medical areas. This author indicates that the breadth and depth of biosensor applications has resulted in a 25% increase in the world biosensor market prediction made as early as August 2002 by Kalorama Information for the year 2005 of \$2.3 billion to \$2.9 billion. The revised estimate was made in the year 2006. Heffner (2006) further indicates that earlier predictions indicated that the medical application was around 90% of the total biosensor market, which includes clinical applications for diabetes, cholesterol, and coagulation monitoring. However, medical applications of biosensors are now down to around 60%. Heffner (2006) attributes these changes to the rapidly advancing technologies and to some serious business challenges.

Newman and Setford (2006) in their very recent review of enzymatic biosensors indicate that the biosensor field has expanded dramatically in the last 45 years or so since the first demonstration of a biosensor application in 1962. These authors emphasize that the self-diagnosis of blood glucose levels by diabetes sufferers, however, still remains the dominant area of biosensor application. In their recent review they provide a historical development of the biosensor field and also provide a commercial perspective.

Business Wire (2005a,b) mentions that Frost and Sullivan (http://www.sensors.frost.com) indicate that the revenue from the World Biosensors market was \$2.34 billion in 2004. This is projected to increase by 87.1% to \$4.38 billion by the year 2008. This report emphasizes that a couple of key issues that are restraining the biosensor market are a small number of research laboratories and a low level of commercialization of biosensors that are developed in research laboratories. The increased investment in new laboratories is essential to permit the application research required for determining the commercial applicability of biosensors.

Market Research.com (2005) has projected the medical biosensor applications and market to the year 2008. This 81-page report was published by Takeda Pacific, and provides (at least the authors claim this) critical business and competitive intelligence required for strategic planning and market research in the biosensor area. The report claims to cover the United States, United Kingdom, Europe, Asia, and global markets.

The report estimates that global biosensor market for medical purposes to about $7.1 billion in the year 2004. For the years 2005, 2006, 2007, and 2008, the estimated biosensor market is in dollar billion, 7.79, 8.54, 9.37, and 10.28, respectively. This report analyzes medical biosensors and applications, diabetes and clinical applications, and medical biosensors for artificial organs. The report emphasizes the development of innovative medical biosensor devices that would work as a part of artificial organs such pancreas, liver, or kidney. The report also examines the activities at key biosensor companies, and selected academic centers. Finally, it also provides a future outlook for the development of biosensors.

Yurish *et al.* (2005) have recently analyzed the trends in world sensors and microelectromechanical systems (MEMS) markets. These authors emphasize that fixed and durable medical sensors are gradually being replaced by disposable biosensors that cost less and are more hygienic. They indicate that low cost and high volume sales are becoming the key components for remaining competitive in the medical biosensor market. In order to enhance biosensor market penetration into non-medical areas of application, increasing investment is recommended for biosensor R & D (Infoshop.com, 2005; Frost and Sullivan, 2005). Yurish *et al.* (2005) cite a Business Communication Company, Inc. (Norwalk, CT, USA) report (RG-116N Fiber Optic Sensors) that the global revenues for fiber optic sensors was projected to be $304.3 million for the year 2006. This was projected to increase by 4.1% to $371.8 million by the year 2010.

Fraser (1995) indicated that in 1995 the control of diabetes or the determination of blood glucose levels was the only high-volume market for biosensors. However, even at that time this author indicated that there were opportunities available in a variety of lower volume niches in the medical diagnostics, pharmaceuticals, and other areas. This author emphasized that the future expansion of biosensor technology would be dictated by a balance between market opportunities and technical and/or financial obstacles.

Menon (2004) in a short review on optical biosensors mentions two models of commercial case studies for biosensors. He indicates that the first biosensor was commercialized in 1991. The two models are the razor blade model and the OEM model.

In the razor blade model the technology platform for the biosensor is such that the revenue is generated by a consumable commodity rather than by an actual instrument. The razor/razor blade is a good example, wherein the revenue is generated from the sale of razor blades. This author emphasizes that photonics based biosensors, such as the glucose monitor is based on this principle. In the OEM model, the biosensors such as the SPR biosensor made by Biacore provides high throughput. These are expensive instruments, and costs of hundreds of thousands of dollars, and can only be purchased by industrial organizations and well-heeled university research laboratories.

An example or two of the razor/razor blade model as it applies to biosensors and diagnostics is perhaps useful here. The diabetic test strip market is very large and profitable. These strips cost a few cents to make, and sell for 60–75 cents each in boxes of 100 for $60–75 (DiabetesStore.com, 2007; Diabetic Test Strips, 2007). Small Times (2007) emphasizes that these strips are the profit center. It is estimated that there are 12 million diabetics who use these strips eight or more times a day. This means that over close to 100 million strips are used each day in the United States alone. Using the low value of 60 cents per strip this leads to a revenue of \$60 million from diabetic strips in the United States alone. The number of diabetics is estimated to climb to 14.5 million by the year 2008 (Diabetic Test Strips, 2007). The number of diabetics was estimated to be 12 million in the year 2003 (Smalltimes, 2003). One may anticipate errors in the estimate of the actual number of diabetics made by different authors, nevertheless, the numbers of diabetics is very large, and in all probability, this number is bound to increase due to increasing obesity and poor nutritional intake by quite a large fraction of the population, in general. Diabetic Test Strip (2007) indicated that the sales of diabetic strips was \$1.6 billion in the year 2002. In the year 2008 this diabetes glucose monitoring market was estimated to be \$3 billion, and is/was expected to climb by 7.2% annually in the year 2003–2008, period.

The glucose testing meters in accord with the razor/razor blade model are either available at no cost, or at a minimal cost and their price is decreasing. A very recent check on the prices for glucose meters on the internet (Test Medical Symptoms@Home, Inc., 2007) reveals that one can get a glucose meter plus 10 test strips for a price range of \$15–66. As mentioned above, most companies will supply meters free of charge, and this is also evident from the recent TV advertisements, at least in the United States, especially if one is of age (65 and over, generally), and under Medicare. Testing strips and meters are also available for monitoring and determining cholesterol levels. Some strips like the Cardiochip PA Cholesterol Plus Glucose Test will check cholesterol and glucose in multi-panel test strips (Test Medical Symptoms@Home, Inc., 2007). One may also obtain meters to check for cholesterol, low-density lipoproteins (LDL) and high-density lipoproteins (HDL).

Axela Biosensors (2006) has recently indicated that it will incorporate Beckman Coulter's Universal Linker Capture Technology in its dotLab™ biosensors. It further indicates that Beckman Coulter reported a 2005 annual sales of \$2.44 billion. Out of this 71.5% of this revenue was generated by recurring revenue from supplies, test kits, services, and operating lease agreements. Once again, reinforcing the razor blade–razor model. Finally, one may also have noticed the razor–razor blade model in other non-biosensor business/marketing areas such as the printers and printer cartridges, where the major profit lies in the sale of the printer cartridges vis-à-vis the cost of the printer itself.

Radke and Evangelyn (2002) analyzed the biosensor market for the pathogen detection industry. These authors combined the military, food, medical, and the food industry together. In the year 2002, their estimate for the pathogen detection systems was \$56 million, with an anticipated compounded annual growth rate of 4.5%. By the year 2005 these authors estimated this market to grow to \$192 million with an estimate of 34 million tests. This works out to about \$5.64 per test. Perhaps, under present day circumstances this number could be estimated to be \$10 per test. These authors further state that as legislation creates new standards for microbial monitoring, especially after a pathogenic outbreak or two that seems to occur regularly after periodic intervals, there will be pressure to commercialize biosensors for food industry application that are increasingly sensitive to these pathogenic organisms.

Patel (2006) in a recent review of affinity biosensors for food analysis applications estimates the projected global affinity biosensor market to grow from \$6.1 billion in 2004 to \$8.2 billion in the year 2009 for the major industrial sectors such as Pharma, Medicare, and Food. This represents a 34.4% increase in about 5 years. Note that, and as expected, the projected estimates for the worldwide biosensor market are different by different authors. The projections are dependent upon who is making these projections, on what assumptions are these projections based on, and the reliability of these assumptions and projections, and the individual or organization making these projections.

Kalorama Information (2006) has recently published a report entitled "Medical and Biological Sensors and Sensor Systems: Markets, Applications, and Competitors Worldwide, 2nd Edition." In this 290 page, $3500 report the authors analyze the commercial challenges for biosensors. They also discuss the advantages of different biosensors such as the surface plasmon resonance (SPR) biosensor, the use of monolithic semiconductor processing, the application of thin and thick films for sensor fabrication, the applications of polymer films in sensors and fiber-optic biosensors. These authors delineate biosensor applications, and also present their estimates of the biosensor markets by geographic locations such as North America, Europe, Japan, Latin America, China and Japan, and the rest of the world. These authors emphasize that the commercial medical biosensor market is dynamic in the sense that reports that are a year or two old are out of date due to the rapid penetration of new technology. Finally, the report emphasizes the potential markets and emerging technologies in the different biosensor areas of application.

These authors also attempt to analyze the different variables that may influence the size of biosensor markets, or how the biosensor revenues may be affected. Some of these variables include (a) the high costs of biosensor development, (b) there are manufacturing and product problems, (c) technologies that are synergistic with biosensors are maturing and competing, and (d) nanotechnology and miniaturization is bound to influence biosensor development costs and economics in general. Interestingly enough, these authors also examine the role that small companies with niche markets play vis-à-vis large companies with mass markets.

The Center for High-Rate Manufacturing (CHM) in Boston, Massachusetts is funded by the National Science Foundation in Arlington, Virginia and includes Northeastern University in Boston, the University of Massachusetts in Lowell, the University of New Hampshire in Durham, and Michigan State University in East Lansing (Center of High-Rate Manufacturing, 2007). One of the thrusts of this center is nanotemplate-enabled high-rate manufacturing. The intent here is to guide nanobuilding blocks to self-assemble over large areas in high-rate, scaleable, commercially viable processes such as injection molding, and extrusion. The authors emphasize that their template process will permit the fabrication of a single carbon nanotube (CNT) electromechanical switch. The authors estimate that the market for these switches at $100 billion. The center has combined with Nantero, a company that manufactures CNT computer memory switches. These nanotemplates are also to be used in a biosensor for the rapid (8–10 min detection time) analysis of antibodies. The authors emphasize that they have developed technical cost models for each step of their nanoprocess to help analyze the tradeoffs between economic and environmental tradeoffs. This is in-line with the current line of thinking of the possible hazards in nanomanufacturing processes. Also, fiscal feasibility and economics of scale-up of each nanoprocess is being carefully analyzed.

13.2 COLLABORATION BETWEEN COMPANIES, UNIVERSITIES, AND STATE AND GOVERNMENTAL AGENCIES: TRENDS IN COLLABORATION

Different trends and models for collaboration between different entities are being noticed at the present time. Some of these trends are presented here. Incubators are springing up wherein scholarly research at universities may be gradually turned into or groomed to becoming successful business ventures. The University of Connecticut at Storrs is one such example (UConn Advance, 2005). Zangari, the executive program director of the Technology Incubator Program (TIP) indicates that the incubator nurtures startup high-tech companies. The intent is to bring the University of Connecticut's new ideas to the market place. The incubator provides newly minted entrepreneurs laboratory and office space, and ready access to UConn's facilities and equipment. TIP fits in with the broader perspective of UConn's larger Office of Technology Commercialization. In essence, the TIP program provides enterpreneurs with assistance to get rolling for less than it would cost them to rent office space, etc. Zangari further underscores the importance of the TIP program in that (a) half of small businesses fail within 4 years, and (b) 87% of companies that used the incubator facilities stay within the community.

Kent State University's (Kent, Ohio) Office of Technology and Economic Development also assists university faculty to startup new companies (Kent State University: Research and Graduate Studies News, 2006). The biosensor program at Kent State University is a good example of collaboration between a university and a company. Two startup companies, Origen and Pathogen Detection Systems have emerged from the co-licensing of liquid biosensor technology developed jointly by Kent State University and NEOUCOM. The liquid biosensor technology used by these startup companies can detect bioterrorism agents, pathogens in food and water, and it has potential for military, environmental, and medical applications. Kent State University's Office of Technology and Economic Development also provides (a) leads to venture capital, (b) management talent, (c) office and laboratory space, and (d) leads to design, manufacturing, and new technologies. The primary intent is to facilitate regional economic growth for Northeast Ohio.

University of Oxford News (2004) indicated that Isis Innovation, Oxford's technology company will collaborate with Mitsui to take the University's Intellectual Property into Japanese markets. Mitsui is one of Japan's biggest companies, and its Europe affiliate (Mitsui, Europe) had recently made an investment in Oxford BioSciences Ltd. (a company spun out of the University of Oxford in the year 2000). The intent, of course, is to eventually introduce advanced products into new markets.

Business Wire (2005) indicates that Advance Nanotech, Inc. in New York is a premier provider of financing and support services to help drive the commercialization of nanotechnology devices. This company has recently announced the financing of BiMAT. This is new technology that should assist in the early detection of Avian Influenza (Bird Flu) in humans and in animals. BiMAT is a partnership between Advance Nanotech, Inc. and The Center for Advanced Photonics and Electronics (CAPE) at the University of Cambridge, United Kingdom, Alps Electric Company, Dow Corning Corporation, and the Marconi Corporation. CAPE is an integrated research facility for electrical engineering. The BiMAT technology should assist first responders such as medics, emergency medical technicians (EMTs), and doctors to analyze microscopic biological material quickly for pathogenic diseases on site. This procedure will eliminate the need to send the testing material to laboratories offsite, and thereby eliminate the problems and errors associated with this type of transfer.

Business Wire (2005) indicates that the sensor BiMAT is developing is an integrated, low cost, and disposable biosensor and sensor arrays. They may be used for point-of-care (POC) diagnostics, clinical monitoring, and for biomolecular research. These biosensors will incorporate thin film polysilicon transistors deposited on lightweight, inexpensive substrates. Advance Nanotech, Inc emphasizes that these low-cost thin film (TFT) devices will have a wide range of application for disposable biosensors that they plan to develop.

At this juncture it is perhaps appropriate to provide an example of collaboration between two companies. Axela Biosensors, Toronto, Ontario, Canada, has acquired the license to use the patented A^{2R} universal linker capture technology from Beckman Coulter (Axela Biosensors, 2006). This technology will be incorporated into Axela's dotLab™ Sensors, and provide researchers the unique ability to create user-defined multiplex biomarker panels. Axela Biosensors indicates that this will permit the researchers using this new launched dotLab system to create multiplex assays without the need for specialized pipetting or spotting techniques. Also, this technique will permit researchers to build combined assays with micromolar and picomolar sensitivity requirements.

Phylogica (2006), a drug discovery company in Subiaco, Western Australia has recently indicated that Axela Biosensors will be testing its novel screening technology on Phylogica's large pool of small protein fragment (peptides). These peptides, Phylogica claims act as drugs by blocking a disease process at the protein level. Phylogica indicates that the initial step in the collaboration involves the validation of Axela's diffractive optics technology (DOT) so that it can effectively analyze Phylomer^R (small protein fragments) candidates. The diffracted light in the DOT technology will be used to analyze the interactions between proteins, antibodies, and other smaller molecules.

Tekes teknologiaporssi (2006) located in Finland recently indicates that a United Kingdom company is looking for assay development partners to help develop and manufacture its highly sensitive and low cost biosensor assay that greatly reduces sample preparation and incubation time. Their ligand-binding assay uses standard immunoassay techniques. Since their biosensors are robust, they are able to quantitatively detect analytes in complex aqueous-based systems with generally no sample preparation. This biosensor is apparently easy to use, reduces incubation time, and is also highly sensitive.

The patented technology uses a proprietary method of growing a polymer layer on a single-use electrode. This acts as the transducer in the biosensor system, and reduces sample time. The UK company indicates other advantages that include easy and quick sample preparation, can work with milk, plasma, and blood samples, only very small samples are required, and it is very sensitive (for example low femtomole detection of large proteins). Furthermore, the development platform has a low capital as well as a low consumable cost. Besides, the biosensor is able to detect a wide range of low, medium, and high molecular weight analytes. The quantitative results obtained during the detection process may be used for both monitoring and screening applications.

13.3 FACTORS THAT COULD HELP INCREASE/DECREASE BIOSENSOR MARKETS

Fraser (1995) as early as 1995 indicated that as far as biosensor markets are concerned, there are problems waiting for biosensor solutions. In effect, the future expansion of biosensor markets will depend on the balance between market opportunities and technical/or financial obstacles. A Kalorama Information (2006) report indicates the following four factors that would help expand the biosensor market: greater quality of life expectation, innovative new technologies, research at Industrial/Governmental/Academic Laboratories, and the gradual acceptance of biosensor methods to help improve healthcare and other applications.

Business Wire (2005) indicates that a few technical problems are slowing down the commercialization of biosensors. These include selectivity, reproducibility, reliability, and the cost of manufacturing. These types of considerations not only delay new product launches but also prevent the large-scale production of a wide variety of newly developed biosensors. This author emphasizes that the manufacturing aspects need to be streamlined. Miniaturization and the application of nanotechnology and nanobiotechnology principles should assist the biosensor field to make inroads into hitherto untapped areas of application. These authors emphasize that the high costs of R&D for a new biosensor remains significantly high (around $40–50 million, and involves 8–10 years development time). This coupled with the lack of financial support for these types of endeavors limits the development of biosensor to larger companies, and excludes the smaller ones. Smaller companies (less than 50 personnel perhaps; an arbitrary number) can, however, develop biosensors in niche and specialized areas of application (Business Wire, 2005).

Heffner (2006) estimates that it can take about 5 years and $40 million to get a biosensor to the market. The author emphasize that poorly capitalized sensor companies can go out of business before reaching commercialization. This is particularly true if the companies are not careful and frugal especially in the development of medical sensors where regulatory approval is necessary. Fortunately, however, this author indicates that due to similarities in the detection of pathogens and human diseases, and for environmental applications, some of the technical know how and resources may be diverted from these types of areas where regulatory approval is not a strict as is for human application. It is important to point out that for medical purposes, at least, the newer products (biosensors) must go through developmental as well as regulatory steps (or process).

Further collaboration between University–Industry partners should assist in driving down the costs for biosensor development and assist in their commercialization. Non-biosensor technologies continue to pose a threat to biosensors, especially since they are inexpensive compared to the biosensors. However, end users of biosensors should be educated on the fact that although other methods of analysis are available, it is the ease of use of the biosensors that makes them the correct choice of technique of application for the detection of most analytes of interest.

Parce (2006) of Nanosys, Inc., located in Palo Alto, California delivered an interesting seminar at Columbia University, New York, NY entitled "Research in Academia and Start-up Companies: Different Goals, Different Focus," emphasizes the difference in goals, funding, and external forces whilst doing research in the academia and in industry. This is in spite of the fact that methods and tools for research are the same. The author highlights these differences for products over the last 20 years, which followed the path from an R&D process to a start-up company. Examples of biosensor and microfluidic (lab-on-a-chip) systems were presented. The author emphasizes that the fundamental research discovery is just the tip of the iceberg with regard to the series of engineering efforts required to effectively bring product to the market. He predicts that similar processes will also apply to nanotechnology products as well.

The National Science Foundation, NSF 05-526 (2005) Program Solicitation document entitled, Sensors and Sensor Networks (Sensors) has also identified some of the problems that hinder the development and subsequent commercialization of biosensors. These include required advances in enhanced sensitivity, selectivity, speed, robustness, and fewer false alarms. Besides, these biosensors should be able to function unattended, and in unusual, extreme, and complex environments. The above-mentioned document emphasizes the need for improved methods for sensor fabrication and manufacture, advances in signal processing, and integration with electronics on a chip. Lower power consumption would also be of assistance. This document further emphasizes economic manufacture and the delivery of biosensors. The NSF document does, however, indicate that emerging technologies (nanotechnology quickly comes to mind) do exhibit the potential to provide biosensors of lower cost, greater robustness, and increased lifetime and reliability.

In a subsequent National Science Foundation Program Solicitation, 06-566 entitled Biophotonics with a proposal submission window: August 15, 2007 — September 15, 2007, the NSF requested for proposals that use innovative basic research in biomedical photonics which would then lay the foundation for new techniques and applications in medical diagnostics and therapies. The program solicitation indicated that molecular-specific sensing, imaging, and monitoring systems with high optical sensitivity and resolution would be of considerable assistance in biology and medicine. The program solicitation further states that low cost diagnostics will need the integration of photonics, microbiology, and material science.

13.4 EXAMPLES OF BIOSENSOR COMPANIES, THEIR PRODUCT, AND THEIR FINANCIAL BACKERS

Different examples of biosensor companies along with their product and what analyte it detects will be presented here. Information on their financial backers wherever available is also presented. Most, if not all, of the examples have been selected at random to provide an overall perspective of the biosensors that have been or are in the process of being commercialized.

Biophage Pharma, a Canadian biopharmaceutical company develops new therapeutic and diagnostic products based on phage technology against bacterial contamination (Biophage Pharma, Inc., 2007). This company has developed a portable PDS^R biosensor, and provides service in immunogenicity and immunotoxicity. Furthermore, its MELISA^R testing device may be used for the detection of sensitization to more than 20 different allergens and pollen from a single blood sample. The company specializes in immunogenicity and immunotoxicity, as mentioned above, and in the sensitization to metals.

The company has recently indicated that it has recently received orders from clinics in California and in Quebec for its MELISA^R detection device (Biophage Pharma, Inc., 2006). The company indicates that the MELISA^R test is the first scientifically validated metal allergy test, and this should help consolidate their product development activities, and focus on the commercialization of its new PDS portable biosensor. The MELISA^R detection test is able to simultaneously screen for metals such as nickel, mercury, lead, silver, gold, zirconium, titanium, and manganese. This should assist in metal-sensitive patients suffering from ailments such as Psoriasis, Eczema, systemic lupus erythematosus (SLE), Sjorgen's disease, etc. The company recommends using the MELISA^R testing procedure prior to any invasive procedure such as pacemakers, silicone breast implants, insertion of screws, knee prosthesis, and cochlear implants, etc. This will assist in identifying alternate solutions, if necessary, to these invasive procedures.

Axela Biosensors (2006) referred to earlier in this chapter is located in Toronto, Ontario, Canada, commercializes products that assists in the validation of protein biomarkers from discovery into routine clinical practice. Its DOT^{TM} permits the real time detection and making quantitative of protein binding events in complex media. The DOT is an efficient and low cost tool. The DOT sensor is an optical sensor that uses microfluidics and photonic technology. Axela Biosensors is a privately held company whose major investor is Ven Growth Private Equity Partners, which is one of Canada's premier private equity managers.

Babcock (2007) in an interesting talk entitled "Adventures in Technology Development: How an SB Startup is Commercializing The Next Big Biosensor" mentions the different facets involved in setting up a start-up company for a MEMS-based biosensor. He indicates that this is the world's most sensitive detector of mass in fluid systems. The biosensor is designed to monitor the progression of HIV in developing countries (a highly important and much needed topic and area of use). He also outlines the different types of resources available at the local business/university/technology interface. Needless, to say the path that he took from a University of California postdoctoral student in physics to a small business CEO is anticipated to be rather circuitous. Most such frequent lectures at meetings of interest where scientists convene to analyze and discuss technical topics would help encourage university faculty members, especially those who have never thought of such matters, to venture off on their own also.

IST results (2006) have recently indicated that a Cypriot startup company, SignalGeneriX founded in 2004 is using digital signal processing (DSP) to assist in the detection of prostate cancer and three-dimensional positioning. The company has obtained a broad portfolio of intellectual property rights for core DSP algorithms (for example, security biometrics, speech recognition, and image compression). The intent of the company is to specialize in solutions for the medical, telecommunications, environmental monitoring, security, and military applications.

The initial financing for SignalGeneriX was from personal funds, research grants, and from bank loans. Their project entitled TAMIRUT includes eight consortium partners. The company aims to develop a new ultrasound biosensor for the early detection of prostate cancer with the 4 million Euros (equivalent to \$5.19 million: exchange rate €1 = \$1.2976, January 25, 2007) that they have raised. The company emphasizes that prostate cancer is one of the highest cancer risks for men in Europe, and is apparently curable only at an early stage, Thus, the need for early detection. This will compete with the prostate specific antigen (PSA) test that is already being used to screen for prostate cancer.

The TAMIRUT approach is based on an approach already being used for ultrasonic imaging of fluid flow in the heart and in the liver. The CEO of SignalGeneriX, Professor Anthony Constantinides of Imperial College, London emphasizes that the microbubbles used in TAMIRUT are capable of binding to specific molecular structures found in prostate cancer. Lipid coating on the microbubbles reflects the ultrasound. The analysis of the local concentrations of the microbubbles reveals the existence and grading of prostate cancer tumors. The company has developed along with Genoa University signal processing methods that distinguish microbubble echoes from tissue echoes. This technique is apparently consistent with the Program Guidelines on Biophotonics issued by the NSF in 2007 (National Science Foundation, PD 07-7236, Program Solicitation, Biophotonics, 2007) for proposals that are, and we quote, "very fundamental in science and engineering to lay the foundation for new technologies beyond those that are mature and ready for use in medical diagnostics and therapies."

Ohmx is a privately held company in Evanston, Illinois (Ohmx, 2006). It was founded by Professor Thomas Meade of Northwestern University in Evanston, Illinois. Very recently, the Illinois Technology Enterprise Center (ITEC) has invested $25,000 in Ohmx Corporation. Meade indicates that ITEC has been a tremendous source for Ohmx in help identifying initial investors, and federal and state grant funding agencies. The company is developing portable, electrical detection devices for use in diagnostics, drug development, and biodefense, food, water and environmental applications. The company was founded in the year 2005, and has raised more than three million dollars since then on completing a Sales of Preferred Stock financing in February 2006.

The basis of the research on biosensors at Ohmx is a new, simple, and easy way to detect a wide variety of bacteria, viruses, and molds. The company plans to develop a low cost, simple to use, reusable biosensor reader that uses a disposable biosensor chip to detect a wide variety of harmful analytes in a fluid system. The size of the biosensor is to be like a personal digital assistant (PDA). The company's research is at the interface of molecular engineering, analytical chemistry, and biology.

Applied Nanotech, Inc. a subsidiary of Nano-Proprietary, Inc. (Applied Nanotech, Inc., 2006) indicates that miniaturized enzymatic biosensors are in demand in the medical, environmental, and chemical analysis arena. They emphasize that the rapid commercialization of biosensors is hindered by cost, manufacturing complexity, and single use of a specific analyte. Besides, the sensitivity of the analyte to be detected is limited by the substrate that the enzyme is immobilized on. This company has developed a miniaturized biosensor array based on enzyme-coated nanotubes (ECNT). They indicate that their ECNT tongue can analyze quite a few analytes from a single drop of blood, urine, or saliva for metabolic analysis. They also emphasize that their sensor manufacturing process is scaleable to produce hundreds of sensors on a miniaturized chip.

Furthermore, their sensor uses low power, and can also be operated by a battery. The ECNT biosensor is able to detect glucose, phenol, formaldehyde, hydrogen peroxide, lactose, uric acid, alcohols, and ascorbic acid. Furthermore, the ECNT biosensor can also be used for toxic gas detection and for chemical warfare sensing.

Innovative Biosensors, Inc. (IBI) located in College Park, Maryland develops and manufactures rapid testing systems for the detection of pathogens (Innovative Biosensors, Inc., 2006). Some of the pathogens that its biosensor can detect/or will be able to detect includes (a) a rapid test for bovine spongiform encephalopathy (BSE). A grant has been awarded by the National Institutes of Health (NIH), National Heart, Lung, and Blood Institute (NHLBI), and Small Business Innovation Research (SBIR) program; (b) detection of *E. coli* 0157:H7 in foods. IBI has also entered into a cooperative R&D agreement with the U.S. Army Medical Research Institute of Infectious Diseases (USAMRIID) for the development of a Severe Acute Respiratory Syndrome (SARS) test using the company's CANARY™

technology. A highly sensitive, portable and rapid detection system is to be developed for SARS. The cellular analysis and notification of antigen risks and yields (CANARY^{TM}) was originally developed by Rider *et al.* (2003) in the Lincoln Laboratory at Massachusetts Institute of Technology (MIT) in Boston, Massachusetts. IBI has obtained an exclusive license for the CANARY technology. IBI (2006) indicates that in May 2005 the company had raised $3.5 million in Series A financing. In October 2006 the company, IBI announced that it had raised a total of $6.25 million in its A round of financing with the help of additional investors.

Professor Evangelyn Alocilja and her colleagues of the National Center for Food Protection and Defense (NCFCD) at Michigan State University, East Lansing are also developing biosensors for the rapid detection of microbial pathogens in foods and products (National Center for Food Protection Defense, 2006). These researchers are using polyclonal antibodies as biological sensing elements. Polyaniline is used as the molecular nanotransmitter and molecular bridge. The NCFPD plans to develop an electrochemical biosensor prototype disposable unit for the detection of *Bacillus anthracis* in food products in less than 15 min per sample. The sensitivity of the proposed biosensor will be in the 10 to 100 cfu/ml range. This NCFPD estimates should be applicable for the field-testing of real-time diagnosis of pathogenic contaminants.

West Virginia's Lane Department of Computer Science and Electrical Engineering is developing a microchip sensor (the size of a postage stamp) that can detect bacteria in a water sample to improve its quality (Wilson, 2001). The biosensor may be used for both municipal and home-based water systems. Multi-Sense, a start-up company is collaborating with West Virginia University to provide part of the research support and prototyping the sensor, as well as leasing laboratory space on campus for its activities. For a start-up company all of this can be very expensive, and this way the university acts as an incubator, which is of value to the start-up company. This should assist Multi-Sense in its commercialization efforts.

Nanosensors, Inc. located in Santa Clara, is a nanotechnology company and its principal business is to develop and market sensors for the detection of explosives, chemicals, and biological agents (Sensors Speciality Markets, 2006). It uses its recently licensed silicon-based biosensor to detect *E. coli*. Their biosensor consists of two different parts: (a) a disposable housing unit on which the actual sensor is mounted, and (b) an external, but separate acquisition unit. The signal in the acquisition unit is converted to an appropriate format so that results can be displayed. The company plans to field test their initial units in early 2007. The company indicates that it hopes that feedback information from the field trials will help improve, and also help them to assess the commercial viability of their biosensor.

Oxford Biosensors is located in Oxfordshire, United Kingdom and was formed in the year 2000 from technology that emanated from the University of Oxford. Its primary intent is to make a new category of diagnostics for Primary Health Care (Oxford Biosensors, 2007). The company would like to make low complexity detection devices that are suitable for POC by health workers, with a minimal of training. Each of these strips will measure health-related variables that are associated with different diseases. This is to be made possible from a single finger prick blood sample. The company has its own pilot manufacturing facility, and does not need to outsource. This aspect has a major impact on biosensor cost, since companies often outsource.

Tormey (2007) in a guest editorial entitled, "New economics drive product development in biotech" indicates that specialization in design and manufacturing has changed the product design and development business. He indicates that specialized, smaller, and leaner design and manufacturing units provide companies with economies of scale that they cannot achieve by themselves. These smaller specialized manufacturing companies may provide a significant time-to-market advantage. This is important, since they indicate that for example, if your product has a 6-month delay, this could result in a 33% less life cycle profit.

Large pharmaceutical companies use biosensors to screen for drugs. Business Resource Software, Inc. (2007) indicates that Nepkar located in Oxford, United Kingdom plans to revolutionize drug screening. It will use yeast genetic engineering for the discovery of new drugs. Nepkar plans to engineer human cell targets in yeast such that the yeast can respond to molecules, both natural and unnatural, that match the target. Screens for new drugs are possible that either block the target function (antagonists) or mimic the action of the natural ligand (agonists).

Nepkar indicates that the initial sets of targets are G-protein-coupled receptors (GPCRs). They are a major focus of pharmaceutical drug discovery. Apparently, according to Business Resource Software, Inc. (2007) of the top 100 drugs, 18 are directed at GPCRs. Furthermore, apparently an estimated 60% of all commercial drugs act on GPCRs.

Nepkar requires £1.5 million (equivalent to US \$2.939 million, exchange rate £1 = \$1. 9592, exchange rate January 29, 2007) to build a staff base and for operating expenses for 2 years (Business Resource Software, Inc., 2007). After 2 years the company expects to have a neutral cash flow based on its rapidly growing contract drug discovery operation, and a greatly expanded intellectual property portfolio. A seed capital £250,000 (US \$489,800, exchange rate January 29, 2007) has been provided by British Biotech, and a further £250,000 will be provided by British Biotech to help develop a yeast-based GPCR screen. For its financial backing of Nepkar, British Biotech will retain a 15% stake in Nepkar.

Finally, it is perhaps appropriate to mention three companies that make the SPR biosensor. The biggest one is Biacore that has recently been bought by General Electric (GE) in the USA (Biacore, 2006). Biacore is now a part of GE healthcare. The intent of the acquisition is to create a center of excellence wherein a wide range of solutions to the life science community will be offered. GE emphasizes that products from both companies may be used from early research in academia all the way up to manufacturing and quality control in the pharmaceutical and biotechnology industries. GE emphasizes that together with Biacore they can provide unbeatable solutions to help elucidate disease mechanisms and also help provide novel therapeutics. The SPR systems provide protein interactions in real time. The software that comes along with it provides values of the binding and dissociation rate coefficient values, along with affinity values. However, the analysis does not take into consideration the influence of external diffusional limitations as well as the heterogeneities that exist on the biosensor surface. This is one of the major themes of this book to help rectify the errors made in the estimation of the binding and dissociation rate coefficients, and in the affinity values.

The error arises due to the fact that the commercially available programs that come along with the biosensor equipment from the different manufacturers do not take into account either the diffusional limitations that are present in these types of systems, and the heterogeneities that are inherently present on biosensor surfaces. The aim is that if the biosensor is run properly these diffusional effects will not be present. In effect, if these diffusional effects are present, then one is really observing 'diffusion-disguised' kinetics, and not inherent kinetics, as is claimed. Furthermore, the effects of heterogeneities on the surface are completely ignored. As mentioned thorough out in different chapters in the book, the influence of heterogeneities on the biosensor surface can be quite severe on the values of the binding and dissociation rate coefficients, and affinity values estimated.

Biosensing Instruments (2006) located in Tempe, Arizona also provides high-performance SPR instruments for research and analysis. Both slow and fast kinetics of analyte–receptor interactions may be determined, as well as the size of the analytes involved in these interactions.

Genoptics (2006) located in Cedex, France uses surface plasmon resonance imaging (SPRi) to determine biomolecular interactions in the life science laboratory in a multiplex fashion. Multiplex implies the simultaneous measurement of all of the phenomena observed on a biochip surface for an entire reaction.

Sensortec was founded in 1996, and is in St. Helier, Jersey, United Kingdom (Sensortec, 2006). It has developed a robust and flexible platform for disposable biosensors. Immunoassay techniques are used for detection purposes. Low cost materials are used to produce the disposable cartridge. Blood samples may be manipulated by fluidics.

The company also has a new platform technology for food quality assurance and environmental applications at a competitive price. Their technology platform can be automated, multiplexed, and miniaturized. This makes it suitable for point-of-use applications. Pharmaceutical applications are also possible. The company indicates that due to the simple and novel potentiometric measurements involved, their UTS^{TM} biosensor is more sensitive than current amperometric and optical biosensors. Furthermore, the company emphasizes that their UTS^{TM} technology requires minimal sample preparation. Additional advantages include rapid detection times (less than 15 min), for single analyte detection, and a wide dynamic range (4 to 5 orders of magnitude). The design of the biosensor is such that it may be operated as a one time disposable diagnostic, as a panel for diagnostic tests, and even as a small sensor array.

The company is financially backed by Chordcapital (Chord Capital, Sensortec, 2006). Sensortec has licensed its technology to DxTech LLC, a private company located in Melbourne, Florida. DxTech will use the Sensortec technology for the development of a multi-analyte microfluidic cartridge for use in its diagnostic platform. DxTech (2006) hopes to initiate the transition between centralized to distributed diagnostics with its unique, fluid-based medical diagnostic platform. The company claims that real-time diagnostics may be obtained at POC that will be of assistance to clinicians. A drop of blood is only required for a credit card size disposable cartridge.

VTT Technical Research Center with its corporate headquarters in Espoo, Finland is the biggest contract research organization in Northern Europe (VTT, 2006). It provides high-end technology solutions and innovation services. The VIT organization was established in 1942, and last year its turnover was 225 million euros or 291.3 million US dollars (€1 = US $1.2948, exchange rate January 29, 2007). The company is using its strengths in biotechnology, information technology, and in electronics to develop and manufacture an inexpensive disposable biosensor using the results obtained from different research institutes. VTT promotes networking between companies and also helps create new businesses.

One of VTT's area in biosensor research is the development of advanced systems for sensing phenomena at the molecular level. Tailored coatings with functional molecules are placed/immobilized/adsorbed on appropriate substrates to promote molecular recognition. This sensing is aided by case-specific transducers. The company's immuno- and DNA-based sensors, as well as their chemical sensors find applications in the areas of biomedical engineering, clinical diagnostics, and in the monitoring of different types of processes.

It is of interest to note that even a university would, besides its many roles, also like to present itself with available technology (in this case biosensor) platforms, and is seeking potential collaborators in the biosensor and other areas. Biopartner (2006) indicates that the University of West England at Coldharbour Lane in Bristol, England has developed a portfolio of biosensors that may be used in the detection of volatile organic compounds (VOCs) based on heated ceramics and conducting polymers. These biosensors find applications in the indoor environment, agri-food, and in the biomedical areas. UWE indicates that its screen-printed carbon electrodes (SPCEs) may be used for the detection of environmental pollutants in air, water, and contaminated land. Also these SPCEs may be used for the detection of clinical metabolites, hormones, bacteria, and trace heavy metals in water. UWE is also actively pursuing collaborators (companies) in the clinical diagnostic and environmental areas requiring R&D to identify target marker compounds.

Phillips (2005) located in Eindhoven, The Netherlands has developed a biosensor that provides high analytical performance with simplicity in use and is also of low cost. One biosensor is based on magnetic particles (for magnetic biosensors), and the other is based on Raman spectroscopy (for optical biosensors). Phillips emphasizes that their biosensors should improve on the different biosensor performance parameters such as sensitivity, speed, and reliability for applications such as protein and pathogen monitoring, near patient testing in medical centers and at home-testing for variables such as blood, urine, and saliva tests. The company is also looking at developing a disposable biosensor that would be manufactured at low cost. This is for possible use in a hand-held reader.

Phillips emphasizes that the current trend is to be able to predict diseases before the symptoms actually manifest themselves. This is possible if one has very sensitive biosensors to be able to detect the relevant biomarkers for the different diseases. The company indicates that their biosensors are able to detect specific molecules (for example, these biomarkers for different diseases) at very low concentrations (for example, 10^{-13} moles per liter, and at lower concentrations).

The goal of the Nanoscale Science and Engineering Center (NSEC) at Ohio State University, Columbus, Ohio is to create devices that will make diagnostics, treating, and managing diseases easier, less expensive, and more effective. The NSEC center collaborates with the Center for Mutlifunctional Polymer Nanomaterials and Devices (CMPND), Center for Advanced Polymer and Composites Engineering (CAPCE), and Integrative Graduate Education and Research Training Program (IGERT). The CMPND center has an on-going program for the use of conjugates of biomolecules and nanoparticles or conductive polymers for biosensors and lab-on-a-chip devices. This center indicates that for the future of health care low-cost and highly sensitive biosensors and lab-on-a-chip devices are essential. This center further states that the present-day biosensors that rely on optical and radiation methods are either too expensive or are not sensitive enough for single-molecule detection and analysis. In order to facilitate single-molecule analysis, the CMPND center plans to use nanoparticles and conductive polymers to form conjugates with biomolecules for advanced proteomic biosensing applications. The CMPND center is also exploring ligand–protein– nanofiber interactions with the eventual goal of developing high-performance electrical signal based biosensors for the early detection of chemical and biological agents.

13.5 CONCLUSIONS

This last chapter attempts to provide a perspective of what is required to setup a biosensor industry. At the outset it is appropriate to indicate that the author has spent his entire life in the academia with a few summers spent at private companies and at National Laboratories. Five years were also spent at the National Chemical Laboratory in India. Also, some consultancy with biosensor and other types of companies has also been undertaken in the US.

After providing an introduction to the market for biosensors in this chapter, the trends in collaboration between companies, universities, and state and governmental agencies are presented, followed by the factors that may help increase or decrease the biosensor markets. Examples of real life biosensor companies are given. The biosensors may be at different stages of development on getting the biosensor to the market. Some companies may already have their biosensor on the market. This is besides the ones used for measuring sugar levels for managing diabetes.

Besides the monitoring of sugar levels for diabetics, the biosensor can be a valuable tool for diagnostic purposes as well as for the detection of pathogens in the environment. Except for the monitoring of sugar levels in diabetic patients (which is a huge market, and it is increasing due to the rise in obesity levels in individuals worldwide coupled with poor food and exercise habits), biosensors are finding it difficult to make any major headway in other possible areas of application. This is in spite of the fact that they can be of tremendous use to diagnose early quite a few intractable and insidious autoimmune diseases, for example. Needless to say, quite a few of these diseases may be managed better if they were diagnosed during their early stages of incidence.

The information presented in this chapter is difficult to get from the classical open literature sources, such as journals. Thus, most of the information presented in this chapter is obtained from internet sources. There may be a reliability factor here. Needless to say, private companies will guard this type of economic information very carefully.

Biacore in Sweden, which manufactured and marketed the SPR biosensor was one of the more successful, if not the most successful biosensor company. It has recently merged with General Electric in their Life Sciences division. The reason for this is not clear; however, if this author is permitted to speculate, it could be that a biosensor company as strong as Biacore was unable to stay strong and independent for long. If this is true, and there may be quite a few reasons behind this, this does not bode well for the biosensor industry as such; though this may be just a single and isolated example.

Examples of biosensor companies seeking financial backers or collaborators are also presented in this chapter. Due to the relatively small biosensor markets, and the time and resources required to bring a biosensor to the market hinders venture capitalists as well as possible biosensor entrepreneurs to help develop biosensors for newer applications, where reasonable amounts of profits or return on investment (ROI) may be obtained. Hopefully in the not to distant future with advances in biosensor technology, and in nanotechnology, especially the manufacturing aspects, and the lessons that the companies and the individuals have learned in bringing biosensors (successfully or even unsuccessfully) to the market will all come together in the future to let companies make reasonable amounts of profit by manufacturing and marketing biosensors for different areas of application. This may begin to attract more venture capitalists to support and develop biosensors for newer applications.

REFERENCES

Applied Nanotech, Inc., Enzyme coated carbon nanotubes (ECNT) platform, www.nano-proprietary.com, 2006.

Axela Biosensors, Axela biosensors to license patented A^{2R} universal linker capture technology from Beckmann Coulter. November 28. http://www.axelabiosensors.com, 2006.

Babcock, K, Adventures in technology development: How an SB startup is commercializing the "next big biosensor," CEO Affinity Biosensors, http://www.physics.ucsb.edu/~phys260g/KenBabcock.htm, 2007.

Biacore, 2006, http://www.biacore.com/lifesciences/ge/Acquisition/index.html?

Biophage Pharma, Inc., Biophage pharma receives important orders from Canada and U.S., November 3, 2006, press release, http://www.biophagepharma.net/index.php?option=com_content&task=view&id=38&item, 2006.

Biophage Pharma, Inc., Home page, Welcome, http://www.biophagepharma/.net/index.html, 2007.

Biopartner, Company profile. http://www.globalwatchservice.com/pages/, 2006.

Biosensing instrument. http://www.biosensingusa.com, 2006.

Business Resource Software, Inc., Nepkar: Executive Summary, http://www.businessplans.org/Nepkar/Nepkar00.html, 2007.

Business Wire, Emerging applications and greater health consciousness drive world biosensor markets. http://www.findarticles.com/articles/mi_mOEIN/is_2005_May_23/ai_13773210, 2005a.

Business Wire, Advance nanotech announces financing for nanotechnology devices to detect infectious disease: BiMAT program holds promise for early detection of bird flu. http://www.findarticles.com/p/articles/mi_m0EIN/is_2005_Oct_3/, 2005b.

Center for High-Rate Manufacturing; http://www.nano.neu.edu/research/, 2007.

Chord Capital, Sensortec. www.chordcapital.co.uk/Investment portfolio/Sensorttec.aspx, 2006.

CMPND, Center for multifunctional polymer nanomaterials and devices: biomedical devices, http://www.capce.ohio-state.edu/CMPND/.

DiabetesStore.com, Glucose Test Strips. http://www.diabetesstore.com, 2007.

Diabetic Test Strips, Diabetic Test Strips—Diabetic testing supplies—The medSupplyGuide.com. http://themedsupplyguide.com/diabetic-test-strips, 2007.

DxTech, Igniting the change from centralized to distributed diagnostics. http://www.dxtech.com/index.html, 2006.

Fraser, D, Biosensor markets: opportunities and obstacles. *Medical Devices and Technology*, 1995, **6**(2), March, 32–37.

Frost and Sullivan, World Biosensors Markets, April 18. 2005, http://www.frost.com/prod/servlet/report-brochure.pag?id=A142-01-00-00-00.

Genoptocs, http://www.genoptics-spr.com, 2006.

Heffner, S, Biomarket trends: Biosensor markets outperform expectations. *Genetic and Engineering News*, September 15 http://www.genengnews.com/articles/chitem.aspx?aid=1864&chid=0, 2006.

Infoshop.com, Biosensors and Bioelectronics, April. http:www.the-infoshop.com/study/bc29282_bioelectronics.html, 2005.

Innovative Biosensors, Inc., Diagnostic technology for pathogen detection. http://www.innovativebiosensors.com/.pressrelease_pathogen_detection.html, 2006.

IST Results, Start-up signals new processes for cancer detection and local positioning. January 30, 2006. http://istresults.cordis.europs.eu/index/cfm/section/news/tpl/article, 2006.

Kalorama Information, Medical and Biological Sensors and Sensor Systems: markets, Applications, and Competitors Worldwide, 2nd edn., May 1. http://www.kaloramainformation.com/sitemap/product.asp?productid=1083189, 2006.

Kent State University, Research and Graduate Studies News, Licensing of liquid crystal biosensor technology at Kent State, NEOUCOM, leads to new startup companies. 103(1). http://www.kent.edu/enews/rags/Fall2006NewsI.cfm, 2006.

Market Research.com, Medical biosensor applications and market to 2008_Diabetes management, Point-of care and related applications, published by Takeda Pacific, May 1. http://www.marketresearch.com/map/prod/1100483.html, 2005.

Menon, N, Optical biosensors: applying biosensors: applying photonics products to the biomedical diagnostics markets. *Optical Society of America*, menon@chromologic.com, Chromologic LLC, Corning, New York, 2004.

National Center for Food Protection Defense, A Homeland Security Center of Excellence. Division of University Relations, Michigan State University, East Lansing, 2006.

National Science Foundation, NSF 05-526, Program Solicitation, Sensors and Sensor Networks (Sensors), Proposals due March 03, Arlington, VA, USA, 2005.

National Science Foundation, PD 07-7236, Program Solicitation, Biophotonics, Full Proposal Window: August 15, 2007—September 15, 2007.

Newman, JD and SJ Setford, Enzymatic Biosensors. *Molecular Biotechnology*, 2006, **32**(3), 249–268.

Ohmx Corporation, ITEC Pre-seed fund makes award to Ohmx Corporation, http://www.ohmxbio.com, 2006.

Oxford Biosensors, Welcome to oxford biosensors! http://company.monster.co.uk/oxbiouk/, 2007.

Parce, JW, Research in academia and start-up companies: different goals, different focus. Nanosys Inc., lecture delivered at Columbia University, Interschool Laboratory, New York, NY, March 8, 2006.

Patel, PD, Overview of affinity biosensors in food analysis. *J AOAC International*, 2006, **89**(30), 805–818.

Phillips, Phillips demonstrates biosensor technologies for high-sensitivity molecular diagnostics. http://research.phillips.com/newcenter/archive/2005/, 2005.

Phylogica, Phylogica signs international deal with Canadian biotech. http://www.phylogica.com, 2006.

Radke, S and A Evangelyn, Market analysis of biosensors for food safety. *American Society of Agricultural and Biological Engineers Technical Library: Abstract*, http://asae.frymulti.com/abstract.asp?aid=9233&t=2, 2002.

Rider TH, MS Petrovic, FE Nargi, JD Harper, ED Schoebel, RH Matthews, DJ Blanchard, LT Bortolin, AM Young, J Chen and MA Hollis, A cell-based sensor for rapid identification of pathogens. *Science*, 2003, **301**, 213–215.

Sensors Speciality Markets, Nanosensors to test *E. coli* biosensor prototype. http://speciality.sensorsmag.com/sensorsspeciality/article/, 2006.

Sensortec, Universal electrochemical sensor: sensitive, simple, economical. http://www.sensortec.uk.com/news, 2006.

SmallTimes, Agamatrix equates its algorithms with pain-free test for diabetes. http://www.smalltimes.com/Articles/, 2003.

Tekes teknologiaporssi, A highly sensitive, low cost biosensor assay development system that dramatically reduces sample preparation and incubation time. http://www.tekes.fi/partner/fin/search/, 2006.

Test Medical Symptoms@Home, Inc., Diabetes Tests. http://www.tetssymptomsathome.com, 2007.

Torney, P, Guest editorial: new economics drive product development in biotech. Vie President, Silicon Valley Instruments, Morgan Hill, CA 95037, USA, email: Pete.Torney@SV Instruments.com, 2007.

UCONN Advance, Incubator helps turn scholarly research into viable business ventures. http://advance.uconn.edu/2005/05022/05022208.htm, 2005.

University of Oxford news, Oxford establishes collaboration with Japanese company Mitsui. Enquiries to webmaster, http://www.ox.ac.uk/webmaster.html, 2004.

VTT, Biosensors. http://www.vtt.fi/palvelut/cluster4, 2006.

Wilson, J, Fulfilling the mission: West Virginia University Alumni Magazine, Summer. http://www.ia.wvu.edu/~magazine/issues/summer2001/htmlfiles/mission/html, 2001.

Yurish, SY, NV Kirianaki and IL Myshkin, World Sensors and MEMS markets: Analysis and trends. *Sensors and Transducers Magazine (S&T e-Digest)*, 2005, **62**(2), December, 456–461.

Index

www.ingramcontent.com/pod-product-compliance
Lightning Source LLC
Chambersburg PA
CBHW081546031225
36271CB00005B/308

* 9 7 8 0 4 4 4 5 3 0 1 0 3 *